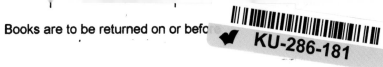

KU-286-181

Plants at the Margin

Ecological Limits and Climate Change

Plants at the limits of their distribution are likely to be particularly affected by climate change. Biogeography, demography, reproductive biology, physiology and genetics all provide cogent explanations as to why limits occur where they do. The book brings together these different avenues of enquiry, in a form that is suited to students, researchers and anyone with an interest in the impact of climate change. Margins are by their very nature environmentally unstable – does it therefore follow that plant populations adapted for life in such areas will prove to be pre-adapted to withstand the changes that may be brought about by a warmer world? This and other questions are explored concerning the changes that may already be taking place on this planet. Numerous illustrations are included to remind us that knowledge of the existence of plants in their natural environment is essential to our understanding of their function and ecology in a changing world.

R. M. M. CRAWFORD has taught and researched at the University of St Andrews since 1962, pursuing the study of plant responses to the environment in a wide range of habitats in Scotland, Scandinavia, North and South America and the Arctic. He is a Fellow of the Royal Society of Edinburgh, a Fellow of the Linnean Society and an associate member of the Belgian Royal Academy.

Plants at the Margin

Ecological Limits and Climate Change

R. M. M. Crawford

Professor Emeritus, University of St Andrews, Scotland

CAMBRIDGE UNIVERSITY PRESS
Cambridge, New York, Melbourne, Madrid, Cape Town, Singapore, São Paulo

Cambridge University Press
The Edinburgh Building, Cambridge CB2 8RU, UK

Published in the United States of America by Cambridge University Press, New York

www.cambridge.org
Information on this title: www.cambridge.org/9780521623094

First published 2008

Printed in the United Kingdom at the University Press, Cambridge

A catalogue record for this publication is available from the British Library

ISBN 978-0-521-62309-4 hardback

For Barbara

and all who inhabit, study and
value marginal lands

Fortunatus et ille, Deos qui novit agrestes Virgil, Georgics: Book II

Contents

Preface   *page* xiii
Acknowledgements   xv

PART I   THE NATURE OF MARGINAL
         AREAS

1   **Recognizing margins**   3
    1.1   Defining margins   5
    1.2   Margins and climate change   5
    1.3   Limits to distribution   8
          1.3.1   Physiological boundaries   9
          1.3.2   Resource availability   9
          1.3.3   Resource access and
                  conservation in marginal areas   15
    1.4   Genetic boundaries   17
    1.5   Demographic factors   17
          1.5.1   Limits for reproduction   19
    1.6   Relict species and climate change   19
          1.6.1   Evolutionary relicts   20
          1.6.2   Climatic relicts   20
    1.7   Endangered species   23
    1.8   Agricultural margins   24
    1.9   Conclusions   26

2   **Biodiversity in marginal areas**   29
    2.1   Biodiversity at the periphery   31
    2.2   Assessing biodiversity   31
          2.2.1   Definitions of biodiversity   31
          2.2.2   Problems of scale and
                  classification   34
          2.2.3   Variations in assessing
                  genetic variation   35
    2.3   Variation in peripheral areas   36
    2.4   Disturbance and biodiversity   36
          2.4.1   Grazing   37

	2.4.2	Fire	42
2.5	The geography of marginal plant biodiversity		43
	2.5.1	The South African Cape flora	45
	2.5.2	Mediterranean heathlands	48
	2.5.3	Mediterranean-type vegetation worldwide	50
	2.5.4	The Brazilian Cerrado	51
2.6	Plant diversity in drylands		52
2.7	Plant diversity in the Arctic		57
2.8	Conclusions		59

PART II   PLANT FUNCTION IN
MARGINAL AREAS

3	Resource acquisition in marginal habitats		**63**
3.1	Resource necessities in non-productive habitats		65
3.2	Adaptation to habitats with limited resources		68
	3.2.1	Capacity adaptation	69
	3.2.2	Functional adjustment	70
	3.2.3	Adverse aspects of capacity adaptation	72
	3.2.4	Climatic warming and the vulnerability of specific tissues	74
3.3	Habitat productivity and competition		77
	3.3.1	Plant functional types	78
3.4	Life history strategies		81
	3.4.1	Two-class life strategies	81
	3.4.2	Three-class life strategies	83
	3.4.3	Four-class life strategies	83
3.5	Resource allocation		84
3.6	Resource acquisition in marginal areas		85
	3.6.1	Competition for resources in marginal areas	85
	3.6.2	Deprivation indifference	86
	3.6.3	Deprivation indifference through anoxia tolerance	87
	3.6.4	Avoiders and tolerators	89
3.7	Alternative supplies of resources		90
	3.7.1	Light	90
	3.7.2	Precipitation	91
	3.7.3	Ground water	92
	3.7.4	Carbon	96
	3.7.5	Nitrogen	98

	3.7.6	Phosphate	100
	3.7.7	Phosphate availability at high latitudes	101
3.8	Mycorrhizal associations in nutrient-poor habitats		102
	3.8.1	Mycorrhizal associations in the Arctic	102
	3.8.2	Cluster roots	103
3.9	Nutrient retention in marginal areas		103
3.10	Changes in resource availability in the Arctic as a result of climatic warming		106

4	Reproduction at the periphery		**109**
4.1	Environmental limits to reproduction		111
4.2	Sexual reproduction in marginal habitats		111
	4.2.1	Pre-zygotic and post-zygotic limitations to seed production	111
4.3	Germination and establishment in marginal areas		114
4.4	Phenology		116
	4.4.1	Reproduction in flood-prone tropical lake and river margins	116
4.5	Hybrid zones		118
	4.5.1	Transient and stable hybrids	118
	4.5.2	Hybrid swarms	120
	4.5.3	*Spartina anglica* – common cord grass	122
	4.5.4	*Senecio squalidus* – the Oxford ragwort	123
4.6	Genetic invasion in marginal areas		126
	4.6.1	Invasion and climatic warming	127
	4.6.2	Climatic warming, disturbance and invasion	130
	4.6.3	Theories on habitat liability to invasion	131
4.7	Reproduction in hot deserts		131
	4.7.1	Diversity of plant form in drought-prone habitats	131
	4.7.2	Desert seed survival strategies	134
4.8	Flowering in arctic and alpine habitats		135
	4.8.1	Annual arctic plants	140

4.9   Mast seeding   142
4.10   The seed bank   146
       4.10.1 Polar seed banks   147
       4.10.2 Warm desert seed banks   148
4.11   Biased sex ratios   148
4.12   Clonal growth and reproduction in
       marginal habitats   153
       4.12.1 Asexual reproduction   153
4.13   Longevity and persistence in
       marginal habitats   155
4.14   Conclusions   158

PART III   MARGINAL HABITATS –
            SELECTED CASE
            HISTORIES

5   Arctic and subarctic treelines and the
    tundra–taiga interface   161
    5.1   The tundra–taiga interface   163
           5.1.1   Migrational history of the
                   tundra–taiga interface 163
    5.2   Climatic limits of the boreal forest 166
           5.2.1   Relating distribution to
                   temperature   166
           5.2.2   Krummholz and treeline
                   advance   169
    5.3   Climatic change and forest migration 174
           5.3.1   Boreal migrational history 174
    5.4   Fire, and paludification at the
           tundra–taiga interface   178
           5.4.1   Post-fire habitat degradation 178
           5.4.2   Treelines and paludification 179
           5.4.3   History of paludification 181
           5.4.4   Bog versus forest at the
                   tundra–taiga interface 183
    5.5   Homeostasis and treeline stability 185
    5.6   Boreal forest productivity at high
           latitudes   187
           5.6.1   Physiological limits for tree
                   survival at the tundra–taiga
                   interface   188
           5.6.2   Carbon balance   190
           5.6.3   Carbon balance versus tissue
                   vulnerability at the treeline 191
           5.6.4   Winter desiccation injury 191
           5.6.5   Overwintering photosynthetic
                   activity   191

    5.7   Future trends at the tundra–taiga
           interface   193

6   Plant survival in a warmer Arctic   197
    6.1   Defining the Arctic   199
    6.2   Signs of change   199
    6.3   The Arctic as a marginal area   204
           6.3.1   Mapping arctic margins 204
    6.4   Pleistocene history of the arctic
           flora   205
           6.4.1   Reassessment of ice cover in
                   polar regions   205
           6.4.2   Molecular evidence for the
                   existence of glacial refugia
                   at high latitudes   211
           6.4.3   Evidence for an ancient
                   (autochthonous) arctic flora 213
    6.5   Habitat preferences in high arctic
           plant communities   213
           6.5.1   Incompatible survival
                   strategies   214
           6.5.2   Ice encasement and the
                   prolonged imposition
                   of anoxia   214
    6.6   Mutualism in arctic subspecies 215
    6.7   Polyploidy at high latitudes   216
    6.8   Arctic oases   219
    6.9   Phenological responses to increased
           temperatures   221
    6.10   Conclusions   224

7   Land plants at coastal margins   225
    7.1   Challenges of the maritime
           environment   227
           7.1.1   The concept of oceanicity 228
           7.1.2   Physical versus biological
                   fragility   231
    7.2   Northern hemisphere coastal
           vegetation   235
           7.2.1   Foreshore plant communities 235
           7.2.2   Dune systems of the North
                   Atlantic   238
           7.2.3   Arctic shores   240
    7.3   Southern hemisphere shores   246
           7.3.1   Antarctic shores   246
           7.3.2   New Zealand   248
    7.4   Global shore communities   250
           7.4.1   Salt marshes and mudflats 250

7.4.2　Rising sea levels and
mudflats　251
7.5　Hard shores　252
7.5.1　Cliffs and caves　252
7.5.2　North Atlantic cliffs　254
7.6　Trees by the sea　256
7.6.1　Mangrove swamps　256
7.7　Physiological adaptations in coastal
vegetation　263
7.7.1　Drought tolerance　263
7.7.2　Nitrogen fixation　264
7.7.3　Surviving burial　264
7.7.4　Flooding　267
7.8　Conservation versus cyclical
destruction and regeneration in
coastal habitats　269
7.9　Conclusions　271

8　Survival at the water's edge　273
8.1　Flooding endurance　275
8.1.1　Life-form and flooding
tolerance　277
8.1.2　Seasonal responses to
flooding　281
8.2　Aeration　281
8.2.1　Radial oxygen loss　281
8.2.2　Thermo-osmosis　282
8.3　Responses to long-term
winter flooding　284
8.3.1　Surviving long-term oxygen
deprivation　285
8.4　Flooding and unflooding　286
8.4.1　Unflooding – the post-anoxic
experience　286
8.5　Responses to short-term flooding
during the growing season　287
8.5.1　Disadvantages of flooding
tolerance　289
8.6　Amphibious plant adaptations　290
8.6.1　Phenotypic plasticity in
amphibious species　290
8.6.2　Speciation and population
zonation in relation
to flooding　291
8.7　Aquatic graminoids　292
8.7.1　*Glyceria maxima* versus
Filipendula ulmaria　295

8.7.2　Sweet flag (*Acorus calamus*)　295
8.7.3　Reed sweet grass
(*Glyceria maxima*)　298
8.7.4　The common reed (*Phragmites
australis*)　300
8.7.5　Amphibious trees　301
8.8　Tropical versus temperate trees in
wetland sites　301
8.9　Conclusions – plants with wet feet　305

9　Woody plants at the margin　307
9.1　Woody plants beyond the treeline　309
9.2　Woody plants of the tundra　311
9.3　Montane and arctic willows　314
9.4　Mountain birches　318
9.4.1　Biogeographical history of
mountain birch　322
9.4.2　Current migration　323
9.5　Dwarf birches *Betula nana* and
B. glandulosa　323
9.5.1　Biogeographical history of
dwarf birch　323
9.6　Ecological sensitivity of woody
plants to oceanic conditions　324
9.7　Juniper　326
9.8　Heathlands　329
9.8.1　Relating heathlands to
climate　329
9.8.2　Possible migration behaviour　332
9.8.3　Historical ecology of
heathlands　334
9.9　New Zealand: a hyperoceanic
case study　334
9.10　Conclusions　337

10　Plants at high altitudes　339
10.1　Altitudinal limits to plant survival　341
10.2　Mountaintop isolation　343
10.2.1　Inselbergs – isolated
mountains　345
10.2.2　African inselbergs　347
10.3　Aspects of high-altitude habitats　348
10.3.1　Geology and mountain
floras　350
10.3.2　Adiabatic lapse rate　352
10.3.3　Mountain topography and
biodiversity　352

10.4	Physiological implications for plant survival on high mountains	355
	10.4.1 Water availability at high altitudes	355
	10.4.2 Adapting to fluctuating temperatures	355
	10.4.3 Protection against high levels of radiation at high altitudes and latitude	356
	10.4.4 Effect of UV radiation on alpine vegetation	359
	10.4.5 Oceanic mountain environments	360
	10.4.6 Phenological responses of mountain plants	361
10.5	Alpine vegetation zonation – case studies	363
	10.5.1 Temperate and boreal alpine zonation	364
	10.5.2 Tropical and subtropical mountains – East Africa	364
	10.5.3 South America	365
10.6	The world's highest forests	366
	10.6.1 The Peruvian Highlands	369
10.7	High mountain plants and climate change	369
	10.7.1 Indirect effects of increased temperature on alpine vegetation – reduction in winter snow cover	372
	10.7.2 Effects of increased atmospheric CO_2 on high mountain vegetation	372
10.8	Alpine floral biology	376
10.9	Conclusions	379
11	**Man at the margins**	**381**
11.1	Human settlement in peripheral areas	383
11.2	Past and present concepts of marginality	384
	11.2.1 Agricultural sustainability in marginal areas	386

11.3	Man in the terrestrial Arctic	389
	11.3.1 Acquisition of natural resources at high latitudes	391
	11.3.2 Future prospects for the tundra and its native peoples	395
11.4	Man on coastal margins	396
	11.4.1 Human acceleration of soil impoverishment in oceanic regions	398
	11.4.2 Sustainable agriculture in oceanic climates: Orkney – an oceanic exception	403
11.5	Exploiting the wetlands	404
	11.5.1 Coastal wetlands	404
	11.5.2 Human settlement in reed beds	405
	11.5.3 Agricultural uses of wetlands	406
	11.5.4 Recent developments in bog cultivation	409
	11.5.5 Future uses for wetlands	411
11.6	Man in the mountains	411
	11.6.1 Transhumance	411
	11.6.2 Terrace farming	413
11.7	Conclusions	417
12	**Summary and conclusions**	**419**
12.1	Signs of change	421
12.2	Vegetation responses to climate change	422
12.3	Pre-adaptation of plants in marginal areas to climatic change	423
12.4	Physical fragility versus biological stability and diversity	424
12.5	Marginal areas and conservation	426
	12.5.1 Regeneration and the role of margins	426
12.6	Future prospects for marginal areas	430
References		433
Author index		461
Species index		465
Subject index		471

Preface

Margins have long provided key questions for ecological investigation. Today with climatic warming becoming ever more apparent margins as regions of ecological change invite an assessment of their responses to environmental alteration. The purpose of this book is therefore to examine how marginal plant communities in different parts of the world are responding to climate change. Practically every aspect of modern biological enquiry can be used to address the nature of margins. Biogeography, demography, reproductive biology, physiology and genetics all provide cogent explanations as to why limits occur where they do. The aim of this book is to bring together, wherever possible, different avenues of enquiry in relation to explaining the existence of limits to plant distribution. Each of these disciplines can contribute to our understanding of the biological consequences of climatic warming.

Marginal areas have a number of features in common. These can be seen in demographic limits to population renewal, in adaptations to shortness of the growing season, in problems of access to resources, and impediments to reproduction. To avoid repetition an attempt is made therefore to discuss these common features before moving on to individual case studies.

Part I examines the nature of margins and their effects on biodiversity. Part II is functional, and explores how plants in marginal areas overcome the shortness of the growing season and other physical limitations in acquiring resources and reproducing. The remaining chapters look at individual examples of marginal areas which have been selected on the supposition that they may be sensitive to climatic change.

In a scenario of a warmer world it is highly probable that changing climatic conditions will have a particularly marked effect on human exploitation of marginal areas. The history of human settlement in peripheral areas is therefore discussed in relation to our use of plants in marginal areas. Climatic change will also create problems for conservation particularly in relation to the interactions between human beings, their livestock and the environment. The consequences of both higher temperatures and greater human populations create a worldwide problem with particularly serious consequences for marginal regions.

In this book an attempt is made to compare the sensitivity of different margins with climate change and to explore the question of whether or not all peripheral areas are equally likely to suffer losses in biodiversity as a result of climatic change. The converse situation is also considered. Margins are by their very nature environmentally unstable. Does it therefore follow that plant populations adapted for life in areas of climatic uncertainty will prove to be pre-adapted to withstand the changes that may be brought about by a warmer world?

Numerous illustrations have been included as a reminder of the place of plants in their habitats and that whatever may be learnt from the application of sophisticated methods of investigation it is the existence of the plant in its environment that has prompted our initial curiosity.

Acknowledgements

This book would never have been finished if it were not for the many colleagues and friends who have been willing to give me the time and benefit of their specialist knowledge. I am especially grateful to colleagues who have read particular chapters, Professor R. J. Abbott (St Andrews), Professor R. Brändle (Berne), Professor F.-K. Holtmeier (Münster), Professor Ch. Körner Bale, Professor D. Tomback (Colorado), Dr L. Nagy (Glasgow), Professor S. Payette (Québec) and Dr C. Vassiliadis (Paris). They may have saved me from error; if not, the fault is entirely mine. Many others have provided invaluable help in sourcing data and providing illustrations from all corners of the globe.

I am particularly grateful for detailed documentation as well as access to extensive collections of images from distant places to Professor R. Cormack (St Andrews), Dr A. Gerlach (Oldenburg), Professor F.-K. Holtmeier (Münster), Professor R. Jefferies (Toronto), Dr L. Nagy (Glasgow), and Professor J. Svoboda (Toronto). The privilege of using these images is acknowledged in the legends.

My own opportunities for studying plants in different parts of the world have been greatly aided by generous assistance from the Natural Environment Research Council, the Carnegie Trust for the Universities of Scotland, the Leverhulme Foundation and the Erskine Trust of the University of Canterbury (New Zealand).

This work would never have been undertaken had it not been for the stimulation and encouragement provided by the Cambridge University Press and I am particularly indebted to Dr Alan Crowden for the initial imaginative prompting that made me attempt this task, and to Dr Dominic Lewis and the production staff for bringing it to completion.

Part I
The nature of marginal areas

Fig. 1.1   A marginal population of mountain pine (*Pinus mugo*) colonizing stabilized scree slopes in the Vercors Regional National Park (France).

1 · Recognizing margins

1.1 DEFINING MARGINS

All species have limits to their distribution, and populations that demarcate margins demonstrate an end-point in adaptation to a changing environment. Margins are therefore of particular interest as they represent limits to survival that may alter with climatic change. Plants are ideally suited for the study of peripheral situations as their sedentary nature facilitates mapping and historical recording. Many plant atlases record limits to plant distribution both past and present (Hultén & Fries, 1986; Meusel & Jäger, 1992). Boundaries can also be observed between biomes (vegetation formations characterized by distinct life-forms) as in the latitudinal and altitudinal limits to tree growth. The interface between one vegetation type and another can vary as to whether it is abrupt and easily visible even at a distance (*limes convergens* or *ecotone*), as in the natural treelines of the *Nothofagus* forests in the Andes (Fig. 1.2), or whether it is diffuse, as one vegetation zone gradually merges into another (*limes divergens* or *ecocline*) as at the interface between the southern limits of the boreal forest and the northern limits of the deciduous broad-leaved forest. In this latter case, a more quantitative approach is needed for monitoring change, which may be ecologically just as significant as the movement of discrete boundaries (Fig. 1.3).

Significant plant migrations are to be expected as a response to climatic change. However, care is needed in distinguishing climate-induced changes from the current effects of widespread alterations in land use. Environmental change is likely to create diffuse boundaries as one community gradually replaces another but this can also result from changes in land use. Many alpine pastures are no longer grazed in summer with their former intensity and in many areas this has allowed a gradual uphill advance of tree cover even on steep mountain sides (Fig. 1.1).

In terms of plant distribution, living at the edge of any habitat or community poses the question: why do these plants grow there and no further? Numerous biological disciplines have applied their respective methods in the investigations of limits. Biogeography, demography, reproductive biology, physiology and genetics all provide cogent explanations as to why limits occur where they do. Each discipline is correct in its own particular way and can provide adequate answers to the questions that are asked. Whether or not the different disciplines provide an answer as to a prime cause for any particular limit depends entirely on human perceptions of the problem. Where there are no geographical obstacles, physiological failure might be expected to account for the inability of a species to survive. On the other hand a geneticist might explain the boundary as due to a lack of variation and failure to adapt (see Section 2.2), or merely dismiss the importance of boundaries with the comment that 'plants are static and it is their genes that migrate'. By contrast a demographer would claim that boundaries are no more than the place where recruitment finally fails to balance mortality. The recognition of boundaries can depend therefore in large measure on what the observer is capable of seeing.

Recognizing a margin and explaining the reasons for its existence are interconnected processes. What is observed as a margin can depend on the manner of observation. Life-forms present obvious boundaries that are immediately visible, as with treelines, or when rising water tables cause meadows to become marshes and bogs or changes in coastal topography cause dunes to give way to flood-prone dune slacks. Other boundaries may be discernible only when species or populations are examined in detail for morphological, demographic or genetic characteristics. Understanding the causes of these wide-ranging limits requires an equally comprehensive approach in recording the nature of variation in plant populations. The negative effect of warm winters on the carbon balance of certain arctic species can explain their southern limits, while their northern boundary is more likely to depend on the time available for growth and resource utilization rather than resource acquisition (see Section 5.6.2). Plants with adequate reserves and low resource demands may be able to survive a negative balance for many years and thus ensure continuance of peripheral populations during temporary episodes of climatic deterioration. The study of marginal areas therefore requires perceptive recording of relevant biogeographical data for matching with possible causes of limits to distribution from what is known about demography, physiological requirements and genetic variation.

1.2 MARGINS AND CLIMATE CHANGE

Despite varying concepts as to what constitutes a margin, boundaries provide an opportunity for observing

Fig. 1.2   *Limes convergens* as seen in two natural treelines in Patagonia at the frontier between Chile and Argentina (40° 30′ S; 70° 50′ W). Below the snow-covered peaks can be seen an upper limit to tree survival with the deciduous southern beech (*Nothofagus pumilio*). Below is the upper limit for the evergreen *Nothofagus dombeyi*.

limits to plant survival. Consequently, it is the condition of plants at the approach to a boundary and how they may respond to environmental changes that provides much of the subject matter of this book.

Comparisons between geographically different areas need to be made with caution. Although some boundaries such as treelines and scrub zones may appear similar, and contain species with comparable functional types, they may have geographically different evolutionary and ecological histories and be controlled by diverse environmental factors. Isomorphism – the occurrence of similar forms in unrelated taxonomic groups – is common in plants. It can therefore be dangerous to make global generalizations without careful examination. Even within a single biome, boundary positions may differ in their response to climate change.

In the interface between the arctic tundra and the boreal forest (see Chapter 5), the intuitive pre-diction would suggest that the tundra will retreat and the boreal forest should expand in response to climatic warming. However, in Alaska, the treeline is currently at its most northerly Holocene extent, while in north-eastern Canada there has been a retreat since the mid Holocene (Edwards & Barker, 1994). A forest retreat southwards (see Section 5.3.1) has been observed in the Siberian Lowlands (Kremenetski *et al.*, 1998). In Finland an inland rather than a northward migration of the bulrush (*Typha latifolia*; Fig. 1.4) has been attributed to reduced ice cover rather than just warmer temperatures at higher latitudes (Erkamo, 1956).

Some well-defined communities contain groups of species sharing a common boundary (e.g. salt marshes and wetlands; Figs. 1.5–1.6). In these cases it is possible to define such specific plant communities in terms of their present-day species composition. Nevertheless, in terms of geological time these assemblages are only temporary, and species and population aggregations have no biological permanence. The late-glacial floras of Great Britain and Denmark and Russia consisted of communities which contained species which are now dispersed as plants of tundra, steppe, mountains and dunes. In the same way, the northern European forests during the Holocene had an ever-changing species composition (Huntley, 1990).

Fig. 1.3   *Limes divergens* as seen in autumn in northern Vermont, USA showing deciduous trees gradually interspersing and replacing coniferous forest.

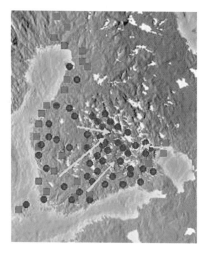

Fig. 1.4   Eastward spread of *Typha latifolia* in Finland 1900–50. Symbols: red, at or before 1900; blue, new records 1926–50. (Adapted from Erkamo, 1956.)

Species survival in peripheral regions is not controlled merely by the impact of average probability of adverse climatic conditions. Survival is also profoundly affected by competition from other species and by the frequency of extreme events, such as drought, flooding, freezing and disturbance. These periodic events, sometimes measured in terms of decades or centuries, can create a complicated pattern for limits to plant existence both geographically and at a microsite level. Marked differences can also be found between populations of the same species. One of the most widespread and ancient of arctic flowering plants is the early flowering purple saxifrage (*Saxifraga oppositifolia*; see Section 6.4.2). Plants of this species from the High Arctic (Spitsbergen) can be deprived of oxygen for months when encased in ice and yet survive. Such a well-developed tolerance of anoxia is not found in more southerly populations of this species (Crawford *et al.*, 1994).

Fig. 1.5   Community limits imposed by a regular flooding regime. A very distinct boundary is visible between the non-flooded dune vegetation and the upper limit of flooding in a salt marsh marked by a zone of *Suaeda vera* near Romney, south-east England.

Boundaries can also be found at a local level, particularly in regions with marked seasonal variations in climate. Many local factors modify the degree of exposure of plants to the adverse season with the result that boundaries can be found in relation to topography, geology, and soil type. Changes in bedrock, particularly at outcrops of chemically basic rocks such as limestone can produce an abrupt boundary that extends over considerable distances (Fig. 1.7). Flooding frequency and depth is also a powerful discriminator between plant communities and creates easily visible boundaries (Figs. 1.5–1.7).

1.3   LIMITS TO DISTRIBUTION

In many demographic studies, edge effects or boundaries have traditionally been regarded as merely 'nuisances'. Nevertheless, they can be modelled and investigated for the dynamic properties of populations at the edge of their distribution. Such studies can be used to provide graphic demonstrations that the distribution patterns of individuals at population edges have distinctive properties due to the influence of space on population margins. Processes occurring at the margin of populations are likely to have an influence

Fig. 1.6   Aerial view of wetland plant community zonation.
Boundaries between plant communities on the shores of
Loch of Kinnordy, Angus, as photographed in 1969.
(Photo J. K. S. St Joseph.)

on the extension and distribution, particularly when
populations are small and localized (Antonovics
et al., 2001).

1.3.1   Physiological boundaries

Figure 1.8 is an attempt to divide the physiological
limitations to species distribution into a number of
discrete categories. Physiological requirements for
plant survival are considered under the heading of
resource availability, which involves biological properties and can therefore be separated from purely
physical limits to viability such as heat and cold tolerance. As modelling studies suggest, individuals at the
margins of populations may be exposed to special
spatial features resulting from a different relationship
with their neighbours which can be influenced by a
variety of factors including seed production, pollen
dispersal, gene flow and the availability of potential
sites for establishment (Antonovics *et al.*, 2001). The
margin is also a region where environments oscillate.
For sedentary organisms, particularly if they require

several years to develop before they can reproduce, this
poses a particular problem. First and foremost, the
long-term survival of individuals will depend on their
physiological capacity to survive in a fluctuating
environment.

Whatever their location in a population, be it
marginal or at the core, all individuals if they are to
survive have to attain a positive carbon balance and
secure adequate supplies of nutrients, light, and water.
This does not have to be steady-state existence. For
some species access to certain resources can be interrupted for prolonged periods without causing any
serious injury. Consequently, in the study of limits to
distribution, particular attention has to be paid to the
timing of environmental stresses and the frequencies of
extreme events. Ultimately, the survival of any individual or population is related to the relative resource
needs of the species in question as compared with their
competitors. This, however, cannot be observed in any
one growing season. Short-term experiments (e.g. 3–4
years), which can include raising the ambient temperature with shelters or giving additional nutrients,
therefore have limited value. Species that succeed in
capturing resources to the detriment of a less vigorous
competitor, in what for these dominant plants may have
been a series of favourable growing years, may ultimately be excluded from the habitat if they cannot survive at other times when resource levels are reduced.
The species with the smaller demand may have therefore a greater long-term viability (Section 3.6.2).

1.3.2   Resource availability

In ecophysiological studies it is often considered
desirable to have some common unit for assessing the
relative viability of plant populations as they approach
limits to their distribution. Given that plants may be
limited by different resource deficiencies at various
stages in their life cycle, carbon balance is commonly
used as an appropriate currency for measuring success
in resource acquisition as it is the investment of carbon
that makes possible the acquisition of all other plant
resources. In recent years the acquisition of resources as
affected by environmental factors has been much
studied, possibly because carbon acquisition either in
individual plants or whole communities is readily
monitored by recording carbon dioxide flux from

Fig. 1.7   Boundary zone marking the interface between limestone and acid Torridonian sandstone in the Scottish north-west Highlands. View looking north to Elphin with cliffs of Durness limestone on the right (east) supporting a calcicole flora that contrasts with the bog and acid mountain vegetation of the Torridonian sandstone and Lewisian gneiss lying to the north-west.

individual leaves, whole plants, or even forest canopies (Lee *et al.*, 1998). In this approach, limits to plant distribution can be viewed in terms of carbon balance with the potential theoretical physiological limit for any species or community being reached only when carbon gain is no longer greater than expenditure. Current interest in climate change and the desire to be able to predict through modelling future limits to plant distribution makes the use of a general metabolic currency, such as carbon, a potentially attractive proposition.

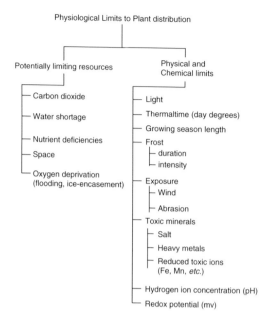

Fig. 1.8   A selection of physiological factors that can impose limits on plant distribution.

If such a generally applicable method could be found to monitor energy resources it might serve to detect potential carbohydrate impoverishment before the plants in question showed any other signs of decline in viability.

In some cases it is possible to demonstrate a relationship between carbon balance and limits to distribution and physiological viability. Drought and cold both reduce the potential for carbon acquisition. In addition, certain stressful environmental situations can further deplete carbon reserves by causing an increase in carbohydrate consumption. The depletion of carbohydrate reserves by high respiratory activity at warm temperatures has often been suggested as a factor that limits the southward extension of northern species (Mooney & Billings, 1965; Stewart & Bannister, 1973). Various case histories relating plant distribution to carbon balance are discussed in Chapter 3. Although in certain cases carbon starvation is associated with failure to survive it is not a universal method that can be applied crudely to all plant forms in any situation for determining their distribution limits as there is remarkably little evidence that carbon starvation is the primary cause for either woody or herbaceous species failure at the cold end of their ecological distribution.

The most thoroughly examined aspects of the ecology of plants in relation to temperature are the altitudinal and latitudinal limits for the survival of trees. Many studies have sought to determine whether or not the low temperature regimes of high latitudes and altitudes cause trees to come into a carbon balance deficit. Physiologically, this would appear a simple and logical explanation of the effect of low temperatures on tree survival. It might be expected that as woody plants devote a considerable part of their resources to the formation of non-productive trunks and stems they may be unable to support such a growth strategy when growing seasons are cool and short. However, an extensive worldwide study of the carbon balance in trees at their upper altitudinal boundaries has shown the converse, namely, that tree growth near the timberline is not limited by carbon supply (Fig. 1.9) and that it is more probable that it is sink activity and its direct control by the environment that restricts biomass production of trees under current ambient carbon dioxide concentrations (Körner, 2003).

A worldwide study looking at numerous thermal indicators found that a growing season mean soil temperature of 6–7 °C provided the best generalized indicator of montane treelines from the tropics to the boreal zone (Körner & Paulsen, 2004). The soil temperature provides an approximate indicator of thermal conditions at the treeline and suggests that an edaphic thermal summation of growing season conditions could replace the older Köppen's Rule that the limit to tree growth coincided with the 10 °C isotherm of the warmest month of the year (Fig. 1.10). This modification from a measurement indicating maximum warmth to a temperature mean for the entire growing season reflects a realization that the altitudinal limits to tree growth are not directly related to the ability to make a net carbon gain. Instead, the treeline is more likely to be related to the length of growing season that is needed for the production and development of new tissues (Körner, 1999). However, in any discussion between cause and effect it is necessary to remember that mean temperatures do not exist in nature and therefore should be considered merely as indicators and not causal factors. The same is true for mean soil temperatures (Holtmeier, 2003).

Although overall carbon starvation is not a feature of large woody plants at their upper limits of distribution, it is nevertheless possible that certain organs

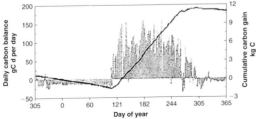

Fig. 1.9 (Above) The Swiss stone pine (*Pinus cembra*) growing as a timberline tree in the Alps with the Matterhorn in the background. The stone pine is an example of a tree which shows no signs of suffering a carbohydrate shortage even at its highest locations. (Below) Seasonal course of daily carbon balance and the net cumulative carbon exchange of an experimental *Pinus cembra* tree. The data were collected between 5 November 2001 and 31 December 2002. (Reproduced with permission from Wieser *et al.*, 2005.)

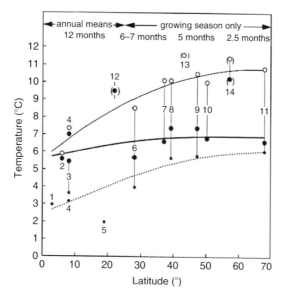

Fig. 1.10   Temperature at alpine treelines for various latitudes. Upper line: mean temperature for the warmest month. Middle heavy line: mean temperature for the growing season. Lower line: mean temperature for the growing season for patches of forest from very high altitudes. Treeline altitudes (and extreme tree limits): 1, Kilimanjaro (3950 m); 2, Mt Wilhelm, New Guinea 3850 m (4100 m); 3, Bale Mts, Ethiopia 4000 m (4100 m only approx. estimates of temperature); 4, Venezuelan Andes 3300 m (4200 m); 5, *Polylepis* record, northern Chile (4900 m); 6, Khumbu Himal (Everest region) 4200 m (4420 m on sunny slopes, 200 m less on shaded slopes); 7, White Mts, California 3600 m; 8, Colorado 3550 m; 9, Central Alps 2100 m (2500 m); 10, Rocky Mts, Alberta 2400 m; 11, Northern Scandinavia (Åbisko) 680 m (750 m). Island mountains not considered in lines fitted to data: 12, Mauna Kea, Hawaii 3000 m; 13, Craigieburn Range, New Zealand 1300 m; 14, Cairngorms, Scotland 600 m. (Data compiled by Körner, 1999, from various sources and reproduced with permission.)

can be seriously carbon deficient under specific conditions. When carbon balance is viewed in relation to plants with a large biomass such as trees it is easy to overlook the necessity of examining the vitality of specific organs. The fact that one summer day's photosynthetic activity near the timberline of the Swiss stone pine (*Pinus cembra*) is equivalent to the total respiration for the entire winter (Wieser, 1997) can easily lead to the conclusion that a carbon deficit is not likely to be a limiting factor. Unfortunately, this whole-plant approach overlooks the fact that in many cases it is certain vulnerable key tissues and not whole plants that are likely to suffer from carbon deficits. The root tips in Sitka spruce (*Picea sithchensis*) become depleted of carbohydrate in winter, particularly when the soil is flooded or merely water saturated. This is not immediately harmful to the roots, but when air returns in spring, when the water table subsides, the roots show dieback as a result of post-anoxic injury. Reduced carbohydrate availability can also be expected to cause a

deficiency in antioxidants and thus render the root tips vulnerable to membrane damage when suddenly re-exposed to oxygen. Again this is not immediately harmful as there are usually sufficient upper roots to keep the tree alive. However, when repeated over a number of seasons the resultant shallow rooting of trees in wet soils makes them highly vulnerable to wind-throw. The spruce tree as a whole is never carbon limited (Jarvis & Leverenz, 1983), yet this type of injury, which can be traced to a localized carbohydrate deficiency, is one of the severest limitations to Sitka spruce cultivation in oceanic conditions. Despite the fact that Sitka spruce is capable of making a net increase in biomass in every month of the year in the British Isles, it is the carbon depletion of the root meristem in warm, water-saturated soils that brings about the eventual collapse of trees (Crawford, 2003). Similarly, intermittent mild winter periods can deplete carbohydrate levels in shoots of *Vaccinium myrtillus*, and the progressive respiratory loss of cryoprotective

Table 1.1. *Definitions of ecological concepts used in discussion of the nature of margins*

Concept	Definition	Reference
Biome	A biological division of the Earth's surface that reflects the ecological and physiognomic character of the vegetation.	Allaby, 1998
Competition	*Consumptive competition*: Exploitative competition – the simultaneous demand by two or more organisms or species for a resource that is actually or potentially in limited supply.	Calow, 1998 Harper, 1977
	Interference competition: the detrimental interaction between two or more organisms or species.	Grubb, 1977
	Pre-emptive competition: the ability to occupy an open site and this pre-empted space.	
Deprivation indifference	The ability to survive a temporary absence of an essential resource.	Crawford, 1989
Ecological release (competitive release)	The expansion of the range of a species when competition for its niche is removed.	Wilson, 1959 Allaby, 1998
Ecotone	A narrow transition zone between two or more different communities. Such edge communities are usually species rich.	Allaby, 1998
Fitness: immediate and long-term	The ability of a given genotype to contribute to subsequent generations. Populations that maintain a high degree of variability can be considered to have long-term fitness as they are pre-adapted to environmental alteration.	Lincoln *et al.*, 1998 Crawford, 1999
Isomorphism	Apparent similarity of individuals of different race or species.	Lincoln *et al.*, 1998
Köppen's Rule	The low temperature limit for tree growth is reached when the mean temperature of the warmest month of the year is less than $10\,°C$.	Köppen, 1931
Life history traits / strategies	Major features of an organism's life cycle most directly related to birth and death rates.	Calow, 1998
Limes convergens	A well-defined boundary zone (Latin *limen*, threshold defining the boundary between two major habitat types)	Allaby, 1998
Limes divergens	A diffuse boundary zone (cf. above) where one major habitat changes gradually into another.	Allaby, 1998
Metapopulation	A network of subpopulations isolated in habitat patches. The long-term persistence of the species depends on local (patch) extinction and recolonization and on net gene flow between subpopulations.	Levins, 1970 (see Hanski, 1999)
Montgomery effect	The ecological advantage conferred by low growth rates for survival in areas of low environmental potential.	Montgomery, 1912
Niche	The requirements that the environment has to meet to allow the persistence of a species or population.	Calow, 1998

Table 1.1. (cont.)

Concept	Definition	Reference
Pre-adaptation	An organism may be described as pre-adapted to a new situation where pre-existing morphological structures or physiological adaptations have been inherited from an ancestor for a potentially unrelated purpose.	Allaby, 1998
Refugium (plural refugia)	An area that has escaped major climatic changes typical of a region as a whole and acts as a refuge for biota previously more widely distributed.	Lincoln *et al.*, 1998
Relict species (climatic and evolutionary)	Persistent remnants of a formerly widespread fauna or flora existing in isolated areas or habitats.	Lincoln *et al.*, 1998
Resource	Any component of the environment that is consumed by an organism.	Section 3.1
Ruderal species	A species of open, disturbed conditions usually resulting from human activity (Latin *rudus*, *rudera*, rubble, rubbish dump)	Grime, 2001
Tolerance: physiological and ecological	The range of environmental factors in which an organism can survive. In absence of competition: physiological tolerance; in natural surroundings and exposed to competition: ecological tolerance.	Walter, 1960

sugars renders the shoots sensitive to frost damage, a phenomenon also observed in *Picea abies* (Ögren, 1996).

The desire to have a common currency for measuring potential distribution limits in physiological terms can obscure the realization that even along a simple environmental gradient many factors operate on plant survival that cannot be quantified under the common currency of carbon balance. In the case of the treeline cited above, time for resource utilization in growth rather than carbon acquisition may be the overruling limit to survival. A further disadvantage of using resource acquisition in general as a measure of potential limits to distribution comes from the fact that most species have broadly similar resource requirements (Grubb, 1977). Where plant species do differ is in their relative tolerance of adverse factors. Earlier works on plant ecology in the last century stressed *ecological tolerance* (see Table 1.1) as one of the most important factors in plant ecology (Walter, 1960). When the objective of the study is assessment of the productivity of a community such as a stand of forest trees, then no objection can be made to assessing carbon balance. However, if the aim of the study is to deter-

mine the causes of distribution limits, then the investigation should be free of any concept of yield or productivity and concentrate on the ultimate evolutionary criterion, namely survival.

1.3.3   Resource access and conservation in marginal areas

Limits to distribution are determined not just by the ability to acquire resources, but also by their utilization and conservation. Prolonged snow cover, flooding, drought or disturbance can impede the ability of plants to access and conserve resources. Many of the adaptations that allow plants to live in marginal areas have evolved as means of overcoming limited access to resources. One of the commonest solutions for survival on minimal resources, including carbon, is to reduce demand, which has similarities with the concept of *deprivation indifference* mentioned above. In an earlier study of North American cereals the concept of the ecological advantages of low growth rates was encapsulated in what has become known as the *Montgomery effect*, which states that 'in areas of low environmental potential ecological advantage is conferred by low growth rates' (Montgomery, 1912).

Species with widespread distributions such as the circumpolar polar arctic–alpine species *Oxyria digyna* and *Saxifraga oppositifolia* achieve their geographical spread in part by altering their size and form, both phenotypically and genotypically, and thus reducing the need for resources as measured in absolute terms of carbon sequestration. Some arctic species are so adapted to conserving resources that they can survive under continuous snow cover for two to three years in succession. Populations that live in areas where

Fig. 1.11   Examples of species that can survive prolonged and continuous snow cover for more than one growing season. (Above) The dwarf buttercup (*Ranunculus pygmaeus*; scale divisions = 1 cm). An extreme dwarf species of *Ranunculus* of circumpolar arctic and subarctic distribution that grows among moss and beside streams and often beside glaciers and snow drifts and therefore in some years remains buried for an entire growing season. (Below) The blue heath (*Phyllodoce caerulea*), a species in danger of disappearing from the mountains of Scotland due to the lack of winter snow.

this hazard is relatively frequent usually consist of diminutive specimens of certain widespread species such as *Polygonum viviparum*, *Oxyria digyna*, *Salix polaris*, and *Ranunculus pygmaeus* (Fig. 1.11).

Plants in marginal habitats, as in coastal habitats, semi-deserts and in the Arctic, frequently show a high degree of polymorphism. No one form or ecotype is continually favoured as environmental conditions are continually oscillating, causing populations to consist of a mixture of different forms. Such polymorphic populations aid survival in fluctuating environments by providing a number of ecotypes usually with slightly different habitat requirements (see Table 1.1). The existence of stable assemblages of interbreeding, yet distinct ecotypes, is usually described as a *balanced polymorphism*. Balanced polymorphisms confer *immediate fitness* by increasing the ecological tolerance of the species as a whole. They also enhance *long-term fitness* as the existence of a range of adaptations in different yet interfertile ecotypes can pre-adapt a species to long-term climatic oscillations (Crawford, 1997a). Examples of this phenomenon are readily visible at high latitudes (see Chapter 2) where the constant risks of disturbance and environmental stress maintain a pronounced degree of polymorphism in arctic plant populations. These polymorphisms can be seen in the exposed ridge and snow bank forms of *Dryas octopetala* as well as in the different forms of *Saxifraga oppositifolia* found in High

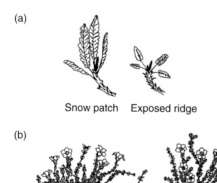

(a)

Snow patch   Exposed ridge

(b)

Fig. 1.12 Ecotypic variation related to growing season length in arctic plants. Snow patch (left) and exposed ridge (right) forms of (a) *Dryas octopetala* and (b) *Saxifraga oppositifolia*. (The *D. octopetala* figure is reproduced with permission from McGraw & Antonovics, 1983, and *S. oppositifolia* from the drawing by Dagny Tande Lid, in Lid & Lid, 1994.)

Arctic coastal sites on the beach ridges and low shores (Crawford, 2004). Both these species have evolved forms that are adapted to short and long growing seasons in the same geographical area (Fig. 1.12).

1.4 GENETIC BOUNDARIES

The use of molecular genetic markers has now opened up an entirely new dimension in our ability to trace the migration and development of plant populations, especially during the post-glacial vegetation re-advance into marginal areas. We now know that Britain received oaks, shrews, hedgehogs and bears from Spain, and alder, beech, newts and grasshoppers from the Balkans (Hewitt, 1999).

In relation to marginal oceanic areas the post-glacial history of oak is of particular interest. The distribution of some of the principal lineages of white deciduous oak is shown in Fig. 1.13. It is noteworthy that the post-glacial population that re-established itself in the west would have had a long period of residence in the more oceanic regions of Iberia (Brewer *et al.*, 2005). Differentiation of the cpDNA haplotypes of this lineage as it now occurs within Britain and Ireland (Fig. 1.14) reveals a further differentiation between the oceanic regions of Ireland and western Britain as opposed to the more continental regions to the east, with one particular lineage being more predominant in the more oceanic regions (Lowe *et al.*, 2005).

Further examples of the use of genetic markers in tracing plant migrations are discussed in Chapter 6 in relation to the long-term survival of the purple saxifrage (*Saxifraga oppositifolia*).

1.5 DEMOGRAPHIC FACTORS

Demographic limits to plant distribution include those factors that adversely affect recruitment or increase mortality. Demography therefore involves many factors that are not the property of species or populations but are instead a function of habitat and location (Antonovics *et al.*, 2001). Thus geographic barriers to dispersal as well as the provision of microsite space within a habitat both come under this heading. The principles of *Island Biogeography* (MacArthur & Wilson, 1967) relate species numbers on islands to an equilibrium that becomes established between the rates

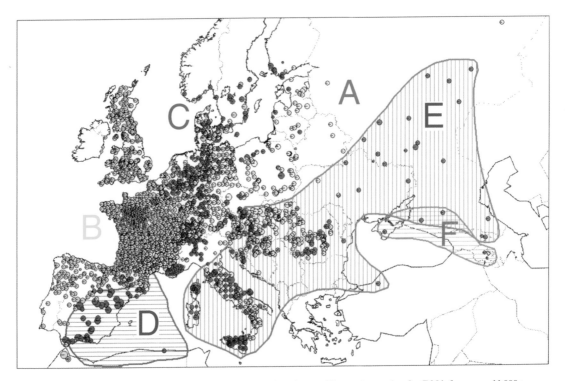

Fig. 1.13   Distribution of five lineages of deciduous oak (A–E) based on a wide-ranging study of cpDNA from over 11 000 trees. The spatial patterns reflect regions of glacial refugia and subsequent migrations. The distribution of lineage (B) indicates the distinct nature of these oceanic marginal populations. (Reproduced with permission from Petit *et al.*, 2002.)

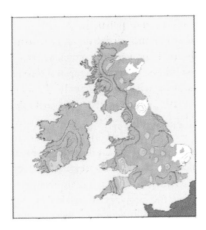

Fig. 1.14   Geographic differentiation of the distribution of cpDNA haplotypes for British and Irish oaks after applying a Kriging average where green are areas with no overall dominance, and white, orange and yellow represent areas of dominance for the three most dominant haplotypes. (Reproduced with permission from Lowe *et al.*, 2005.)

at which species colonize an island as compared with the rates at which they become extinct. For any particular island, species richness depends on the area, topography, number of habitats on the island, as well as accessibility to a source of colonists and the species richness of that source. These principles clearly illustrate the necessity in demography to consider the nature of the habitat and how it modifies recruitment and mortality rates, particularly at population margins where opportunities for recruitment and the hazards of extinction are different from core locations. Patchiness that occurs in marginal situations can frequently be found to accelerate population decline and decrease species richness (Eriksson & Ehrlen, 2001). However, the effect is not consistent as there are numerous instances where local species richness can be increased as fragmentation can aid seedling recruitment.

The soil seed bank is also a space with a demographic dimension. The number of embryo plants that lie dormant in the soil until there is an opportunity for them to germinate adds both numbers and variation to nascent populations. Disturbance can also be included under the heading of demography as it affects both recruitment and mortality. Ruderal plants (Latin *rudera*, broken rocks) are defined as plants that can withstand frequent and severe physical disturbance (Table 1.1). The disturbance factor can be discussed as a negative phenomenon, as are all processes that destroy plant biomass (Grime, 2001). Nevertheless, from a demographic point of view, disturbance also has a positive aspect in providing fresh space for colonization, as does rejuvenation of communities in aiding diversity through limiting the extent to which any one species can permanently dominate a habitat. At the subspecies level, metapopulation development on a larger scale is facilitated in sites that are frequently disturbed and where extinctions create opportunities for new migrants (Hanski, 1999).

Other biological factors influencing recruitment and mortality are those characteristics which are often referred to as *life history traits* or *life history strategies* (for definitions see Table 1.1). These include size, growth pattern, resource storage, as well as reproductive strategies such as whether a plant is an annual or a perennial, male, female or hermaphrodite, a rapid reproducer with many seeds, or a slow reproducer with few seeds. Theoretically, species are expected to select for an optimal set of characteristics that will produce the highest population growth rate in a particular environment (Sibly & Antonovics, 1992). The search for such optimal sets of characteristics can be studied experimentally under controlled conditions as well as with numerical models, but in the field where environments fluctuate from one year to another the attainment of an optimal solution is elusive and will probably always remain so.

1.5.1 Limits for reproduction

Compared with animals, plants are endowed with a broad and versatile range of reproductive mechanisms. Consequently, there is a continuing debate about the nature of plant species and whether they differ from animal species as independent lineages or 'units of evolution', or are merely arbitrary constructs of the human mind, which result from the never-ending activities of over enthusiastic taxonomists who insist on giving specific names to the numerous subspecies of dandelions, blackberries and oaks. An analysis of the phenetic and crossing relationships in over 400 genera of plants and animals has indicated that although discrete phenotypic clusters exist in most genera (>80%), the correspondence of taxonomic species to these clusters is poor (<60%). The lack of congruence as perceived by botanists, it is argued, may be caused by polyploidy, asexual reproduction and over differentiation by taxonomists, but not by contemporary hybridization (Rieseberg *et al.*, 2006). This same study pointed out that crossability data indicated that 70% of taxonomic species and 75% of phenotypic clusters in plants correspond to reproductively independent lineages (as measured by post-mating isolation), and thus represent biologically real entities. It can be argued that, contrary to conventional wisdom, when a wide spectrum of plant species is considered and not just the celebrated horror stories of plants such as dandelions, plants are more likely than animal species to represent reproductively independent lineages (Rieseberg *et al.*, 2006).

Despite the findings of the above study, fertile hybrids between species are common and chromosomal barriers to fertility where they arise can be removed as a result of polyploidy. Consequently, it can be argued that plants differ from animals in that there is much greater gene flow between species. Plants also differ from animals in being sedentary and are therefore strongly selected for particular habitats and as a result ecotypic variation contributes to genetic diversity. Polymorphisms and polyploidy, together with the genetic memory provided by the seed bank, are all powerful means for augmenting genetic variation and provide the flowering plants with a facility for rapid adaptation to change that is rarely found in higher animals. As is argued throughout this book, plants from marginal habitats, are seldom lacking in genetic diversity and in some cases are even more diverse in peripheral than core habitats, with the former frequently being colonized by hybrid species (for references see Chapters 2 and 4).

1.6 RELICT SPECIES AND CLIMATE CHANGE

Climatic change lies at the base of the restricted range of most relict species. Species that were once widespread and now exist in isolated 'islands' provide examples of a marginal existence that in many cases can be related to specific causes (Milne & Abbott, 2002). Not all relict species are in danger of extermination as some of the colonies that still exist, although only covering a fraction of their former distribution, frequently contain populations that are both numerous and viable in terms of regeneration. Endangered relict species are those that exist only in small populations throughout their remaining distribution and which could be eliminated easily by exceptional climatic events or human disturbance. Relict species represent a challenge to the view that for species to survive in a changing world they either have to migrate or adapt. Their continued survival excites curiosity and notice as they are frequently remarkable examples of ancient and now isolated biota that have been fortunate enough to find a habitat where survival has been possible despite a changing world. To encompass the disparate biogeographical histories of relict species it is convenient to consider them under two headings, namely evolutionary relicts and climatic relicts.

1.6.1 Evolutionary relicts

Evolutionary relicts include groups such as the cycads (Cycadaceae) comprising about 100 species in nine genera, all with limited distribution and confined to tropical and subtropical regions. In a sense, these species, which are sometimes referred to as *living fossils*, could also be considered as climatic relicts, but from a time that has long since passed. These slow-growing cycad species were once dominant features of the vegetation in Mesozoic times when the temperature gradient between the Equator and the poles lay between only 10 °C to 20 °C and much of the world had a tropical climate. Many evolutionary relicts are also found in geographically isolated areas where their secluded existence protects them from excessive competition and predation. Two extreme examples are the Chinese maidenhair tree (*Ginkgo biloba*) and the dawn redwood (*Metasequoia glyptostroboides*).

Ginkgo biloba (Figs. 1.15–1.16) is truly a relict species that is apparently identical to 200 million-year-old fossil specimens. It had long been thought that the species was extinct until the German botanist and physician Engelbert Kaempfer (1651–1716) discovered the tree in Japan on a visit to Buddhist monks in Nagasaki in February 1691. Apparently, ginkgo trees had survived in China and were mainly found in mountain monasteries where they were cultivated by Buddhist monks. From China they were introduced to Japan and Korea. Ginkgo seeds were brought to Europe from Japan by Kaempfer. Whether ginkgo still exists in the wild in China is not certain. Some trees can be seen growing naturally in two small mountainous areas on the border between Zhejiang and Anhwei provinces (Tian Mu Shan Reserve) (Fig. 1.15), in central China in Hubei province and in western China in Guizhou and Sichuan provinces. They may, however, have been seeded there from cultivated trees in temple gardens (Kwant, 2006).

Other notable discoveries of relict species include a deciduous member of the Taxodiaceae, the dawn redwood (*Metasequoia glyptostroboides*, discovered in 1941 and first collected in 1947 (Fig. 1.17). More recently, the Wollemi pine (*Wollemia nobolis*, Araucariaceae) was discovered in 1994 by intrepid abseilers into a ravine in the Australian Blue Mountains in a remote area of the Wollemi National Park (New South Wales, Australia). With only around 40 adult trees known to be growing in the wild, it is one of the world's rarest plant species, and this relict population shows no genetic variation between individual trees.

The genus *Araucaria* itself, formerly of worldwide distribution in Triassic, Jurassic and Cretaceous times, now has a relict distribution with the natural distribution of the classic xerophytic tree, the monkey puzzle (*A. araucana*), now confined to between 37° S and 40° S in the arid zone of Argentina and Chile. Similarly, the Norfolk pine (*A. heterophylla*) is confined to Norfolk Island, and *A. bidwillii* is the tallest tropical tree to occur naturally in Queensland, Australia. A combination of plate tectonics and climatic change has reduced these species to isolated areas, usually in warmer climates which resemble the more widespread conditions of uniform warmth that existed on Earth before the Cenozoic climatic decline.

1.6.2 Climatic relicts

Climatic relict is a term usually applied to plants that have had their distribution curtailed more recently and in which a direct link with more recent climatic change is readily observed. Coastal environments now provide the only refuge for some of the coniferous trees that once had a worldwide distribution. The Californian redwood (*Sequoia sempervirens*), in common

Fig. 1.15 Mature tree of *Ginkgo biloba* growing in Lichuan county, Hubei province (China). In this province are many ginkgo trees, possibly growing in the wild. (Photo Wei Gong, reproduced with permission from the web pages of Cor Kwant, 2006.)

with several species of *Chamaecyparis* and *Taxodium*, had a transcontinental distribution during the Tertiary period but are now found only in very restricted coastal habitats. Apparently the oceanic niche, with its diminution of climatic extremes, together with a reduction in competition from more recently evolved tree species, provides many relict tree species with an environment in which they are still viable. As with many of the

other dominant species of these coastally restricted forests, survival appears to be due in part to the longevity of these species, which can exceed 1200 years for *S. sempervirens* and 3000 years for the Alaskan cedar (*Chamaecyparis nootkatensis*) (Laderman, 1998a).

Many of the tree species in these maritime forests share characteristics that have similarities with the ericoid species of the oceanic heathlands of north-west

Fig. 1.16   Foliage of the maidenhair tree (*Ginkgo biloba*). A relict species in China and apparently identical with fossil specimens 200 million years old.

Europe. In form they are usually evergreen, and possess sclerophyllous foliage. Ecologically they inhabit nutrient-poor soils, and have a requirement for high levels of atmospheric moisture. They also illustrate the limiting effects of such oceanic environments, which have been described as 'success through failure' in that these coastally restricted forests survive in regions where other species have found it impossible to survive (Laderman, 1988b).

The genus *Rhododendron* subsection *Pontica* is remarkable among Tertiary relict groups for its disjunct distribution (Section 4.5.2). *Pontica* lineages have survived the Quaternary in south-west Eurasia, south-east North America and north-east Asia with little or no subsequent speciation (Milne, 2004).

It is not necessary, however, to have to search the shores of the Pacific Ocean or the mountains of Asia for examples of arborescent refugia species with an ericoid growth form. The strawberry tree (*Arbutus unedo*) is a member of the Ericaceae that has long been a member of the Irish flora. There has been debate as to whether the strawberry tree is truly indigenous to Ireland or whether monks introduced it during their pilgrimages in the early Christian period to the Iberian Peninsula to shrines such as Santiago de Compostela. However, investigations of the pollen record have now shown it to have been present for over 3000 years in County Kerry and 2000 years in County Sligo and this tree can therefore be added to the worldwide list of ericoid trees that have survived as relict forests on islands (Mitchell, 1993). Whether or not this tree survived in Ireland or in some neighbouring area at the time of the Late Glacial Maximum (18 000 BP) has not yet been determined.

Many species that once had a wide distribution at middle latitudes during post-glacial times retain a scattered and fragmented occurrence on mountains. These climatic relicts are particularly noticeable in Europe due to the predominantly east–west orientation of the mountains as compared with the more favourable corridors for adjustment to climatic change that are found in the north–south orientated mountains such as

Fig. 1.17   Botanic garden specimen of the dawn redwood (*Metasequoia glyptostroboides*). This deciduous conifer was only discovered in 1945 and seeds collected in 1947; it has subsequently been introduced widely to western gardens.

Fig. 1.18   The snake's head fritillary (*Fritillaria meleagris*) growing in Magdalen meadow (Oxford). Probably a naturalized introduction as it is not recorded in Britain before 1732.

the Andes and the Rockies in the New World, and the Altai mountains of Asia, and the Urals at Europe's eastern fringe. Some montane species are currently in retreat due to climatic warming at a rate that can be observed over decades (see Chapter 10).

1.7   ENDANGERED SPECIES

Many countries publish lists as *red data books* of species that are in danger of becoming at least locally extinct. Such lists have prompted many studies into the ecology of these species in an effort to ensure their survival. In most cases the marginality of the species is due to the destruction of their habitat by human activity. In some cases the species have very narrow climatic and habitat limits which makes them highly dependent on the preservation of a restricted set of environmental conditions. In the British Isles, one such example is *Primula scotica* which is restricted to the north coast of Scotland and parts of the Orkney archipelago and survives mainly in maritime sedge heath that occurs just to landward of cliff edges. Not only is this species limited now to less than 30 sites, it also appears to have

a narrow climatic window for reproduction, with seedling production dependent on mild winters and absence of storms at flowering time (Fig. 2.3).

Other endangered species may have a wider ecological niche but their habitats may be in areas that have been subjected to large-scale changes in land use. The disappearance of water meadows has reduced the distribution of many orchid species. The snake's head fritillary (*Fritillaria meleagris*), once widespread in damp meadows and pastures in southern England (Fig. 1.18), has suffered from modern techniques of pasture improvement and is restricted to areas where it is especially protected. Magdalen College Meadow (Oxford) is claimed to be the major refuge of the British population. Cleansing of beaches with removal of litter is also destroying the regeneration niche for many coastal species. Sea kale (*Crambe maritima*) is an endangered species found at the foot of cliffs where it

Fig. 1.19   Sea kale (*Crambe maritima*) growing on shingle at the base of a cliff in south Fife, Scotland. A species which reaches its northern limit of distribution in Fife and which could be expected to migrate northwards if there is sufficient climatic warming and its habitat is not excessively disturbed.

grows on sites with decaying seaweed (Fig. 1.19). Disturbance of shingle beaches, possibly aggravated by climatic warming, is causing a retreat northwards of another relatively rare shingle species, the oyster plant (*Mertensia maritima*).

The reclamation of hill land for improved pasture has resulted in a drastic reduction in moorland throughout the Atlantic seaboard of Europe. Particularly severe in terms of species loss has been the removal of many wetlands through drainage and peat extraction. On a European scale, sedges, rushes, irises, orchids and gladioli are all groups of species which have particularly suffered through habitat destruction.

1.8   AGRICULTURAL MARGINS

A striking feature of many early human settlements is the early date at which farmers settled in areas that would have been climatically peripheral for agriculture even under climatic conditions that were warmer than at present. Mesolithic and early Neolithic sites are to be found in northern coastal sites in Scotland and northern Scandinavia that must have been subject to a substantial risk of crop failure. Here and elsewhere

Fig. 1.20   Early Neolithic settlement at the Scord of Brouster, Shetland. The settlement appears to have been first occupied about 4500 BP and shows evidence of cultivation of barley as well as hunting. The end of occupation approximates to the beginning of peat growth at the site around 3500 BP. The end of occupation may also have coincided with the soil profile having become podzolized and many house sites in similar marginal areas may have been abandoned at this time (Whittle *et al.*, 1986).

the practice of agriculture can be viewed as an early attempt at risk reduction in securing an adequate food supply. If marginality in relation to agriculture is quantified by an assessment of risk in terms of frequency of crop failure then nowhere is this risk more easily demonstrated than in areas with limited growing seasons. In oceanic habitats in north-western Europe high lapse rates (the rate at which temperature decreases for each unit increase in height in the atmosphere) coupled with wind exposure magnify the risk of crop failure with increasing altitude. It is therefore always surprising that many early settlements were at considerable altitudes on hillsides. Up until the Bronze Age the climate was probably warmer, which increased the upland area where crops such as barley could be grown on what are now exposed hillsides with impoverished soils (Fig. 1.20). Striking examples of such Neolithic and Bronze Age settlements are still visible in the remoter regions of the British Isles such as the Orkney and Shetland archipelagos (Section 11.2).

Current interest in climatic change has prompted extensive research into future limits for specific crops in comparing where they presently reach a limit in their viability and where these limits will lie, given a specific degree of climatic warming. Areas where wheat and potato are now grown with satisfactory yields may become marginal as future productivity may be limited by water shortages. Conversely, other areas with more oceanic climates may benefit from increased temperatures and become more productive.

The development of agriculture in Asia Minor and the Levant was once thought to have been due to increased drought as the pluvial period of the Pleistocene at low latitudes gave way to more arid climates in the Fertile Crescent (Childe, 1928). However, more recent examination does not suggest that this area suffered a sudden climatic change at the Late Glacial/ Holocene transition that occurred in the North Atlantic region (Bell & Walker, 1992). The synchrony in the multiple origins of agriculture in both the New and the Old World has been discussed in relation to a number of environmental factors including increased population density, greater social complexity and climatic change. The increase in atmospheric carbon dioxide levels from 200 to 270 μmol mol^{-1} with the retreat of the Pleistocene ice sheets has also been proposed as an important contributing factor in the development of cereal cultivation (Sage, 1995). Among other climatic factors, increased seasonality, global temperature rise, and a reduction in rainfall may all be possible contributing causes. To identify any one cause is probably an impossible task as in different regions of the world different factors may have stimulated experimentation with crop cultivation.

Unfortunately, the invention of agriculture has in many areas been a non-sustainable solution in maintaining human population growth. The occupation of the land by annual crops that complete their growth before the onset of summer drought reduces total vegetation cover. As a result, when populations increase, and grazing intensifies, desertification has all too often been an inevitable consequence of initial agricultural success. The extensive use of irrigation by one of the earliest civilizations, the Sumerians of southern Mesopotamia (third millennium BC), is thought to have led to salination and their eventual decline. Many areas which once supported flourishing civilizations with the early development of agriculture have now either been abandoned or become agriculturally impoverished. In some cases this may have been a result of climate change, but more frequently it is as a result of landscape degradation to an extent that the region has become marginal for successful crop production.

Similarly, the introduction of Neolithic agriculture to hyperoceanic regions such as Ireland, western Scotland and Iceland, resulted in excessive soil leaching, podzolization, peat growth and consequent failure of tree regeneration. Landscape deterioration by human over exploitation in a land with a highly oceanic environment has particularly been the case in Iceland. The extensive deforestation and overgrazing that occurred from when Iceland was first populated (*c.* AD 874, the *landnám*) have been suggested as the primary external force causing soil erosion and land degradation. The ability in oceanic environments to winter livestock out of doors, together with an absence of shepherding, may have contributed more to land degradation rather than absolute numbers (Simpson *et al.*, 2001). The sum total of all these changes is environmental impoverishment, both for the agricultural community and natural ecosystems (see Chapter 10). The large-scale abandonment of the early upland sites mentioned above may have been due as much to soil impoverishment as to climatic deterioration (Fig. 1.21).

Fig. 1.21   A satellite image of Iceland – a largely treeless land as a result of over grazing. Note the marked erosion patterns around river courses in the mountain areas in the northern part of the island (see also Fig. 11.30). (Photo Dr E. G. Duncan and reproduced with permission from Group for Earth Observation.)

1.9   CONCLUSIONS

Marginal areas can be defined superficially in terms of geographical isolation. However, the range of periph-eral situations increases when biological boundaries are substituted for the frontiers of physical geography. The sedentary nature of vegetation and the facility with which distribution can be mapped makes the study of limits to plant survival an unending subject of enquiry and speculation. Habitats become marginal for plant populations whenever survival is threatened. Conse-quently, peripheral areas for plant survival can be defined biologically in terms of demography, physio-logical requirements, and genetic variation. Each of

these perspectives makes its own assumptions as to why plant populations are limited in their distribution and what constitutes a marginal area. In seeking explan-ations in terms of cause and effect it is therefore pos-sible that more than one explanation may be valid, as in different situations the same species may be limited by contrary causes.

Physiological limitations are a fundamental aspect of survival but are not always readily observed in nat-ural situations due to interference from other species competing for space or resources and oscillations in environmental conditions. If long-term observations are possible, demographic data may discern the causes for failure of populations to maintain their numbers.

Genetic studies, especially when combined with historical information concerning sites and past distributions, can provide equally cogent explanations for distribution limits. Any discussion of marginal areas has therefore to accept that in any particular case study it is probable that there will be more than one kind of limitation. In short, the recognition of boundaries presents many possibilities and depends in large measure on what the observer is capable of seeing.

Fig. 2.1   Biodiversity in a marginal area as seen in a Norwegian
mountain pasture.

2 · Biodiversity in marginal areas

2.1 BIODIVERSITY AT THE PERIPHERY

Biodiversity has now become an integral element in environmental monitoring. Much attention has been given in recent years to the species-rich areas of the Earth, as there is an obvious concern that regions that contain so much of the world's evolutionary heritage do not become biologically impoverished. There is, however, a strong case for giving attention to marginal areas in the preservation of biodiversity even if the numbers of species that they contain cannot compare with the biological hotspots of the world.

Marginal areas, as has already been pointed out, are areas where climatic change is liable to cause disturbance either in location or the nature of the vegetation that survives in these potentially labile localities. Historically, they are areas that will have experienced climatic change in the past and therefore the species that live in these areas may be pre-adapted to climatic change and should therefore be considered particularly relevant in the study of species responses to fluctuating environments. It has been argued (Safriel *et al.*, 1994) that peripheral populations have to be genetically more variable than those from core areas, since the variable conditions induce fluctuating selection, which maintains high genetic diversity. Alternatively, due to marginal ecological conditions at the periphery, populations there are small and isolated: the within-population diversity is low, but the between-population genetic diversity is high due to genetic drift. It is also likely that peripheral populations evolve resistance to extreme conditions. Thus, peripheral populations rather than core ones may be more resistant to environmental extremes and changes, such as global climate change induced by the anthropogenically emitted 'greenhouse gases'. They should therefore, it is argued, be treated as a biogenetic resource used for rehabilitation and restoration of damaged ecosystems. Climatic transition zones are often characterized by a high incidence of species represented by peripheral populations and hybrids, and therefore should be conserved now as repositories of these resources, to be used in the future for mitigating undesirable effects of global climate change (Safriel *et al.*, 1994).

2.2 ASSESSING BIODIVERSITY

2.2.1 Definitions of biodiversity

Biodiversity is commonly understood to describe all aspects of variation in living organisms. To be scientifically meaningful, however, diversity has to be both qualified and quantified. The first requirement, qualification, refers to the level of assessment whether it be ecosystem complexity, species richness, or genetic variation. This last category, genetic variation, is the fundamental issue as it represents the only means of assessing the heritable properties that result from DNA variation upon which all biodiversity ultimately depends. Unfortunately, information at the molecular level is not usually available in sufficient detail for ecological field studies. Assessing biodiversity is therefore frequently carried out using various visible approximations such as ecosystem complexity, variation in life-forms and strategies within communities, numbers of species, subspecies and ecotypes or other appropriate estimations of population variation. This can also include the reproductive biology of species, whether or not they are polymorphic, inbreeding or outbreeding, or reproducing entirely vegetatively. Any or all of these characteristics can be useful when comparing differences in plant populations between regions.

A flexible and simple definition of biodiversity that can be readily applied in the field is the number of different taxonomic items that are found in any specific area (or volume). However, when using even this basic measure, some consideration has to be given to the method of sampling and not least to the *species accumulation curve* or *species area curve* as it has long been known in plant ecology. This, the oldest known pattern in ecology, describes the relationship between the area under examination and the number of species present and in most cases fits the equation:

$$S = cA^z$$

where S is the number of species, A the area sampled and c and z are constants. The exponent z varies between 0.15 and 0.30 independently of the biome and taxa under study (Calow, 1998). As with many mathematical relationships in ecology the concept of the species accumulation curve has been refined for comparative use under strictly defined ecological

Table 2.1. *Definitions of biodiversity*[a]

Alpha diversity   The diversity of species within a particular habitat or community (*S*) where (*S*) = species · unit area

Beta diversity   A measure of change in species along a gradient from one habitat to another. One possible measure of beta diversity (*B*) is

$$(B) = (g(H) + l(H))/2S$$

where $g(H)$ is the number of species gained along the gradient H and $l(H)$ the number of species lost along the gradient H.

Regional richness $(Sr) = S(B + 1)$

Gamma diversity   The richness in species of a range of habitats in a geographical area and dependent on the alpha diversity of the habitats it contains and the extent of beta diversity between them. Thus

$$Sr = (Bn + 1)(Bm + 1),$$

where *Bn* is beta diversity and *Bm* is gamma diversity.

---

[a] There are many mathematical assessments of biodiversity with varying properties. The above example was chosen as it was used in the chapter by Cowling *et al.* (1992) on the biodiversity of Fynbos discussed below.
Source: (Lincoln *et al.*, 1998).

conditions. For comparisons of different species, accumulation curves have to be meaningful and therefore should be based on areas which have what is termed continuum vegetation, where there is continuous plant cover and only one individual at any one point. Each individual therefore has a finite area and, when viewed from above, the ground is everywhere obscured by vegetation (Williamson, 2003). Convenient as this might be in simple agricultural situations, or hypothetical models, these necessary rigorous conditions are not readily found in the field. The extent to which the vegetation is patterned and made up of patches of varying size also has an enormous influence on the species accumulation curve (Greig-Smith, 1983). It follows therefore that in comparing biological diversity between different areas it has to be recognized that both the nature of the vegetation and the method of recording can cause significant alterations. A variety of statistical tests are available which allow an assessment to be made as to when species accumulation curves can be considered as defining a uniform sampling process for a reasonably stable situation (Colwell & Coddington, 1995). When this can be established it is possible to record species richness for specific sites in terms of *alpha diversity* (species per unit area; Table 2.1).

In marginal areas where vegetation cover tends to be irregular or distinctly patterned, simple estimations of alpha biodiversity may not be suitable (Fig. 2.2). More appropriate for marginal areas and fundamentally different from alpha diversity is *beta diversity* (Table 2.1). Here vegetation is compared along a gradient from one habitat to another. Quantification of variation between habitats is not a direct count of taxonomic richness but is a comparison of the variation between habitats using a suitable index of similarity giving a statistic that is inversely related to diversity. In a heterogeneous region with much interhabitat variation the similarity coefficients will be low and the communities of that region will be considered ecologically as showing high beta diversity. This represents a diversity that depends on the heterogeneity of the area under consideration.

Other units of diversity sometimes used in ecological studies include gamma and delta diversity. Gamma diversity is usually defined as richness of species over a range of habitats in a geographical community and is dependent on both the alpha diversity of the habitats and the extent of beta diversity between them. Some authors also attempt to differentiate between gamma and delta diversity in which gamma diversity denotes variation between different locations within a community and delta diversity refers to differences between landscapes. However, this level of distinction can confuse elements of environmental and geographical variation (Cowling *et al.*, 1992) and the finally derived statistics are not clearly related to any physical reality.

Fig. 2.2   An extreme case of naturally patterned vegetation at Ny Ålesund, Spitsbergen. A polar semi-desert community in Svalbard, Norway (78° 56.12′ N, 11° 50.4′ E) with clonal patches of *Dryas octopetala*, which has been studied in relation to the potential effects of climatic warming and increased nutrient availability (see Wookey *et al.*, 1995).

Assessments of genetic diversity face the same problems as taxonomic diversity. At the gene level distinctions have to be made between allelic richness and evenness in allelic frequencies (Frankel *et al.*, 1995). Formulae for estimating degrees of similarity and difference are numerous and their particular value can depend on the nature of the items being compared either in studies of genetic geography or in assessments of genetic variation (Hawksworth, 1995).

All assessments of biodiversity depend on the particular aspect of biological variation that is being sampled. It does not necessarily follow that this has always to be based merely on species. For some purposes different taxonomic levels are appropriate. This is particularly relevant in relation to plants where species are less rigidly defined than in animals. For plant communities, biodiversity can be usefully assessed and compared between sites in terms of families, genera, species, populations, gene frequencies, or other molecular markers of variation in evolutionary history. If comparisons are required at a global level, there is a case for using families or genera, but if detail is needed in relation to specific sites, attention needs to be given to subspecies and ecotypes as it is at these levels that adaptation takes place and where the potential of populations to evolve and undergo speciation resides. There are even cases where species survival in marginal areas may be enhanced by not undergoing further speciation and maintaining instead a wide range of polymorphic, interfertile populations. Examples of this aspect of biodiversity are discussed in detail in relation to some widespread and ancient diploid arctic species (Section 6.7).

With the development of molecular biology it is now possible not just to identify particular ecological races or ecotypes, but also to detect variation within species or subspecies in terms such as allelic heterozygosity or DNA variation as observed in the nucleus, the mitochondrion and the chloroplast. The extent of the variation at a molecular level can be mapped, and provides a quantifiable assessment of current variability. The application of cladistic methods can also reveal the genetic and biogeographical history of particular species or populations (Section 6.4.2).

2.2.2   Problems of scale and classification

Scientifically, it is difficult to make comparisons of plant biodiversity on a global scale as every region of the world has plant communities that have evolved under different circumstances. Listing of areas with the greatest number of endemic species is an indication of a long uninterrupted history of speciation. However, when as in certain cases many of the endemic species are apomictic, they are not usually contributing actively to further genetic modification and are merely testimony to past evolutionary activity.

Modern biological technology has created the possibility of having a range of different interpretations of biodiversity. Similarly, ecological studies have some degree of choice when comparing diversity in plant communities. Instead of just contemplating the species richness of the tropics as compared with temperate and arctic zones (which were issues that attracted Darwin and his contemporaries) it is now possible to look at specific examples of intrazonal variation at the community, species or subspecies level and make assessments of those factors which either favour or discourage biodiversity. In areas where differences are readily apparent in species frequencies, comparisons are made without further subdivision. In others, as for example in the Arctic, where species numbers are low, comparisons can be made at the subspecies or population level.

Ecological classifications have long been used to create order by grouping plants by their collective and often contrasting responses to defined environments. Calcicoles and calcifuges, halophytes and glycophytes, eutrophic and oligotrophic are descriptive terms which reflect contrasting and frequently incompatible adaptations to diametrically opposed habitats. Adaptation to one set of conditions usually proves maladaptive for the opposing set. Plants that inhabit dry soils need reducing conditions at their root surfaces in order to convert insoluble ferric iron to the soluble ferrous form. However, plants in flooded habitats need to have oxidizing conditions at their root surfaces to precipitate ferric iron from ferrous iron so that iron toxicity does not result from an excessive uptake of ferrous iron. These are opposing strategies and cannot coexist in the same root system.

The development of computer models to imitate plant behaviour has created renewed interest in grouping plants into functional types (Smith *et al.*, 1997). For computer simulations of vegetation responses to changing environmental conditions some degree of functional aggregation, sometimes referred to as *scaling-up*, is obviously essential (Section 3.4). Even within well-defined ecological groups such as halophytes or calcicoles complications can arise from the unending variety of adaptations and evolutionary strategies that exist in response to any particular stress. Halophytes can be divided into osmo-regulators and osmo-conformers, and different calcicole species differ in their responses to the physical and chemical aspects of adapting to calcium-rich soils. Eutrophic habitats can present an equally heterogeneous set of environmental conditions. It cannot even be generalized that all populations of the same species will fall into any one category. Different ecotypes of the same species can be found in relation to drought, salt, flooding and other stresses. Most species have different ecotypes in relation to their phenology, a fact which was already well known to Roman farmers who knew that seed corn could not be transported successfully from one part of their empire to another (White, 1970).

To ignore this level of variation and to confine observations to only the more obvious phenotypic short-term responses of plants to the environment is to ignore the many subtle genotypic responses which will eventually determine the long-term evolutionary responses of species to environmental change. A striking example of the above argument has already been discussed in relation to the purple saxifrage (*Saxifraga oppositifolia*) which in the High Arctic occurs with distinct ecotypes which have divergent strategies in relation to differences in growing season length. This is an example where *scaling-up* would fail to recognize the fundamental mechanisms that allow polymorphic species to adapt readily to changing environments. Short-term phenotypic responses to environmental perturbation may be detected by the process of scaling up and looking at the effects of climate change on forest canopies or grassland productivity. The long-term effects of environmental change will, however, act on the genotype and therefore examination of the physiological and genetic responses of individual plants and populations should not be ignored (see also Section 6.6).

2.2.3 Variations in assessing genetic variation

An example of how different estimations of genetic variation can be obtained within one endemic species, surviving in a very limited area is seen in the subarctic Scottish endemic *Primula scotica* – a possible glacial relict species (Fig. 2.3). A study of five sites in Orkney and nine from the Scottish mainland revealed variation between individuals at only one of 15 enzyme-encoding loci examined. In addition a survey of DNA sequence variation found no genetic diversity, either within or between a subsample of four populations (Glover & Abbott, 1995). In this study of individual genes the

plants exhibited a 'fixed' level of heterozygosity per individual, indicating that it is of allopolyploid origin. Thus, despite the high level of heterozygosity per individual it might be concluded that there is almost a complete lack of genetic diversity both within and between populations. However, in a separate study of the same species, within the same small areas of its occurrence in northern Scotland and Orkney, it was shown from an examination of both growth and form that plants grown from seed collected from different populations showed considerable variation (Ennos *et al.*, 1997). This latter study of character responses looked at aspects of plant variability which

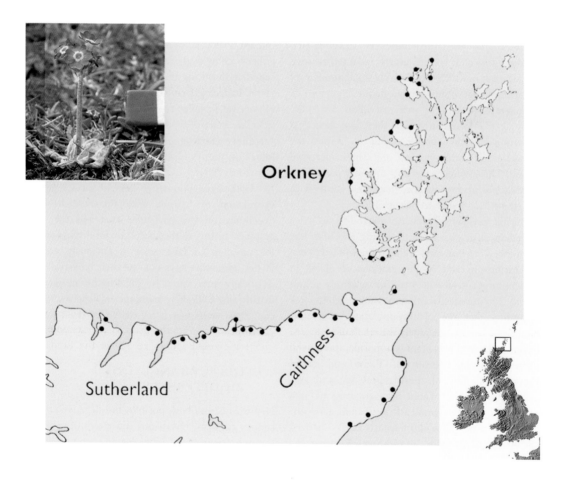

Fig. 2.3   The Scottish primrose (*Primula scotica*). A possible glacial relict species endemic to Scotland and Orkney possessing limited genetic diversity between and within populations when examined for specific alleles yet exhibiting marked genetic variation in multi-gene controlled variation in growth and form (see text inset of *P. scotica*; scale = 1 cm).

are multi-gene controlled and do not appear to be so severely limited as enzymatic variations under control from specific single genes. Assessments of genetic variability can therefore differ depending on what is being assessed, and global statements projected from any one particular facet of variation, if not properly explained, can be potentially confusing.

2.3 VARIATION IN PERIPHERAL AREAS

The special nature of isolated or marginal and peripheral areas with regard to variation and evolution was first discussed in relation to tropical ants and gave rise to the recognition of the phenomenon known as *ecological release* (Wilson, 1959). Areas studied included savannas, monsoon forests, sunny margins of lowland rainforests and salt-lashed beaches. In ants, the expanding dominant species became adapted to such marginal areas, in which the relatively small numbers of (ant) species that occurred had fewer competitors. Consequently, species in marginal areas are more flexible as the places in which they live have less competition and therefore according to Wilson the species are *ecologically released* and show increased variation. This argument was subsequently pursued in the *Island Biogeography* theories developed jointly with MacArthur (MacArthur & Wilson, 1967).

During periods of climatic instability, as have occurred in polar regions both during and after the Pleistocene, it can be expected that there will have been many extinctions and re-immigrations of plants in peripheral areas. Consequently, the relatively recent end of the Pleistocene glaciations has suggested to some investigators that the plant populations that now inhabit arctic regions are largely the result of Holocene migrations from areas that supported peripheral populations during periods of Pleistocene glacial advance. Application of the classical age and area concept (Willis, 1922) would associate areas of high diversity with a long period of occupation and conversely areas which have only recently been colonized would have less diversity. In addition, the many climatic oscillations that have taken place at high latitudes will have resulted in numerous extinctions and re-immigrations. The genetic consequences of these Ice Age climatic fluctuations would have been to cause a reduction of variability, as during phases of northward

expansion recolonization will take place mainly from individuals spreading out from the periphery of southern populations (Hewitt, 1996). For species that survived the Pleistocene glaciations in southern Europe such as grasshoppers, hedgehogs, oak, common beech, silver fir and black alder trees, brown bears, newts, shrews, water voles, and house mice, molecular data confirm a reduction in diversity from southern to northern Europe in the extent of allelic variation and species subdivision (Hewitt, 1999). However, other molecular data have also shown for plants that there is an ancient arctic flora that survived the Pleistocene glaciations at high latitudes (see also Section 6.7). Even in the Arctic, subspecies diversity varies in relation to glacial history. In one of many studies showing this effect clonal diversity in four species of *Carex* was found to be lower in areas that had been deglaciated in the last 10 000 years as compared with areas that had either become deglaciated 60 000 years ago or not glaciated at all during the Weichselian (Stenström *et al.*, 2001). In species that have a long uninterrupted history of survival in polar regions there is emerging a considerable body of evidence for a high level of genetic variability (Section 6.7) even at the most northern limits of species distribution (Grundt *et al.*, 2006; Crawford, 2004).

Both these contrasting views of species variability in relation to latitude may therefore be correct depending on the particular area and the biogeographical history of the species under discussion. Time for variation to take place is obviously necessary. Hence, areas that have not had their biological history drastically interrupted by the Pleistocene glaciations usually have higher numbers of species than in equivalent areas which lay under ice during the last glaciation.

2.4 DISTURBANCE AND BIODIVERSITY

Disturbance has both positive and negative effects on species richness. Disastrous ecosystem destabilization through frequent burning, logging, and overgrazing can be found in every part of the world. The logging of tropical rainforests, the destruction of African savannas by elephants or fire, the denudation of Iceland from overgrazing by domestic livestock, and the destruction of arctic salt marshes by large populations of nesting geese, are but a few examples where repeated disturbance, from whatever cause, eventually prevents the establishment of perennial species and frequently leads to drastic soil erosion.

In the modern world overgrazing is usually due to human intervention and frequently leads to a reduction in species diversity. Despite the obvious calamitous situations where overgrazing and other forms of disturbance can lead to total habitat destruction there are many examples where plant diversity is improved and ecosystems renewed by periodic fires, physical erosion, and grazing. All these forms of disturbance have a history that predates the arrival of human beings as an ecological force. Plants have evolved in dynamic relationship with disturbance both by physical forces and herbivory and therefore it should not be surprising that a lack of disturbance can lead to a reduction in species richness.

Exclusion of grazers from fenced sites in the Arctic can lead to the dominance of mosses to an extent that inhibits the regeneration of flowering plants. Removal of winter grazing from maritime sedge–heath communities in Orkney deprives the rare endemic Scottish primrose (*Primula scotica*) of the bare patches caused by sheep's trotters on wet soils which facilitate seedling establishment (Fig. 2.3). Forest and heathland regeneration can benefit from burning, provided that the frequency and extent of the fires does not entirely remove the possibility of reseeding. Even bog vegetation can be rejuvenated by fire as it removes invading trees, while the carbon particles from the fire wash into the soil, impede drainage and thus help the growth of bog (Mallik *et al.*, 1984).

Whatever the situation, the basic principle underpinning plant species richness is that all persistent species (species recognized as being part of a given plant community) must be able to increase when rare (Crawley, 1997). This is a fundamental property for plant survival which is particularly relevant to the survival of plant communities in marginal areas where windows of opportunity for reproduction may be intermittent, and the maintenance of species diversity can be impeded by a few dominant life-forms. Community impoverishment in this way is particularly noticeable in peripheral areas where a single dominant species or a group of species with similar life-forms can prevail and prevent regeneration and invasion by a more diverse flora. Marginal and oceanic habitats

(see Chapter 7), as well as extensive areas of the arctic tundra (see Fig. 6.13) can become dominated by bryophytes to the exclusion of other species if there is no disturbance from periodic fires and grazing is absent.

2.4.1 Grazing

Grazing is a form of disturbance that has long been recognized as one of the fundamental forces that maintains biodiversity by controlling aggressive and potentially dominant species. As a factor shaping plant communities grazing probably predates human pastoral activity. The highly noticeable impact of Neolithic settlements on the species composition of natural vegetation, and especially on forests, has caused the influence of pre-landnám (pre-agricultural settlement) grazing to be ignored or else assumed to be negligible. Nevertheless, in Europe the early Holocene had an extensive herbivorous fauna. Grazing mammals included aurochs (*Bos primigenius*; Fig. 2.4), tarpan or European wild horse (*Equus przewalski*), bison (*Bison bonasus*), red deer (*Cervus elaphus*), elk (*Alces alces*), roe deer (*Capreolus capreolus*), beaver (*Castor fiber*), and the omnivorous wild boar (*Sus scrofa*), all of which maintained a presence in Europe from the end of the Pleistocene (*c.* 12 000 BP) to the early Middle Ages and some until the present day. In historical times the effects of the larger grazing mammals in creating marginal brushwood regions around forests is well-documented, and there is therefore no reason to

Fig. 2.4   Skeleton of a young aurochs. This specimen, now in Roskilde Museum, dating from 10 000 BP was found in a bog near Himmelev (Denmark).

Fig. 2.5   Musk oxen (*Ovibus moschatus*) in north-east Greenland. One of the last remaining large mammals of the Pleistocene mammoth fauna still found with viable populations in Greenland, Arctic Canada, and Alaska. Grazing is essential for maintaining nutrient levels and floristic diversity especially in the High Arctic.

suppose that grazing disturbance in pre-Neolithic times did not exist and help to create the brushwood interface between forest and open land. It is in these marginal zones, created by grazers between forest and open land, that many tree species, such as pedunculate oak (*Quercus robur*), sessile oak (*Q. petraea*) and hazel, regenerate most readily. It has been suggested (Vera, 2000) that the presence of these tree species in the pre-landnám forest pollen record should be taken as evidence for the influence of the early indigenous grazing fauna on maintaining openings in the ancient forests of lowland Europe and thus contributing to their biodiversity. However, further examination, using as a control the situation in Ireland where ancient forests existed but large mammal herbivores were absent, has shown that oak and hazel (*Corylus avellana*) still regenerate successfully without the disturbance of large herbivores (Mitchell, 2005). Nevertheless, large herbivores, both wild and domesticated, can have a significant impact on forests, as can be seen in the present situation with grazing by excessive numbers of red deer in the Scottish Highlands.

In the Arctic the impact of grazing by wild geese (Anser spp.), lemmings (*Lemmus lemmus*), musk oxen (*Ovibus moschatus*) and reindeer (*Rangifer tarandus*) has a profound influence on the diversity of plant communities and the fertility of the soil (Holtmeier, 2002). The tundra has probably been grazed as long as this vegetation has existed. In northern Eurasia and in Alaska the mammoth fauna of Eurasian horses, the Mongolian ass (kulan), bison, yak, and musk oxen (Fig. 2.5), as well as mammoth, woolly rhinoceros and reindeer, grazed on the disappearing Pleistocene tundra–steppe. The abrupt reduction of their natural habitat started about 12 ka BP. The last mammoths became extinct in northern continental Siberia *c.* 9.6 ka BP. However, a dwarf mammoth population inhabited Wrangel Island from 7.7 to 3.7 ka BP (Kuz'min *et al.*, 2000; see also Fig. 11.7). Caribou and reindeer, and to a lesser extent musk oxen, still remain widespread throughout the High Arctic (Klein, 1999). The length of the Holocene presence of reindeer at high latitudes is particularly notable (Fig. 2.6) and their presence in the past would have contributed, as it does

today, to the turnover of nutrients. When there is little or no grazing the vegetation becomes relatively nutrient poor with an extensive accumulation of dead forage material which can shade the soil surface; this impedes the warming of the upper layers, causing the permafrost level to rise and form ice wedges, which limits the rooting capacity of vascular plants. The end result can be an increase in bryophytes and the suppression of many of the flowering plant species.

Overgrazing, where plant communities can suffer long-term damage, is also found in the Arctic, even in the remoter regions. In polar semi-desert communities of north-west Spitsbergen the reproductive potential of the keystone vascular species, *Dryas octopetala*, is currently constrained by low summer temperatures, resulting in the infrequent production of viable seeds. This is further depressed on the Brøgger Peninsula by the summer foraging behaviour of a numerous population of reindeer (*Rangifer tarandus platyrhynchus*) on the flowering shoots (floral herbivory). Surveys of neighbouring coastal tundra areas with considerably less floral herbivory show just how severe the grazing pressure is on reproductive shoots of *D. octopetala* in this region. This situation is not unique to this area of Spitsbergen and also extends to other species of flowering plants (Cooper & Wookey, 2003).

In salt marshes both in temperate regions and in the Arctic there is a delicate balance between goose grazing and diversity of the salt-marsh vegetation. Overgrazing can be highly destructive and lead to catastrophic changes in the ecosystem. In arctic salt marshes along the western shores of the Hudson Bay large areas have been destroyed as a result of overgrazing by increasing populations of the lesser snow goose (Jefferies, 1997). The North American lesser snow goose has traditionally wintered in the coastal marshes of the Gulf of Mexico. The development of large fields of rice and other winter cereals with large nitrogen inputs has led to a large increase in their population as a result of reduced winter mortality. The high number of birds that now come to breed on the shores of the Hudson Bay has dramatically affected these arctic coastal marshes (Figs. 2.7–2.9). Studies over many years concentrating on the area around La Pérouse Bay (Jefferies *et al.*, 2004) have shown that the breeding colony there has grown from about 1300 pairs in 1967 to an estimated 44 500 pairs in 1997. Their preferred species are the stoloniferous

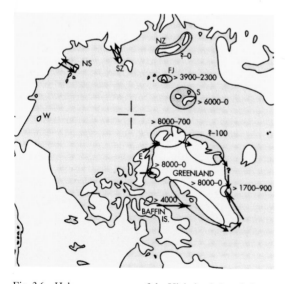

Fig. 2.6   Holocene occupancy of the High Arctic by reindeer (caribou). Island abbreviations: E, Ellesmere; S, Svalbard (Spitsbergen); FJ, Franz Josef Land; NZ, Novaya Zemlya; SZ, Severnaya Zemlya; NS, New Siberian Islands; W, Wrangel Island. (Reproduced with permission from Klein, 1999.)

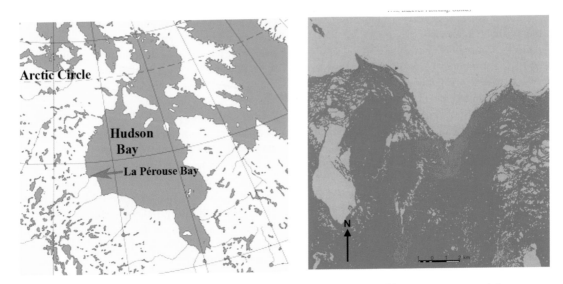

Fig. 2.7   Goose grazing in the Arctic. (Left) Location of studies on the effects of increased lesser snow goose populations on the western shore of the Hudson Bay. (Right) Satellite image of vegetation changes at La Pérouse Bay from 1973 to 1993. Red refers to the areas that have lost vegetation over the period, green indicates no net change. (Reproduced with permission from Jano *et al.*, 1998.)

Fig. 2.8   Damage to coastal stands of willow by overgrazing by lesser snow geese on the shores of the Hudson Bay. (Left) Grubbing of intertidal marsh in early spring. (Right) The birds have removed the insulating mat of graminoid vegetation around the willows. This has exposed the willow roots and as a result of increased evaporation has led to hypersaline soil conditions (3 × oceanic water at 120 parts per thousand) thus causing the death of the willows. (Photos by Professor R. L. Jefferies.)

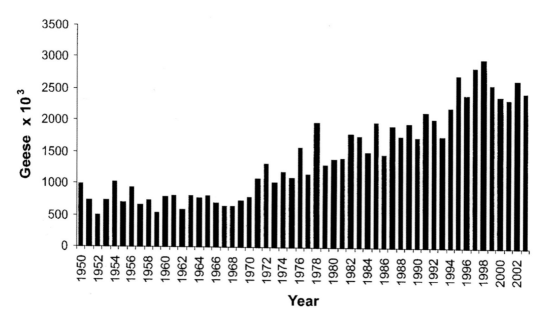

Fig. 2.9   Population index of the mid-continent light geese (mainly lesser snow geese). Data taken from winter counts from 1950 to 2003 as recorded in U.S. Fifth and Wildlife Annual reports. (Reproduced with permission from Jefferies et al., 2004.)

graminoids of the salt marshes, which are rich in soluble sugars and amino acids, of which *Carex subspathacea*, *C. aquatilis* and *Puccinellia phryganodes* are the principal species. The geese can remove 90% of the annual above-ground primary production as well as grubbing up the buried stolons. Once the vegetation has been lost, increased evaporation from the exposed sediment draws inorganic salts to the surface from the underlying marine clays that were deposited under the ancient Tyrell Sea. In summer these salts accumulate and give rise to hypersaline conditions which inhibit recovery of the graminoid vegetation and also kill the willows in the supratidal marshes (Jefferies & Rockwell, 2002).

In other marginal areas, salt marsh vegetation can benefit from moderate goose grazing by arresting succession and preventing the establishment of taller species (van der Wal *et al.*, 2000). High brent goose (*Branta bernicla*) populations can in this way increase the areal extent of grazing lawns (Person *et al.*, 2003). There can also be a benefit from grazing by both mammals and geese. On the Dutch island of Schiermonnikoog, tall-growing shrub species such as sea purslane (*Atriplex portulacoides*) can eventually make the marsh unsuitable

for feeding geese. However, winter grazing by brown hares (*Lepus europaeus*) can retard vegetation succession and facilitate grazing by brent geese by preventing shrubs from spreading into younger parts of the marsh.

In many European marginal areas, free-ranging herbivores are now being used more widely in the interests of conservation, particularly in natural grassland areas, dune heaths and wetlands. Herds of cattle and horses, and to a lesser extent sheep and geese, are proving effective in preserving biodiversity. A mixture of cattle and horses owing to their different feeding strategies is particularly efficient in controlling invasive coarse grasses and woody species. In France, where grazing by horses and cattle is used for conservation management in the Camargue, free-ranging animals have been shown to be highly efficient in preserving biodiversity of this marginal habitat. Horses are particularly successful in plant management because they remove more vegetation per unit body weight than cattle and make greater use of the more productive plant species, especially the graminoids (Menard *et al.*, 2002). The processes contributing to diversity in vegetation structure can therefore be linked to an alternation of positive and negative interactions which leads through competition to the creation of sequences of shifting mosaics.

2.4.2   Fire

Fire, through natural causes, is a worldwide factor with a long history that predates human intervention in vegetation succession and regeneration. Consequently, plants have evolved a dynamic relationship that has operated on a large scale in the evolution of the life history traits of numerous species. The South African fynbos, the Brazilian cerrado and the heathlands of Europe are all examples of vegetation types that have evolved in association with fire and depend on fire for their existence (Figs. 2.10–2.11). The dependence of these species-rich areas on periodic conflagration raises the question of whether biodiversity is positively or negatively related to fire frequency. In common with other disturbance phenomena, a high frequency of fires can eliminate many species while a low frequency leads to the dominance of competitive species and contributes to species loss. In general, within a single fire cycle of destruction, recovery and post-fire succession,

species richness typically peaks in the first year after the fire and then declines as the community ages (Bond & van Wilgen, 1996). Thus, it might be expected that heaths with intermediate fire frequencies will have the greatest biodiversity. In the South African Fynbos fires occur with a frequency that varies between 5 and 40 years, while in the Brazilian Cerrado the fire frequency is between 1 and 3 years (Bond & van Wilgen, 1996). This high frequency appears to be essential for the preservation of the typical cerrado flora (see below), as protection from fire for two to three decades brings about an increase in woody plants and other fire-sensitive species (Moreira, 2000).

In Scotland large areas of moorland are burnt in order to regenerate the heather for sustaining viable populations of birds for game-bird shooting and for sheep grazing (Figs. 2.12–2.13). Depending on whether the management objectives are primarily for shooting or agriculture, muirburn frequency can vary. It is also possible to manage the intensity of the fire depending on the weather conditions at the time of burning and the age and size of the heather plants. Moors that are managed for shooting red grouse (*Lagopus lagopus*

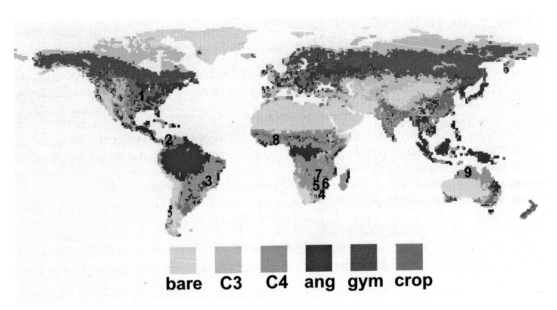

bare   C3   C4   ang   gym   crop

Fig. 2.10   Land cover according to dominant functional types. The colour key represents from left to right: bare ground, C3 grasslands, C4 grasslands, angiosperm forests (deciduous and evergreen), gymnosperm forests, and crop lands. Land-cover types: 1, broad-leaved evergreen forest; 2, broad-leaved deciduous forest; 3, mix of 2 and coniferous forest; 4, coniferous forest; 5, high latitude deciduous forest; 6, wooded C4 grassland; 7, C4 grassland; 9, shrubs and bare ground. (Adapted with permission from Bond *et al.*, 2005.)

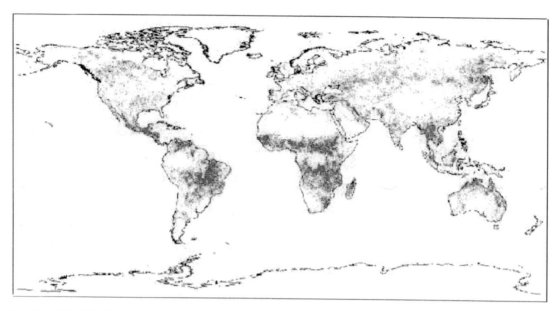

Fig. 2.11   Global distribution of fire in 1988 mapped by World Fire Atlas European Space Agency. Note that most fires occur in C4 grass ecosystems. (Reproduced with permission from Bond *et al.*, 2005.)

scoticus), which are dependent on young nutritious heather shoots as their main food, are best managed by burning small patches at a time, especially if combined with a relatively high frequency of burning. How often these patches of moor are burnt can vary between 8 and 20 years depending on the productivity of the moorland.

In relation to the preservation of biodiversity the extent and frequency of burning, and the habitats in which this management occurs, are contentious issues as burning has both positive and negative effects on the biota and the moorland (Yallop *et al.*, 2006). Blanket bogs lose their conservation value if burnt frequently with hot fires. Wild birds can be adversely affected by too frequent muirburn. Meadow pipit (*Anthus pratensis*) density has been observed to increase with moderate heather cover but falls if it is too high (Vanhinsbergh & Chamberlain, 2001). It appears that a mosaic of heather, bog and grassland may be the optimum habitat for meadow pipits.

Recent changes in upland management and high labour costs have resulted in moorlands no longer being regularly burnt. The disappearance or degradation of heather mosaics (Fig. 2.13) in some areas is having a deleterious effect on the biodiversity of this habitat.

Studies on the time sequence of species diversity in relation to fire on Scottish moorlands have shown that there is a decline in species number after long periods without burning (Hobbs *et al.*, 1984). The maintenance of biodiversity in these marginal heath habitats will therefore depend in the future on a reassessment of the how and when muirburn is practised as climate changes and the use of the moorlands also alters. Upland sheep grazing is proving less profitable than it was in the past and an ecosystem that can produce one red grouse per hectare per annum may prove financially more rewarding, with a faster cash return than traditional methods of upland farming.

2.5   THE GEOGRAPHY OF MARGINAL PLANT DIVERSITY

In terrestrial plant communities species richness in terms of alpha diversity generally decreases from the Equator to the poles. It is therefore understandable that biodiversity studies are frequently concerned with species assemblages at low latitudes. Conditions that limit plant distribution and create marginal areas can also be expected to influence biodiversity. Globally, tropical rainforests are commonly considered to be the

Fig. 2.12   Heather burning on a Scottish moor. Burning mature heather (*Calluna vulgaris*) in early April on the Hill of Alyth, Angus, Scotland.

outstanding regions for plant biodiversity and as such currently command much attention. Understandably, if there is to be a political will to preserve rainforests there is a need to increase public awareness of the species richness of these tropical plant communities. However, anxiety about the fate of the tropical rainforests tends to reduce awareness that there are many other regions where biodiversity is also likely to be at risk, not just from human activities, which are at least partially capable of being controlled, but principally from global climate change which is probably beyond control as long as human populations continue to grow.

There are several notable areas outwith the tropics that are marginal in terms of resources for plant growth and yet are remarkable for their diversity of plant life. Such areas are commonly referred to as biodiversity hotspots (Myers, 2003). They may not all be as rich in species as the tropical regions, but relative to their own area they stand out, either in terms of species numbers or for richness in subspecific variation. Notable

Fig. 2.13   The Lowther Hills, Southern Uplands, Scotland, showing an example of muirburn as practised when the object is to create a moor suitable for game-bird shooting. The burning is carried out frequently (every 8–20 years) and in small patches. This creates the maximum of young nutritious heather which is ideal feeding for grouse. The frequency and intensity of burning depends on the object in view as any burning regime can have positive or negative effects on different aspects of the moorland ecosystem (see text).

examples of such biodiversity hotspots include the Cape Peninsula of South Africa (and other regions with a similar long-term history of Mediterranean-type climatic stability), and the central Brazilian Cerrado. In addition there are numerous climatically favoured thermal oases in both hot and cold deserts. Even in the High Arctic there are localized floristic hotspots which in comparison with surrounding areas have a high level of species diversity.

2.5.1   The South African Cape flora

The Cape Peninsula with its outstanding flora of over 8500 species of flowering plants has a greater diversity of plant species at a regional level than the richest tropical rainforest (Cowling *et al.*, 1992; Linder & Hardy, 2004). Particularly famous for species richness within the Cape region is the unique *Fynbos*. The term Fynbos, derived from the Dutch *Fijnbosch*, is used to describe woody vegetation with thin or fine branches which is a characteristic feature of the varied collection of heath and scrub communities that survive on ancient, nutrient-deficient, acid soils (see Figs. 2.14–2.17). The Cape

region as a whole has a rugged landscape with sharply differentiated habitats in close proximity to one another coupled with variable and complex precipitation patterns (Goldblatt & Manning, 2000). In the Fynbos in particular, the soils are particularly nutrient deficient, being developed mainly on nutrient poor quartzite sand. The Fynbos alone has a flora numbering 2285 species and infraspecific taxa, including many endemic species in an area of 471 km^2. The flora is also remarkable for the large number of species that are found in just a few genera, with the genus *Erica* alone having 650 species (Goldblatt & Manning, 2000). The Fynbos can be grouped under five broad types depending on their dominant species (Table 2.2).

Considering the Cape flora as a whole, the long Pliocene–Pleistocene history of climatic stability, like that of southern Europe, North America or southern South America, has probably been a major factor in contributing to the unique species richness of this region (Goldblatt & Manning, 2000). The extraordinary species richness of the Fynbos is due in part to topographic and geological diversity and this has contributed to both beta and gamma diversity. Using the

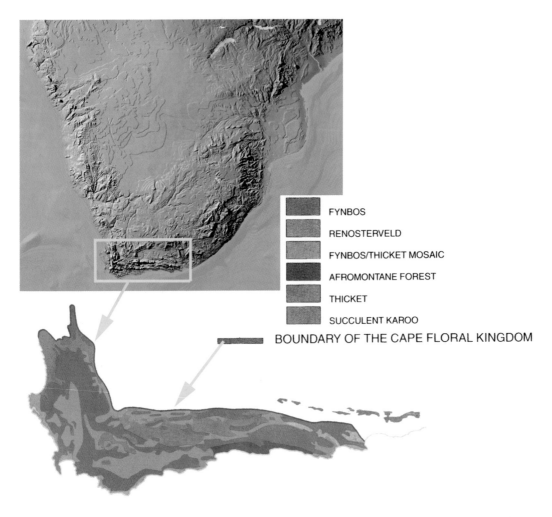

FYNBOS

RENOSTERVELD

FYNBOS/THICKET MOSAIC

AFROMONTANE FOREST

THICKET

SUCCULENT KAROO

BOUNDARY OF THE CAPE FLORAL KINGDOM

Fig. 2.14   Location of the Fynbos and Succulent Karoo within the Cape Floral Kingdom. (Reproduced with permission from Cowling & Richardson, 1995.)

definitions given in Table 2.1 the Fynbos can also be claimed to be richer in species per unit area (greater alpha diversity) than most tropical rainforests. In the case of the Cape Fynbos vegetation, species richness is attributed largely to a high degree of endemism, which appears to be aided by development of specific ericoid mycorrhizal-mediated nutrient uptake, which explains edaphic specialization and speciation particularly in the Ericaceae and Fabaceae (Cowling et al., 1992).

The Cape region is also a very ancient landscape that has maintained global migration routes along the African mountain chains. Consequently, long-term species migration has resulted in the acquisition of groups with geographical affinities using both the Gondwana track (Antarctic) which includes all southern hemisphere connections as well as African and boreal tracks. The latter includes taxa such as *Ranunculus*, *Anemone*, *Galium*, Dipsaceae, etc., which have had access to South Africa along the East African Highlands (Linder et al., 1992). Comparison of the nature of endemism in Mediterranean heathlands with the South African Fynbos and a similar community in Australia, the Kwongan, shows that in both continents more than 90% of the endemic species are edaphic

Fig. 2.15   Restiod Fynbos vegetation near the South African Cape. The dominant white flowered species is Cape snow (*Syncarpha vestita*, Asteraceae) growing among clumps of Restionaceae. (Photo R. J. Mitchell.)

specialists. There are nevertheless notable differences between the two regions in the biological profiles of the endemics. South African endemic species are more likely to be low shrubs with soil-stored and ant-dispersed seed while Australian endemics are concentrated among low to medium-height shrubs with either canopy-stored or soil-stored seed (Cowling *et al.*, 1994).

The Succulent Karoo communities on the west coast of southern Africa (see Fig. 2.17) are similar to the Fynbos, in that they are floristically part of the Greater Cape flora and considered to be one of the Earth's 25 biodiversity hot spots (Myers, 2003). Like the Fynbos

the species-rich vegetation of the Succulent Karoo (about 5000 species) contains more than 40% endemic species. The rainfall is usually less than 250 mm and falls predominantly in winter. In contrast to the Fynbos the soil is nutrient rich and the succulent vegetation is not fire prone. The Aizoaceae (ice plants) which dominate the succulent vegetation of the Karoo are represented in this region by 1750 species in 127 genera. A study of the possible causes of this extraordinary diversity (Klak *et al.*, 2004) has suggested that this group has diversified both recently and rapidly, with the estimated age for this radiation lying between 3.8 and 8.7 million years ago, yielding a per-lineage

Table 2.2. *Summary of principal types of Fynbos vegetation occurring in South Africa*

Proteoid Fynbos consists of bushy vegetation usually taller than 1.5 m and exhibits much colour in winter due to the large number of *Protea* species in flower. The principal locations are at altitudes lower than 1000 m and are most common at the base of mountains in areas with deep colluvial soils.

Ericaceous Fynbos, with a large number of species from the Ericaceae and resembling the temperate heathlands of Europe; this type occurs in permanently moist and cool environments on seaward-facing slopes and the upper regions of coastal mountains.

Dry Fynbos, found where the soil conditions are too dry to sustain shallow-rooted vegetation and consequently is dominated by small ericaceous shrubs able to extract moisture from lower regions of the soil with deep roots. This vegetation type is common on inland mountains and also on coastal sand dunes and the borders of the Succulent Karoo (see below).

Restiod Fynbos describes communities where the precipitation is not adequate for woody plants such as proteoids and ericoids. The Restionaceae, a southern hemisphere monocot family consisting of rhizomatous xeromorphic herbs with photosynthetic stems, is typical of nutrient-poor soils. There are over 180 endemic Restionaceae species in the South African Cape region where they characterize the *Fynbos* in areas where rainfall is low but predictable and where shallow-rooted shrubs absorb most of the moisture leaving little for deeper rooted plants. Alternatively, restiods also flourish where drainage is blocked by a water-impermeable layer (Fig. 2.17).

Graminoid Fynbos, typical of the eastern Fynbos where rain falls mostly in summer and conditions are particularly suitable for tropical grasses.

Source: Adapted from Cowling & Richardson (1995).

Fig. 2.16 Botanists examining South African Cape heathland. The dominant purple flowering species in the fore- and middle-ground is *Erica melanthera*. (Photo R. J. Mitchell.)

diversification rate of 0.77–1.75 per million years. Diversification of the group appears to be associated with the origin of several morphological characteristics and one anatomical feature, namely the development of wide-band tracheids in the Ruschiodeae clade which

are absent in all sister taxa. These specialized cells are considered as preventing tracheid collapse and therefore provide an adaptation for resisting water stress. The lack of this feature in species-poor clades of the Aizoaceae has been taken to suggest that this characteristic is a key innovation that may have facilitated this rapid radiative evolution (Klak *et al.*, 2004).

2.5.2 Mediterranean heathlands

Mediterranean heathlands, like the South African Fynbos, are also noted for their species richness and high levels of endemism, but differ in having lower numbers of species per genus (Ojeda *et al.*, 1995). As with the Fynbos the biodiversity of these sites is due in part to their antiquity. The European Mediterranean region, owing to its climatic stability, has been a refugium for many species throughout the Pleistocene and still retains a high number of endemic species. Again, like the South African Fynbos, the ancestors of the species and populations found in this region today have been present ever since the pre-Pleistocene climatic cooling period created the summer-dry Mediterranean

Fig. 2.17   The Little Karoo, a South African Cape region near Tradow's Pass, alt. 351 m, with less precipitation than in the Fynbos and a more xerophytic flora including many succulent species. (Photo R. J. Mitchell.)

type climate some 2–5 million years ago and made fire a driving force in the evolution of this distinctive flora.

In the European Mediterranean heathlands a wide variety of imprecise terms are used to describe the fire-prone, dwarf-shrub zone. These include *garrigue*, and *maquis*. Both are spread over vast areas and are derived from the burning and grazing of former forests. The taller form of scrub with species of broom (*Cistus* spp.), tree heather (*Erica arborea*), rosemary (*Rosmarinus officinalis*), and myrtle (*Myrtus communis*) is usually termed maquis while the term garrigue denotes a low-growing vegetation often containing many prickly dwarf shrubs with drought-resistant foliage but with a species composition that varies depending on local conditions. In Spain the *tomillares* is a form of garrigue with several species of thyme, sage and lavender, while in Italy gorse (*Ulex europeaus*) provides a link with the heaths of western Europe and Great Britain.

The more recent history of the Mediterranean garrigue and maquis can be traced from the earliest human occupation of the region. Archaeological botanical studies looking at plant remains in charcoal deposits can detect vegetation changes beginning in the middle Neolithic, when deciduous oaks declined and were replaced by evergreen oaks. The Chalcolithic (Greek *khalkos* + *lithos*, copper + stone) period is a phase in the development of human culture in which the use of early metal tools appeared alongside the use of stone tools (*c.* 6500–5500 BP). This period of human cultural development also witnessed the arrival and spread of meso-Mediterranean and thermo-Mediterranean plant formations, creating first the maquis, which then led to the development of garrigue vegetation during the Bronze Age (Heinz *et al.*, 2004). The degradation stages of this vegetation follow the sequence

deciduous forest → evergreen forest → maquis

→ garrigue → steppe.

The fire-prone nature of the vegetation stems from a combination of a climate with hot dry summers and the evolution of the vegetation to adapt to the dry season with partially desiccated, sclerophyllous vegetation. Many Mediterranean plants are also characterized by being active producers of volatile organic carbons. These monoterpenes and sesqui-terpenes (isoprenes; Fig. 2.18), commonly referred to as essential oils, lend a characteristic odour to the foliage. A long-recognized property of these com-pounds is to act as grazing deterrents. Some act as direct deterrents. Camphor has long been known to discourage egg laying by moths and the volatile oils in peppermint are claimed to discourage geese from grazing this plant. In other cases the mode of action is more subtle. In wild thyme (*Thymus vulgaris*) the terpene composition varies from plant to plant. Slugs appear to distinguish between individual plants and their terpenes and their preferences change with time. This polymorphism diversifies the attacks of the

predators and apparently increases the survival chances of certain individuals (Gouyon *et al.*, 1983). A similar situation is found in thyme-feeding aphids (Linhart *et al.*, 2005). It has also been shown that volatile oils reduce heat injury to the electron transfer system and that isoprene production in certain heat-resistant species increases under conditions of high light and high temperatures (Haraguchi *et al.*, 1997). It is not entirely clear how this protection is achieved. One possibility is that membrane stability may be enhanced by these essential oils altering the liability for phase change at high temperatures. Another pos-sibility is that the excretion of the volatile carbon compounds enables the electron transfer system to function without becoming inhibited by excessive carbon accumulation, an explanation that is analogous to the supposed benefits of light respiration.

A downside of this adaptation appears to be the increased sensitivity that volatile carbon producing plants show to ozone pollution. Such a sensitivity to oxidative damage also indicates that the vegetation is prone to fire damage. In some cases the volatile organic compounds may increase the readiness of the foliage to ignite, hence the rapid burning that then ensues and which can pass through the vegetation with great speed. The swiftness with which the fire passes through the vegetation is, however, a phenomenon which aids survival as the brevity of the fire does not incinerate all the wood. The extreme case is the biblical 'burning bush' (probably *Dictamnus albus* although plants in the genera *Bassia* (Chenopodiaceae) and *Combretum* (Combretaceae) are also sometimes known as burn-ing bush). Benzene has been confirmed among the various volatile compounds produced by *Dictamnus albus* and easily decomposes into chavicol and the very flammable hydrocarbon 2-methyl-1,3-butadiene (isoprene b.p. 34 °C). The secretion of isoprene, which can be especially intense on hot windless days, leads to formation of the isoprene cloud that may inflame the bush without causing too much harm to the source plant (Fleisher & Fleisher, 2004).

2.5.3   Mediterranean-type vegetation worldwide

A worldwide examination of regions with Mediterra-nean-type climates, and this includes the South African Fynbos, and the Australian Kwongan, suggests that in

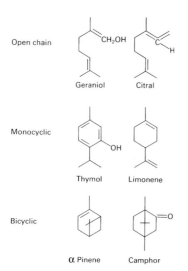

Fig. 2.18   Volatile oils commonly found in plants.

Table 2.3. *Factors which can lead to rapid speciation*

(1) Poor soils favour shrubs that are killed by fire and re-establish from seed. The frequent fires that occur in most Mediterranean heathlands also encourage shrubs that can resprout.

(2) Nutrient-poor soils can favour species that rely on seed production and seed conservation in buried seed banks for reproduction. These so-called seeders adopt a strategy that avoids the extra investment in underground organs that takes place in long-lived plants.

(3) The numerical dominance of seeders on poor soil lowers their extinction rates.

(4) Seeders have relatively short generation times, which leads to increased speciation rate.

Source: Wisheu *et al.* (2000).

addition to having soils with low soil pH and low nutrient levels, there are a number of common demographic features that may account for their floristic diversity. In particular, elevated speciation rates coupled with depressed rates of extinction could lead to enhanced species richness with fire as the predominant promoter of diversity.

Various aspects of this situation have been suggested as responsible for the rapid speciation that has taken place in these regions (Wisheu *et al.*, 2000; Table 2.3).

2.5.4   The Brazilian Cerrado

The Brazilian savanna biome usually referred to as the Campo Cerrado (Portuguese *Campo*, field; *cerrado*, closed, cf. *campo limpo*, clean field – grassland) or just Cerrado, is a vast area of dense scrub and woodlands extending over 2 million km^2 (Figs. 2.19–2.20) and is frequently so dense that vehicle access is impossible. The Cerrado is another example of a very ancient landscape and extends from the southern margins of the Amazonian forest to outlying areas in the southern states of São Paulo and Paraná. The antiquity of the Cerrado even suggests that it existed in prototypic form in the Cretaceous before the separation of the South African and South American continents. During the Pleistocene the Cerrado would have expanded as the extent of the Amazonian forest shrank, and then subsequently contracted as the rainforest spread during the Holocene (Prance & Lovejoy, 1985).

The ancient soils on which the commonly twisted and fire-resistant trees of the Cerrado grow are highly acidic and rich in soluble aluminum (Hueck, 1966). Highly toxic levels of aluminium up to 1000 mg kg^{-1} d. wt have been recorded in the foliage of some species (Geoghegan & Sprent, 1996), which necessitates a high degree of adaptation and probably contributes to the high level of endemism that occurs here at a species level. Considering the hostile nature of the soils the Cerrado has an extraordinarily rich flora. The vegetation comprises about 800 species of trees and large shrubs and several times that number of ground species (herbs and subshrubs).

An important aspect of the Cerrado is that unlike the African savannas it has lost the fauna of large herbivores that would have existed at the time of its creation. The herbivores that remain comprise anteaters, armadillos, opossums, some monkeys and various rodents such as agoutis, picas and capybaras (Ratter *et al.*, 1997). The introduction of cattle and horses is likely to have restored the vegetation to the structure it might have had in its earlier history.

When the flora of gallery forests, mesophytic forests and other habitats occurring in the biome are included, the total number of vascular plant species is estimated to reach about 10 000. In common with other savannas the cerrado flora is resistant to fire and shows all the usual morphological features associated with frequent fires: corky bark, xylopodia (lignotubers) and in grasses tunicate (having a dry paperlike covering) leaf bases (Ratter *et al.*, 1997). The flora is dominated by a few endemic genera with hundreds of species. In the genus *Mimosa* alone there are 189 species, 74% of which are endemic (Simon & Proenca, 2000). The Cerrado also shows high levels of beta and gamma diversity with many subdivisions of recognizable forest and scrub communities (Eiten, 1972).

Fig. 2.19   Open cerrado – a species-rich vegetation maintained by frequent fires. The relative dominance of species is also influenced in certain areas by frost. The dominant trees in the middle and far distance (*Vochysia tucanorum*) suffer significant damage to stems and leaves and are slow to recover when exposed to frost. Brando and Durrigan (2004) consider that the frequency and intensity of frosts can maintain more open forms of cerrado vegetation even in sites where both water and nutrient availability could support denser vegetation. (Photo Dr P. E. Gibbs.)

2.6   PLANT DIVERSITY IN DRYLANDS

The ability to withstand long periods of moisture shortage is not a prerogative of any one plant group or life-form. In trees, shrubs, grasses, ferns, mosses and lichens there are species which are drought tolerant (Figs. 2.21–2.22) and others which are restricted to areas of plentiful water. This evolution of drought-resistant species in all the major life-forms of plants is sometimes overlooked as it is easy to be deceived by the dramatic changes in vegetation that take place when passing from areas of high to low rainfall. In particular, the reduction in tree cover which can make a very vivid

impression on arriving in an arid zone can lead to an underestimation of the range of drought tolerance in certain tree species.

The absence of trees in many dry areas is often as much due to man and his animals as to a reduction in precipitation. When passing into a rain shadow area from one of plentiful precipitation, reduction in water supply produces a greater change on the species composition of the forest than on the existence of the tree form. Tree species native to arid regions do not necessarily consume more water than grassland. In the cool rainforests of Chile and western Patagonia rainfall of 2500–3800 mm is needed to support the growth of the evergreen southern beech (*Nothofagus dombeyii*).

Fig. 2.20   Fire and the Cerrado. (a) Geographical location of the Cerrado. (b) Incidence of fire in South America in September 2003 as recorded from satellite detection of night-time thermal spots (source: *World Fire Atlas* – European Space Agency web page: http://dup. eserin.esa.int/ionia/. (c) Burning cerrado vegetation (Photo Dr P. E. Gibbs). (d) Post-fire cerrado – note the twisted blackened stems from recent fires from which new shoots will emerge.

Fig. 2.21   An ancient specimen of mesquite (*Prosopis juliflora*) in the desert at Bahrain (Persian Gulf). This remarkable specimen is known locally as the tree of life. The species is native to Central America and the southern states of the USA but is now widespread in other arid lands. In some areas, as in the USA, it is considered an invasive bush that is taking over former grasslands as a result of overgrazing and increasing temperatures. In other places, as here in the Persian Gulf, it is valued for its sand-binding properties, shade, edible fruits, and as a source of honey.

Sixty kilometres to the east the rain shadow of the Andes has reduced the rainfall to less than 350 mm, yet in spite of this trees still survive in areas that are not grazed.

The succulents are one of the most characteristic plant groups of arid lands. Despite their restriction to a limited number of families, and a homogeneity of form in adopting the capacity to carry out crassulacean acid metabolism (CAM), they are nevertheless surprisingly variable in their morphology and have growth forms that can be trees as well as shrubs. The well-known Joshua tree (*Yucca brevifolia*) is one of the larger species to show CAM metabolism (Fig. 2.22). As long as there is some moisture, even if it is only coastal fog, there is nearly always a flora adapted to exploit an ecological window, however short. Hot deserts are usually ancient habitats and the plants and animals that live there have long histories of drought adaptation. Consequently, the biodiversity of arid lands at low latitudes should not be a surprise.

Fig. 2.22 Variation in tree forms in drought-prone habitats. (Top) The Cordilleran cypress (*Austrocedrus chilensis*) in the western edge of the Patagonian Desert (Argentina). (Above left) The Joshua tree (*Yucca brevifolia*) in the Mohave Desert – a dendroid member of the Liliaceae and one of the larger species to possess crassulacean acid metabolism (CAM). (Above right) *Ceiba chodatii* (Bombaceae) in Paraguay, a water-storing bottle-tree (Photo Dr P. E. Gibbs.)

Although deserts can be climatically grouped as arid areas there is great variation in the periodicity and duration of the drought period. Examination of any one major desert area reveals considerable climatic variations. The Atacama Desert in Chile and Peru is one of the driest in the world. At Antofagasta measurable rainfall can sometimes be detected in less than 6 years out of 20 and then the precipitation does not exceed 4–6 mm per annum. However, from May to November a dense cold mist, the *Garua* (Peru) or *Camancha* (Chile), rolls in with the sea breeze from the cold Humboldt Current. It is particularly dense at night but is sufficiently thick by day to obscure the tropical sun for days at a time. In the month of August some areas exposed to this mist have only 36 hours of sunshine and an average temperature of 13 °C while less than 800 m up-slope above the fog the temperature rises sharply to 24 °C. This mist supports a lichen-dominated vegetation over extensive areas of the fog-shrouded hills facing the sea.

In certain particularly favoured spots bordering the Atacama Desert there exists a unique forest vegetation, the *lomas*. The precipitation under the trees can be eight times that which is condensed in the open. Typical tree species are *Carica candens*, various species of *Eugenia*, *Caesalpina tinctoria* and *Schinus molle*, and near Lachay, *Acacia macrocantha* (Hueck, 1966). The lomas have a high proportion of endemic species together with a rich flora of bromeliad species.

Also striking in the Atacama Desert are the 14 species of *Tillandsia* with both epiphytic and unrooted, terrestrial representatives (Rundel & Dillon, 1998). All the *Tillandsia* species listed by Rundel and Dillon are epiphytic in the broad sense, but in addition to growing on plants they also grow on rocks, and six species (*T. purpurea*, *T. latifolia*, *T. capillaris*, *T. marconae*, *T. werdermanii* and *T. landbeckii*) have all evolved the ability to survive unrooted on sand (Fig. 2.23). As noted by Rundel, nowhere in the world are bromeliads more dominant or have more biomass than in these coastal species growing on sand. Many of these species grow at the absolute limits of vascular plant tolerance. It is possible to find an entire community consisting of a single *Tillandsia* species.

The Sahara occupies an area of almost nine million square kilometres in Africa and lies in a zone with less than 100 mm mean annual rainfall. Nevertheless, despite this general overall aridity there exists considerable bioclimatic diversity. Between the Mediterranean Sahara in the north, and the Tropical Sahara in the south, there are also the Central Plains Sahara, the Montane Sahara and an Oceanic Sahara, all of which have their own distinctive climatic patterns and biological diversity. Distribution patterns of plants and animals are closely linked with the climatic parameters, particularly in the amount and seasonality of rainfall and temperature. The latter may play as important a role as the former in controlling animal and plant distribution, since Mediterranean species can dominate in communities at higher elevations under tropical rainfall regimes, whereas tropical species intrude into the Mediterranean rainfall regions wherever winter temperatures are sufficiently warm (Lehouerou, 1995).

African deserts also provide astonishing examples of just how little rainfall is necessary to support a species-rich flora, provided the precipitation is regular in its occurrence. The sandy coastal belt between Port Nolloth and Alexander Bay on the north-west coast of Namaqualand in South Africa is part of the biologically diverse Succulent Karoo biome. In an area of 750 km^2, 300 plant species occur with 24% endemicity (Desmet & Cowling, 1999).

Egypt's Sinai Peninsula supports a flora of about 1285 species. The southern region is particularly species rich, supporting 800 species (including infraspecific taxa) with approximately 4.3% endemic species. Beta diversity between different landforms in the St Catherine area reflects a large biotic change between slopes and terraces on the one hand and between terraces and ridges on the other (Ayyad *et al.*, 2000). As yet, molecular genetic studies in deserts are few but those that have been carried out reveal considerable diversity between populations. In some cases this may be associated with the longevity of many desert perennials. An investigation of the genetic substructure of the insect-pollinated desert species *Alkanna orientalis* in the St Catherine desert revealed that the subpopulations growing in three different steep-sided wadis and a central plain area were genetically distinct from each other, but nevertheless showed evidence of gene flow, particularly between two of the wadis and the adjacent plain (Wolff *et al.*, 1997). The high mountain ridges between the wadis of the St Catherine desert make movement by bees across the ridges unlikely, nor do bees forage far beyond the range of a few plants at each wadi. Seed transport by flash floods therefore appears to be the more

Fig. 2.23 *Tillandsia latifolia* (Bromeliaceae) in the Atacama Desert. This is a free-living bromeliad species that survives unrooted and moves about the desert blown by the wind.

likely cause of gene flow between populations and illustrates again that even in an environmentally stressed habitat some benefit can accrue from disturbance. Hot deserts despite their apparent spatial and sometimes temporal emptiness can contain floras that are both species rich and varied and illustrate yet again that marginal areas can have significant biodiversity.

2.7 PLANT DIVERSITY IN THE ARCTIC

Cold climates do not feature frequently in discussions of biodiversity and the Arctic is normally not included in global maps showing sites of high plant biodiversity

(Myers, 2003). Nevertheless, within the Arctic there is ample evidence of diversity within species and of the presence of recognizable subspecies (Table 6.4; Fig. 2.24).

As discussed above (Section 2.1), peripheral populations, provided they are not isolated into populations of minimal size and subjected to genetic drift, can be expected to be more variable, since the changeable nature of marginal conditions induces fluctuating selection, which maintains high genetic diversity (Brochmann *et al.*, 2004; Grundt *et al.*, 2006; Eriksen & Topel, 2006). Genetic drift is probably unlikely in both hot and polar deserts where perturbation and wind erosion can transport seeds or fragments

Fig. 2.24   The early purple saxifrage (*Saxifraga oppositifolia*) growing at 79° N near Ny Ålesund, Spitsbergen. (Left) The tufted form which resumes growth early on dry banks and beach ridges is not flood tolerant and conserves carbohydrate in a taproot. (Right) The prostrate form which inhabits low-lying, late snowmelt habitats has a rapid growth rate and does not conserve carbohydrate.

of clonal species over vast distances and at great speed over frozen surfaces.

If the environmental argument of environmental uncertainty in peripheral areas is coupled with the case for ecological release (Section 2.3) due to a lack of competition, then the High Arctic is an area where substantial subspecific variation should be expected. Evidence to support this view can be found in various studies of diversity at high latitudes. A survey of allozyme diversity in populations of each of the 11 species of the North American angiosperm genus *Polygonella* (Polygonaceae) showed that the two most widespread species had reduced within-population gene diversity with respect to their narrowly endemic congeners (Lewis & Crawford, 1995).

In bird species, comparison of variability in peripheral areas with core areas has found greater variation at the periphery (Safriel *et al.*, 1994). A possible explanation is that the endemic species, several of which inhabit known Pleistocene refugia, have been able to maintain higher levels of diversity because of population stability during glacial cycles. As the authors point out, if this explanation is correct, an important implication for conservation is that, for many genera in eastern North America, the species richest in gene diversity may be those most in danger of extirpation in the next decade. The genus *Polygonella* is not alone in this respect and a number of tree species are also notable for showing the same phenomenon, namely *Pseudotsuga menziesii*, *Abies grandis*, *A. concolor*, *Pinus moticola* and *Thuja plicata*.

Population variation studies have progressed in recent years from classical taxonomic descriptions concentrating on morphological characters to detailed examination of many aspects of molecular genetics. Physiological characteristics are also worthy of attention in relation to metapopulations, not least for the possibility that they may link cause with effect. In any fluctuating environment there can be expected to evolve considerable physiological variation depending on factors such as water availability and length of growing season. Certain physiological variants may be evident only at times of particular stress. Thus, the remarkable capacity for arctic populations of some widespread plant species to withstand prolonged periods of total deprivation of oxygen (see Section 3.6.3) which occurs during periods of ice encasement is found only at high latitudes and does not occur in more southern populations of the same species (Crawford *et al.*, 1994). Some characters, such as those which influence phenology, as in snow-bed and beach-ridge populations in *Saxifraga oppositifolia* (Teeri, 1972) and *Dryas octopetala* (McGraw, 1987), are also recognized as taxonomic variations as they are accompanied by associated morphological adaptations.

2.8 CONCLUSIONS

The examples discussed above from peripheral areas in a variety of biomes illustrate that given time and space, and even the briefest of growing seasons, the capacity of plants to evolve and inhabit even the most extreme habitats is almost unending. Wherever there is light and moisture, even if it is only for a few brief weeks in the year, either in the High Arctic or in some of the Earth's driest deserts, flowering plants will survive, reproduce and diversify.

Part II
Plant function in marginal areas

Fig. 3.1 Luxuriant growth on the tundra. The Arctic lousewort (*Pedicularis dasyantha*) growing as a hemi-parasite on mountain avens (*Dryas octopetala*) at Blomsterstrand, Kongsfjorden, 79° N, Spitsbergen.

3 · Resource acquisition in marginal habitats

3.1 RESOURCE NECESSITIES IN NON-PRODUCTIVE HABITATS

Plant survival in exposed habitats with minimal resources frequently attracts attention for tenacity in an unpromising environment (Fig. 3.2). Despite impoverished soils, harsh physical conditions, and a lack of resources, such areas can provide habitats for a wide range of diverse communities (see Chapter 2). Plants vary enormously in their needs for resources and recognition of this phenomenon has been a driving force in the development of many ecological concepts including competition, mutualism and life history strategies. A study of plants in marginal areas therefore has to give attention to habitats with low levels of resources and examine how the many species that survive in these areas obtain their mineral nutrients, access to water, carbon dioxide and even oxygen for plants that live in wetlands and swamps. In such areas climatic warming is likely to expose the vegetation to further stress, particularly in relation to access to water.

Resources (sometimes termed essential resources) have been defined biologically as any component of the environment that can be utilized by an organism (Lincoln *et al.*, 1998). Ecologically, it would be more precise to replace utilized with consumed as this defines resources that are liable to exhaustion and for which organisms may have to compete. Thus, light, carbon dioxide, water and even oxygen are all essential resources that are consumed. Oxygen is often omitted from discussion of essential resources, partly due to the failure to recognize that many plant tissues frequently have to endure hypoxic conditions, either as a consequence of their morphological development, flooding, or competition with the soil microflora for rapidly diminishing supplies of oxygen. It is now apparent that plants differ in their tolerance of anoxia with profound ecological implications, particularly in amphibious and aquatic habitats. Oxygen will therefore be considered in this discussion as a resource (see also Chapter 8).

In marginal areas low temperatures and short growing seasons can limit plant growth, which raises the question as to whether thermal energy and time should be considered as essential resources. Heat is electromagnetic radiation, and like light can be absorbed by plant tissues. The extent to which plants can absorb or dissipate thermal energy can be maximized or minimized through phenotypic plasticity in relation to shoot morphology, pigmentation, and orientation (Fig. 3.3). Nevertheless, heat is not a resource for which plants normally compete. It is, however, one of the most decisive environmental factors influencing resource acquisition.

Whether or not time should be considered a resource is dependent on the life history of the species under discussion. Time is neither matter nor energy and as it exists into infinity is never totally depleted for perennial species and therefore cannot strictly be considered a resource. There may be a limited amount of time in any one growing season as measured in frost-free days or thermal time (day degrees) and this can impose limits in facilitating access to resources, particularly for annual species that have only one growing season in which to complete their life cycle. Time also affects competition as metabolic rates and timing of growth and development (phenology) determine the efficiency of biological processes and influence the outcome of competition for resources.

Entropy increases with time and as life has been defined as the negation of increase in entropy, the availability of energy is therefore time dependent. Nevertheless, time differs from the consumable resources in that it is in a sense never exhausted. In the Arctic where growing seasons are very short, the development of essential organs for reproduction can be spread over several years, making development less dependent on time. The arctic wintergreen (*Pyrola*

Fig. 3.2 Sea rocket (*Cakile maritima*) surviving with minimal resources. This plant has succeeded in flowering with a sprinkling of sand on a concrete block on the Churchill Barrier, Orkney.

Fig. 3.3   The glacial buttercup (*Ranunculus glacialis*) maximizing utilization of heat. (Left) Pre-fertilization state where the petals are white and reflect light and heat radiation to the centre of the flower. (Right) Post-fertilization stage when the petals close, turn red and absorb heat and protect the developing seeds. This species holds the high-altitude record in Scandinavia (2380 m Galdhøpigen, Norway) and in the Alps at 4725 m Finsterahorn, Bavaria, is only exceeded by *Saxifraga biflora* (see Chapter 10).

grandiflora) which grows in northern Iceland and Greenland can take several years to complete its reproductive cycle from the initial formation of a flowering shoot to the dispersal of seed (Fig. 3.4) and is therefore indifferent to the brevity of the growing season in relation to its eventual capacity to produce flowers and seeds. Similarly, the monocarpic desert century plant (*Agave americana*) can commonly take 20–30 years before flowering (usually well short of a century) and then dying.

A species that is constrained by time in accomplishing its reproductive cycle is the Iceland purslane (*Koenigia islandica*; Fig. 3.5), the only annual species able to survive in the Arctic. A study of this species under controlled environments concluded that a conservative seed germination strategy, a diminutive stature, and rapid development contributed to the success of this species in alpine and arctic environments. In addition, low temperature optima for growth and flowering were also important adaptations. The success of this annual plant, which is found in the High Arctic, the Himalaya, and in the southern hemisphere in the Tierra del Fuego, may be due also to being day-length neutral in relation to flowering (Heide & Gauslaa, 1999).

Nutritionally, plants differ from animals in that their sources of energy are less varied. Most green plants have only one major energy resource and that is light. Carbon dioxide, oxygen, mineral nutrients, and water are also necessary to capture energy, but this list is much less varied than the variety of alternative energy resources that are available to support animal life. For animals, energy is obtained from what they eat, which presents a degree of choice normally denied to plants. Plants, despite their dependence on relatively few basic resources, differ from animals in that they are not readily starved to death once they have become established. Individual resources such as light, water, carbon dioxide and oxygen can be experimentally withheld in many species for prolonged periods without necessarily proving fatal.

Hibernation in animals and dormancy in plants allow some species to withstand adverse periods when access to resources is temporarily hindered by unfavourable environments. Normally, for animals such periods of adversity will be no longer than a few months in winter, and the animals will not necessarily be totally dormant throughout the entire adverse period. For plants, however, resource limitations, whether they are due to cold, drought or flooding, can be endured in some cases for years and in the case of seeds, sometimes even for centuries (see Section 4.1).

The seed bank is a feature unique to plants which allows whole populations to survive many years of resource deprivation. A less extreme condition, common

Fig. 3.4 Arctic wintergreen (*Pyrola grandiflora*), a species not limited by time for successful reproduction, growing on a barren rocky substrate near Mesters Vig (72° N), north-east Greenland. This specimen, approximately 20 cm tall with a corolla 14–18 mm wide, probably took several years to develop the flowering bud before elongating the scape.

Fig. 3.5   Iceland purslane (*Koenigia islandica*), a species limited by time for successful reproduction, growing in patchy moss communities in Iceland. This species, approximately 3 cm high (see 1 cm scale), is the only annual species to survive in the Arctic and for the size of the plant succeeds in producing relatively large seeds in a very short growing season.

in deciduous shade-tolerant trees, is prolonged survival in the non-reproductive state while available light is close to the illumination compensation point (the intensity of illumination at which the oxygen evolved by photosynthesis is equivalent to the oxygen consumed by respiration – usually about 1% of full daylight). At such low light intensities, where energy acquisition from photosynthesis is only just adequate to meet the needs of maintenance respiration, growth is almost negligible, yet many saplings succeed in surviving for long periods in this suppressed state.

Most animals require conditions that permit sexual reproduction if their populations are to remain viable. This is not always the case for plants. In the Arctic and Boreal regions numerous species can survive as vegetative clones for centuries and even millennia. Aspens (*Populus tremuloides*), bracken (*Pteridium aquilinum*) and various arctic sedges (*Carex* spp.) are but three examples where clonal age has been estimated in millennia. In the Siberian Arctic estimates of over 3000 years have been made for the longevity of clones of *Carex ensifolia* ssp. *arctisiberica* (Jónsdóttir *et al.*, 2000). More examples are coming to light as molecular markers present new and remarkable insights into the longevity of many clonal species. The ability to lie dormant as seeds, or for perennating

plants to miss several growing seasons while buried under mud or ice, together with the capacity of many established plants to withstand starvation, demonstrates that plants are often endowed with a certain degree of indifference to a constant and uninterrupted provision of resources, a characteristic which facilitates their survival in marginal habitats (see Sections 1.3.2 and 3.6.2).

3.2   ADAPTATION TO HABITATS WITH LIMITED RESOURCES

Avoidance or tolerance is probably the most apt description of the principal survival strategies that are associated with plants in marginal areas. Either the plants adapt to reduced access to light, nutrients, water or oxygen, or else they have some means of avoiding acute deprivation of these essential resources. Low growth rates and a diminutive size are the commonest responses. Genotypic variation resulting in populations adopting this minimalist approach is often particularly evident in the distinct local races or ecotypes that have evolved in many marginal populations. Phenotypic plasticity can also produce dwarf plants. In some cases this reduction in growth is considered as merely a passive response due to a lack of resources. However, it may also be accompanied by other changes which suggest that even what appears to be a passive response may actually have a genetic basis. The dwarf and pubescent specimens of the sea plantain (*Plantago maritima*; Fig. 3.6) that are often found on sea cliffs, if transplanted to a less stressful habitat not only become much larger, but also usually lose their hairy leaves. However, not all coastal populations of sea plantain are hairy, which suggests these cliff-top populations are genetically diverse.

In relation to reduced light, plants vary in their response depending on whether they tolerate this condition by showing reduced growth or else attempt to escape from shade by the opposing strategy of increased growth as in shoot extension (etiolation). In these cases where the response is an attempt to escape from shade there has to be a high probability that increasing growth will achieve access to improved illumination otherwise the response will be maladaptive and decrease the probability of survival. Extension growth is frequently observed in aquatic plants where rooted macrophytes respond to increased submergence by stem and petiole extension (Voesenek *et al.*, 2004). In submerged aquatics

Fig. 3.6   Hairy diminutive form of the sea plantain (*Plantago maritima*) growing on a cliff top in Orkney. This hairy form occurs along with glabrous forms on areas exposed to salt spray. However, when transferred to the shelter of a glass house the plant can grow larger and lose its pubescence.

the extension is generally hormone controlled. For terrestrial plants shade detection, leading to shoot extension, is mediated by sensing the change in the quality of light, particularly the ratio of red to far red as perceived via phytochrome, or other photoreceptors.

Both phenotypic and genotypic variation can be examined in relation to their physiological responses to environmental conditions under headings which highlight metabolic activity. The first is *capacity adaptation*, in which alterations in specific morpho-logical and physiological properties enable the plants to increase their capacity to carry out existing metabolic reactions under suboptimal conditions. The second is

functional adjustment, which describes changes in behaviour (e.g. phenology) or the adoption of different physiological mechanisms as a response to alterations in environment.

3.2.1   Capacity adaptation

Marginal habitats frequently limit growing season length and reduce metabolic activity due to low temperatures. For many species these limitations can be overcome by the ability to increase metabolic rates at low temperatures by means of increased amounts of active enzymes, an adaptation which is referred to as

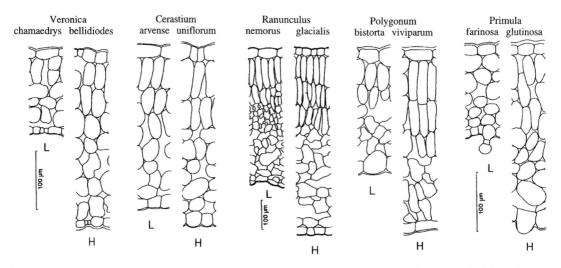

Fig. 3.7   Capacity adaptation as a result of genotypic variation. Contrasting internal morphology of pairs of lowland (L) and upland (H) species in their development of leaf thickness and internal air space (see also Chapter 10). (Reproduced with permission from Körner *et al.*, 1989.)

metabolic or capacity adaptation. As a means of adaptation it is found in cold-water fish (Hochachka & Somero, 1973) as well as in plants. Arctic plants are notable not only for having high levels of the carbon dioxide fixing ribulose bisphosphate carboxylase (RuBisCO) but also for having high densities of mitochondria. The maintenance costs of large amounts of enzymatic proteins in high-latitude plants will be minimized as long as they remain within the low temperature environment and can exploit the long days of the short arctic summer. If as a result of climatic warming temperatures increase, particularly where nights are longer at lower latitudes, high levels of RuBisCO could prove maladaptive due to increased maintenance costs. Physiologically, the 24-hour light regime makes it economic to have higher levels of RuBisCO. This enzyme has a dual function as a carboxylase and an oxygenase and can therefore fix or liberate carbon dioxide. In warm climates this functional duality can cause a significant loss of fixed carbon through light respiration. However, in the Arctic, temperatures are so low that the oxygenase activity is not significant and the 24-h usage of the high investment in carbon dioxide fixing capacity is amply repaid.

Alpine species are noted for their thicker leaves as compared with related species from lower altitudes (Körner *et al.*, 1989). This is also a form of capacity adaptation, as thicker leaves, with a greater development of palisade tissues and a higher internal volume, will facilitate greater assimilation of carbon dioxide in regions where atmospheric carbon dioxide concentrations are reduced, as at high altitudes (Fig. 3.7). This adaptation is similar to the manner in which human beings living at high altitudes are partially adapted to low atmospheric oxygen concentrations, as is evident in Quechua Indians born and raised at high altitudes in the Andes who possess a greater vital capacity (air expired after inspiration; see also Chapter 10).

3.2.2   Functional adjustment

Functional adjustment can take a number of forms in flowering plants but is dependent in all cases on phenotypic plasticity (the ability of an individual to change during its lifetime) as opposed to genetic variation (heritable variation) which varies both between species and within populations of the same species. The crucial difference in terms of plant responses to the environment is that phenotypic plasticity provides an immediate reaction to environmental stress, while genetic variation only allows change from one generation to another as a response to selection. Such an immediate phenotypic adaptation is found in the ability of plants to change the density of stomata in their leaves. Plants of

Fig. 3.8 *Vaccinium myrtillus* (bilberry) growing in a Norwegian mountain birch forest.

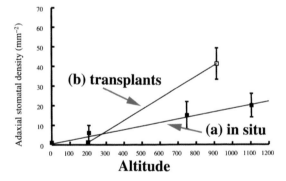

Fig. 3.9 Change in stomatal density with altitude for *V. myrtillus*. (a) *In situ* observations. (b) Transplants from 200–900 m. (Reproduced with permission from Woodward *et al.*, 2002.)

Vaccinium myrtillus show an increase in adaxial stomatal density with increasing altitude (Figs. 3.8–3.9). When individual plants were transplanted from 200 m to 905 m there was an even more marked response demonstrating the ability to respond phenotypically to the effects of increased altitude. Similarly, plants of this species growing in snow patches had mean adaxial stomatal densities of 18 mm^{-2} while those in exposed areas had mean values of 55 mm^{-2}. The signal for altering the development of the leaf in relation to stomatal density

appears to be sensed by mature leaves, which then influence the stomatal density in the leaves of the developing bud (Woodward *et al.*, 2002).

Resource foraging is another aspect of functional adjustment which is seen in the ability of some plants to actively search for resources. This is most readily apparent in stoloniferous plants where surface stems or underground stolons can place shoots in locations favourable for light capture and where subsurface stolons can aid roots in their foraging for mineral nutrients. The purple saxifrage (*Saxifraga oppositifolia*) has two forms (Sections 2.2.4 and 3.2.4), one of which has tufted shoots and a taproot and does not forage, and lives in drier habitats with a longer growing season. The other form adopts a creeping habit which allows it to forage for resources and exploit cracks in rocks and other places for shelter, nutrients and water, and has a greater metabolic activity both in respiration and photosynthesis which compensates for the shorter growing season associated with wet hollows where snow lies late into the growing season. The common bog cotton sedges are an example of two closely related species that share adjacent habitats, with one, *Eriophorum angustifolium*, being a stoloniferous foraging species while *E. vaginatum* is a tussock-forming non-foraging species. The latter is more tolerant of changing depths in the water table while the former is more successful in wetter areas through which the plants spread by means of their flood-tolerant stolons (Figs. 3.10–3.11).

The larger the organism the greater, and frequently disproportionately greater, is the share of resources that it can command to the detriment of competitors. Thus, large trees remove much of the light from any terrain they occupy. In scientific terms, competition for light is size asymmetric, in that a large plant can potentially dominate a competitive relationship and therefore the light resources obtained by the taller plant are disproportionate to its size (Blair, 2001).

In marginal habitats the environment is often uncertain and episodes of drought, flooding, storm exposure or cold, plus physical disturbance, can undermine the tendency of large plants to dominate the landscape. In these situations, size, at least above ground, can be incompatible with long-term survival, and organisms with reduced exposure and lower demands for resources may prevail. It is therefore important to consider the dual nature of plant existence with one part in the air and the other rooted in the ground or

Fig. 3.10   Bog cotton (*Eriophorum vaginatum* and *E. angustifolium*) colonizing a bog from which peat has been recently extracted industrially in Caithness, Scotland.

Fig. 3.11   Divergent strategies in bog cotton growth. (Left) The tufted non-foraging species *E. vaginatum* which favours drier portions of the bog. (Right) The foraging species *E. angustifolium* which grows in the wetter parts of the bog.

submerged in water. In contrast to aerial shoots the below-ground parts of plants live in a potentially more heterogeneous habitat where resources are rarely evenly dispersed. Therefore accessing resources from soil may be more size symmetric as the amount of soil nutrients obtained will be in direct proportion to the size of the foraging organs. Unfortunately, most studies examining below-ground competition use homogeneously distributed nutrient resources and soil homogeneity is not often found in nature.

3.2.3   Adverse aspects of capacity adaptation

Every adaptation to a specific environment has its limitations which arise from an increased dependence on a particular set of environmental conditions that is

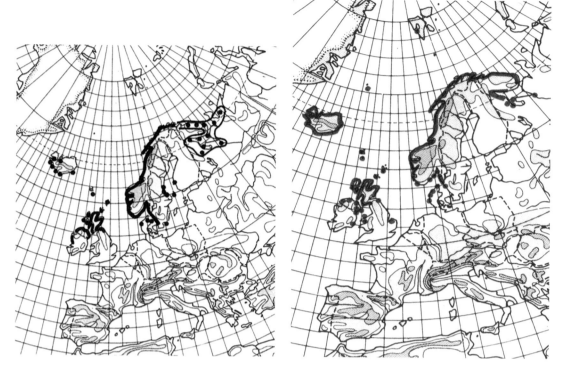

Fig. 3.12   European distribution of two coastal arctic species that reach their southern limits of distribution in the British Isles and Scandinavia. (Left) Scot's lovage (*Ligusticum scoticum*). (Right) The oyster plant (*Mertensia maritima*). (Reproduced with permission from Hultén & Fries, 1986).

created in the adapted plant. In this sense 'adaptation is the first step on the route to extinction' (Crawford, 1989). Thus the use of capacity adaptation by plants of cold climates to compensate for the limitations of low temperatures can place such adapted plants at a disadvantage in warmer environments. Examples of this are found in the carbon deficits that arise in some arctic species as a result of warm environments. The southern distribution of the arctic coastal herbs Scot's lovage (*Ligusticum scoticum*) and the oyster plant (*Mertensia maritima*) are examples of this condition. Both these species occur in the Arctic (Figs. 3.12–3.15) and reach the southern limits of their distribution on the shores of Scotland. They are also remarkable for the speed with which they extend their foliage when growth recommences after winter. Such rapid growth, exploiting the reserves of previous carbon gains, requires high respiration rates. A typical response of *L. scoticum* to temperature in its respiration rate is shown in Fig. 3.14. The rapid increase in dark respiration rate with

temperature is also found in the oyster plant (*Mertensia maritima*) and contrasts with the lower respiration rates of two comparable coastal species of more southern distribution. The depletion of carbohydrate reserves by high respiratory activity at warm temperatures has often been suggested as a limiting factor in the southward extension of northern species (Crawford & Palin, 1981).

An examination of the carbohydrate content of various *Vaccinium* species growing in Scotland (Bannister, 1981) showed that the species which suffered the severest depletion of carbohydrate reserves at warm temperatures, *V. uliginosum*, was the most restricted in its southern range. Thus for some perennial herbs and woody heath species the consequence of having high respiration rates to exploit carbohydrate reserves in short cool growing seasons places the plants at a disadvantage in warmer climates (see also Figs. 3.12–3.15).

Many studies have sought to determine whether or not the total carbon balance of woody plants can be

Fig. 3.13   Coastal species with predominantly arctic and subarctic distribution growing on a shingle beach in Orkney. (Left) Scot's lovage (*Ligusticum scoticum*). (Right) The oyster plant (*Mertensia maritima*).

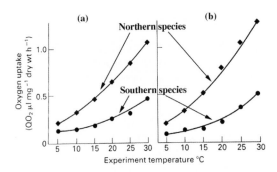

Fig. 3.14   Overwintering root respiration rate at a range of temperatures for four coastal species. (a) *Ligusticum scoticum*, a northern species, and *Crithmum maritimum*, a southern species. (b) *Mertensia maritima*, a northern species, and *Limonium vulgare*, a southern species. (Data from Crawford & Palin, 1981.)

related to their latitudinal or altitudinal distribution. Intuitively, it might be expected that woody plants which devote a considerable part of their resources to the formation of non-productive trunks and stems may be unable to support such a growth strategy when growing seasons are cool and short. However, an extensive worldwide study of the carbon balance in trees at their upper altitudinal boundaries has shown the converse, namely that tree growth near the timberline is not limited by carbon supply (Fig. 3.16) and that it is more probable that it is sink activity and its direct control by the environment that restricts biomass production of trees under current ambient carbon dioxide concentrations (Hoch & Körner, 2005).

Although carbon limitation as measured in the total biomass of woody plants may not be a feature that relates directly to distribution of woody species with temperature, it is nevertheless possible that certain vital organs, such as root tips and buds, and cambial tissue, rather than the whole plant can be seriously carbon deficient under specific conditions (Section 3.2.4).

3.2.4   Climatic warming and the vulnerability of specific tissues

The meristematic region of roots is always anaerobic. Consequently, any diminution of oxygen supply to the roots will increase the amount of root that is hypoxic, which accelerates the drawdown of carbohydrate levels through anaerobic respiration. Translocation of carbohydrates does not appear to be able to replenish the distal regions of roots in mid-winter. Carbohydrate depletion of these tissues reduces antioxidant content and thus renders the roots vulnerable to post-anoxic injury when water tables fall and aeration is restored in

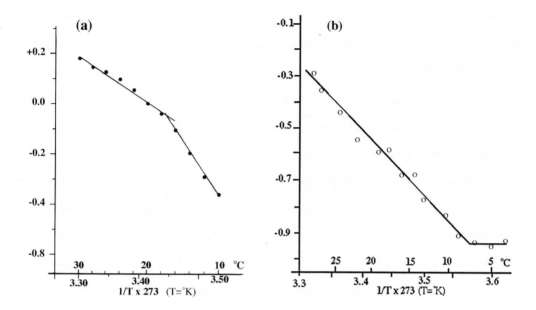

Fig. 3.15   Arrhenius plots of root respiration rate (oxygen uptake – see above) in relation to temperature. (a) *Ligusticum scoticum.* (b) *Crithmum maritimum.* (Data from Crawford & Palin, 1981.)

spring (Crawford, 2003). Observation of Sitka spruce roots (*Picea sitchensis*) has shown that winter warming when combined with flooding can lead to a dieback of the root system when the water table drops in spring and oxygen is suddenly restored to tissues that have endured a prolonged period of oxygen deprivation. This damage is not immediately fatal or even damaging to the tree, but repeated dieback of the deeper roots makes the trees liable to windthrow, which causes the greatest loss of trees in oceanic climates.

The damage caused by warm winter temperatures is unlike heat injury in that no visible signs or tissue lesions are observed. Arctic plants transplanted to the south of their natural range typically grow for a season or two until they have exhausted their carbohydrate reserves. Spring starvation then prevents the resumption of growth and the plants merely disappear.

Limitation in southern distribution will be caused by competition in many species, but there is a significant number of examples where this explanation is not adequate. Coastal habitats with their open communities which frequently have to be renewed after storms give ample opportunity for the establishment of fresh colonists. The continuous nature of coastlines

also provides a habitat which extends through changing climatic zones without imposing any topographical barrier to plant dispersal. As these examples show, starting on the shores of Greenland and Spitsbergen (Fig. 3.17, left panel), there are several species which enjoy an uninterrupted distribution as far south as the British Isles, and in the case of sea lyme grass (*Leymus arenarius*; Fig. 3.17, right panel) even to the coast of northern France, but cannot be found further south.

A striking example of a species confined to high latitudes, even in the apparent absence of competition, is the whiplash saxifrage or spider plant (*Saxifraga flagellaris*). This species which does not occur south of 75° N in either Greenland or Spitsbergen is related to the high altitude whiplash saxifrage species of the Himalaya (e.g. *S. stenophyllla*) of Pakistan and Nepal. In both Greenland and Spitsbergen this saxifrage grows in open shingle habitats and there are no grounds for believing that it is competition with other more aggressive species that limits its southern spread. When visiting the plant at 75° N in Spitsbergen (Fig. 3.17) it is difficult to imagine what there might be in more southern climates that is inimical to the survival of this species.

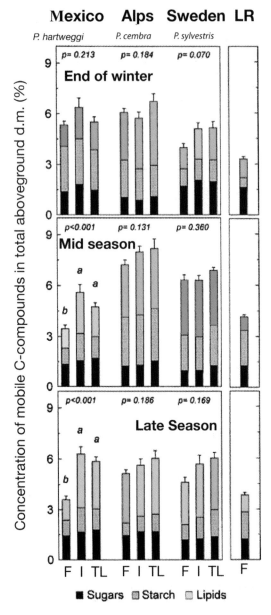

Fig. 3.16   Estimated concentration (% dry matter) of low-molecular weight sugars and starch (NSC) and lipids in the total above-ground biomass of pines along altitudinal transects as compared with a temperate lowland reference site near Basel, Switzerland (LR), at three dates during the local growing seasons. Values are the means (+SE) of 5–15 trees. F, closed forest; I, intermediate stand between closed forest and treeline; TL, treeline stand. (Reproduced with permission from Hoch *et al.*, 2003.)

A regime of continuous polar light and minimal risk of heat injury allows arctic plants to adopt a number of other low temperature compensating mechanisms which would not be viable in warmer climates. Dark colours absorb heat and the predominant hue of much of the vegetation cover of the Arctic is brown not green. This is particularly the case in the ultra-short growing season habitat of some shore communities (Fig. 3.18). In addition, a number of species have the ability to keep their flowers and sometimes also their leaves oriented towards the sun (sun tracking) and by having parabolic flower structures that concentrate the sun's rays on their reproductive organs, which not only hastens their development but also increases their attractiveness to insects (Fig. 3.19; see also Fig. 10.3). A number of woolly species even produce their own mini-greenhouses. Female plants of the arctic willow (*Salix arctica*) can become covered in down which can raise their leaf temperature to 11 °C above the air temperature (see Fig. 4.1).

So important is adjustment of metabolic rate to temperature that even within the High Arctic there are specialized forms of the same species for living in cold or warm microhabitats which may be only a few metres from each other. One striking example of such differing functional adaptation in ecotypes of a widespread arctic species is found in the purple saxifrage (*Saxifraga oppositifolia*), one of the hardiest plants of polar regions. On dry, exposed ridges with warmer temperatures and a longer growing season than the adjacent low shore, the purple saxifrage has a tufted form that starts growing before the snow melts in adjacent low-lying shore habitats. By contrast, on the late, cold shore sites this species is found usually as a trailing prostrate plant that does not give an impression of being particularly robust. However, appearances are misleading and this frail-looking prostrate saxifrage has a capacity adaptation which allows it to outperform the more robust type on the beach ridge in gross photosynthetic capacity, respiration and shoot growth (Crawford *et al.*, 1993). Physiologically, the tufted form on the beach ridge is much more drought tolerant, and conserves carbohydrate for periods of stress. The plants of the shore habitat are by contrast spenders and use a much greater portion of their energy gains immediately for rapid growth (Figs. 3.20–3.21). The two forms have developed opposing strategies which aid their

Fig. 3.17   Examples of species of cold climate coastal plants with southern limits to their distribution in Europe. (Left) The spider plant *Saxifraga flagellaris* ssp. *platysepela* on an arctic shore in Spitsbergen at 75° 30′ N. This species occupies open habitat with apparently no physical barriers to its southern migration yet is not found south of 73° N. (Right) Sea lyme grass (*Leymus arenarius*), a coastal species in dunes or gravel from the Arctic to north-west Spain.

survival in their particular microhabitat but which would disadvantage them if conditions were to change. Plants of the tufted and creeping ecotypes of purple saxifrage maintain their distinctive growth forms in cultivation and although interfertile have been recognized as distinct types (Lid & Lid, 1994). This same phenomenon is found in the mountain avens with different ecotypes occupying exposed ridges and snow hollows (Section 1.3.3).

3.3   HABITAT PRODUCTIVITY AND COMPETITION

A long-standing ecological question of particular relevance to resource utilization marginal areas is whether or not intensity of competition is a direct function of habitat productivity. It can be argued that as productivity rises so will the intensity of competition increase. The obverse situation would be that communities with low productivity are made up of species that have lower growth rates and therefore make lower demands on their habitat for resources and create communities with lower levels of competition intensity (Grime, 2001). However, the contrary effect can also be argued. If competition for light increases there should be a reduction in competition for other resources such as nutrients and water so that the total intensity of competition will be independent of productivity (Tilman, 1987). The nature of this argument is conceptual and depends on how competition intensity is to be measured. Two possible variants exist depending on whether absolute competition or relative competition is a true reflection of the intensity of competition (Table 3.1). Comparison of a number of data sets using these two indices tends to support the view that competition intensity increases with productivity if absolute competition intensity is calculated, while the view that competition intensity is independent of productivity is true if relative competition intensity is calculated (Grace, 1993). This distinction, however, does not apply with equal clarity to all studies. The use of the term performance in many studies introduces an element of ambiguity as it may mean growth or biomass accumulation in one study and flowering or seed production in another. The ambivalence of just how competition should be assessed underlines the principle that it is not a physical reality and that what is observed is highly dependent on how the concept of competition has been defined and how survival is assessed (see also Section 3.6.2).

Fig. 3.18   View of an arctic salt marsh – Longyearbyen (Spitsbergen). The brown colour of the marsh is typical of much arctic vegetation even in summer. The dominant species is the creeping salt marsh grass (*Puccinellia phryganodes*, see inset).

Communities of plants and animals are usually dynamic with constantly changing species assemblages that result from continuing and long-term competitive processes. Unfortunately, it is only rarely that the processes of competition can be observed in action in natural plant communities. Consequently, most competition studies are carried out on crops where variation is minimal, conditions controlled, and the problem is confined either to the effects of density on crop yield, or to the interaction of usually not more than two species in carefully controlled replacement experiments. Such studies can be used in estimating maximal sustainable yields as well as increasing resource-use efficiency in crops. They may also prove helpful in combating pollution by finding the optimal planting densities for nutrient uptake and thus reducing the rate at which

artificial fertilizers need to be applied to agricultural land. Human life spans are usually too short for long-term observation of competition in natural systems and most examples are restricted to monitoring the effects of invasive and aggressive alien species (see Section 4.6).

3.3.1   Plant functional types

The concept of categorizing plants into broad groups has a long history, beginning with Aristotle's disciple and successor Theophrastus of Erosos (*c.* 370–285 BC) who grouped terrestrial plants into three categories: trees, herbs and vines.

This process of categorization has continued with ever-increasing inventiveness, particularly in modern times, as some degree of rationalization is necessary,

Fig. 3.19   The arctic poppy (*Papaver radicatum*). This specimen is growing at sea level at Mesters Vig (Greenland 73° N). The species also holds the second highest altitude flowering plant record for Greenland at 2450 m (Halliday, 2002).

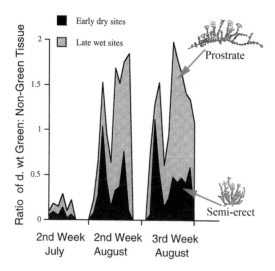

Fig. 3.21   Ratio of d.wt of green to non-green tissues in plants of the semi-erect and prostrate forms of *Saxifraga oppositifolia* inhabiting respectively dry warm ridge sites and late, wet shore sites at Kongsfjord, Spitsbergen. Note the greater proportion of green tissue in the prostrate form that developed during the peak of the growing season which will compensate in part for the shortness of the growing season on the wet shore site.

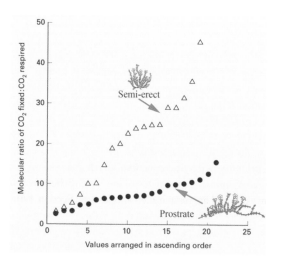

Fig. 3.20   Molecular ratio of carbon dioxide fixed (net photosynthesis) to respiration in shoots of *Saxifraga oppositifolia* of two distinct ecotypes. Triangular symbols, measurements made on semi-erect plants taken from a dry beach ridge; solid circles, measured at same time from an adjacent low beach late, wet site with prostrate creeping plants. Note the semi-erect plants from the early dry site are carbon savers and the plants from the late, wet site are carbon spenders.

particularly when computer-based models are used to predict the effects on plants of environmental change. Different biomes can be typified for comparative studies by grouping plants into different functional types. For arid zones, functional types can be defined by categorizing plants in relation to their responses to drought. In a large-scale study of global distribution of ecosystems in relation to fire, angiosperm and gymnosperm forests can be grouped functionally according to whether or not they are resistant to fire. Similarly, grasslands can be categorized depending on whether their dominant species are C3 or C4 plants. This latter categorization has been suggested as evolving in relation to fire some 6–8 million years ago when fire was already a major factor in the spread of C4 grasslands into former forested areas long before there was any human-induced forest burning (Bond *et al.*, 2005; see Figs. 2.10–2.11).

In other situations different categories can be employed. Studies on climate change responses in arctic plants have found it convenient to consider the various species grouped into deciduous or evergreen shrubs, sedges, and herbs, while in herbivory studies

Table 3.1. *Ecological definitions related to competition studies*

r-selection   Selection favouring a rapid rate of population increase, typical of species that colonize short-lived environments or of species that undergo large fluctuations in population size.

K-selection   Selection producing superior competitive ability in stable unpredictable environments, in which rapid population growth is unimportant as the population is maintained at or near the carrying capacity (K) of the habitat.

Consumptive competition   (Exploitative competition) The simultaneous demand by two or more organisms or species for a resource that is actually or potentially in limited supply.

Interference competition   The detrimental interaction between two or more organisms or species.

Pre-emptive competition   The ability to occupy an open site and thus pre-empt space.

Niche   The ecological role of a species in a community; the multidimensional space of which the co-ordinates are the various parameters representing the species' conditions of existence to which it is restricted by the presence of competitor species.

Fundamental niche   The entire multidimensional space that represents the total range of conditions within which an organism can function and which it can occupy in the absence of competitors or interactive species; pre-interactive niche; ecospace, cf. physiological tolerance.

Realized niche   That part of the fundamental niche actually occupied by a species in the presence of competition or interactive species; realized ecospace; post-interactive niche; cf. ecological tolerance.

Absolute competition   Productivity in monoculture minus productivity in mixed culture.

Relative competition   (Productivity in monoculture minus productivity in mixed culture)/Productivity in monoculture.

species can be usefully placed into functional types in relation to their palatability, the essential factor being that species within each grouping share a combination of physiological, morphological and life-span characteristics. Such an approach is valid if the variability in key characteristics is greater between than within the defined types (Smith *et al.*, 1997). It has to be noted, however, that scaling up or aggregation runs counter to normal scientific method which usually demands that investigations should proceed from the particular to the general, i.e. bottom-up and not top-down. It is nevertheless understandable that the complexities of biology, and in particular ecology, often cause this logical sequence to be reversed or ignored in response to the need for expediency in arriving at decisions necessary for conservation and political action.

The premise that the variability of key characteristics is greater between than within vegetation types cannot always be justified particularly in marginal situations. Ecotypic variability within species and even neighbouring populations of the same species is very common in marginal situations, with different ecotypes frequently having opposing and incompatible adaptations, e.g. flooding versus drought tolerance, shade versus high light requirements etc. In such cases grouping these populations together as one functional type ignores basic ecological differences.

Equally, at the species level, minor variations within one vegetation type can be occupied by closely related species that exploit different survival strategies in relation to their physiological and reproductive processes. Examples of such differences can be found in the use of crucial resources such as water. There are even differences between sexes of the same species, as has been shown in the higher growth rate and greater ability in female plants to conserve water in arctic willows (see Section 4.11; Crawford & Balfour, 1983; Dawson & Bliss, 1989a). Similarly, in male and female boxelder trees (*Acer negundo*) female plants show higher carbon assimilation rates than males, both when well watered and when droughted (Fig. 3.22). The significance of such differences is ignored when scaling-up

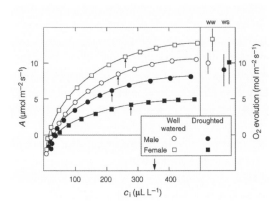

Fig. 3.22   Carbon assimilation rate as affected by sex in male and female trees of box elder (*Acer negundo*) growing under common garden conditions. Plot of assimilation rate (A) as a function of intercellular CO_2 concentration (c_i) and the rate of oxygen evolution (right panel) for male (circles) and female (squares). Plants were well watered or droughted for 14 days then measured. Arrows of the A–c_i curves show c_i at ambient [CO_2] of 353 ppm. (Reproduced with permission from Dawson *et al.*, 2004.)

methods are employed. Only by examining local differences in biodiversity and progressing from the particular to the general can biologically based action plans for conservation and restoration be accurately formulated which reflect the nature of the species under consideration.

3.4   LIFE HISTORY STRATEGIES

The ability of less competitive species to survive when more aggressive species are hindered from reaching maximum productivity by environmental variation or uncertainty has long been realized as fundamental to species diversity. Many attempts have been made to look for combinations of adaptations or strategies that can be used to describe the manner in which different species respond to the pressure of competition for resources or interference from neighbours. The term strategy, when applied to ecology, is divorced from its more common military usage and merely denotes a 'complex of adaptations involving a large number of interacting and coevolving traits' (Calow, 1998). Recently, plant life strategies (for definitions see Table 3.2) have gained new importance in ecological thinking as they can form a basis for conceptualizing the processes that maintain

biodiversity (see Chapter 2). However, when these classifications are being applied, as for instance in computer models of vegetation responses to changing environments, it is important to remember that like the concept of competition they are merely human perceptions used to create order from an infinite variety of plant behavioural responses to the environment and that they are therefore imaginative scenarios rather than biological realities.

3.4.1   Two-class life strategies

The earliest life strategy classifications were based on a two-class division on the use of resources using the terms such as capitalists and proletarians, possibly reflecting radical political thinking at this period (Grime *et al.*, 1988). This and other two-way classifications in essence anticipated the classification of r and K selection strategies where organisms are divided on the basis of whether their lifestyles are adapted to frequently disturbed environments or niches, or whether they are able to exploit continuously highly productive habitats (Fig. 3.23). In r-selected species the typical habitat is one where resources are transient and temporary and thus the population that grows most rapidly prevails. Such populations increase their intrinsic rate of population growth either by increasing their birth rate (seedling production) or more effectively decreasing generation time. Such rapidly growing populations suffer high levels of mortality particularly at the seedling stage. Thus efficient dispersal mechanisms will be favoured, as they will reduce juvenile mortality by distribution of seed to potentially favourable sites. Weed species are sometimes described as being typical r-selected species. However, this is not a universal generalization. Many weed species can be highly competitive and show a degree of ecotype flexibility that allows them to adapt to different situations and operate a successional sequence of their own. This is particularly the case with dangerous perennial weeds such as creeping thistle (*Cirsium arvense*), field bindweed (*Convolvulus arvensis*), perennial sowthistle (*Sonchus arvensis*), couch grass (*Elytrigia repens*), bracken (*Pteridium aquilinum*) and field horsetail (*Equisetum arvense*).

Habitats with long-term stability carry populations of species that are usually close to carrying capacity (K) with little room for further population growth. Such

Table 3.2. *Examples of definitions for classifying plants in relation to function*

Life-form   Classification of species on the basis of morphological features that are associated with their environment. Raunkiaer's scheme (1934), which is based on the height of the perennating bud above the ground during the adverse season, is still the most widely applied system.

Life history traits   Features of an organism's life cycle that are most directly related to birth and death rates and thus to Darwinian fitness. Life history traits can be viewed as the way that individuals allocate their limited resources (Calow, 1998).

Life history strategy   A grouping of similar or analogous life history characteristics which recur widely amongst species or populations and cause them to exhibit similarities in ecology (Grime, 2001).

Functional types   Species or groups of species that have similar responses to a suite of environmental conditions. In practice these groupings are usually related to physiological function as it is affected by different environments.

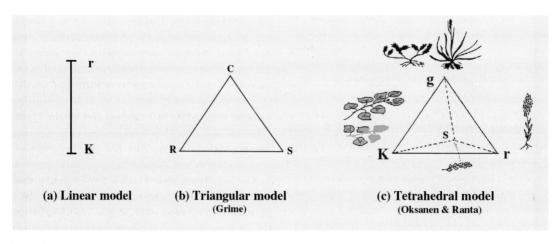

(a) Linear model   **(b) Triangular model**   **(c) Tetrahedral model**
(Grime)   **(Oksanen & Ranta)**

Fig. 3.23   Two, three and four classification strategies for interpreting plant responses to competition and environmental stress. (a) The basic two classification system between r and K selection. (b) The triangular classification model of Grime between competitors (C), stress tolerators (S) and ruderals (R). (c) The tetrahedral modification of Oksanen and Ranta, 1992, with the three primary strategies (K, r and g) on the face of the tetrahedron. The g strategy is typified by ericoid and graminoid species with low root to shoot ratios and low stature. The additional strategy s lies behind the front plane and represents decreasing allocation to support tissues, with the extreme type being a hepatic. (Reproduced with permission from Oksanen & Virtanen, 1997.)

species can have low levels of fecundity which is offset by a longer life span. Selection for the particular competitive strategies makes K-selected species highly specialized and consequently more sensitive to environmental change. They are also less able to recover from low population densities and are therefore likely to be species in danger of extinction. Vegetative reproduction can enhance the probability of survival by conferring almost virtual immortality. Such cases are seen in many northern populations of aspen which have had difficulty in setting seed during the Little Ice Age (a period of cooling lasting approximately between the thirteenth and nineteenth centuries) but nevertheless have survived due to an ability to sucker. Similarly, stands of trees lying outside the main boundaries for forest regeneration are maintained by layering. Some alien species achieve pest status due to their capacity for vegetative reproduction. In Scotland the common rhododendron (*Rhododendron ponticum*) was introduced to the Highlands and owing to its capacity for layering and shade tolerance is able to outcompete much of the native woodland flora.

3.4.2 Three-class life strategies

Although simple binary divisions are sufficient to highlight some of the most contrasting dichotomies in competition behaviour in both plants and animals there is considerable improvement when plant competition is viewed through three-strategy models.

A variety of additional third classes of plant behaviour have been suggested. *Adversity selection* (Southwood, 1988) brings into play the quality of the habitat where stressful conditions can be accommodated at a metabolic cost to the organism but results in impaired growth or reproduction. The notion of a third dimension in competition strategy as developed by Grime recognizes differences in mortality between the behaviour exhibited by juvenile as opposed to adult and established plants. Variations in regenerative strategies exploit differences in resource investment, mobility and dormancy which are all expressed mainly in the regenerative phase. This has led to the suggestion (Grime *et al.*, 1988) that there are two major classes of external factors which affect the life strategies of plants. The first may be described as stress and consists of all those phenomena which can restrict photosynthetic production, such as shortages of light, water and mineral nutrients or suboptimal temperatures.

The second class is referred to as disturbance and covers all phenomena which result in the partial or total destruction of biomass such as herbivory, human disturbance through trampling, mowing and natural physical disturbances from wind damage, frost drought, soil erosion and fire. In the three-class life strategy concept, it is asserted that plants can occupy only two extreme marginal conditions, namely either extreme stress or extreme disturbance. The combination of both extreme stress and extreme disturbance is considered by Grime to result in areas that are not habitable by plants.

3.4.3 Four-class life strategies

There are, however, situations in some marginal habitats that have prompted the suggestion that it is possibly more realistic to visualize plants as having up to four basic life strategies. In some arctic, subarctic and montane habitats it is possible to have intense grazing in a severely resource-limited environment and survive (see Section 10.3). In many instances it can even be demonstrated that this grazing is essential for the survival of flowering plants in order to avoid build-up of litter and shading and the consequent formation of lethal ice wedges in the soil.

Even in relatively unproductive arctic and alpine habitats, grazing has always existed and is frequently accompanied by intense physical disturbance through cryoperturbation and erosion. It has been argued that grazing selects for entirely different kinds of traits from devastating physical disturbance, which leaves no survivors *in situ* (Oksanen, 1990, 1993). Loss of above-ground tissues obviously selects for high root to shoot ratios and for vegetative reproduction by means of horizontal rhizomes, especially if the loss is caused by grazers, which prefer nutrient-rich floral shoots. Moreover, herbivory selects for tough, narrow, finely lobed or scale-like leaves and against broad, mesomorphic leaves. Grazing is particularly detrimental to the latter as partial consumption of broad mesomorphic leaves creates long wounds, exposing the remaining tissues to desiccation and invasion by parasitic fungi.

The concept that plants have evolved certain stable defence strategies suggests that there are two main strategies in relation to withstanding grazing. Plants can either invest large amounts of reduced carbon in defensive compounds, which inevitably results in low growth rates, or else they can defend themselves solely by means of readily available inorganic substances (e.g. silica) or by means of mechanical deterrents (e.g. thick cuticles, thorns), which are relatively efficient even at low levels of resource investment. The two main ways of surviving intense grazing have been described as ericoid and graminoid strategies (Oksanen, 1990). The main advantage of typical graminoids – their ability to produce erect leaves rapidly from basal intercalary meristems – is claimed by Oksanen to be reduced in habitats with persistent, intense grazing, where shoot competition never becomes intense. The above four-class strategies have properties that can be compared with each other by an expansion of Grime's triangular strategy diagram into a tetrahedron (Oksanen & Ranta, 1992). One corner of the scheme represents the K-strategy or competitiveness *sensu* Tilman (Tilman, 1988), i.e. a suite of morphological, physiological and reproductive traits which enables plants to complete their life cycles in an environment where critical resources are depressed to a low level by these plants. Another pole is the classical r-strategy, which can be regarded as identical to Grime's R or ruderal strategy. Grime's competitors or C-strategists, which are adapted to resource-rich environments where interference competition prevails, are regarded as intermediate between K- and r-strategists, as proposed by Grace (Grace, 1990). The third pole is formed by g-strategists, adapted to frequent but small losses of above-ground tissues. The s-strategy represents decreasing allocation to support tissues with the extreme type being a hepatic. This system also envisages further subdivisions with ericoid, graminoid, and *Dryas* strategies, being subsets of the more general concept of g-strategy (Oksanen & Virtanen, 1997).

Despite its conceptual complexities, this discussion of a four-class classification of plant life strategies as compared with the more traditional concept of two- and three-class strategies serves to demonstrate that as conditions become more marginal for plant survival this need not be accompanied by any loss in diversity of either form or function in higher plants. As is argued throughout this book, plants surviving in marginal areas are not necessarily impoverished in the diversity of their adaptations.

3.5 RESOURCE ALLOCATION

Numerous attempts have been made to rationalize the behaviour of plants with respect to their allocation of resources. The concept of plant functional types (see Table 3.2) provides a basis for grouping species or populations that share comparable responses to a particular stress. It is a debatable point as to whether plants that share a similar marginal habitat with regard to the availability of light exhibit the same properties in relation to their allocation of resources. Late successional shade-tolerant trees might be expected to share some common attributes as to how they manage to survive as seedlings on the shaded forest floor. A study of seedlings of five such species (*Fagus sylvatica, Acer pseudoplatanus, Quercus robur, Taxus baccata, Abies alba*) grown in a natural forest understorey in very low and low light microsites and exposed to varying CO_2 concentrations demonstrated that no single species combined all the characteristics traditionally considered as

adaptive to low light conditions and increasing carbon dioxide. At very low light intensities, only *F. sylvatica* and *T. baccata* responded to CO_2 stimulation of seedling biomass with increased leaf-area ratios and decreased leaf dark respiration. At slightly higher light levels, interspecific differences in the biomass response to elevated CO_2 were reversed and correlated best with leaf photosynthesis (Hättenschwiler, 2001). Although the seedlings in these species may all be shade tolerant, this tolerance is achieved in different ways and does not strictly represent a common functional type.

The short growing seasons and low thermal input into arctic habitats requires plants that live in these areas to control carefully the allocation of their carbon resources. In addition to the short period normally available for photosynthesis, the Arctic has also a very variable climate. In cold years the accumulated temperatures in day degrees can be so low as to result in no net photosynthetic gain being made by the plants. In Alaska a fall of just over one degree can be sufficient to result in a reduction of net photosynthesis by 27% (Chapin *et al.*, 1980). Therefore, in maintaining an adequate partitioning of their carbon resources between growth and carbohydrate reserves arctic populations ensure their energy supplies are adequate to last not just from one year to another but for more than one missed growing season. This conservation of carbohydrate may be achieved in several ways depending on the relationship between growth and productivity in the species concerned. In arctic species a high proportion of the carbon dioxide that they fix is allocated to the maintenance of underground organs.

The large root systems of many northern and arctic plants with their considerable carbohydrate reserves, while enabling these plants to survive long periods with little photosynthetic gain, require nevertheless a considerable portion of the available energy input just for their maintenance. The bog rosemary (*Andromeda polifolia*) growing in the Swedish tundra translocates 75% of its fixed carbon below ground (see Fig. 9.17). Similarly in an overall estimation for grass species in Alaska 59% of the carbon dioxide fixed was found to be translocated below ground where half was used just for maintenance respiration and the other half contributed to new biomass (Chapin *et al.*, 1980). By comparison, prairie grasses translocate 57% of the carbon fixed below ground but use 85% of this portion to contribute to new growth and only 15% for maintenance. The strategy of the northern plants in maintaining adequate reserves to combat the thermal uncertainties of their environment places them at a disadvantage in terms of growth potential with less well-insured species. Thus, it is not just in their photosynthetic capacity that arctic plants differ from those further south but also in the manner in which they allocate their reserves.

3.6 RESOURCE ACQUISITION IN MARGINAL AREAS

Strategy theories are devised to explain different modes of plant life in a manner which aids human understanding. Convenient and useful as these theories can be (see Chapter 2) they merely reflect human perceptions and inevitably discard information in the desire to produce conformity with a set ecological philosophy. One such dangerous simplification can be seen in the concept of environmental stress which is usually conceived as a permanent feature of particular habitats. Thus plants that inhabit areas that are considered chronically unproductive are classified as stress tolerant (Grime, 2001). There is, however, an increasing realization that resources are not constantly available to plants and that there are instead pulse periods when resources are high and most growth and resource accumulation occurs, as well as interpulse periods when resources are too low to be taken up and when resource-deficit mortality is most likely to take place (Goldberg & Novoplansky, 1997). Even when resources are available, conditions may be such that they are temporarily inaccessible. Prolonged flooding, drought and low temperatures frequently induce lengthy, quiescent periods with regard to resource acquisition during which competition for resources is minimal and survival depends on the possession of stress-tolerance adaptations. These quiescent periods can then be followed by periods of rapid growth when competition for resources comes into play as a factor in determining survival.

3.6.1 Competition for resources in marginal areas

Acquiring resources demands more from plants than just being in the right place at the right time. There is also the need to be able to acquire resources in the face of competition both from other species (interspecific competition) and from other individuals of the same species (intraspecific competition). When natural phenomena are discussed in terms of concepts such as competition there is always the possibility of arriving at conflicting conclusions depending on how the concepts are interpreted. Resource acquisition has stimulated considerable debate as to whether competition is more intense in areas that are rich in resources as compared with areas where resources are scarce. There has been therefore much research on whether the intensity of competition changes along environmental gradients with changing productivity. Many studies demonstrate that competition is more intense in productive habitats (Grime, 2001; La Peyre et al., 2001). Nevertheless, there are situations in marginal areas where despite very reduced productivity and low biomass accumulation competition may still be a factor in determining the species composition of plant communities. The biased sex ratio in favour of female willows that is frequently found in both arctic and subarctic areas appears to be linked to the ability of the female plants to outcompete the males for space in favourable habitats (Crawford & Balfour, 1990). It is possible therefore to question the frequent assumption that as resources become constrained all plants are similarly restricted in their supply and the factors that control plant distribution become physical rather than biological. In marginal areas such as deserts and bogs as well as arctic and montane regions, environmental fluctuations, some seasonal, others at longer intervals, are demographically decisive and determine how competition through the presence or absence of neighbours will affect recruitment, growth and mortality. As will be discussed in the following sections of this chapter, the nature of competition, particularly in marginal areas, will vary between the adverse and favourable periods with different populations and species within any one area profiting or losing depending on their relative (competitive) susceptibilities to environmental oscillations.

Stress-tolerant species are usually described in terms of the particular resource (the stressor), which is in limited supply and yet does not adversely affect their long-term survival. Plants that are characteristically found in shade or in arid conditions are described respectively as shade or drought tolerators. However, such classifications, even with the recognition of intermediates, do not take into account the fact that plants can have opposing strategies at different stages in their life

Fig. 3.24   The grass state of the southern pine (*Pinus palustris*) at the edge of a forest in South Carolina, USA. The plant will remain in this dwarf stage until exposed to the heat of a forest fire.

cycle and at different seasons of the year. An example is *Glyceria maxima* which has a high tolerance of anoxia early in the growing season but by early summer is markedly less tolerant due to a reduction in carbohydrate reserves (Braendle & Crawford, 1999). *Pinus palustris* seedlings (Fig. 3.24) remain in a genetically controlled dwarf condition – the 'grass state' – with a deep taproot, which permits seedlings to lose foliage and survive low intensity fires (an adaptation also found in various other pines, notably *P. engelmannii* in Arizona) as part of the understorey shade-tolerant plant community; they maintain this state until exposed to the heat of a forest fire when the seedlings elongate to become part of the forest canopy (Nelson *et al.*, 2003).

Resource acquisition and resource deprivation are not just positive and negative aspects of the same phenomenon. The ability to exist for an ecologically meaningful period without a certain resource is a vital characteristic, genetically controlled, and with substantial survival importance. Surviving without a specific resource for a certain period of time does not preclude the eventual demand and active use of this resource when it becomes available. One of the most powerful competitive strategies available to plants in marginal areas is to be able to dispense temporarily with a resource that is in continual demand by other species. This ability for the adapted species to be able to endure deprivation more easily than a potential competitor has been termed *deprivation indifference* (Crawford *et al.*, 1989).

3.6.2   Deprivation indifference

In natural situations most plants exist under suboptimal conditions due to the constant and ubiquitous effects of competition, predation and climatic oscillations. Survival in such situations depends on a combination of relative tolerance to extreme conditions and the extent of fluctuations in resource availability. The severity and duration of these periods of adversity are highly variable and plants differ greatly in their ability to withstand periods of deprivation. In addition, an ability to survive deprivation of a particular resource, whether it be light, oxygen or nutrients, alters with the seasons and the stage of development of the plant and is not necessarily a fixed characteristic. The ability to endure temporary resource limitations has been described as deprivation indifference (Crawford *et al.*, 1989).

The ability of plants to endure resource deprivation is seen in those species that can endure long periods of drought, darkness, and even in some cases total oxygen deprivation. Adaptation to these conditions has created the specialized floras that have settled in many marginal areas such as deserts, lakes, marshes and periglacial polar regions. Consequently, long periods of drought or flooding rarely extinguish the natural vegetation of marginal areas even though they may be disastrous for agriculture. Desert plants can lie dormant in the seed bank for decades yet spring to life with rapid growth after periods of rain. There are even species of flowering plants that can live where it never rains and survive by extracting enough water to complete their life cycles from fog and dew (Section 2.6, Fig. 2.23). In the Arctic, long winters and ultra-short growing seasons provide striking examples of just how little in the way of resources are necessary to sustain plant life. Some arctic species can even survive total encasement in ice without access to oxygen for many months and still emerge and flower (Section 3.6.3).

This approach to resource limitations differs from the more classical studies where plants are classified as shade, drought or flood tolerant as permanent characteristics. A disparity between species in their ability to withstand long periods of deprivation, particularly in adverse seasons when resource capture is usually minimal, can determine the eventual outcome of competition rather than resource capture during a non-stressful period. By definition, a resource is something that is consumed (Table 1.1). Therefore oxygen, as it is consumed by plants, should also be considered as a resource. Plants that can withstand long periods of flooding, particularly those that overwinter under mud and are therefore cut off from the usual aeration mechanisms, may have to survive without oxygen for months. The degree to which they can endure deprivation of this particular resource can determine the eventual outcome of competition. No plant can endure a permanent absence of oxygen, but the ability to do without this essential resource longer than a competitor may eventually determine the outcome of competition in specific habitats. Such is the range of deprivation indifference and ecological tolerance shown by many species of flowering plants that it can be argued that it is the ability to survive under suboptimal conditions that lies at the base of species diversity.

Examples of deprivation indifference to a number of resources, including water, light, carbon dioxide and oxygen, can be found in many species. As mentioned above, saplings of many deciduous forest trees can survive near the light compensation point for years. Once a gap occurs the long-suppressed saplings can grow rapidly in response to the greater availability of light. Typical examples are found in oak seedlings, and in shade-tolerant evergreen species which minimize biomass losses through long life spans and reduced respiration rates coupled with low leaf/mass ratios (Walters & Reich, 1999). Different tree species can share the same species of mycorrhizal fungi and thus be connected to one another by a common mycelium which permits transfer of carbon and mineral nutrients. Field studies on whether or not these movements of resources are bidirectional are limited. However, laboratory experiments using isotope tracers have shown that the magnitude of a one-way transfer can be influenced by shading of 'receiver' plants and fertilization of 'donor' plants, indicating that movement may be governed by source–sink relationships (the internal movement of resources within plants from where they are acquired to where they are consumed). Shading has been found to have an important influence and the transfer of carbon by shaded seedlings of Douglas fir (*Pseudotsuga menziesii*) from paper birch (*Betula papyrifera*) accounted on average for 6% of carbon isotope uptake through photosynthesis. This magnitude of net transfer is influenced by shading of *Pseudotsuga menziesii* and strongly suggests that such source–sink relationships regulate carbon transfer under field

conditions and contribute significantly to the shade tolerance of the understorey trees while in the suppressed condition (Simard *et al.*, 1997).

Carbon dioxide is a resource that can be recycled during periods of drought. Orchids and cacti rarely wilt and during extreme drought continue to photosynthesize by recycling the carbon dioxide released from respiration and tissue and protein breakdown. Plants with C4 and CAM improve their access to limited supplies of carbon dioxide by using the initial non-light-dependent CO_2-fixing enzyme phospho-enol-pyruvate carboxylase. A number of drought-tolerant species have large reserves of stored water and this together with morphological adaptations to reduce water loss enables the plant to recycle internally respired carbon dioxide.

Bromeliad phreatophytes are also highly efficient CAM plants, particularly those that are obligate epiphytes such as *Tillandsia usneoides* and *T. brachycaulos* (Fig. 3.25). This latter species is capable of making a net diurnal gain of carbon even after a continuous drought of 30 days (Graham & Andrade, 2004).

3.6.3 Deprivation indifference through anoxia tolerance

Flood-tolerant plants which can survive without oxygen longer than their competitors are able to maintain their

Fig. 3.25   The obligate epiphytic bromeliad *Tillandsia brachycaulos*, a native of Mexico and central America. In common with most other epiphytic bromeliads the leaf rosette acts as a well for trapping water. This species can continue to fix carbon dioxide even after 30 days of continuous drought (Graham & Andrade, 2004).

Fig. 3.26   Use of an anaerobic incubator to test the anoxia tolerance of pseudo-viviparous plantlets of alpine meadow grass (*Poa alpina* var. *vivipara*). (Left) Anaerobic incubator where plants are held in the dark under an atmosphere of 90% oxygen-free nitrogen and 10% hydrogen continually circulated over a platinum catalyst. (Right) The plantlets in the glass vessel have just been taken from the incubator and found to have suffered no injury even though they had been kept under total anoxia for three weeks at 20 °C in total darkness.

presence in an extreme habitat as they can endure a stress that has eliminated those species that require an uninterrupted supply of oxygen. Rhizomes of wetland species such as the bulrush (*Typha latifolia*) and the common club-rush (*Schoenoplectus lacustris*) can survive for months under total anoxia and even produce new shoots without having to wait for access to oxygen. Mature green overwintering leaves of *Acorus calamus* and *Iris pseudacorus* have been shown to survive up to 75 days and 60 days respectively of anoxia in the dark. During this period of anaerobic existence there is a down-regulation of metabolism and carbohydrate reserves are conserved (Figs. 3.26–3.27). On return to air *A. calamus* leaves rapidly recover full photosynthetic activity. This well-developed ability to endure prolonged periods of oxygen deprivation in both these species is associated with a down-regulation in metabolic activity in response to the imposition of anaerobiosis (Schlüter & Crawford, 2001) for other examples of anoxia tolerance (see Figs. 3.28–3.29 and Crawford, 1992).

Many arctic species can in a sense dispense with time and survive being deprived of one or even more entire growing seasons. Diminutive forms of species such as *Polygonum viviparum*, *Salix polaris*, *Ranunculus*

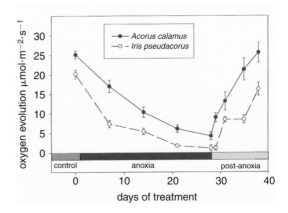

Fig. 3.27   Down regulation of photosynthetic activity under anoxia in two wetland species that show a high tolerance of oxygen deprivation. The leaves of the sweet flag (*Acorus calamus*) and the yellow flag iris (*Iris pseudacorus*) were able to recover most of their original photosynthetic activity after 28 days under anoxia as observed during a 10-day post-anoxic recovery period (Schlüter & Crawford, 2001).

pygmaeus, *Sibbaldia procumbens*, *Potentilla hyparctica* and *Saxifraga oppositifolia* can survive under snow banks for one or more growing seasons (Pielou, 1994) and some can even withstand many months of total

ANOXIA TOLERANT

Saxifraga caespitosa
S.   oppositifolia
S.   foliosa
Ranunculus sulphureus
Cardamine nymanii
Eriophorum scheuchzeri
Juncus biglumis
Carex misandra
Luzula arctica
Dryas octopetala
Puccinellia vahliana
Deschampsia alpina
Alopecurus borealis

ANOXIA-INTOLERANT

Saxifraga hieracifolia
S.   cernua
Ranunculus pygmaeus
Oxyria digyna
Pedicularis hirsuta
Cochlearia groenlandica
Polygonum viviparum
Draba oxycarpa

Anaerobe Jar

Fig. 3.28   Experimental testing of anoxia tolerance in some High Arctic perennial species. The experiments were carried out at Ny Ålesund (Spitsbergen, 79° N). Intact plants were placed in anaerobe jars (as shown above) and examined for their ability to survive 7 days of total anoxia in the dark at ambient arctic temperatures. In all the anoxia-tolerant species the entire plant including leaves and flowers survived undamaged except for *Salix polaris*, which lost its leaves but otherwise remained viable (Crawford *et al.*, 1994).

encasement in ice yet emerge and become metabolically highly active with high rates of photosynthesis when the ice and snow eventually melt. When examined experimentally many arctic species show a remarkable ability to withstand total anoxia (Fig. 3.28), with the whole plant including leaves and sometimes flowers surviving an anaerobic incubation in the dark at 20 °C for over a week (Crawford *et al.*, 1994).

The ecological consequences of deprivation indifference can be considered in relation to the processes that create renewal gaps in plant communities. The similarity of the resources required by plants has prompted the suggestion that differential use of resources is unlikely to be an important ecological discriminator between plants because all species have the same basic needs. Instead, it is the ability of species to regenerate in gaps, where individuals have died, that is viewed as the ultimate cause of coexistence (Grubb, 1977). The causes of the gaps will be associated frequently with adverse periods which take place irregularly; consequently the presence or absence of a species in any particular community will be governed by

competition that takes place only at irregular intervals, as for example when regeneration of the community takes place either after a natural catastrophe or when dominant individual plants die. It follows, therefore, that as gaps occur, as opposed to total devastation, not all species are equally affected. This ability to occupy an open site and thus pre-empt space can be described as pre-emptive competition or founder control (Werner, 1979). Species which can survive resource deprivation are therefore in a position of being able to maintain their presence in these areas despite the risks that arise from flood, fire, drought or storm and will be in a favourable position to pre-empt spaces from other species.

3.6.4   Avoiders and tolerators

A further useful ecological concept in relation to resource deprivation is that of avoiders and tolerators, a view that has been developed particularly in relation to stress physiology in plants (Levitt, 1980). The adaptive responses of being able to either avoid rather than

Fig. 3.29   Sea club-rush (*Bolboschoenus maritimus*) growing on the shore of the island of Gotland (Baltic Sea – reduced salinity). (Inset) Extension growth of dormant bud while kept in the dark under total anoxia for 12 days.

tolerate stress appears to be a more flexible attribute in terms of plant physiology than the concept of competitor or tolerator.

In evolutionary terms every adaptation carries a certain negative element in that it increases habitat specialization. Species that manage to avoid exposure to the stress do not have to resort to the more specialized adaptations that are needed to confer an ability to survive. Stresses that impinge at a cellular level, as in tissue desiccation, or oxygen deprivation, require specialized physiological adaptations which can involve a metabolic cost and consequently may reduce the competitive status of the species in the unstressed condition. Plants avoid stress most commonly either by reducing demand or by not being active during the adverse season. Thus drought avoiders either reduce the rate of water loss, or adjust their phenology to grow at times when water is available. The commonest

mechanism for plants avoiding low-oxygen stress is to have aeration mechanisms that supply oxygen to their roots when they are flooded.

3.7   ALTERNATIVE SUPPLIES OF RESOURCES

3.7.1   Light

There is no real alternative to light for flowering plants unless it is parasitism on other species. The dependence of parasitic plants on their hosts ranges from hemi-parasitism in chlorophyllous species such as the louseworts (*Pedicularis* spp.) and mistletoes (*Viscum* spp.). An arctic example is *Pedicularis dasyantha* shown in the frontispiece to this chapter (Fig. 3.1). Achlorophyllous species are obligate parasites entirely dependent on soluble sugars from their hosts for their

Fig. 3.30 Purple toothwort (*Lathraea clandestina*), a widespread European species growing parasitically on willow.

energy supply. Such species remain entirely under-ground except when flowering. A particularly colourful example of this life strategy is the purple toothwort (*Lathraea clandestina*) which is a widespread parasite on poplars and willow and is even cultivated in gardens, as is the specimen flowering in March in the Botanic Garden at St Andrews, Scotland (Fig. 3.30).

3.7.2 Precipitation

Fortunately for many terrestrial flowering plants the upper soil layer is not the only place where water can be found. Dew, deep underground reserves, and parasit-ism are resources that specially adapted plants can exploit in the search for water. Hemi-parasitism is an immediate alternate source of water for adapted species that is exploited by many species. The genus *Pedicularis* is notable for having numerous species mainly in the northern hemisphere, and especially in drought-prone alpine and arctic habitats. In areas where water stress is frequent these plants transpire freely, largely at the expense of water obtained from host plants (Fig. 3.1).

Apart from parasitism there are two other main sources which can supply plants with water in the absence of precipitation. The first and commonest is ground water, or more exactly the *phreatic zone* (Greek *phrear*, a well) – the region of the soil or rock zone below the level of the water table and where all voids are saturated. The second is dew. For dew to be condensed in any quantity and provide water that can be taken up by plants it has to be protected from re-evaporation. Thus dew will contribute to the water needs of plants only when it condenses inside some protected system such as underneath leaf scales as in some bromeliad species or else trapped in the wells formed by leaf rosettes on many epiphytic species (Fig. 3.25). The forest canopy can also serve for trapping moisture as is commonly found in the trees that inhabit mountain cloud zones. A cryptic form of dew is provided by the upward movement of water vapour at night from the warmer water table to the cold surface soil in porous soils. This is particularly important for the grasses that colonize sand dunes (see below).

3.7.3 Ground water

Plants that can access the phreatic zone are termed *phreatophytes*, and can reach water that is not available to shallower rooting species. The creosote bush (*Larrea divaricata*) has a root system that can extend many metres down to the phreatic zone. A study of the root systems of some Chihuahuan Desert plants found that the roots of 11 shrub or shrub-like species could be traced through various soil horizons to depths of 5 m (Gibbens & Lenz, 2001). Desert phreatophytes, and also some riparian species, combine the possession of long-lived water-seeking roots with more ephemeral surface root systems which supply the plant with nutrients as well as water when the soil is in a humid condition. Samples collected from the native Australian tree *Banksia prionotes* over 18 months indicated that shallow lateral roots and deeply penetrating tap (sinker) roots obtained water of different origins over the course of a winter-wet/summer-dry annual cycle. During the wet season lateral roots acquired water mostly by uptake of recent precipitation (rainwater) contained within the upper soil layers, while taproots derived water from the underlying water table. The shoots therefore obtained a mixture of these two water sources. As the dry season approached, dependence on recent rainwater decreased while that on ground water

increased. In high summer, shallow lateral roots remained well-hydrated and shoots well-supplied with ground water taken up by the taproot. This enabled plants to continue transpiration and carbon assimilation and thus complete their seasonal extension growth during the long (4–6 month) dry season. Parallel studies of other native species and two plantation-grown species of *Eucalyptus* all demonstrated behaviour similar to that of *Banksia prionotes* in that shallower resources were accessed by lateral roots in the upper soil layers (Dawson & Pate, 1996).

This potential to tap an alternative source of water has contributed to the invasive success of several southern European species of the genus *Tamarix* spp. (saltcedars) which were first introduced to North America in the 1800s. Many of the species escaped from cultivation and by 1987 were estimated to have invaded at least 600 000 ha. The saltcedars benefit from a deep water supply but are only facultative phreatophytes as they are often able to survive under conditions where deep ground water is inaccessible. The high evapotranspiration rates of salt cedars can lower the water table in heavily infested areas and even alter the ancient composition of the underground water reserves. Mature plants are tolerant of a variety of stress conditions, including heat, cold, drought, flooding and high salinity. Saltcedars are not obligate halophytes but survive in areas where ground-water concentrations of dissolved solids can average 8000 ppm or higher. In addition, the leaves of saltcedars excrete salts that are deposited on the soil surface under the plants, inhibiting germination and growth of competing species (Di Tomaso, 1998).

Variation in the sources of water used by tree species has significant implications for forestry. Riparian forests in their association with river banks are further examples of marginal plant communities. In a study of tree transpiration in a riparian forest in south-eastern Arizona containing *Populus fremontii*, *Salix gooddingii* and *Prosopis velutina* it was found that *Salix gooddingii* did not take up water in the upper soil layers during the summer rainy period, but instead used only ground water, even where the depth of the ground water exceeded 4 m. *Populus fremontii*, a dominant 'phreatophyte' in these semi-arid riparian ecosystems, used mainly ground water, and at an ephemeral stream site during the summer rainy season derived only 26–33% of its transpiration water from upper soil layers. By contrast, at another ephemeral stream site during the summer rainy period, *Prosopis velutina* derived a greater fraction of its transpiration water from upper soil layers than at a perennial stream site where ground-water depth was less than 2 m (Snyder & Williams, 2000). These results show the flexible manner in which phreatophytes and riparian trees can use alternative sources of water depending on availability.

The ability of phreatophytes to access deep water also has consequences for the water status of the upper layers of the soil due to a phenomenon described as *hydraulic lift*. During the night when transpiration is less, the upward movement of water in the roots of the phreatophytes nevertheless continues. As supply can exceed demand, the deeper-rooted species release some of the water they are transporting upwards into the top layers of the soil, where it becomes available to shallow-rooting species (Williams *et al.*, 1993). Shallow-rooted species can then profit from the presence of deep-rooted plants by the provision of indirect access to the ground-water reserves that would otherwise be beyond their reach. The deep-rooted plants in turn benefit from the growth of the shallower rooted plants as many of them can fix atmospheric nitrogen (bird's foot trefoil, clover, medick) and thus contribute to the nutritional status of the entire plant community.

In addition to the phenomenon termed 'hydraulic lift', where water is redistributed from a depth to dry topsoil, recent studies have detected a process of hydraulic redistribution which includes downward transfer of water when the surface layers of soils with low permeability become wet after rainfall. A comparison of sap flow in vertical and lateral roots of *Grevillea robusta* trees growing without access to ground water at a semi-arid site in Kenya showed that a reversal of sap flow occurred when root systems crossed gradients in soil water potential. Reverse flow in roots descending vertically from the base of the tree occurred, while uptake by lateral roots continued, when the top of the soil profile was wetter than the subsoil. This transfer of water downwards by root systems, from high to low soil water potential, was termed *downward siphoning* and represents the reverse of hydraulic lift. This downward siphoning was induced by the first rain at the end of the dry season and by irrigation of the soil surface during a dry period. It has been suggested that by transferring water beyond the reach of shallow-rooted neighbours, downward

siphoning may enhance the competitiveness of deep-rooted perennials (Smith *et al.*, 1999).

Dew

Dew is another major source of water that can have profound ecological consequences. In times past there was much discussion as to whether 'dew descended or rose', causing the remark that dew is 'nowhere until it is formed'. Meteorologically this may be correct, but from an ecological point of view it is nevertheless possible to indicate the source of the water that contributes to the dew that can be used as an alternative source of water by vegetation. Dew is formed by the distillation of water either upwards or downwards and provides another potential source of moisture which in certain circumstances can benefit vegetation. Cool surfaces will condense dew from the atmosphere as well as from water vapour in the soil profile. Many species do not have roots capable of reaching the phreatic zone or even that region about 40 cm above the water table from which water may be expected to rise through capillary action.

In sand dunes dew can distil upwards from the deeper layer as the temperature there is relatively constant and at night remains warmer than the upper dune layers which cool and therefore condense water vapour coming from the phreatic zone. Sand dunes have a low retention of water and grasses that inhabit dunes such as marram grass (*Ammophila arenaria*) are not phreatophytes and cannot reach the depths where these water reserves lie. Because of these daily oscillations of temperature in sand dunes the upper layers are able nevertheless to have their water content replenished on a daily basis, even during periods of drought, provided the phreatic zone is not too far below the surface of the dune. Unfortunately, in many places the local needs for farming and golf courses can lower the water table to such a depth that it renders the internal circulation of dew ineffective. In such situations dunes erode rapidly, as has been found in the Netherlands where coastal dune systems have been extensively exploited as a source of fresh water (Maarel, 1979).

Distillation of dew in semi-desert soils is a phenomenon that has been recognized since antiquity. In the Negev Desert there are stone mounds called *teleilat el einab*, literally the hillocks for grapevines, which are believed to have been constructed for the cultivation of vines or other small fruits by the Nabatean communities in the first and second centuries AD (Glueck, 1959). Even larger structures up to 10 m high and 24–30 m in diameter are to be found at Theodosia in the Crimea (Hitier, 1925). These may be expected to condense up to 300 litres of water as dew in one night (Monteith, 1963).

This trapping of dew with stones is generally described as *lithic mulching* and is an agricultural strategy that has been practised for more than one thousand years both in the Old and the New World. The earliest lithic-mulch plots are associated with ancient Nabatean sites in the Negev of southern Israel mentioned above. Roman viticulturalists and olive growers in Italy and nearby Mediterranean regions used stone mounds between 100 BC and AD 400 and perhaps a century earlier or later (White, 1970).

Fog

In the Atacama Desert site of Chilca (Peru) approximately 1500 stone-lined pits were built in the lomas or fog oases (see below) for the retention of subsurface moisture and possibly the collection of dew from fog moisture may have allowed both maize and potatoes to be grown in these pits (Lightfoot, 1994). The ecological importance of fog (which is just a large body of water-saturated air at the dew-point temperature where water droplets condense), is also an alternative source of water for plants and deserves serious discussion. The coastal redwood forests of California are one example, and the pine forests of the Canary Islands are another, of remarkable plant communities that are sustained by fog (Figs. 3.31–3.32). In a study of the heavily fog-inundated coastal redwood (*Sequoia sempervirens*) forests of northern California it was found on average that 34% of the annual hydrological input was from fog-drip from the redwood trees themselves. When trees were absent, the average annual input from fog was only 17%, demonstrating that the trees significantly influenced the magnitude of fog-water input to the ecosystem. Stable-isotope studies were used to characterize the water sources used by the plants. In summer, when fog was most frequent, 19% of the water within *S. sempervirens*, and 66% of the water within the understorey plants, came from fog after it had dripped from tree foliage into the soil. For the redwood trees this fog water input comprised 13–45% of their annual transpiration (Dawson, 1998).

In the Canary Islands, the Canary pine (*Pinus canariensis*) maintains a relatively high timberline at 1900–2000 m due to its ability to intercept moisture from the north-west trade winds on the enormous surface area of the pine needles in the forest canopy (Fig. 3.31). The needles of the Canary pine are borne in threes on densely crowded clusters up to 270 mm in length (Fig. 3.32). This great needle length increases the surface area of foliage available for the condensation of dew. Rain gauges placed within the forest can record annual precipitation levels of as much as 2030 mm while those on deforested parts of the mountain record only 510 mm. Quantitative studies of the interception of fog water in relation to tree density have revealed the initially surprising result that nine years after a light thinning in a 48-year-old *Pinus canariensis* plantation the throughfall of water in the forest increased to about twice that measured in the unthinned stand. The long period that elapsed between thinning and re-estimation of throughfall had resulted in an increase in the leaf area index (LAI, the total leaf surface exposed to incoming light expressed in relation to the ground surface beneath the plant) of lightly thinned plots, and this together with surface roughness of the forest canopy (tree height variability) aided interception of the fog and enhanced the amount of throughfall. Continuous severe thinning, however, had the reverse effect, and led to a gradual decline in throughfall values (Aboal *et al.*, 2000).

The Japanese cedar (*Cryptomeria japonica*) resembles the giant redwood (*Sequoiadendron giganteum*) both in its foliage and being a dominant tree in mountain cloud forests. Like the giant redwood the *Cryptomeria* forests enhance their rainfall through intercepting moisture from fog. Unfortunately, in the modern polluted world, fog now transports to mountain forests increased amounts of ammonium, nitrates and sulphates above the level that would be supplied by rain. In a study of the *Cryptomeria* forests at Mt Rokko, Kobe, Japan, it was found that in addition to increasing precipitation, fog also increased the deposition of the atmospheric pollutants SO_4^{2-}, NO_3^-, H^+, and NH_4^+ to six to twelve times the corresponding deposition via rain. These values are equal to, or exceed, the maximum deposition reported for Appalachian forests in the eastern United States (Kobayashi *et al.*, 2001). Deposition of mineral nutrients to forests is, however, probably neither a new nor entirely unnatural phenomenon. At a remote site in southern Chile it has been noted that organic nitrogen

Fig. 3.31   View looking down on a forest of Canary pine (*Pinus canariensis*) emerging through the cloud zone at 1800 m on Tenerife, Canary Islands.

was the dominant component of the mineral deposition from cloud water and was being deposited at a greater rate than in more human-impacted sites in north-eastern USA at up to 9 kg ha^{-1} and contributing substantially to the N-economy of Chilean coastal forests. It was suggested that it was the adjacent ocean, where biological productivity was high, that may have been the major source of the nitrogen in Chilean cloud water. This proposed marine–terrestrial flux via cloud deposition had not previously been identified as a significant source of nutrients but nevertheless could be taken as an example of the ocean feeding the forest (Weathers *et al.*, 2000).

One of the most spectacular examples of a fog-supported vegetation is to be found in the Atacama Desert in Chile and Peru. The desert here is one of the driest in the world. At Antofagasta measurable rainfall can sometimes be detected in less than 6 years out of 20, and even then the precipitation does not exceed 4–6 mm per annum. However, from May to November a dense cold mist, the Garua (Peru) or Camancha (Chile), rolls in with the sea breeze from the cold Humboldt Current. It is particularly dense at night but is sufficiently thick by day to obscure the tropical sun for days at a time. In the month of August some areas exposed to this mist have only 36 hours of sunshine and an average temperature of 13 °C, while less than 800 m above, on the neighbouring hills, the temperature rises sharply to 24 °C. This mist supports a lichen-dominated vegetation over extensive areas of the fog-shrouded hills facing the

Fig. 3.32 The Canary pine is particularly adapted for capturing water from dew due to its very long needles (up to 270 mm), which provide a large surface area giving the forest the capacity to quadruple the annual precipitation by condensing dew from the north-west trade winds.

sea. In certain favoured spots there even exists the unique forest vegetation, the *lomas* (see also Section 2.6). The precipitation under the trees can be eight times that which is condensed in the open. Typical tree species are *Carica candens*, various species of *Eugenia*, *Caesalpina tinctoria* and *Acacia macrocantha* (Hueck, 1966). The lomas communities have a high proportion of endemic species, both trees and ground flora, and were probably much more widespread in the past. Fog-dependent ecosystems are very fragile. The damage that has been done by cutting for fuel and grazing by goats has reduced this vegetation to a few isolated pockets, but some restoration success has been achieved by planting imported *Eucalyptus* and *Casuarina*, which has led to a partial recovery of the local flora (Section 2.6).

3.7.4 Carbon

Compared with animals plants lack variety in their basic resources, namely carbon dioxide and water. However, this does not mean that there is no element of choice available as to where these resources can be found. Given the relatively low levels of carbon dioxide now present in the atmosphere compared with some previous geological epochs, such as the early Carboniferous, and the Jurassic

(when it might have been as high as 3000 ppm), it is not surprising that many species in marginal habitats have developed the facility to access the alternative sources of carbon dioxide that reside in both the soil atmosphere and water. The development of *Sphagnum magellanicum* hummocks (as studied in wet conditions in the laboratory) has been found to depend on high carbon dioxide concentrations in the water in the upper layer of peat in which organic matter decomposes aerobically (the *acrotelm*). *Sphagnum* plants grown in a low dissolved carbon dioxide treatment appeared to be carbon-limited, with increased mineral nutrient concentrations and decreased ratios of carbon to other nutrients. These results demonstrate that, at least in wet conditions, atmospheric carbon dioxide alone is not sufficient to enable *S. magellanicum* to develop its normal hummock growth pattern. It was therefore concluded that substrate-derived carbon dioxide is an important carbon source for *Sphagnum* spp. (Smolders *et al.*, 2001).

In this same context it is relevant to note that when carbon dioxide enrichment was applied in the atmosphere as in the Free-air Carbon Dioxide Enrichment (FACE) experiments on ombrotrophic peat bog lawns in Finland, Sweden, the Netherlands and Switzerland there were no significant effects on sphagna or vascular plant biomass at any of the sites even after a three-year treatment with an increased atmospheric carbon dioxide concentration at 560 ppm (Hoosbeek *et al.*, 2001). This suggests that, just as with other nutrient-poor ecosystems, increased atmospheric carbon dioxide concentrations will have a limited effect on bog ecosystems. However, given the ability of *Sphagnum magellanicum* to utilize the carbon dioxide in the soil water, it is hardly surprising that carbon dioxide enrichment of the atmosphere has little effect on wetland mosses.

The season of the year is also important in assessing the role of soil respiration in contributing to the carbon resources of flowering plants. In the short growing season of the High Arctic flowering plants are metabolically active for a relatively short time and achieve the bulk of their growth and development within a few weeks of snowmelt. This episode of early growth coincides with the period that soil respiration releases carbon dioxide, much of which, like the arctic nitrogen pool, may have been derived from the photosynthetic activity of cyanobacteria growing on the soil surface. Measurements made in Spitsbergen show that the carbon dioxide flux progresses from a net carbon dioxide source of 0.3 g C m^{-2} d^{-1} during late snowmelt to a mid-summer net carbon dioxide sink of -0.39 g C m^{-2} d^{-1}, returning to a net CO_2 source of 0.1 g C m^{-2} d^{-1} in the early autumn (Lloyd, 2001). The low-growing, spreading and cushion plants of the Arctic are particularly well adapted to profit from this augmentation of the atmospheric carbon dioxide supply.

An estimation of the use made by these plants of soil carbon can be obtained from examination of the stable isotope ratio of the carbon sequestered in the flowering plants. The relative abundance of ^{13}C versus ^{12}C (expressed as $\partial^{13}C\%$) is frequently used as a measure of relative water efficiency over extended periods. It can also be used to distinguish the source of carbon dioxide used for photosynthesis. Carbon dioxide derived from soil respiration often has a $\partial^{13}C$ value of around -19%. In a sample of plants of *Saxifraga oppositifolia* growing in Spitsbergen a mean $\partial^{13}C$ value of $-19.51\% \pm 0.32$ ($n = 8$) suggests that carbon dioxide derived from soil respiration supplied a considerable portion of their photosynthetic needs (Crawford *et al.*, 1995).

Forest plant communities can also benefit from carbon dioxide emitted from soils particularly where there are substantial soil reserves of organic carbon as in the boreal forest. Currently the carbon balance of these northern forests attracts much interest with regard to whether or not they will provide a sink or source for atmospheric carbon dioxide as the climate becomes warmer. Although some authors consider that the northern forests represent potentially a large sink for carbon (Myneni *et al.*, 2001) there are other studies that demonstrate that these same regions are not only very heterogeneous in soil types but also vary from year to year and are therefore unlikely to moderate the rise in atmospheric carbon dioxide during the next century (Schlesinger & Andrews, 2000).

Comparison of the day and night vertical profiles of atmospheric carbon dioxide concentration in the lower portion of forest communities (Buchmann *et al.*, 1996) have long been used to demonstrate the extent to which soil respiration augments the supply of atmospheric carbon dioxide to the forest plant community as a whole (Fig. 3.33). The organic matter of the forest floor provides the low-growing early spring

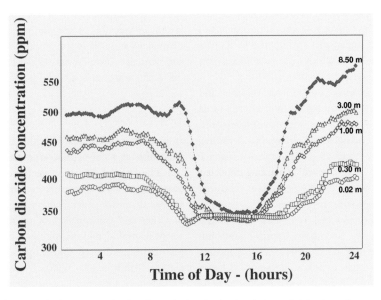

Fig. 3.33 Changes in atmospheric concentration in a forest during the day in relation to height above the ground. Observations taken from a mixed forest of box elder (*Acer negundo*) and *A. grandidentatum* in Red Butte Canyon 1400 m a.s.l., Utah, USA. Note the marked uptake of carbon dioxide before midday, especially in the region under 1 m above the soil surface, indicating the use made by the vegetation of carbon dioxide emitted from the soil. (Figure from Professor Nina Buchmann; see also Buchmann *et al.*, 1996.)

forest flora with luxury levels of carbon dioxide as compared with plants of taller status in open habitats. Rosette-shaped plants, which carpet the ground with actively photosynthesizing leaves in early spring, are therefore able to compensate for the shortness of their growing season through their ability to utilize high concentrations of carbon dioxide emanating from soil respiration.

Aquatic and amphibious plants in many instances are also able to access high levels of carbon dioxide from submerged soils due to the efficacy of their ventilation mechanisms. The aerenchyma in roots and rhizomes and the hollow stems of such species as *Typha latifolia* facilitate the supply of oxygen from the shoot to the root. Gaseous diffusion of carbon dioxide also operates in the reverse direction. The partial pressure of carbon dioxide (pCO_2) in these aerenchyma gas spaces has been estimated to be more than 10 times atmospheric pCO_2 and it appears that the source of this carbon dioxide is from microbial (soil) respiration. Over 50% of total leaf volume in *T. latifolia* is occupied by gas spaces and most of the total gas-space volume in plants is in the shoot. Photosynthetic rate in C3 plants, such as cattail, can increase with increasing carbon dioxide concentrations under natural conditions. For this reason, cattail and other emergent wetland plants possessing continuous gas-space pathways appear to have a significant carbon supplement as compared with other C3 plants growing in well-aerated soils (Constable *et al.*, 1992; Constable & Longstreth, 1994).

The warming of leaves and petioles in water lilies and reeds (*Phragmites australis*) by solar energy leads to a net downflow of air from younger to older tissues. This process, sometimes called thermo–osmosis, depends on younger tissues having a higher resistance to diffusion than older leaves. Consequently, air travels downwards from younger leaves through submerged rhizomes and ventilates upwards through older organs. The process ventilates submerged organs and alleviates a potentially hypoxic condition in rhizomes (Armstrong *et al.*, 1994) and in its upward movement through mature stems and leaves delivers carbon-dioxide-enriched air to aerial photosynthetic tissues.

In aquatic species a further enhancement of carbon dioxide utilization is found in those species which possess CAM and C4 metabolism. There is evidence for the occurrence of the CO_2-concentrating

mechanism crassulacean acid metabolism (CAM) in five genera of aquatic vascular plants, including *Isoetes*, *Sagittaria*, *Vallisneria*, *Crassula* and *Littorella* (Keeley, 1998). CAM is most frequently thought of as an adaptation that increases water use efficiency in drought-resistant species. However, in some aquatic habitats dissolved carbon dioxide levels can be low and CAM contributes to the carbon budget by increasing both net carbon gain and carbon recycling. Aquatic CAM plants tend to be found in shallow pools that experience low carbon dioxide levels by day and higher levels at night. CAM plants are therefore able to take advantage of the higher night-time carbon dioxide levels (Keeley, 1998).

Littorella uniflora (shoreweed) is an amphibious species which forms a shallow water turf down to a level of 4 m but flowers only when the plants are exposed above the waterline (Fig. 8.7). A study of photosynthesis in this species (Robe & Griffiths, 2000) has shown a remarkable facility for optimizing the ability to sequester carbon. When submerged the plant shows crassulacean acid metabolism and high leaf lacunal CO_2 concentrations which are maintained as it emerges above the water level, suggesting a continued carbon dioxide uptake from sediments. Shore weed is an excellent example of a truly amphibious species in that it is phenotypically plastic both in morphology and physiology and is poised at any time to emerge from the submergence onto dry land without suffering any noticeable water deficit.

3.7.5 Nitrogen

It has often been assumed that ammonium is the predominant form of inorganic nitrogen nutrition in tundra soils and that it is preferentially absorbed in comparison with nitrate. Recent investigations have shown that this generalization neglects both habitat variation and the capacity of some species to access various soluble and insoluble sources of inorganic nitrogen both directly and through mycorrhizal association (Atkin *et al.*, 1996). Although NH_4^+ is the predominant form of inorganic nitrogen in most arctic soils there is evidence that nitrification takes place in certain situations and producing substantial quantities of NO_3^- even exceeding NH_4^+. In both mesic and drier soils higher temperatures can increase the presence of nitrate (Nadelhoffer *et al.*, 1992). Faeces

Fig. 3.34   Nitrogen and phosphate enriched vegetation on bird cliffs near Grumantbyen, 78° 10′ N, Spitsbergen.

deposits also provide sources for nitrate production. Consequently, as soils become warmer with earlier snowmelt at high latitudes it is probable that nitrate availability will increase. Nitrate is more mobile in soils than ammonium and the relatively luxurious vegetation that is found in the lowest regions of vegetation below bird cliffs is a consequence of the flushing of nitrate downslope from ammonium rich soils (Fig. 3.34). A possible limiting factor remains, and that is the low rates of nitrate reductase activity commonly found in arctic vegetation even when fertilized with nitrate (Atkin, 1996). However, the degree of induction of nitrate reductase activity is not uniform in arctic plants and in certain ancient bird cliffs in Spitsbergen which have been identified as having a particularly long history of being free of ice (Fig. 6.13) a surprisingly high level of induction of nitrate reductase activity has been recorded (Odasz, 1994).

The ability of arctic plants to absorb soluble organic nitrogen has now been recognized. In some arctic soils, particularly under moist or wet conditions, the concentration of free amino acids can exceed that of inorganic nitrogen. In particular glycine, aspartic and glutamic acids can be found in organic-rich soils

(Atkin, 1996). Wet and mesic soils have been particularly noted for their high concentrations of soluble nitrogen compounds. However, even in dry heath soils, water-extractable amino acids can be found. Isotope studies using C^{14}-labelled free amino acids showed that *Eriophorum vaginatum* preferentially absorbed free amino acids such as glycine (Chapin *et al.*, 1993). It has been calculated that the uptake rates of free amino acids may account for between 10% and 82% of the total N uptake in the field, depending on species and community. Deciduous shrubs had higher uptake rates than slower growing evergreen shrubs, and ectomycorrhizal species had higher amino acid uptake than did non-mycorrhizal species.

A study of the sources of nitrogen for plant growth in a coastal marsh grazed by snow geese in Manitoba, Canada, showed that amounts of nitrogen, primarily ammonium ions, increased in the latter half of the growing season and over winter, but fell to low values early in the growing season. Free amino acid concentrations relative to ammonium concentrations were highest during the period of rapid plant growth in early summer, especially in soils in the intertidal zone, where the median ratio of amino acid nitrogen to ammonium

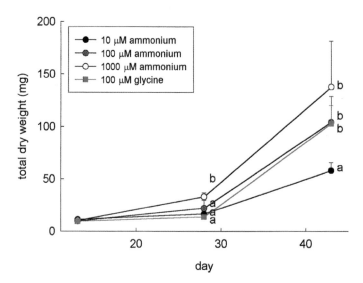

Fig. 3.35 Comparison of increase in total dry weight at fixed nitrogen concentrations using ammonium or glycine for plants of *Puccinellia phryganodes* grown in hydroponic culture in the field at La Pérouse Bay, Manitoba. (Reproduced with permission from Henry & Jefferies, 2002.)

nitrogen was 0.36 and amino acid concentrations exceeded those of ammonium ions in 24% of samples. Amino acid profiles, which were dominated by alanine, proline and glutamic acid, were similar to goose faecal profiles. In a continuous flow hydroponic experiment conducted in the field, growth of the salt-marsh grass *Puccinellia phryganodes* on glycine was similar to growth on ammonium ions at an equivalent concentration of nitrogen (Fig. 3.35). Thus, when supplies of soil inorganic nitrogen are low, amino acids represent a potentially important source of nitrogen for the regrowth of plants grazed by geese, and amino acid uptake may be as high as 57% that of ammonium ions (Henry & Jefferies, 2002).

These studies demonstrate that in a nutritionally marginal region such as the tundra, plants short-circuit the mineralization decomposition step by directly absorbing amino acids. This implies that in the organic soils of these tundra systems the following conditions occur:

(1) inorganic nitrogen is an inadequate measure of plant-available soil nitrogen
(2) mineralization rates underestimate nitrogen supply rates to plants
(3) the large differences among species in capacities to absorb different forms of N provide ample basis

for niche differentiation of what was previously considered a single resource
(4) by short-circuiting N mineralization, plants accelerate N turnover and effectively exert greater control over N cycling than has been previously recognized.

3.7.6 Phosphate

The essential role of phosphorus in plant metabolism might suggest that the demand for this element in the nutrition of plants would be relatively great. However, an examination of plant ash shows that the amount of phosphorus in the typical plant is lower than calcium, potassium or magnesium. Phosphates in the mineral component of soils are relatively insoluble and less readily leached than nitrate or sulphate especially if there is a good supply of calcium resulting in a high pH and the formation of insoluble calcium phosphate. In acid soils phosphoric acid combines with iron, aluminium or magnesium to form colloidal hydrates which are precipitated. Leaching is therefore minimized and phosphorus can accumulate in areas where there is or has been human habitation. This is particularly noticeable in phosphate-deficient soils formed from slow-weathering rocks as in the Northern Isles

Fig. 3.36   Nutrient retention in an oceanic landscape. Remains of prehistoric settlement on the island of Papa Stour (Shetland). Note the retention of nutrients aided by preferential grazing which has retained greater fertility than in the surrounding nutrient-deficient and phosphate-impoverished landscape.

and Outer Hebrides (Scotland; see photo of abandoned prehistoric settlement on the Shetland island of Papa Stour, Fig. 3.36). The organic component of soils can contain appreciable amounts of organic phosphates usually as inositol phosphates and in particular phytic acid (inositol hexaphosphate). Despite the retention of phosphorus in insoluble mineral and organic combination phosphate leaching is not entirely prevented. In natural plant communities phosphate is generally more limiting than nitrogen and many species which in the past were considered as classic nitrophiles have been shown experimentally to respond more to phosphorus than nitrogen. Thus nettles (*Urtica dioica*), dog's mercury (*Mercurialis perennis*) and tufted hair grass (*Deschampsia caespitosa*) show only a limited response to nitrogen unless they are given additional phosphate (Pigott & Taylor, 1964). In areas with hard acid rocks and high rainfall such as on the Lewisian gneiss of the Outer Hebrides (Scotland) phosphate deficiency is a problem for agriculture.

Although extreme examples of phosphate deficiency are not common, the ability of plants to access insoluble phosphate varies and has important ecological consequences. One of the most important properties of mycorrhizal systems is the marked influence they have on phosphorus uptake in areas with low phosphate availability. The most noticeable effect of vesicular-arbuscular mycorrhizae (VAM) is the growth improvement provided by the supply of phosphorus. The fungal hyphae not only render phosphate soluble but forage for supplies well beyond the zone of depletion that can normally be reached by the roots and root hairs without fungal associates. VAM are the most abundant of endo- and ectomycorrhizae and are generally the most effective in stimulating growth in the host plant. Plant species without ericoid mycorrhiza consistently show low inherent relative growth rate (RGR), low foliar N and P concentrations, and poor litter decomposition, while plant species with ectomycorrhiza had an intermediate RGR, higher foliar N and P, and intermediate to poor litter decomposability, plant species with arbuscular mycorrhizae showed comparatively high RGR, high foliar N, and P, and fast litter decomposition (Cornelissen *et al.*, 2001).

3.7.7   Phosphate availability at high latitudes

The arctic and boreal lands with their mineral-poor, gravel soils and equally nutrient-deficient wetland peats provide a biological challenge for plant nutrition which has attracted much investigation particularly with reference to phosphorus. Phosphate uptake has also been shown to vary greatly between high latitude species. In a study which compared 15 tundra species near Toolik Lake, Alaska, the potential for phosphate absorption alone was found to vary 20-fold between species (Kielland & Chapin, 1992). Within this tundra ecosystem deciduous shrubs had the highest potential for phosphate absorption and contrasted with the evergreen shrubs which had the lowest, the difference being presumably due to their large annual nutrient requirement for producing new leaves every year. In many species mycorrhizal associations appear to enhance phosphate uptake as they do elsewhere. In arctic Canada a survey of root colonization and spore populations of arbuscular mycorrhizal fungi revealed that of 197 plant-root systems and soil rhizospheres examined, 28% were associated with arbuscular mycorrhizae (Dalpe & Aiken, 1998).

3.8 MYCORRHIZAL ASSOCIATIONS IN NUTRIENT-POOR HABITATS

The role of mycorrhizal associations in plant nutrition increases with vegetation succession. As ecosystems develop and mineral soils give way to soils with increasing organic matter so does the importance of mycorrhizal associations increase for plant nutrition. There are three main types of mycorrhizae: (1) endomycorrhizae, (2) ectomycorrhizae and (3) vesicular-arbuscular mycorrhizae (VAM). In endomycorrhizae, the fungus penetrates and lives within the cells of the root cortex and external growth is limited to finely branched haustoria (arbuscules). These are the most widely distributed of mycorrhizae and are found on herbaceous species, especially orchids, as well as woody ericaceous plants such as heathers and rhododendrons. This type of mycorrhizal association is particularly beneficial to flowering plants that grow on nutrient-deficient soils. In the ectomycorrhizae a mycelium mantle is developed on the outside of the root. The hyphae penetrate the cortex of the root an form an intercellular meshwork termed a Hartig net, the outer mantle of which replaces the piliferous (outer) layer of the root. This type of association is very common in temperate trees and appears to be essential for the proper growth of the trees. Vesicular-arbuscular mycorrhizae (VAM) are a form of ectomycorrhizae where the fungus lives between the cells of the cortex,

forms finely intertwined, dendroid hyphae and penetrates the cells with temporary hyphal projections which may be swollen vesicles or finely branched hyphae called arbuscules. These are probably the most ancient forms of fungal association with higher plants and are widespread throughout the world.

3.8.1 Mycorrhizal associations in the Arctic

It has been a long-standing assumption that fungal associations are less common in cold climates. Arctic soils lie under snow and ice for the greater part of the year and even in the growing season have a tendency to remain cold and often wet. A combination of anaerobic conditions and low temperatures have therefore been considered as presenting obstacles to the development of mycorrhizal associations particularly in plant communities in the High Arctic (Kytoviita, 2005). However, it is always dangerous to make generalizations about an area that is as heterogeneous as the Arctic especially now that many regions are showing clear signs of climatic warming.

In the subarctic soils of northern Sweden it has been shown that on the drier fellfield prolonged shading, warming, and fertilization treatments of hybrid willow stands (*Salix herbacea* × *Salix polaris*) more than doubled the above-ground biomass and shoot growth, but decreased the number of root tips per unit root mass with no long-term changes in total ectomycorrhizal colonization. The net result was therefore an increase in the density of the potential host plant tissues, which increased the intensity of fungal colonization in this ecosystem. An analysis of the fungal species also suggested that there was a shift from drought-stress-tolerant fungi towards a dominance of minerogenic fungi, which may take place if nutrient availability increases substantially because of anthropogenic disturbances (Clemmensen & Michelsen, 2006).

In the High Arctic on Axel Heiberg Island (approximately $80°$ N) VAM have been found in three genera of the Asteraceae (*Arnica*, *Erigeron* and *Taraxacum*). The Axel Heiberg Island is an example of a thermal oasis and is therefore ideal for functional fungal root endophyte development (Allen *et al.*, 2006).

These examples are taken from favoured well-drained and warm sites. Climatic warming at high latitudes may make these sites more favourable for the development of mycorrhizal associations but in other regions where the climate is more oceanic, and where prolonged ice encasement and bog growth may be enhanced by increased nitrogen input from anthropogenic sources and higher temperatures, the conditions may remain adverse for the development of active mycorrhizal associations.

3.8.2 Cluster roots

Root clusters are an alternative to mycorrhizae for acquiring nutrients in impoverished soils. These have been defined as densely packed groupings of determinate rootlets which develop, grow, then cease growth and undergo a synchronous burst which liberates an exudation of organic acids. These exudations last for about two days but in this time achieve a localized reduction in soil pH creating an increase in the pool of phosphate and iron available for plant uptake (Skene, 1998, 2003). Cluster roots were originally observed in the Proteaceae where they occur in almost all species and were thought to be a particular characteristic of this family; they were therefore termed proteoid roots. They are now known to be much more widespread both taxonomically and geographically and are particularly prevalent in the non-mycorrhizal plants that inhabit nutrient-deficient soils as in the South African Fynbos and Australia. Recently, they have been shown to occur on temperate species (Fig. 3.37) such as lupin (*Lupinus albus*) (Hagström *et al.*, 2001), bog myrtle (*Myrica gale*), alder (*Alnus incana*, *A. rubra*, *A. glutinosa*), and sea buckthorn (*Hippophae rhamnoides*). The spatial clustering of the roots and the synchrony in their development and the large exudative burst are remarkable and have been suggested as analogous to a cavalry charge, where a concentrated and sudden attack prevents bacteria from metabolizing the exudate completely before it can enhance root nutrient uptake (Skene, 1998).

In some Australian sites there appears to be a definitive association in habitats rich in Proteaceae between zones of root proliferation and ferricrete layers in lateritic soils. This has led to the intriguing hypothesis that certain Australian lateritic and related oligotrophic soils may have been partly derived biologically from soluble iron-rich complexes generated following secretion of low-molecular weight organic acids by phosphate-absorbing specialized cluster roots of proteaceous plants (Pate *et al.*, 2001). Subsequent precipitation of the iron is postulated as leading to development of laterites after consumption of the organic components of the complexes by soil bacteria.

3.9 NUTRIENT RETENTION IN MARGINAL AREAS

Bryophytes and in particular the various species of *Sphagnum* that are common in wetland and oceanic regions have long been recognized as chelators of mineral ions. The rapid drop in soil pH after the colonization of wetlands by *Sphagnum* spp. is well known, Similarly, in many nutrient-deficient landscapes abandoned human settlements frequently stand out for the verdure of their immediate surroundings.

The fifteenth-century abandoned Norse settlements in Greenland (*c.* 1450), Bronze Age and Neolithic homesteads in Scotland (Fig. 3.36), and winter hunting camps in the Arctic can still be recognized from a distance from their vegetation. Particularly striking in the High Arctic are the winter encampments of the Pomors, Russian hunters who crossed the sea (Russian *po more*, on the sea) from the Kola Peninsula to hunt for furs in Spitsbergen in the seventeenth and eighteenth centuries or possibly earlier. Most of the encampments fell into disuse two or three centuries ago but are still marked by the lush bryophyte communities (Fig. 3.38).

Despite these many well-documented examples of nutrient retention in unproductive habitats it was an astonishment to the scientific community when the Russian nuclear reactor at the Ukrainian site at Chernobyl exploded in 1986 that the resulting pollution with radiocaesium (^{137}Cs) that spread westward was retained for so long in the upland oceanic grazing lands of the British Isles. As was pointed out at a subsequent government enquiry (Grime, 2001), the predictions made at the time to allay the fears of British farmers and the general public were made on productive and intensively managed lowland pastures and no account was taken of the upland situation where the 'slow dynamics' of unproductive land would inevitably lead to sequestration and slow release of radiocaesium from both living and dead components. The impoverished upland soils of oceanic lands experience high levels of precipitation combined with mild wet winters, which had led to a false assumption that constant leaching of mineral ions would be inevitable in these conditions. The erroneous belief that this leaching would rapidly

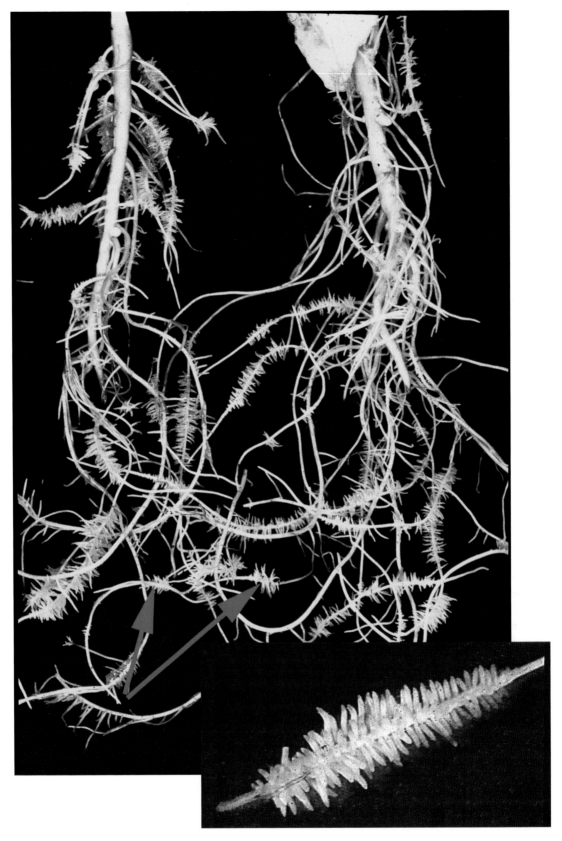

Fig. 3.37   Cluster roots developing on *Lupinus albus*. The plants were grown in a regime of low internal phosphate and iron. Inset shows a close-up of a cluster root nearing its full development. (Photos by courtesy of Dr K. R. Skene.)

Fig. 3.38   Site of long-abandoned high arctic hunting station at Krossfjorden, Spitsbergen. This site was used by Russian trappers (Pomors – see text) probably in the early eighteenth century. The encampment with its residual wooden posts is clearly distinguished by the lush bryophyte growth.

cleanse the soil of radioactive pollution was made in ignorance of the capacity of oligotrophic vegetation for nutrient retention in these marginal sites.

Nutrient retention in upland vegetation depends on both physical and physiological properties. Over large upland areas of the western regions of the British Isles from Shetland to Wales the soils are predominantly thin or else covered with layers of peat of variable depth. In shallow mineral soils the entire soil profile is usually occupied from top to bottom by a dense mat of roots. This total occupation of the soil profile maximizes nutrient retention and minimizes leaching. When the soils are organic the upper layer of peat which becomes aerated during the growing season (the acrotelm) is also densely occupied by living roots which maximize nutrient retention. The lower region (the catotelm) is composed of dense anaerobic peat which both impedes drainage and decomposes extremely slowly and will therefore retain any nutrients that escape from the acrotelm.

Localized areas that have higher nutrient status than the surrounding area as a result of past settlements, or from herding together of domestic animals at certain times of the year, still attract intense grazing. Although the grazing intensity over the area as a whole may be light, the greater proportion of verdant palatable herbage, particularly in winter and early spring, in these more fertile areas attracts many animals. The manuring that results from this behaviour facilitates nutrient cycling in these patches and perpetuates this vegetation pattern in a manner that reflects the past usage of the land even though it may have been abandoned for centuries or even millennia (Fig. 3.38).

Physiologically the vegetation in many arctic and subarctic deciduous species demonstrates a strong facility for retaining nutrients from one season to another. In a study of mineral nutrient economy in shrub tundra in northern Sweden, dwarf birch (*Betula nana*) showed the greatest capacity for the reabsorption of N, P and K from the leaves into the body of the plant before leaf fall (Jonasson, 1992). Similarly, in cotton grass (*Eriophorum vaginatum*) 90% of the phosphorus can be withdrawn from single leaves relative to their total nutrient content so that only about 10% of the annual demand to maintain the biomass needs to be absorbed annually from the soil. Similar patterns for nitrogen reabsorption have been found in *Carex bigelowii* (Jónsdóttir & Callaghan, 1990) and for phosphorus in *Lycopodium annotinum* (Headley *et al.*, 1985). Other more general features of the vegetation of infertile areas also aid the retention of nutrients in upland areas and include the presence of living foliage (including lichens and bryophytes) and active roots throughout the year, as well as longevity, slow turnover rate and a close association with mycorrhizal fungi (Grime, 2001).

Experimental addition of nutrients in arctic sites makes it possible to provide some quantification of the capacity for nutrient retention. In a prolonged study documented for 3–10 years and carried out at 13 sites in Alaska on *Eriophorum vaginatum*, *E. angustifolium* and *Carex aquatilis* the effects of fertilizer on flowering were in some cases still significant after six years (Shaver & Chapin, 1995). These three species belong to the ecological group often referred to as graminoids, which in comparison with some other species show the greatest capacity to respond to additional nutrients. They are also less sensitive to temperature in their absorption of phosphate and ammonium. All three species examined, namely *Carex aquatilis*, *Dupontia fischeri* and *Eriophorum angustifolium*, can absorb phosphate at 1 °C at 20–60% of the rate at 20 °C (Kielland & Chapin, 1992). As already mentioned, graminoid species are also able to absorb free amino acids which occur in the high soluble nitrogen pool of many tundra soils. Thus, despite the slow rates of nitrogen mineralization in tundra soils and low concentrations of inorganic N these soils have considerable concentrations of both structural and soluble organic nitrogen.

The addition of nutrients can fundamentally alter the effects of competition between species. In a nine-year study of environmental perturbations at Toolik Lake (Alaska) nutrient addition increased biomass and production of deciduous shrubs but reduced growth of evergreen shrubs and non-vascular plants (Chapin *et al.*, 1995). Certain species (e.g. *Dryas octopetala*) when growth is stimulated by the artificial addition of nutrients can become frost sensitive and as a result decline in presence. Similarly, in the Arctic when palatable vegetation is stimulated to grow with extra nutrients it can be selectively grazed out of existence by unrelenting visits from reindeer and lemmings. If flowering is promoted this can also attract floral herbivory from reindeer with adverse effects on regeneration (Cooper & Wookey, 2003).

In a sense many manipulation or perturbation experiments run the risk of not being realistic as they can produce short-term changes in the vegetation that would be unlikely to take place under natural conditions. Short-term manipulation experiments such as adding water or nutrients and placing transparent shelters around experimental plots produce an artificial forcing of phenotypic plasticity. The long-term natural response to environmental change would be through genotypic alteration which could select for a totally different set of environmental adaptations with greater Darwinian fitness.

3.10 CHANGES IN RESOURCE AVAILABILITY IN THE ARCTIC AS A RESULT OF CLIMATIC WARMING

No discussion of resource utilization by plants in marginal areas can be considered complete without some speculation as to the possible consequences of climatic warming. Particular attention has been given recently to this aspect of environmental change in relation to polar regions, which may be subject to substantial climatic warming. One of the most likely consequences of any substantial change will be an alteration in mineral nutrient availability (Chapin & Shaver, 1996; Wookey & Robinson, 1997). In general, experiments on the effects of climatic warming are difficult both to carry out and interpret. Generalized conclusions are elusive as the responses of plants vary between species and from one part of the Arctic to another. Populations of arctic heather (*Cassiope tetragona*) from the subarctic regions (Fig. 3.39) are known to respond positively to additions of nutrients while those in the High Arctic are little affected (Havström *et al.*, 1993). Experiments on nutrition in cold climates are also complicated by the need for prolonged observations. Factorial environmental manipulation of growing season temperature, soil nutrient and water status conducted over three years at a polar semi-desert community in Spitsbergen have shown that *Dryas octopetala* can respond rapidly to an amelioration in the availability of inorganic nutrients (N, P and K) by an expansion in leaf area and biomass. Sexual reproduction and seed viability were also markedly improved by elevated temperatures or soil nutrient availability (Wookey *et al.*, 1995).

Although short-term observations often record immediate nutrient-induced growth stimulation, as

Fig. 3.39   Arctic heather (*Cassiope tetragona*) – an evergreen species of acid soils capable of surviving under prolonged snow cover.

in *Dryas octopetala*, in the long-term this may serve only to increase frost sensitivity. It follows therefore that the eventual outcome of additional nutrients is unlikely to depend solely on nutrient supply. The ultimate consequences of greater nutrient availability will depend on competitive interactions between plants, which may be modified substantially by other factors such as herbivory, disturbance, and overwintering survival.

In the Arctic the problem of nutrient supply is compounded by questions of low input rates and accessibility. Nutrients may be present in the ecosystem in organic deposits, but in short, cold, growing seasons they are not as readily available as they would be in soils with an active microbial flora. In addition to low input rates, the Arctic has the lowest nutrient cycling rates of any ecosystem with typically only 1% of the ecosystem's nutrients occurring in the living biomass. In this 'locking up' of nutrients in the non-living material, the Arctic reaches yet another extreme end-point in comparison with other world ecosystems. Where there is an adequate supply of moisture as on wet-mossy sites there is usually an active growth of blue-green algae which accounts for the bulk of nitrogen fixation in most arctic habitats. Although some nitrogen comes into the system through snowfall

and rain this is more than lost by the run-off that takes place with snowmelt in the spring. In the low temperature conditions of the High Arctic, nitrogen fixation, which is largely carried out by cyanobacteria, is severely limited by temperature and the input even in favourable coastal sites in Alaska is only 5% of the nitrogen that cycles through the vegetation annually (Chapin *et al.*, 1980).

In the High Arctic plants are usually of dwarf stature with low growth rates and diminutive size. The frequent assumption that nutrients are limiting in arctic sites is suggested by the common observation that plant growth is often stimulated when additional nutrients are available, as around a rotting carcass or where they have been recently added to experimental plots (Wookey *et al.*, 1995). However, this same stimulation can be found in temperate and lowland plant communities and it is always a false assumption to equate greater productivity with an increase in fitness, without corroborating demographic data. Although the concentrations of readily available nutrients in arctic soils are low there is no consistent evidence from analyses of leaves, either in the Arctic or in alpine regions, that there is any general deficiency in the major nutrients. Addition of nutrients can even have an adverse effect. With some arctic species additional nutrients can reduce fitness in the Darwinian sense (i.e. reducing the ability of an individual to contribute genetically to a subsequent generation) by causing detrimental phenological changes such as promoting early spring growth and delaying the onset of dormancy in autumn, consequently risking frost damage. Frequent augmentation in carbohydrate levels from extra nutrients may also be a contributing cause to increased fungal infections (Körner & Larcher, 1988).

Whether or not low growth rates are caused by a lack of resources in terms of carbon and mineral nutrients or are due to other environmental constraints is therefore open to question. Arctic soils are usually low in available nutrients but whether demand exceeds the ability to liberate nitrogen and phosphorus from organic sources is still a debatable point. As long as the axis of the Earth maintains its tilt in relation to the sun, the long, dark winters of the Arctic will persist and nutrient availability may be only a minor aspect of plant survival.

Fig. 4.1 A female plant of the arctic willow (*Salix arctica*) growing at Mesters Vig, East Greenland (73° N). Generally in the Arctic female willows outnumber the males. In *Salix arctica* (the only arctic willow with measurable growth rings) the females also outperform the males in radial stem growth (see Section 4.11).

4 · Reproduction at the periphery

4.1 ENVIRONMENTAL LIMITS TO REPRODUCTION

Limits to plant distribution can arise either from a failure to grow or an inability to reproduce. In many cases a failure to reproduce may be a more common response to environmental limitations than a failure to grow, probably because reproductive success requires more than just the development of viable seed. Reproduction is accomplished only when there is successful establishment of a new generation of reproductive individuals. The continued existence of viable populations in marginal areas is therefore dependent on accomplishment of flowering, fertilization, viable seed production, germination, and the establishment of new individuals in regions where the environment is uncertain and variable. These basic requirements for completion of the reproductive cycle illustrate the appropriateness of measuring genetic fitness in the Darwinian sense as *the ability of an individual to contribute genetically to the next generation.*

Fortunately, at least for some perennial plants, the arrival of the next generation is not as urgent as it is in animals with their genetically fixed and discrete lifespans. For plants in marginal situations failure to reproduce sexually due to climatic deterioration does not necessarily result in abandonment of the habitat.

'Adapt or migrate' is sometimes presented as the only possible choice open to organisms that live in fluctuating environments where adversity may make sexual reproduction not possible for long periods. Reasonably regular sexual reproduction may be necessary to perpetuate populations of most animals, but many plant species are able to survive for centuries and even millennia without reproducing sexually. There are even situations where plants have survived considerable climatic change without migrating. Examination of the disjunct distribution of some relict plant populations frequently finds populations with a long history of inhabiting marginal habitats with very low rates of reproductive success. The Norwegian mugwort (*Artemisia norvegica*) is an example of a species which in both Norway and Scotland survives in isolated populations. In Norway the colonies manage some reproduction by seed while in Scotland regeneration appears to be mainly vegetative (Fig. 4.2). A detailed examination using specimens from Scotland and southern and central Norway showed that when grown in the Bergen Botanical Garden these populations on either side of the North Sea differed from each other in lipid composition, as well as in leaf and flower morphology. From the general similarity of the populations, it was argued that the differences that have arisen between the Scottish and Norwegian populations may have evolved over post-glacial time (Ovstedal & Mjaavatten, 1992). This and other examples already discussed in Chapter 1 illustrate the well-developed capacity of flowering plants to survive and evolve under seemingly adverse conditions.

4.2 SEXUAL REPRODUCTION IN MARGINAL HABITATS

4.2.1 Pre-zygotic and post-zygotic limitations to seed production

Inability to reproduce sexually may arise from numerous causes, which can be grouped under two headings, namely *pre-zygotic* and *post-zygotic limitations* (i.e. before or after fertilization). Flower development may be prevented or delayed due to unfavourable environmental conditions. Even if a flower and receptive ovules are produced, pre-zygotic limitations may occur due to a lack of suitable vectors for pollination. If pollination succeeds, fertilization of the ovule may still not be achieved, either because the pollen grain does not germinate or else the pollen tube fails to reach the ovary. A brief exposure to low temperatures may be sufficient to prevent fertilization. The small-leaved lime (*Tilia cordata*) is limited in its distribution in the northern parts of the British Isles as pollen fails to germinate below 15 °C, which is sufficient to prevent fertilization, especially when coupled with a short period of stigmoid receptivity (Pigott, 1991).

This restriction in the northern limit to the natural distribution of *T. cordata* in the British Isles is in sharp contrast to its higher latitudinal distribution in Finland (Fig. 4.3a). However, despite the warmer midsummer temperatures of the more continental Finnish climate and therefore avoiding the pre-zygotic limitation of fertilization, the subsequent rapid fall in temperature in late summer causes a post-zygotic restriction by preventing embryo and endosperm development (Pigott, 1981). Thus, in both Britain and Finland a failure to produce seed defines the northern boundary of *T. cordata*, but in the British Isles the cause

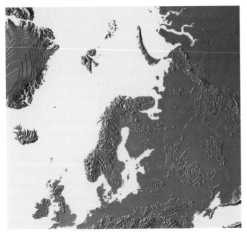

Fig. 4.2 *Artemisia norvegica*; an extreme example of a disjunct distribution and a species that maintains its presence in small isolated populations (data from Hultén & Fries, 1986). (Top) European distribution. (Centre left) Norwegian plant from a population that readily produces seedlings. (Centre right) Scottish plant, from a population that reproduces mainly vegetatively. (Below) Scottish habitat (Norwegian photos Dr A. Moen; Scottish photos Professor R. M. Cormack.)

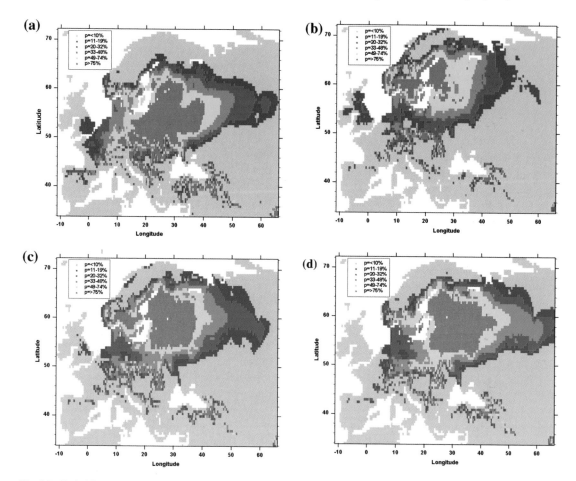

Fig. 4.3   Probability density plots of the potential range of small leaved lime (*Tilia cordata*) modelled on probability distribution under January and July temperatures from European Gridded Climatology for 1961–1990 (Hulme *et al.*, 1995). The four maps show the consequences of an increase in mean annual temperature of 2 °C obtained by distributing the relative amounts of summer and winter warmth in different ways to obtain three contrasting amounts of additional seasonality. (a) The modelled probability distribution under the current European (1961–1990) climate. (b) The summer temperature is increased by 3 °C and the winter temperature by 1 °C. (c) The temperature increase is applied equally, 2 °C each to summer and winter temperatures. (d) The winter temperature is increased by 3 °C and the summer temperature by 1 °C. These maps were made available by courtesy of Dr C. E. Jeffree.

is pre-zygotic and in Finland post-zygotic. Climatic warming can therefore be expected in both cases to permit a significant northward advance of *T. cordata*.

The above reproductive studies on *T. cordata* have shown that specific temperature conditions at certain times of the year can be crucial for successful reproduction. It is therefore essential to include seasonal sensitivity in modelling the consequences for species responses to climatic change. When this is done clear seasonal effects can be demonstrated in relation to the geographic distribution of both insects and plants

(Jeffree & Jeffree, 1996). When these methods are applied to *T. cordata* (Fig. 4.3) it is possible to examine the distribution of *T. cordata* in Europe based on the limits to distribution under current (1961–90) temperatures with the modelled potential distribution under three seasonally different climate change scenarios in which the mean annual temperature is increased by 2 °C in each case, but with contrasting seasonal differences (see figure legend). The marked contrasts in the direction of potential migration of the species between the scenarios shown in Fig. 4.3b–d emphasize the importance of differences in seasonal temperature. The more continental seasonality with a 1 °C rise in winter and a 3 °C rise in summer (Fig. 4.3b) shows an expansion into Ireland and many areas of Britain, except the far north of Scotland, and a very marked northward advance in Scandinavia to northern Norway, as well as through the Kola Peninsula to the shores of the White Sea (Crawford, 2000). It is also of interest to note that in scenario (d) where the winter temperature is increased by 3 °C and the summer temperature by only 1 °C the maps indicate a marked reduction in the probability of the occurrence of *T. cordata* throughout its current British range and in the neighbouring oceanic regions of Western Europe.

North American dwarf birch (*Betula glandulosa*) is an example of a species which can reproduce both sexually and asexually. At the northern limits of the species, less than 0.5% of the seeds (samaras) are viable and populations are maintained by asexual reproduction (Weis & Hermanutz, 1993). The potential role of pollination dynamics in the loss of sexual reproduction in North American dwarf birch has been compared at Tarr Inlet on Baffin Island (64° N), near the northern limit of the species, with populations at Kuujjuaq, Québec (58° N) at the centre of its distribution where sexual reproduction is the primary mode of reproduction. It was found that plants at the peripheral site at Tarr Inlet attained only 15–30% of the pollen production achieved at the more southerly site due to a combination of a lower density of staminate catkins and less pollen per catkin. Potential seed productivity appeared to be further constrained at the northern limit with pistillate catkins producing 50% fewer flowers in the north than in the south. While stigmatic pollen loads were similar at both sites, the northern site had lower pollen viability (68% versus 93%). There was also at the northern site a reduction in cross-fertilization due to clonal growth, which increased the probability of geitonogamous pollination (pollen from different flowers on the same plant and as a result self-sterility, a pre-zygotic limitation) and became the main cause of reduced seed production.

Post-zygotic causes of reproductive failure encompass all the varying aspects of ecology that control species survival. These may start with low or short-lived seed viability, inadequate seed dispersal, as well as destruction by predators and pathogens, and failure to be able to compete with better adapted species.

4.3 GERMINATION AND ESTABLISHMENT IN MARGINAL AREAS

Marginal areas present both advantages and disadvantages for seedling establishment as compared with more central situations. In many forests regeneration is often episodic as it is dependent on some form of disturbance or catastrophe to create space for the next generation. In the southern beech forests (*Nothofagus dombeyii* and *N. pumilio*) of south-central Chile the beeches at lower elevations are dependent on large-scale disturbance by landslides, blowdown, or even rooting by wild pigs. By contrast, in stable regions they become replaced by more shade-tolerant, late-successional species, e.g. *Aetoxilon punctatum* and *Lauerelia phillipiana*, which can be found with ages from 300 to over 635 years old. At the upper peripheral margins of the forest the more open nature of the canopy allows a more suitable habitat for *Nothofagus* regeneration (Pollmann & Veblen, 2004). At subalpine elevations the frost-tolerant *N. pumilio* is particularly successful and has no need in this relatively open forest for any disturbance as an aid to regeneration (Fig. 4.4).

The availability of suitable sites is not sufficient on its own to ensure reproduction. There has also to be an adequate supply of propagules. The various methods used by plants to disperse seeds are well known. However in marginal areas there can also be found highly specialized biological adaptations that have received less attention. Stone pines (five-needled *Pinus* spp. subsection *Cembrae*) are subalpine and timberline species both in North America and Eurasia. The European arve or stone pine (*Pinus cembra*) has long been known to have a mutualistic association with the European spotted nutcracker (*Nucifraga caryocatactes*). The pine seeds are large and wingless and remain within the cones and are therefore maladapted for wind dispersal. Their large size makes them rewarding fodder for the nutcracker birds which collect and distribute the pine seeds in hoards or caches. These hoards are buried, creating a bird-collected soil seed bank (Tomback *et al.*, 2001).

Although this mutualistic relationship has long been noted in Europe it is only recently that detailed studies have revealed a similar connection between the North American whitebark pine (*Pinus abicaulis*) and Clark's nutcracker (*Nucifraga columbiana*; Figs. 4.5–4.6). Detailed studies have shown that this relationship goes

Fig. 4.4   A stand of mature trees of the lenga beech or lenga (*Nothofagus pumilio*) surviving with little evidence of regeneration at the upper limit for tree survival in southern Patagonia.

far beyond being merely a local aid to seed dispersal but has far-reaching long-term effects on the ecology, timing of regeneration, and population genetics, of the Cembrae pines (see reviews in Tomback, 2001, 2005).

The quantity of seed transported by the nutcrackers is impressive. In their specially adapted sublingual pouch they can hold 100 or more pine seeds. It has been estimated that one bird collecting seeds for about

Fig. 4.5   Whitebark pine (*Pinus abicaulis*) growing in the Tioga Pass, California. (Photo Professor Diana Tomback.)

Fig. 4.6   Clark's nutcracker (*Nucifraga columbiana*) sitting on a limber pine (*Pinus flexilis*). The ability of this nutcracker to cache pine kernels – estimated to be as much 98 000 in a growing season – plays a vital role in forest regeneration. (Photo Michael Bowie.)

80 days and working for 9 hours per day will transport as many as 98 000 seeds in a season into caches with 1–15 seeds (mean 3.2) which represents more than 30 600 caches per bird. As has been pointed out (Tomback,

2005), where the whitebark pines will grow is determined first by the cache-site preferences of the nutcrackers and secondly on aspect, substrate, water availability, and climatic conditions of the site. The birds also have a homogenizing effect on genetic differences between populations. The tendency for the nutcrackers to cache seeds both above and below the timberline enables the pines to migrate rapidly with respect to climatic warming or cooling trends.

4.4   PHENOLOGY

Examples of the necessity for appropriate timing of reproductive processes can be found in marginal areas from the poles to the tropics. The brief growing seasons of polar and alpine regions contain many examples of the necessity for prompt and precise phenological control in relation to flowering. Similarly, marginal areas in the tropics, such as lake and river margins, also provide examples where seedling establishment faces phenological problems due to fluctuating environments.

4.4.1   Reproduction in flood-prone tropical lake and river margins

The annual flood pulses of the central Amazonian rivers inundate marginal forests to a depth of 10 m or more every year (Fig. 4.7). The regularity of this flooding has created a complex relationship between forest plants and the fish that migrate into the flooded forests and consume the seeds which are shed during the peak of the flooding season. Dispersal by floodwaters (hydrochory) and dispersal by fish (ichthyochory) are both possible means of seed dissemination. In a study of seed dispersal by the Amazonian catfish (*Auchenipterichthys longimanus*) digestion of the seeds was never found to be harmful for germinability and in many cases caused an improvement. Ichthyochory has been related to the striking ability of the vegetation of these flooded Amazonian lakes to recover even after extensive industrial habitat destruction. For example, in Amazonia, Lake Bata has suffered severe silting from the effluent of bauxite mine tailings being deposited in a layer more than 2 m deep over one third of the lake, destroying the lake vegetation and that of the neighbouring forest. However, despite this massive devastation, almost 80% of the former plant species recolonized the area 10 years after the impact of

Fig. 4.7   The understorey of a seasonally flood-tolerant Amazonian forest. These trees are flooded annually to depths of 4–6 metres for up to 6 months. The red arrow shows the flood-line mark on the foreground tree.

silting (Mannheimer *et al.*, 2003). The arrival of species that were not in the surrounding forest suggests that the catfish, which is the most abundant opportunistic fruit eater in the region, may have played a significant role in the rapid re-establishment of the vegetation.

A constant risk for seeds that are shed into floodwaters is the high probability of not reaching a site where they can become established. *Carapa guianensis* is a hardwood tree from the Brazilian Amazon with large recalcitrant seeds that can germinate both in flood-free (*terra firme*) and flood-prone forests (*várzea*). This particular species fruits throughout the year. Consequently, some seeds fall onto water and are dispersed by the currents, while in the dry season others fall on the ground where they may either be dispersed by rodents or float away when the forest eventually floods. For the seeds that land in water, in addition to the risk of not finding a suitable landing site, there is always the probability that even if they germinate on dry ground only a few months later they may have to withstand the dangers of a full submergence for six months or more.

An experimental study of the effects of flotation on seed germination in this species has shown that floating can delay germination in some of the seeds whereas others germinate while they are floating. Seeds that germinate while floating develop both shoots and roots and are able to continue to float and maintain their reserves until they reaching a suitable landing site. The seeds that delayed germination while floating showed a decline in germinability which was particularly abrupt after 2–2.5 months of floating and was probably due to depletion of endosperm reserves. However, all the seeds that germinated did so without showing any injury as a result of a prolonged floating period. This seed polymorphism with regard to dormancy is a feature of this species which under natural conditions gives a degree of ecological plasticity that enables the species to adapt to different seasons and thus contribute to the ecological success and wide geographical range of the species (Scarano *et al.*, 2003).

4.5   HYBRID ZONES

A hybrid zone is created when hybrids between two taxa flourish in the marginal area (hybrid zone) that lies between two parents and is more common with plants where fertility barriers are weaker than in animals

(see below). Marked environmental gradients such as those that occur in relation to exposure (e.g. near the sea, or in zones of varying flooding frequency, or across altitudinal zones on mountains) create marginal sites where hybrids can be more frequent than either parent. Examples of such zones can often be found in populations of willow, birch, poplar and pine, as well as in many herbaceous species where genetically differentiated populations meet and produce offspring of mixed ancestry. These zones of hybridization can be considered as marginal regions as they may lie along regions of environmental transition, with one taxon being dominant at one end and the other parent better suited to conditions at the other end.

Marginal coastal areas are commonly the location of records of hybrid species. Crosses between creeping willow (*Salix repens*) with *S. cinerea* and *S. caprea* are found in coastal areas, just as are many other willow hybrids that occur in alpine regions. Similarly, sand couch grass (*Elytrigia repens*) forms numerous hybrids with other closely related species, e.g. *E. juncea* and *E. aetherica* (Stace, 1997), which occupy specific zonations in the salt marsh.

The hyperoceanic Orkney Islands to the north of Scotland provide a striking example of a hybrid becoming the dominant form in a marginal area. The partially drained pasture lands of these islands have a saturated soil profile for a large part of the winter and are too wet for the overwintering survival of the common ragwort (*Senecio jacobea*) which is not tolerant of prolonged flooding. The pastures are too dry in summer for the closely related *S. aquaticus* (Forbes, 1976). They are, however, highly suited to the hybrid between these two species. In the hyperoceanic conditions of the Orkney Islands stands of pure *S. jacobea*, as recognized morphologically, are relatively rare and confined to the drier areas of a limited number of coastal sand dunes while the other parent *S. aquaticus* is widespread along lakeside and stream margins. The area occupied in the Orkney Islands by the hybrid between these two species, namely the partly drained pastures, is therefore much greater than that occupied by either parent and the hybrids might be described as a hybrid swarm (Fig. 4.8).

4.5.1   Transient and stable hybrids

A constant question that arises in relation to the hybrids that inhabit the marginal zones between their parents is

Fig. 4.8   Pasture in Orkney (Northern Isles, Scotland) with pasture infested by the hybrid between the common ragwort (*Senecio jacobea*) and the marsh ragwort (*S. aquaticus*). Inset shows detail of hybrid.

whether or not the hybrids are a transient phenomenon or whether they represent an evolutionary advance producing populations with greater fitness than their parents. Two points of view exist in relation to hybrids. The first is that hybrids are essentially transitory but manage to exist in a stable tension zone where selection against them is balanced due to a constant gene flow between the parent species, possibly aided by the vigour that is associated with heterosis in F_1 and other early hybrid generations (Barton & Hewitt, 1985). The second viewpoint asserts that selection depends on the environment favouring opposite traits in the two parental habitats or favouring well-adapted hybrids within a bounded region (Campbell & Waser, 2001). Experimental studies to determine whether or not selection of hybrids is environment dependent have been carried out on a number of species. A study in the Netherlands where *S. jacobea* occurs abundantly in sand dunes and *S. aquaticus* only infrequently at lake edges showed that maternal effects played a role in the fitness of experimentally produced F_1 hybrids, with offspring from *S. jacobea* mothers exhibiting higher fitness than those from *S. aquaticus* mothers (Kirk *et al.*, 2005). This is the reverse of the situation described above for Orkney where *S. aquaticus* is the more frequent parent. This study concluded that the natural hybrids were not distributed in zones where they were most fit with respect to nutrient and water regimes. It was also found that different hybrid generations differed in fitness. It was therefore concluded it was the greater heterosis in early generation hybrids which created their temporary superior fitness and which may be contributing to hybrid swarm stability.

Support for the second model can be found in studies of progressive introgression where the hybrids have gradually spread and replaced the parent plants over wide areas, as has happened in central Florida with hybrids between two taxa in the *Piriqueta caroliniana* complex (Cruzan, 2005). It is not yet clear which model is most appropriate for the vast expansion and maintenance of the *S. aquaticus* × *S. jacobea* hybrid in Orkney where the frequency of the parent species differs from the study in the Netherlands.

4.5.2 Hybrid swarms

The term *hybrid swarm* is often used to describe transient populations that arise from time to time and then disappear, as can be found in the hybrids between wood avens (*Geum urbanum*) and water avens (*G. rivale*). The hybrid (*G. intermedium*) forms swarms that link the parents with a range of intermediate populations. Even in arctic habitats, fertile hybrids are common in several well-studied genera including *Draba* and *Saxifraga* (Brochmann *et al.*, 1992).

Hybrid zones are often brought about by habitat disturbance or environmental change. In particular, human disturbance can lead to the breakdown of ecological isolation. In plants, the static nature of populations results in many species being closely associated with one particular habitat and therefore adaptation to a particular set of local conditions is ecologically advantageous. The resulting divergent evolution between adjacent populations, which still possess some degree of interfertility, is referred to as *parapatric speciation*. The difference between *allopatric* and *parapatric* speciation is

that while the former requires isolation to take place, parapatric speciation takes place despite a certain amount of gene flow between populations. The selective forces favouring a particular biotype in its own region must be sufficiently great to prevent it being swamped by neighbouring biotypes. One of the best-known examples of such a hybrid zone is the hybridization between *Iris fulva* and *I. giganticaerulea* in the Mississippi delta where human disturbance of the natural vegetation by drainage schemes has created a marginal habitat for the parents but where the hybrid between these two species predominates. As long as the intermediate zone lasts, the hybrid will survive and can be looked upon as a stable hybrid (Emms & Arnold, 1997).

On the mountain slopes of Tiryal Dag, north-east Turkey (Fig 4.9–4.12), there is a clear example of habitat-mediated selection for an F_1 hybrid in the ecotone zone between two species of *Rhododendron*. The parents are, respectively, the low-altitude (mostly < 2000 m) *R. ponticum* and the high-altitude (mostly > 2000 m) species *R. caucasicum*, which form extensive hybrid populations wherever their distributions overlap, on mountain slopes between 1800 and 2200 m. Molecular techniques confirmed that the hybrid individuals occurring on the mountain were F_1 hybrids. The hybrids produce viable seed as does the F_2 generation and their backcrosses. However, all non-F_1 hybrid derivatives appear to be eliminated in the hybrid zone at Tiryal Dag as a result of post-germination selection. Thus on this mountain all three taxa are shrubs, and are dominant components of the vegetation within their respective altitudinal habitat ranges. The F_1 hybrid *Rhododendron* × *sochadzeae* is often the commonest shrub within the centre of its habitat range, but at the upper and lower edges of this range it occurs mixed with *R. caucasicum* and *R. ponticum* (Milne *et al.*, 2003).

Fig. 4.9   Location of Tiryal Dag in eastern Turkey.

Fig. 4.10   Tiryal Dag, eastern Turkey – a mountain with a rich flora of *Rhododendron* species. In the foreground *Rhododendron smirnowii* (pink), and *R. luteum* (yellow). (Photo Dr R. I. Milne.)

Fig. 4.11   Parent species and hybrid rhododendrons identified on Tiryal Dag. *Rhododendron caucasicum* (left), *R.* × *sochadzeae* (centre) and *R. ponticum* (right). (Photo Dr R. I. Milne.)

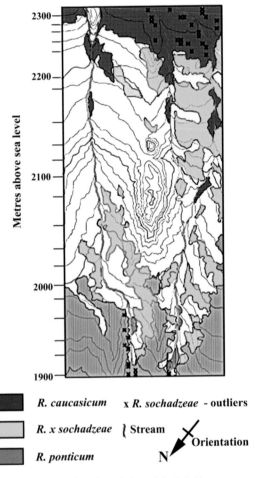

Metres above sea level

2300—
2200—
2100—
2000—
1900—

■ *R. caucasicum*   x *R. sochadzeae* - outliers

□ *R. x sochadzeae*   ⌇ Stream

■ *R. ponticum*   N ⤬ Orientation

Fig. 4.12   Distribution of populations of the hybrid *Rhododendron × sochadzeae* and its parent species on Tiryal Dag in eastern Turkey. (Reproduced with permission from Milne *et al.*, 2003.)

4.5.3   *Spartina anglica* – common cord grass

One of the most striking examples in Europe of hybridization leading to the creation of a new species with an ecological amplitude that is greater than either parent is that which arose between European cord grass (*Spartina maritima*) and the American species (*S. alterniflora*) producing the hybrid *S. townsendii* which was first recorded in Southampton Water in 1870. Chromosome doubling subsequently restored sexual fertility to the autotetraploid, which is now named *S. anglica*, and this facilitated a massive expansion into many estuaries and salt marshes (Fig. 4.13). Similar evolutionary scenarios have also taken place elsewhere. In the Basque region of

south-west France a hybrid has arisen from the same parents that created *S. anglica* but is named differently as *S. × neyrautii* (Salmon *et al.*, 2005).

The vigorous growth of *S. anglica* and its physiological tolerance of salt flooding, anoxia, and the presence of high sulphide contents in estuarine soils has allowed this species to inhabit regions of the foreshore where previously no flowering plants other than the sea grasses (*Zostera* spp.) had ever survived. The perceived benefits of this species for coastal protection and land reclamation caused it to be introduced widely in Europe, North and South America, South Africa, Australia, and New Zealand. The increase in sediment stabilization caused by *S. anglica* growth raises the level of the foreshore and leads to the development of new salt marsh areas. The high productivity of *S. anglica* also results in a large amount of energy and organic matter entering the ecosystem and provides a food source for grazers as well as being an opportune species for coastal protection against rising sea levels. Despite these benefits *S. anglica* is often regarded as an undesirable invasive species in that it leads to the loss of feeding areas for wildfowl and waders in estuary margins and can threaten economic interests such as commercial oyster fisheries. Unfortunately for coastal protection, *S. anglica* has been nominated as among the 100 of the 'world's worst' invaders (Table 4.1).

Despite the many remarkable properties that *S. anglica* possesses in relation to its ability to colonize coastal mudflats there are limits to both its reproductive capacity and its long-term viability. Flowering takes place relatively late in the growing season (July to September) and imposes a northern phenological limit to the spread of the fertile autotetraploid *S. anglica* in Europe; little seed is set in the northern regions of the British Isles. The current range of *S. anglica* is from 48° N to 57.5° N in Europe, from 21° N to 41° N in China and from 35° S to 46° S in Australia and New Zealand (Gray & Raybould in Patten, 1997). However, it remains to be seen whether or not climatic warming will encourage the northward spread of this species. Even within the geographical limits of successful establishment for *S. anglica* fertile seed production is variable both temporally and spatially. Seed does not set in most years, resulting in periods of spread by clonal expansion. Even when sexual reproduction is taking place, less than 5% of the spikelets are likely to produce viable seed. Seeds that fail to germinate in

Fig. 4.13   Common cord-grass (*Spartina anglica*) colonizing mudflats at Bosham (Sussex). Bosham is one of the possible places for the origin of the Legend of the Waves and King Canute the Great (1017–1035). The King, having tired of the flattery of his courtiers, one of whom suggested that the King could even command the obedience of the sea, was provoked to prove him wrong by practical demonstration. It was either here or at Southampton or near his palace at Westminster, that he showed (contrary to the popular version of this legend) that even a king's powers have limits. Where Canute failed, the common salt-grass has succeeded in trapping mud and silt, raising the level of the shore and pushing back the waves.

their first season do not remain viable and consequently there is no soil seed bank (Gray *et al.*, 1991). Furthermore, as salt marshes mature there is a gradual decline in vigour with dieback being a common phenomenon, particularly in shores with fine sediments. This phenomenon is similar to the dieback that is also found in the common reed (*Phragmites australis*; see Chapter 8) and is possibly induced by prolonged anaerobiosis and toxic sulphide levels. Like many other hybrids *Spartina anglica*, despite its initial vigour as a colonizer, depends for its ecological success on having access to marginal areas.

4.5.4   *Senecio squalidus* – the Oxford ragwort

The cosmopolitan genus *Senecio* is rich in species that inhabit marginal areas. The potential for adaptation in this genus to peripheral areas is remarkably high as the genus as a whole crosses ecological boundaries with the same ease with which it has spread geographically. Habitat specific species of *Senecio* can be found from the New Zealand Alps (*S. bellidioides*), to the bogs of Chile (*S. smithii*), as well as on the arid volcanic ash of Mt Etna in Sicily where different species occupy distinct altitudinal zones. It is from this latter location that the Oxford ragwort (*S. squalidus*) has evolved over the past 300 years and provided one of the best-elucidated botanical evolutionary histories in this highly adaptable genus (Harris, 2002; James & Abbott, 2005).

It has long been thought that the British populations of *S. squalidus* came from the escape of introductions made directly to the Oxford Botanic Garden sometime after the garden opened in 1621 and before the end of the seventeenth century. However, a recent review of

Table 4.1. *A selection of species commonly described as invasive and frequently found in marginal habitats*

Species	Common name and notes
Acacia longifolia	**The Sydney golden wattle:** initially planted to stabilize sand dune systems in Portugal, now an extensive and escalating invasive species. Also a widespread invader of water systems in South Africa.
Azolla filiculoides	**Water fern:** an American species introduced as a plant for garden ponds from which it has escaped into watercourses. The plant has become very abundant in recent years possibly due to warmer winters.
Cabomba caroliniana	**Carolina fanwort:** an aggressive invader of nutrient-rich freshwaters. The species is a native of eastern North America and South America but it is now widespread in North and South America. As a fully submerged aquatic plant that outcompetes native freshwater plants it can pose a risk of entanglement and drowning to swimmers.
Crassula helmsii	**Australian swamp stonecrop:** *C. helmsii* originates from Australia and New Zealand and was introduced into Europe as an aquarium plant. Main problems are at present in the British Isles. Vigorous vegetative growth through most of the year, without any period of dieback in winter, blocks ponds and drainage ditches with dense mats that outcompete the native flora and impoverish the ecosystem.
Eichhornia crassipes	**Water hyacinth:** this floating aquatic plant, a native of Brazil, has been for over 100 years a worldwide invasive species which jams rivers and lakes with uncounted thousands of tons of floating plant matter. An acre of water hyacinths can weigh up to 200 tons.
Elodea canadensis	**Canadian waterweed or common elodea:** a submerged aquatic species in which only the small white flowers emerge at the water surface. Much used as an aerating species in aquaria and accidentally introduced into Ireland from North America. The first report of its occurrence anywhere in Europe was in Co. Down in 1836. It has subsequently spread across Ireland and Great Britain and is now common in lowland lakes, ponds, canals, and slow rivers, and became a troublesome aquatic weed in Britain despite the fact that only female plants occur in the British Isles. It is now considered a noxious weed in parts of Europe, Australia, Africa, Asia, and New Zealand.
Fallopia japonica	**Japanese knotweed:** a native of Japan, Korea, Taiwan and China, but it appears that all plants grown in European gardens derive from a single Dutch import from Japan made in the 1820s. Almost all plants in the British Isles are female octoploids and almost all seed set is from hybrids with other species of *Fallopia*, e.g. *F. baldschuanica*. Originally a garden escape and first found in the wild in 1886. Its invasive spread has been from garden rubbish containing rhizome fragments.
Rhododendron ponticum	**Rhododendron:** in parts of western Britain and Ireland it is an invasive species in oakwoods, spreading vegetatively by suckering from the tips of its procumbent branches. First introduced to the British Isles in 1761 from south-west Spain. Now extensively hybridized with subsequent introductions of the cold hardy species *R. catawbiense* and *R. maximum*, thus increasing its range particularly, in colder, montane regions (Milne & Abbott, 2000).

Table 4.1. (cont.)

Species	Common name and notes
Salvinia molesta	**Giant salvinia:** an aggressive and competitive species and possibly one of the world's worst aquatic weeds. Excessive growth results in complete coverage of water surface which degrades natural habitats. Mats of floating plants prevent atmospheric oxygen from entering the water and decaying remains consume dissolved oxygen necessary for fish and other aquatic life.
Spartina anglica	**Common cord grass:** a recent hybrid salt marsh grass first recorded in 1870 (see text) has been used worldwide for coastal protection. It has a high capacity for invasion of mudflats and can displace native estuarine species of *Zostera* and *Salicornia*. It can also lead to the loss of feeding habitat for wildfowl and waders. However, with rising sea levels its role as a coastline protector needs to be reassessed and set against the negatively perceived invasive properties of the species.
Tamarix spp.	**Tamarisk or salt cedars:** introduced from Asia. Invaders of riparian areas throughout the American West where they accumulate salt in their tissues, which is later released into the soil, making it unsuitable for many native species. Because of their tolerance to alkaline and saline conditions, tamarisks are valuable as shade and ornamental plants. However, in many regions they have become a serious problem because they have formed extensive stands and cause great water losses as they take in much more water than the native plants, causing severe desiccation in desert wetland oases.

herbarium and literature records has concluded that material morphologically similar to British *S. squalidus* (2n = 20), a short-lived, herbaceous perennial plant, was collected from Mt Etna, Sicily, and first grown in the British Isles, not as commonly thought in the Botanic Garden at Oxford, but at the Duchess of Beaufort's garden at Badminton and then later transferred to the Oxford Botanic Garden sometime before 1719 (Harris, 2002).

Subsequent taxonomic studies have been unable to recognize in Sicily the taxon *S. squalidus* as found in Britain, and it has been suggested that British *S. squalidus* is a product of hybridization between the Sicilian diploid species *S. aethnensis* and *S. chrysanthemifolius* (Crisp, 1972 in James & Abbott, 2005). Both Sicilian species occur on Mt Etna, with *S. chrysanthemifolius* present up to approximately 1000 m altitude and *S. aethnensis* from approximately 1600 to 2600 m altitude with a series of hybrid swarms occurring between these two species at 1300 m ± 300 m, where certain plants can be found that bear a close morphological resemblance to British *S. squalidus*. It has therefore been suggested that the material introduced to the British Isles in the seventeenth century was collected most likely from these hybrid populations.

Following a century of cultivation in the Oxford Botanic Garden, stabilized derivatives of this hybrid material escaped at the end of the eighteenth century and subsequently spread to many parts of the British Isles.

Further molecular studies of *S. squalidus* in Britain have shown that all the individuals examined were of mixed ancestry and that the taxon is indeed a hybrid derivative of *S. aethnensis* and *S. chrysanthemifolius* (James & Abbott, 2005) and that this species must have originated within a period of 90 years, i.e. between the time hybrid material was first grown in the British Isles in the early eighteenth century and when the species was first reported growing on walls in Oxford away from the Botanic Garden in 1794 (Sibthorp, 1794).

Subsequent to the escape of *S. squalidus* from Oxford, and in its continuing spread northwards through England reaching central Scotland in the mid 1950s, the process of hybridization has continued. Crossing with the native tetraploid species, *Senecio vulgaris* L. (2n = 40), led to the recent origin of three

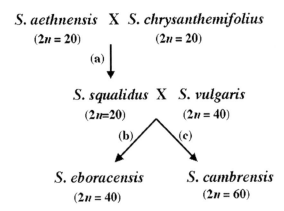

S. aethnensis X *S. chrysanthemifolius*
(2*n* = 20)   (2*n* = 20)

(a)

S. squalidus X *S. vulgaris*
(2*n*=20)   (2*n* = 40)

(b)   (c)

S. eboracensis   *S. cambrensis*
(2*n* = 40)   (2*n* = 60)

Fig. 4.14   Recent hybrid speciation in *Senecio*. Origin of the (a) diploid hybrid species *S. squalidus*, (b) recombinant tetraploid hybrid species *S. eboracensis*, and (c) allohexaploid hybrid species *S. cambrensis*. (Reproduced with permission from Abbott *et al.*, 2005.)

recognized, sexually reproducing, hybrid taxa: an introgressant form of *S. vulgaris*, *S. vulgaris* var. *hibernicus*, a recombinant tetraploid hybrid species, *S. eboracensis* and the allohexaploid *S. cambrensis* (Figs. 4.14–4.15) (James & Abbott, 2005). All these species are inhabitants of marginal areas and are at risk from extinction. Despite extensive searches, the species *Senecio cambrensis* has not been found growing at any of its previously recorded sites in Edinburgh or at other potential sites in the area since 1993. The lineage was present in Edinburgh from at least 1974 and therefore survived in the wild for a minimum of 19 years. The species remains well established in parts of North Wales (Abbott & Forbes, 2002).

4.6   GENETIC INVASION IN MARGINAL AREAS

Genetic invasion can be said to have taken place when hybridization leads to the substitution of the genes in a

Fig. 4.15   The recently evolved *Senecio* hybrid species *S. cambrensis* (2n = 60) flanked by its parents, *S. vulgaris* (2n = 40) and *S. squalidus* (2n = 20). (Photo Professor R. J. Abbott.)

native species by those from an alien taxon. Given time, insidious hybridization by continuous backcrossing can fundamentally alter the nature of an authochthonous (ancient indigenous) species and can therefore be regarded as a genetic invasion. In common with non-genetic invasions, it is marginal areas that are again most susceptible. Approximately 25 years ago smooth cord grass (*Spartina alterniflora*) was introduced into Francisco Bay where it has since hybridized with the native Californian cord grass (*S. foliosa*). A molecular study of the parental chloroplast DNA (cpDNA) genotypes of the parental species has shown that hybridization proceeds in both directions, assuming maternal inheritance of cpDNA. The native species produces little pollen with low viability whereas the hybrids produce abundant and virile pollen. Thus, it is anticipated that the spread of hybrids to other *S. foliosa* marshes could be an even greater threat to the native species than introductions of alien *S. alterniflora* (Anttila *et al.*, 2000).

The common bluebell (*Hyacinthoides non-scripta*) can be considered to occupy a marginal habitat on an international scale as it is confined to the oceanic regions of Europe from Spain to the Netherlands, with over half the world population occurring in the British Isles. A recent survey carried out by the wild plant conservation charity Plantlife International (2004) reported that one in six of the broad-leaved forest populations of *H. non-scripta* in the British Isles now show evidence of hybridization with the Spanish bluebell (*H. hispanica*) and that in boundary areas such as hedgerows and road verges and urban areas the proportion is even higher. The future viability of the native bluebell is therefore at risk as a result of hybridization, in addition to the dangers of habitat loss and illegal collection (Fig. 4.16).

4.6.1 Invasion and climatic warming

Invasion is much discussed in plant ecology in relation to the possible consequences of climatic warming. The use of the term invasion is perhaps unscientific as it carries an embedded negative-value judgement which should be

Fig. 4.16 Bluebells (*Hyacinthoides non-scripta*) in a wood near Oxford (England). This scene, which is typical of woodlands throughout the British Isles in spring, is likely to alter. Plantlife International (2004) has reported that one in six of the broad-leaved forest populations of *H. non-scripta* in the British Isles now shows evidence of hybridization with the Spanish bluebell (*H. hispanica*). The proportion of hybrids is even higher in hedgerows and road verges (see text).

superfluous in the hopefully politically neutral world of plant migration. The question of where natural migration with changing climate ends and invasion begins is not easily answered. If invasion is used to describe the movement of exotic species then migration could describe species that may have been present in the area in question under a past climatic regime. It is probably more realistic to use a quantitative rather than a qualitative judgement in separating invasion from migration. The cases that are most usually described as invasions are those where large sections of the landscape are taken over by one or more species that were not there previously. Migration is possibly better reserved to describe more gradual changes. However, differentiation between these two scenarios will probably always remain elusive. Dramatic migrations which can be described as invasions are most often seen in areas where competition is minimal and where levels of resources fluctuate. In disturbed marginal sites or areas of open ground there can be

rapid colonization by species that are new to the area, particularly if it is combined with some sudden and fortuitous supply of nutrients. It is fortunate given the extensive cultivation of exotic plants in many countries over many centuries that such situations are not globally overwhelming.

In many countries there is increasing concern about the ability of certain alien species to invade and repress or replace the indigenous flora. Exotic species that have the ability to reproduce and spread outside cultivation when an opportunity or space presents itself are usually described as invasive or alien components of the vegetation.

In New Zealand the rapid spread of introduced heather (*Calluna vulgaris*; Fig. 4.17) has been possible as it regenerates more rapidly after fire than the native upland scrub species (Wardle, 1991). In the British Isles and Ireland the number of alien species (or their hybrids with native species) that are able to survive outside

Fig. 4.17   Invasion of Scottish heather (*Calluna vulgaris*) in New Zealand. The heather has invaded a frost-flat at the northern edge of the Tongariro National Park. The trees in the distance are various species of *Nothofagus*, with those at the lower level being a mixture of red and silver beech (*N. rubra*, *N. menziesii*). There are some isolated stands of *N. solandri* at higher altitudes up to 1200 m. (Photo Dr Simon Fowler, Landcare Research, Lincoln, New Zealand.)

cultivation number 1326, which is a high figure when compared with a native flora which consists of only 1486 species (Preston *et al.*, 2002). Most of these alien species inhabit disturbed or waste ground on the fringes of cultivation. Aquatic habitats due to their open nature are also highly susceptible to invasion.

Despite this presence of a large number of introduced species it is remarkable that in Europe in general there has not been a substantial alteration in the species composition of the major native plant communities. Nevertheless, in the Scottish Highlands changes in landuse over the past century have facilitated the spread of bracken (*Pteridium aquilinum*), a native species, and also rhododendron (*Rhododendron ponticum*), which is commonly considered as an introduced invasive plant species notwithstanding the fact that it was present in the Tertiary flora and eliminated by the Pleistocene glaciations.

For a long time the origins of the present 'invasive' population of *R. ponticum* was not known. A molecular study of the naturalized material, chloroplast DNA (cpDNA) and nuclear ribosomal DNA (rDNA) from 260 naturalized accessions of *R. ponticum* throughout the British Isles found that 89% of these accessions possessed a cpDNA haplotype that occurred in native material of *R. ponticum* derived almost entirely from Spain, while 10% of accessions had a haplotype unique to Portuguese material. These results therefore indicated an Iberian origin for a substantial sample of British material which could possibly permit the assertion that the present invasion is just a re-establishment of the ancient pre-Pleistocene European stock (Milne & Abbott, 2000).

The term invasion is often used in describing the changes that are taking place in a range of habitats from undergrazed prairies to upland and alpine pastures. In the case of alpine pastures, the advance of forests that is now taking place in much of Europe from the mountains of Scandinavia to the Alps is more in the nature of re-establishment rather than an invasion. Given time, however, migration can take place into any community. The above invasions or migrations are highly visible as they occur in open landscapes such as alpine meadows. Slow migrations are insidious and can take place so gradually that they may escape notice unless careful long-term records are kept and handed down from one generation to another.

There are already examples in Europe where species are moving from cultivated parks and gardens into the native plant communities due to longer growing seasons and a reduction in the risk of periodic exposure to intermittent very cold periods. One of the warmest areas in central Europe is southern Switzerland with a mean annual temperature of 12 °C and a generous mean annual precipitation of 2000 mm. Current climatic warming trends have enabled many exotic evergreen species introduced to this region in the eighteenth century, such as the southern magnolia of eastern America (*Magnolia grandiflora*), and the Nepal camphor tree (*Cinnamomum glanduliferum*), to become established in native plant communities (Walther, 2002).

In southern Switzerland, *Quercus*, *Tilia* and *Fraxinus* species dominate the present native forest, together with sweet chestnut (*Castanea sativa*) which was a Roman introduction. Over the past four decades the length of the growing season in the region has increased on average from 290 ±13 to 350 ±8 days and, what is probably more significant, the number of frost days has decreased from approximately 75 days per year to fewer than 10 in mild years. Comparisons of previous vegetation surveys between 1960 to 1975 with more recent surveys in 1994 and 1998 have also shown a striking increase in the presence of oceanic species, with holly (*Hedera helix*) climbing into the upper tree layer together with a more vigorous growth of ivy (*Ilex aquifolium*). Exotic broad-leaved evergreen tree species that have invaded from cultivation now include cherry laurel (*Prunus laurocerasus*) and the camphor tree (*Cinnamomum glanduliferum*). The overall change in the forest species composition has been described as *laurophylization*. This term has been suggested (Walther, 1999) as it portrays the spreading of evergreen broad-leaved (laurophyllous) species into deciduous forest, which represents a biome shift from deciduous to evergreen broad-leaved forest. Thus the margin between these two biomes is migrating and the ancient Tertiary forests of central Europe which were eliminated from the region by the Pleistocene glaciations may possibly be restored in coming decades.

In contrast to the above insidious invasions or re-establishment of past vegetation communities, there are regions of the world where the native plant species have been almost extinguished by invading species from another hemisphere. In New Zealand, introduced

Fig. 4.18   Invasion of a New Zealand beach near Christchurch (South Island) with the alien marram grass (*Ammophila arenaria*) replacing the native sedge (see inset) 'pingao' (*Desmoschoenus spiralis*).

European species have largely transformed the vegetation of the coastal plains. Even the sand dune plant cover has been drastically altered (Fig. 4.18), with the native dominant graminoid, a sedge, pingao (*Desmoschoenus spiralis*), being replaced almost totally by the alien marram grass (*Ammophila arenaria*). It is probable that the geographical isolation of New Zealand has left its flora vulnerable to artificially introduced plants. It is also somewhat surprising that it is possible to raise the altitudinal level of the treeline through the introduction of exotic northern hemisphere conifers (Wardle & Coleman, 1992). Fortunately for conservation, this ability of invaders to

outperform the native species in their own habitat and persist without human assistance is not universal, and individual colonies of so-called *aliens* in disturbed ground remain at present the commonest situation due to their inability to reproduce and persist in a foreign environment.

4.6.2   Climatic warming, disturbance and invasion

Detailed examination of the distribution of invasive species, as described in an analysis of the changes in the British flora in the interval between two extensive mapping surveys (1987 and 2004), has revealed disturbance and habitat change as major factors influencing the distribution of invasive species (Braithwaite *et al.*, 2006). One of the most outstanding changes has been the spread of halophytic vegetation along road verges due to the large quantities of salt that are used for road de-icing. Notable examples include the grass-leaved orache (*Atriplex littoralis*) in eastern England and the lesser sea spurrey (*Spergularia marina*) in Scotland. These cases stand out as in the past these species have been restricted to the more southern parts of the British Isles and here climatic warming is probably an important factor in their advance northwards. The spread of wall barley (*Hordeum murinum*) which has increased by 20% (Braithwaite *et al.*, 2006) mainly in built-up areas in the south may also be due to climatic warming.

A coastal species that has been extinct in the Channel Islands (last recorded in the Scilly Isles in 1936) and is now very rare in Ireland, occurring only in two sites in County Wexford (Preston *et al.*, 2002), is the cottonweed (*Otanthus maritimus*). Given that coastal species migrate readily and this species is still found along the west coast of France and the Iberian Peninsula it might be hoped that climatic warming could aid the restoration of this species to the British flora (Fig. 4.19).

Mapping does not always record the true extent of the reproductive capacity of a species. Bracken (*Pteridium aquilinum*; Fig. 4.20) occurs in many plant communities (Marrs & Watt, 2006). It is limited by frost and waterlogging, and although the latter is increasing it is likely that the elevation of the frost line will allow bracken to colonize higher altitudes on mountain slopes. Changing land management in upland areas is failing to keep the spread of this species under control. Thus merely mapping in 10 km squares for presence or absence does not give true indication of the ability of this species to invade pastures and impinge on plant communities with a high conservation interest.

4.6.3   Theories on habitat liability to invasion

A general theory of invasibility in relation to fluctuating resources in plant communities has suggested particular characteristics that make certain environments susceptible to invasion (Davis *et al.*, 2000). The theory

proposes the following predictions in relation to the probability of invasions:

(1) strong fluctuations in resource supply
(2) coincidence of increased abundance of invader propagules with a greater availability of resources
(3) increased availability of resources due to disturbance and other effects which reduce resource capture by the resident vegetation, e.g. disease, and pest outbreaks
(4) a lengthy interval between an increase in the supply of resources and the eventual capture or recapture of the resources by the resident population.

The above-mentioned study (Davis *et al.*, 2000) also points out that there are no observed relationships between either diversity or productivity and invasibility. This is a resource-based theory rather than an assessment of Darwinian fitness as measured by reproductive success. There is nevertheless a connection in that greater growth will usually lead to greater reproductive success. Fluctuating resources are a feature of marginal habitats and therefore likely to have a significant effect particularly in pioneer terrestrial communities or in aquatic habitats where variable environments are a regular phenomenon.

4.7   REPRODUCTION IN HOT DESERTS

4.7.1   Diversity of plant form in drought-prone habitats

The flowering plants of warm deserts include every variety of life-form (Figs. 4.21, 2.22). Desert floras contain trees and shrubs, as well as annual and perennial herbaceous species. Although none of these groups is excluded as a result of drought not all are visible at any one time (see also Section 3.6.2). The relative contribution of each life-form to desert communities differs depending on the particular conditions of each desert in relation to rainfall, seasonality, and temperature extremes. The flowering of perennial herbaceous species, succulents, cacti and trees is less affected by variation in the rainfall than annuals. Above all, it is the annual species that provide the 'desert in bloom' events that attract public attention due to their sporadic nature and the visual transformation of the landscape

Fig. 4.19 Cottonweed (*Otanthus maritimus*), a common species on the Atlantic shores of south and western Europe but now almost extinct in the British Isles. This species might become more common with climatic warming. (Photograph taken in south-west Spain.)

Fig. 4.20   Hillside infestation of bracken (*Pteridium aquilinum*) in the English Lake District. The upper limit to the spread of bracken is the frost line and the lower limit the grazing efficiency of the upland farms. The former is likely to rise and the latter to deteriorate, which will result in an increase in the area of ground covered by bracken.

that takes place with the mass flowering of a wide range of species.

Desert annuals are highly dependent on autumn and winter rains. Consequently, they are by far the most variable group both in terms of their presence or absence in the landscape and for their floral display. In many deserts a massive flowering of the winter annuals in early spring is a rare event and may take place only every three or four years.

In the more climatically variable deserts, massive flowering events can be as much as 20 years apart. The occurrence of peak flowering episodes depends on a number of environmental variables and cannot always be predicted with any certainty. The spring-flowering annuals are dependent on a substantial amount of rain in the previous autumn for germination. The timing of this autumn rain and its intensity, as well as temperature and soil type, all interact in various ways with different species. Usually, heavy rain in autumn followed by regular periods of rain throughout the winter will lead to massive spring-flowering provided there are no unexpected or prolonged cold periods.

Many desert species require a period of 'after ripening' for the successful germination of their seeds. Such seeds will not germinate even when there is sufficient moisture in the soil. Thus, in winter annuals,

when the seeds are shed in early or mid summer there is no germination until the seeds have dried further. In the autumn when soil moisture rises again germination takes place rapidly. In hot deserts only 10% of species have non-dormant seeds compared with over 60% in tropical rainforests (Baskin & Baskin, 1998).

It is one of the marvels of nature that a minute dormant seed, sometimes only a few millimetres in length, is able to sense and respond to a wide range and time sequence of environmental signals and thus optimize its germination responses for maximum survival. In many desert species the seeds are able to sense the difference between sporadic rainfall and precipitation of sufficient intensity and duration to provide sufficient reserves of water for the successful completion of their life cycle. This is achieved in a variety of ways. Post-maturation dormancy mechanisms may respond to the quantity of rain, relative humidity, temperature, light, and soil salinity through germination inhibitors which influence the percentage of seeds that are ready to germinate. In some cases it is simply a certain minimum rainfall that is necessary to remove soluble germination inhibitors. In other cases a period of high relative humidity is needed before the micropyle pore will open sufficiently to allow the ingress of liquid water to initiate germination.

Fig. 4.21   Sagebrush steppe in Nevada in spring. Among the tussocks of the dominant sagebrush (*Artemisia* spp.) can be seen (inset) the scarlet flowers of the Indian paintbrush (*Castilleja linariaefolia*).

In certain leguminous species, e.g. the cutleaf medick (*Medicago laciniata*), water enters the seed through the hilar fissure which is coated with a water repellant. The degree of opening is hygroscopically controlled. Initially it is only wide enough for the entry of water vapour, as free water is excluded by the water-repellant coating. After a set time at a critical water-vapour pressure the hilar fissure opens wide enough for the entry of free water, which then initiates rapid germination. Thus, it is not merely the state of moisture in the soil that triggers germination, but rather the relative humidity of the soil atmosphere and the length of time that it remains saturated that allows seeds to estimate not only the intensity of precipitation but also its duration (Baskin & Baskin, 1998).

Temperature fluctuations, as well as the presence or absence of light, also provide the seed with information as to the depth to which they are buried. Once these conditions have been met, and they appear to act in a similar way on a wide variety of annual species, then rapid germination, growth and flowering then follow, with subsequent renewal of the long-lived soil seed bank (Koller, 1969). A variety of other means, some involving complex structural adaptations, occur in other leguminous species in hot drought-prone habitats. One example is the central and southern American wild tamarind tree (*Leucaena leucocephala*), an invasive thornless tree forming dense thickets; the seeds have specialized cells in the region of the hilum and micropyle which under favourable conditions for germination allow water to enter through a thin palisade layer (Serratovalenti *et al.*, 1995).

4.7.2   Desert seed survival strategies

The desert survival strategies that are most frequently apparent are the various forms of diversification that serve to obviate the dual dangers of predation by herbivores and drought injury. Predation losses and desiccation injury are usually high risk factors in deserts, especially if a high proportion of the seed production germinates at one time. The more extreme the desert, the more unpredictable the minimal and annual precipitation, and the more important become these various adaptations and survival strategies. One prevalent adaptation is delayed germination in a proportion of the seed produced by any one plant. With such adaptations the history of each seed can differ sufficiently, through development and maturation and eventual seed wetting, so that there is a subsequent effect on the phenotypic plasticity of seed germination. In this way the seeds from any one plant avoid the risk of all germinating at one time. This phenomenon can be described as *somatic polymorphism* where the time of the germination of seeds is governed by a maternal influence which varies with the position of the seeds in the plant, inflorescence, or capsule (Gutterman, 2000). In the Fabaceae it has been shown that when seeds are observed separately in relation to their position in the pod (numbering the seeds from the base of the pod) then those in position one (nearest to the point of attachment to the parent plant) will germinate after a shorter period of weathering than those further from the point of attachment to the parent plant. In extreme cases, this somatic polymorphism can determine the probability of germination, even 30 years after seed maturation (Gutterman, 2002).

In common with other marginal areas the maintenance of variation in desert plants appears to be a common strategy for long-term survival. In the lesser sand spurrey (*Spergularia diandra*), a plant native to the Mediterranean region of Europe and Africa (and now a widespread weed), the seeds that develop on the first flowers are black and are usually heavier and germinate more promptly than brown seeds which are produced later on lateral branches. In addition, in some of the *S. diandra* populations there are three distinct genotypes, producing seeds with hairs, glabrous seeds and partially hairy seeds. Thus in one population there can be nine types of seeds differing in colour, size, dispersal and germinability. This strategy of variation within populations is also much evident in other areas where reproduction has to contend with short and variable growing seasons, as is discussed below in relation to arctic and alpine habitats.

Rodents and other seed-eating animals are plentiful in deserts and many species of desert plants do not rely on buried seeds but instead retain their seeds in an aerial seed bank where the accessibility of seeds to rodents may be reduced. This is especially the case with lignified plants and inflorescences (serotiny) which adds a further protection against herbivory by small rodents. In these species the seeds can be retained for many years and constitute an aerial seed bank through their retention in serotinous cones or pods. The seeds have a delayed dehiscence with portions of the aerial seed bank being released when wetted by rain. In investigations of their subsequent germination (Gutterman, 2000) it was found that in many cases the degree of dormancy in the seed also depends on their position in the dispersal unit, which is yet another example of somatic polymorphism.

4.8 FLOWERING IN ARCTIC AND ALPINE HABITATS

Like deserts after rain, arctic and alpine habitats are also noted for their remarkable floral displays. The diminutive nature of plant vegetative organs in these marginal areas is not accompanied by a proportional reduction in the size of the flowers. Consequently, summer flowering displays are highly visible. Despite the abundance of flowers that is often found in montane and arctic habitats, it is necessary to determine the effectiveness of sexual reproduction, first in producing viable seed and secondly in establishing viable populations, under the potentially inhibiting combination of cold and short growing seasons that are found at high latitudes and altitudes.

In a detailed study in Greenland of flowering success, as estimated through seedling production in *Dryas integrifolia*, *Silene acaulis* (Fig. 4.22) and *Ranunculus nivalis*, it was found that, despite prolific flowering in all three species, the production of mature seeds was very uncertain due to the unpredictability and quality of the growing season. Few seedlings survived and the population structure of the established species indicated that the input of new individual plants was episodic (Philipp *et al.*, 1990).

In cold climates annual species are few. The rapid flowering that takes place in the tundra after the melting of snow and ice is achieved by long-lived perennial plants. This prompt flowering, often only

Fig. 4.22   Cushion pink (*Silene acaulis*). The species is predominantly gynodioecious (hermaphrodite and female flowers on separate plants). This cushion is a female plant and the species is an example of the early flowering *pollen risking* phenological strategy. In some sites only females and hermaphrodites can be found while in others female, male and hermaphrodite individuals have been recorded (see Alatalo & Totland, 1997).

days after the plants have become free of snow and ice is a result of two partially related adaptations. The most immediately observable aspect is the short pre-flowering period at the beginning of the growing season which takes place in the long days of midsummer at high latitudes. However, this ability to produce flowers quickly is dependent on a second adaptation, namely the possession of preformed flowering buds. When the flowering requirements of ice grass (*Phippsia algida*), a high-arctic and high-alpine snow-bed grass species, were studied in controlled environments (Heide, 1992) seedlings flowered rapidly in continuous long days at temperatures ranging from 9 to 21 °C, provided they had previously had a short-day period for the initiation of flower primordia at the same temperatures. This snow-bed grass species thus has the

characteristics of a regular long-day plant for the actual act of flowering but the initiation of flowering requires short days. This dual response to day length is found in other species similarly adapted to short growing seasons that occur in the vicinity of snow banks. The control of flowering of *Carex bigelowii* from the Rondane Mountains in southern Norway has also been shown to require short days and moderate temperatures for at least 10 weeks followed by long days for optimum flowering (Heide, 1992). Similarly, a number of arctic–alpine *Carex* species (*Carex nigra*, *C. brunnescens*, *C. atrata*, *C. norvegica* and *C. serotina*) all had a dual induction requirement for flowering (Heide, 1997).

In other cases a more extreme situation exists where the initiation and development of plant organs in cold, short growing season habitats often takes several years in order to bring both leaves and flower buds to maturation. A study of *Polygonum vivipara* (Diggle, 1997) has shown that four years are required for each leaf and inflorescence to progress from initiation to functional and structural maturity. Due to the protracted duration of leaf and inflorescence development, five cohorts of primordia, initiated in successive years, are borne simultaneously by an individual plant, which can limit the number of flowers and asexual bulbils that are produced in any one year.

The short growing and unpredictable seasons that are characteristic of most arctic and alpine habitats carry specific risks in terms of the timing of sexual reproduction. Several studies of flowering in these cold habitats have sought to define strategies in relation to the advantages and risks involved with varying phenological patterns.

In a study of three Nordic *Pinguicula* species growing in Swedish Lapland (Åbisko), *P. alpina* (Fig. 4.23) was examined as an example of an early-flowering outbreeder, *P. vulgaris* as an opportunistic late-flowering inbreeder, while *P. villosa* was taken to represent an intermediate species (Molau, 1993b). The species were found to be sympatric and reproductively isolated due to different ploidy levels. In Molau's view, the length of the growing season combined with the different breeding patterns is critical for reproductive success and

Fig. 4.23   The alpine butterwort (*Pinguicula alpina*), an early flowering outbreeding species of this insectivorous genus.

cannot be simply explained by the classical survival strategy concepts such as r and K selection. Instead, he hypothesized that arctic and alpine flowering plants are more realistically divided into two opposing categories, namely *pollen-riskers* and *seed-riskers*. The pollen-riskers are the early flowering plants which are outbreeding but show high rates of ovule abortion due to pollination failure (Fig. 4.24). Nevertheless, some seed is produced every year and as it is cross-pollinated will also provide some genetic dispersion. The late-flowering species are at the other extreme. The flowers are largely self-pollinating and there is little or no ovule abortion. However, in unfavourable years these late flowering species risk losing their entire seed crop (Molau *et al.*, 2005).

Another arctic–alpine species that exhibits a dual strategy in relation to flowering phenology is the purple flowering saxifrage (*Saxifraga oppositifolia*). This is one of the earliest flowering species both in alpine and arctic habitats and has also a strong tendency to outcrossing (Fig. 4.25). However, the actual time of flowering depends on the thawing of snow banks and those populations that live on cold coastal areas where snow lies late may have their flowering dangerously delayed. Late flowering populations show accelerated phenology but nevertheless there is always a high risk of failure to produce seed before the end of the growing season (Stenström & Molau, 1992). High-Arctic populations of this species, however, have early flowering, tufted, ecotypes that inhabit dry ridges, sometimes only a few metres distant from prostrate late flowering populations growing on cold, wet, coastal sites. These ecotypes have a genetic basis and retain distinct features when cultivated in cold climate growth chambers (Crawford, 2005). Even at its extreme northern limit of distribution in Peary Land (83° 39′ N; Fig. 4.25) the purple saxifrage is recorded as having both the prostrate and tufted forms as well as intermediates which are less specific in their habitat preferences (Dr J. Balfour, pers. obs.). It would appear that these two distinct forms aid the continued long-term presence of the species as they provide a mutual support mechanism for survival in an environment with variable and uncertain growing seasons. In unfavourable years the plants in the low-lying shore habitat are unable to set seed but nevertheless produce pollen which fertilizes some of the adjacent plants on the beach ridge. This pollen transfer from the late microsite to the earlier ridge sites ensures a genetic input of the late flowering population to the

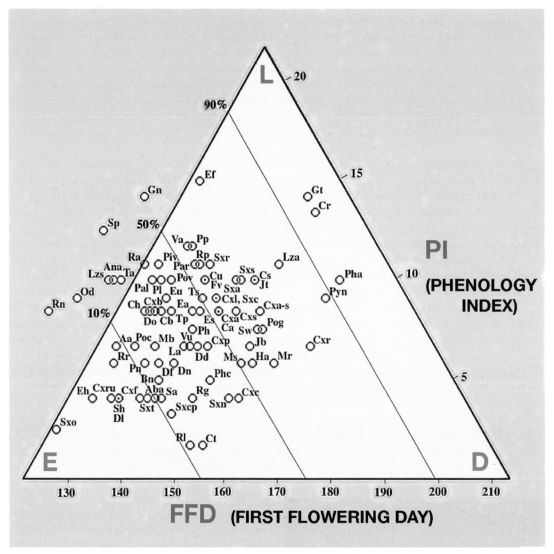

Fig. 4.24   A triangular ordination of flowering phenology in 80 species from Latnjajaure, Swedish Lapland. FFD, first flowering day (Julian day number); PI, phenology index; E, early; L, late; and D, delayed. The dates when 10%, 50%, and 90% of the valley is snow free are inserted as lines. The dotted circles represent two species each. Abbreviations: Aa, *Arctostaphylos alpina*; Aba, *Arabis alpina*; Ana, *Antennaria alpina*; Bn, *Betula nana*; Ca, *Cerastium alpinum*; Cb, *Cardamine bellidifolia*; Ch, *Cassiope hypnoides*; Cr, *Campanula rotundifolia*; Cs, *Calamagrostis stricta*; Ct, *Cassiope tetragona*; Cu, *Campanula uniflora*; Cxa, *Carex atrofusca*; Cxa-s, *Carex aquatilis* ssp. *stans*; Cxb, *Carex bigelowii*; Cxc, *Carex capillaris*; Cxf, *Carex fuliginosa*; Cxl, *Carex lachenalii*; Cxp, *Carex parallela*; Cxr, *Carex rariflora*; Cxru, *Carex rupestris*; Cxs, *Carex saxatilis*; Dd, *Draba daurica*; Df, *Draba fladnizensis*; Dl, *Draba lactea*; Dn, *Draba nivalis*; Do, *Dryas octopetala*; Ea, *Eriophorum angustifolium*; Ef, *Euphrasia frigida*; Eh, *Empetrum hermaphroditum*; Es, *Eriophorum scheuchzeri*; Eu, *Erigeron uniflorus*; Fv, *Festuca vivipara*; Gn, *Gentiana nivalis*; Gt, *Gentianella tenella*; Ha, *Hierochloe alpina*; Jb, *Juncus biglumis*; Jt, *Juncus triglumis*; La, *Lychnis alpina*; Lza, *Luzula arcuata*; Lzs, *Luzula spicata*; Mb, *Minuartia biflora*; Mr, *Minuartia rubella*; Ms, *Minuartia stricta*; Od, *Oxyria digyna*; Pal, *Poa alpina*; Par, *Poa arctica*; Ph, *Pedicularis hirsuta*; Pha, *Phippsia algida*; Phc, *Phyllodoce caerulea*; Piv, *Pinguicula vulgaris*; Pl, *Pedicularis lapponica*; Pn, *Potentilla nivea*; Poc, *Potentilla crantzii*; Pog, *Poa glauca*; Pov, *Polygonum viviparum*; Pp, *Poa pratensis*; Pyn, *Pyrola norvegica*; Ra, *Ranunculus acris*; Rg, *Ranunculus glacialis*; Rl, *Rhododendron lapponicum*; Rn, *Ranunculus nivalis*; Rp, *Ranunculus pygmaeus*; Rr, *Rhodiola rosea*; Sa, *Silene acaulis*; Sh, *Salix herbacea*; Sp, *Sibbaldia procumbens*; Sw, *Silene wahlbergella*; Sxa, *Saxifraga aizoides*; Sxc, *Saxifraga cernua*; Sxcp, *Saxifraga cespitosa*; Sxn, *Saxifraga nivalis*; Sxo, *Saxifraga oppositifolia*; Sxr, *Saxifraga rivularis*; Sxs, *Saxifraga stellaris*; Sxt, *Saxifraga tenuis*; Ta, *Thalictrum alpinum*; Tp, *Tofieldia pusilla*; Ts, *Trisetum spicatum*; Va, *Veronica alpina*; Vu, *Vaccinium uliginosum*. (Reproduced with permission from Molau *et al.*, 2005.)

Fig. 4.25 *Saxifraga oppositifolia* flowering profusely at the limit of its most northerly distribution 83° N in Peary Land (Greenland) and showing the tufted arctic form that is common at high latitudes. (Photo Dr Jean Balfour.)

next generation, even though they may have failed to produce seed (Teeri, 1973).

A comparative study of the alpine sedges *Carex curvula* and *C. firma* has shown that like *Saxifraga oppositifolia*, both *Carex* species are highly dependent on the date of snowmelt for the timing of their reproductive processes (Wagner & Reichegger, 1997). Both species can compensate for late snowmelt by shortening

the time that is required for the emergence of their flowers to as little as 8 days with *C. firma* and 14 days with *C. curvula*. Despite this flexibility, both these species exhibited the characteristic impoverishment of seed production common to early flowering species with the proportion of ovules that ripened to mature seeds (S/O ratio) being only 0.3 in *C. firma* and 0.64 in *C. curvula*. Seed germinability from these marginal sites was poor with none for the *C. curvula* seeds and only 20% of the *C. firma* seeds germinating during a two-year period of observation. In the more sunlit sites seed germination was more successful than from the shaded northern slopes with up to 28% of *C. curvula* seeds germinating over two years. Clearly these studies show that length of growing season rather than temperature makes the sites marginal for sexual reproduction. Predictions as to the effects of climatic change on these alpine species will not depend directly on temperature but on depth of snow banks and the length of the potential growing season.

It is therefore not surprising that in relation to flowering and seed production, the arctic flora shows strong correlations between flowering phenology and snow cover duration, combined with particular reproductive strategies for early and late flowering species. Gynodioecious and dioecious breeding systems are abundant only among early flowering species, whereas apomixis and vivipary are restricted to the late flowering species. The variation in ploidy levels among species increases from early to late flowering. As an overall generalization, these differing reproductive strategies characterize the early flowering groups as preferring the risks that are associated with pollen production and dispersal while late flowering groups face the risks of reproduction as seed producers (Molau, 1993a).

Molau describes such early flowering species as *pollen-riskers*. Although pollen may be wasted in these species the seeds that are produced have ample time for ripening. The converse situation of late flowering has a greater probability of achieving pollination but a deterioration in growing conditions at the end of the growing season can bring about a failure of seed ripening and such species are described as *seed-riskers*. A ten-year study of time of flowering in Swedish Lapland (Molau *et al.*, 2005) has shown that there is a regular pattern in the timing of flowering between the different species, and the seed-riskers are identified as the group most likely to benefit from climatic warming (Fig. 4.24).

4.8.1 Annual arctic plants

In the Arctic the reproductive season is so short that most species are perennials and annual species are few. Probably the largest annual plant in the Arctic is the mastodon plant *Tephroseris palustris* subsp. *congesta*

(syn. *Senecio paludosus*; Figs. 4.26–4.27). In the early stages of growth, the leaves, stem, and flower heads are covered with translucent hairs, producing a 'greenhouse effect' close to the surface of the plant, essentially extending the growing season by a few vital days by allowing the sun to warm the tissues, and preventing the heat from escaping. It is claimed that the species is '*semelparous*' (flowering once), and dies completely after having flowered and produced seed. However, in high latitudes, plants in which the flowers of the first year are killed by early frost may flower and fruit the following year (Aiken *et al.*, 1999). The species is widespread, occurring throughout the northern hemisphere from Canada across Beringia to Eurasia. The most northerly occurrence is at 72° N on Prince Patrick Island, one of the most inaccessible islands in the Canadian Arctic and surrounded by sea ice all year round (Figs. 4.26–4.27).

Other annual species found in the Arctic are in the Polygonaceae and include Ray's knotweed (*Polygonum oxyspermum* subsp. *rayii*) and the widespread and ancient Iceland purslane (*Koenigia islandica*). Ray's knotweed is a declining Atlantic species which extends its distribution to the arctic coasts of Norway and the Kola Peninsula but occurs sporadically, particularly in the more northern shores of the British Isles (Fig. 4.28).

Koenigia islandica (Fig. 3.5) deserves special discussion for our understanding of how it achieves its remarkable ability to complete its life cycle in a single cool, short growing season at high latitudes. The genus as a whole comprises six species that exemplify their derivation from montane ancestors of a small group of high mountain dwelling species displaying adaptive radiation into diverse areas with short growing seasons. Five out of the six species are now confined to high mountain areas in south-eastern Asia, primarily in the Himalayas, whereas the sixth, *Koenigia islandica*, has spread to arctic and alpine areas in the northern hemisphere and even penetrated to the fringes of Antarctica in the Tierra del Fuego (Fig. 3.5).

Iceland purslane (*Koenigia islandica*) shows a progressive reduction in size in regions with very short summers and severe climatic conditions (Hedberg, 1997). Despite the reduction in size of the vegetative parts the seed is notably large for such a small plant. This is, however, a general characteristic of many annual species that use a plentiful maternal endowment in seed

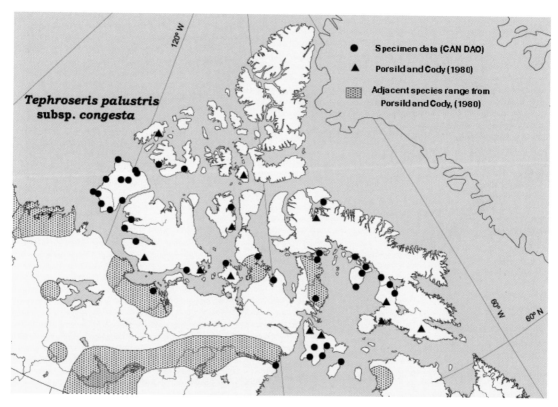

Fig. 4.26   Canadian distribution of the mastodon plant (*Tephroseris palustris*), the largest annual plant to be found in the Canadian Arctic with its most northerly location at 77° N on Prince Patrick Island. (Reproduced with permission from Aiken *et al.*, 1999.)

Fig. 4.27   Mastodon plant *Tephroseris palustris* var. *congesta* the largest annual plant to be found in the Canadian Arctic. (Reproduced with permission from Aiken *et al.*, 1999.)

reserves to facilitate prompt growth and the avoidance of an initial lag phase in their development. A study of the germination, growth and flowering of this arctic–alpine annual (Heide & Gauslaa, 1999) found that seeds germinated poorly at temperatures up to 18 °C, with an optimum at 24 °C (89% in 10 days). Scarified seeds germinated rapidly, and reached 100% germination in 3 days at 21 °C. Flowering was extremely rapid and independent of day length, even in a high-arctic population from 79° N. In full summer daylight anthesis was reached 24 days after germination and seeds ripened after a further 12 days at 15 °C. It was concluded that a conservative seed germination strategy, diminutive size and rapid development together with low temperature optima for growth and reproduction, and indifference to day length for flowering are all important adaptations for the success of an annual plant in high-arctic and high-alpine environments.

Fig. 4.28   Ray's knotweed (*Polygonum rayii*) growing as an annual plant on the north-west coast of Orkney at the end of the growing season in October. This species can grow as an annual or sometimes as a biennial or short-lived perennial and has a distribution range that reaches arctic shores (see text). (Inset) Detail of flowering shoot.

4.9   MAST SEEDING

Mast seeding in trees, with a synchronous super abundance of fruit and seed production by certain species over an extensive geographic area at irregular intervals, is a familiar concept. Less well known is the fact that *masting* (or *mast fruiting*) behaviour can also be found in marginal areas in a number of herbaceous species. Despite the striking floral displays that can be observed in early summer in most alpine habitats not all alpine

Fig. 4.29   A stand of the tussock snow grass *Chionochloa rubra* in autumn at the Lewis Pass, Southern Alps, New Zealand. This genus is noted for flowering at irregular intervals and having mast years with extensive synchronous flowering.

plant species flower with predictable regularity. The high-altitude snow tussock grasses of New Zealand (*Chionochloa* sp.) are highly variable in their flowering and some species, e.g. *C. pallens*, may not flower for a number of years and then have a year with synchronous flowering across extensive populations (Fig. 4.29). Synchronous flowering at irregular intervals is also found in other New Zealand alpine genera such as *Aciphylla* and *Celmisia* (Mark, 1970; Fig. 4.30). The essential biological feature of mast years is that they occur synchronously at irregular intervals. This then leads to the synchronous production of large quantities

of seed; hence the use of the word *mast* (Germanic root *maat*, food). Masting is most commonly observed in wind-pollinated, long-lived plants, and the oaks, beeches and pines of the northern hemisphere have been the most studied species. Early explanations of massive synchronous flowering were based largely on the *resource-tracking* hypothesis emphasizing the need for resources, and in particular carbohydrate reserves, for sustaining a large reproductive effort. Further discussion of the ecological advantages of mast years has given more prominence to the concept of predator *satiation* that comes from having a year of super abundant seed

Fig. 4.30 *Aciphylla horrida*, a New Zealand upland dioecious umbelliferous species which flowers at irregular intervals in mast years with extensive synchronous flowering.

production followed by varying periods with lower starvation levels when the amount of fruit and seed production is only sufficient to support a smaller population of potential predators (Harper, 1977). However, in the last decade a number of additional hypotheses have been suggested. When the various possible causes of masting are considered collectively they can be divided into two classes: first, proximate causes such as weather, stored levels of carbohydrate or environmental signals, and secondly, ultimate causes

which include several possible evolutionary advantages of variable seed output.

A study of mast fruiting in the tropical ectomycorrhizal tree *Dicymbe corymbosa* (Caesalpiniaceae) has drawn attention once again to a possible link between ectomycorrhizas (EM) and mast fruiting (Henkel *et al.*, 2005). There is no proof of a universal association between masting and EM symbiosis but nevertheless many tree species which exhibit this irregular phenological pattern in fruiting are mycorrhizal, as is seen in the Pinaceae, Fagaceae and Betulaceae of temperate forests, and also for several prominent tree families in the tropics, including the Dipterocarpaceae in South-east Asia and Caesalpiniaceae in Central Africa and northern South America. The argument for a role for EM in masting is largely circumstantial and is based on the need for mineral resources, particularly phosphorous. In tropical regions the forest soils are relatively nutrient poor. In Guyana, where the rainforest canopy tree *Dicymbe corymbosa* occurs, the amount of litter that falls at the end of mast year has been measured as nearly 3 t ha^{-1}, which implies that a long intermast period is needed to recover from such a loss as well as an efficient ectomycorrhiza-mediated nutrient cycling mechanism to replenish phosphorous and other minerals lost to mast flowering and fruiting (Henkel *et al.*, 2005).

One of the more compelling explanations that still deserves attention for the evolution of mast seeding has been the suggestion that it is associated with wind pollination (Kelly & Sork, 2002). Many of the well-known masting species are wind-pollinated trees. Less well known, however, are the wind-pollinated grasses that also exhibit the masting habit. Synchronous mass flowering at irregular intervals in polycarpic species, or as a terminal event in monocarpic species, will ensure economies of scale in total reproductive effort and ensure fertilization and seed production, particularly where there is a significant resource investment in the female flower. A striking feature of mast years is not just the greater production of flowering but also the more successful production of mature fruits and viable seeds.

Wind-pollinated species differ from animal-pollinated species, as the latter risk lowering their reproductive success by satiating their pollinators if too many flowers are produced at once. This is not the case with wind pollination, as seen in the mast seeding of outcrossing trees, such as oaks and pines where it maximizes pollination efficiency. By flowering at the same time, trees maximize their chances of pollination and minimize waste, an assertion that presupposes that wind-pollinated species are indeed pollen limited, for which there is increasing evidence (Koenig & Knops, 2005).

The synchronization of flowering, followed by increased seed production, is one of the most intriguing aspects of mast seeding, especially as large scale surveys indicate that conifer genera at sites as far as 2500 km apart spatially synchronize their seed production. As a result of a study of Californian oaks over a period of 11 years it has been possible to relate the geographically widespread synchronization of flowering to climatic signalling. In the case of the blue oak a synchrony of acorn production was detected that appeared to operate almost universally on an astonishingly large population of 100 to 200 million individuals (Koenig & Knops, 2005). Such widespread synchrony appears to rule out direct signalling between plants and also local effects such as those that might be attributed to mycorrhizal associations. Instead it would now appear that mast seeding is another example of the synchronization of population dynamics by environmental fluctuations through the operation of the '*Moran effect*', so named from work in the 1950s by Patrick Moran, an Australian statistician. Moran showed that an external factor, such as weather, that acts across separated populations with similar ecophysiology would tend to produce correlated changes in their abundance and hence synchronize their population cycles (Post, 2003). Moran effects appear to be particularly noticeable in marginal situations and have been detected in reindeer population dynamics in Spitsbergen (Aanes *et al.*, 2003) as well as in subarctic winter moth populations in coastal birch forests (Ims *et al.*, 2004). In Norway, the timing of flowering in three species of flowering plants in 26 populations spanning several hundred kilometres were found to be synchronized with the Arctic Climatic Oscillation over distances up to 500 km (Post, 2003). It appears also that in California, mean temperature in April is spatially synchronized with acorn production by blue oaks as a result of warm, dry conditions during flowering which generally correlate with a larger acorn crop in several oak species (Koenig & Knops, 2005).

Whatever the ultimate causes for the evolution of the masting behaviour some of the proximate causes have a particular relevance for plants in marginal situations. In Austria spruce mast years are reported as occurring at favourable sites every 3–5 years, whereas at higher altitudes the frequency is reduced to every 6–8 years and at marginal sites at the timberline only every 9–11 years (Tschermak, 1950, see Tranquillini, 1979). In an extensive study of flower and seed production in the mountain beech of New Zealand (*Nothofagus solandri* var. *cliffortiodes*; Figs. 4.31–4.32) it was also found that mast years declined in frequency with increasing altitude (Allen & Platt, 1990). The situation at the polar timberline, however, is more extreme and

Fig. 4.31 *Nothofagus solandri* var. *cliffortiodes* festooned with *Usnea* growing at an altitude of 870 m near the Lewis Pass, Southern Alps, New Zealand. Mast years in this species become less frequent with increasing altitude (see Fig. 4.32).

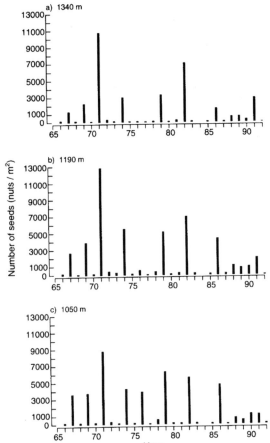

Fig. 4.32 Mast seeding in *Nothofagus solandri* var. *cliffortiodes*. (Reproduced with permission from Allen & Platt, 1990.)

appears to be changing. Prior to the 1920s, mast years for *Pinus sylvestris* at its northern limit in Finnish Lapland and neighbouring Norway were every 60–100 years. Since then, as a result of the climatic improvement that was first noticed in Finland in the early years of the twentieth century (Erkamo, 1956), the frequency of mast years has increased to about every 10 years (Tranquillini, 1979).

Seed weight and viability is also observed to decrease with altitude. Where this has been studied in spruce seed in Austria weight decreases from $900\,mg\,1000^{-1}$ seeds at 1000 m. a.s.l. to $500\,mg\,1000^{-1}$ seeds at 1700 m (the treeline) while germination capacity also falls to below 5% at the timberline (Holzer, 1973, 1950, see Tranquillini, 1979). A similar effect is seen in rowan (*Sorbus aucuparia*). In the oceanic climate of western Scotland, the high lapse rate and large increase in exposure with increasing altitude are probably relevant factors in the very marked reduction in rowan seed viability with altitude from 67% at sea level to 33% at the relatively modest elevation of 567 m (Barclay & Crawford, 1984).

In *Nothofagus* spp. in New Zealand there is a marked improvement in seeding in the year following favourable summer temperatures. This is reflected in a number of unrelated genera (Ogden *et al.*, 1996) and presents a convincing argument that climatic variables are at least a dominant proximal cause of mast years in many species, particularly in the more peripheral regions. Whether or not the climatic cause is due to resource variability or to a specific climatic signal that is sensed by a number of unrelated species is still open to debate.

4.10   THE SEED BANK

The seed bank is one of the most successful means employed by flowering plants to maintain populations in a wide range of habitats. Therophytes (annual plants) are entirely dependent on their seed bank for survival from one growing season to another. Many perennial species also have considerable banks of buried and viable seed which, depending on the species, have different expectations of longevity. Variation in the persistence of viable buried seed has prompted a classification of seed banks into three categories (Thompson *et al.*, 1997):

(1) *transient* – species with seeds which persist in the soil for less than one year, often much less
(2) *short-term persistent* – species with seeds which persist in the soil for at least one year, but fewer than five years. Such populations can aid the maintenance of plant populations after periods of poor seed setting due to environmental stress such as occurs after drought or intermittent disturbances
(3) *long-term persistent* – species with seeds that persist in the soil for at least five years and are capable of contributing to the regeneration of destroyed or degraded plant communities.

This last group of persistent species is the component that has the greatest interest for the study of survival in marginal areas. A number of studies have followed the natural restoration of disturbed communities whether as a result of fire, ploughing, volcanism, or other major disturbances. The above classification, however, has its contradictions, with some investigators finding particular species in categories that are not in agreement with others. There is some evidence that position in the soil may influence longevity. Some of the longest-lived seeds have been recorded from lake bottoms. The Chinese water lotus (*Nelumbo nucifera*) has frequently been reported as yielding viable seeds after centuries of burial in anaerobic mud. A lotus fruit (China antique) from Xipaozi, Liaoning Province, China, has been reliably carbon dated and is the current holder of the world's record for long-term seed viability at 1300 years (Shen-Miller, 2002). Five offspring of this variety, from 200–500-year-old fruits (C^{14} dates) collected at Xipaozi, have recently been germinated, and are the first such seedlings to be raised from directly dated fruits. These lotus offspring were found to be phenotypically abnormal. Most of the lotus abnormalities resembled those of chronically irradiated plants exposed to much higher irradiances. It was suggested that the chronic exposure of the old fruits to low-dose gamma radiation may have been responsible in part for the notably weak growth and mutant phenotypes of the seedlings (Shen-Miller *et al.*, 2002).

Survival may also have been aided by the anaerobic conditions under which they lay buried by providing protection from the dangers of membrane oxidation, which is also a limiting factor in seed longevity. In general, however, there is only a limited number of species where longevity in the seed bank exceeds 100 years. In a study of the seed banks of north-west Europe (Thompson *et al.*, 1997) only 14 species were recorded as surviving more than 100 years. In a similar study in North America, W. J. Beal in 1879 initiated the longest-running experiment on seed longevity by burying the seeds of 21 different species in unstoppered bottles in a sandy hilltop near the Michigan Agricultural College in East Lansing in 1879 (Telewski & Zeevaart, 2002). After 100 years, only one species, moth mullein (*Verbascum blattaria*) remained viable.

Apart from the soil conditions there are certain aspects of the seed that appear to be correlated with longevity in the seed bank. Compact, low-weight seeds are more persistent in the seed bank than seeds which are flattened and elongated. Seeds that weigh less than 3 mg are also usually persistent. Ecologically, the consequences of this differential survival in the seed bank results in pioneer species, with small easily distributed seeds being better represented than the larger seeded species of climax woodlands and shaded habitats.

4.10.1 Polar seed banks

The role of the seed bank in maintaining population diversity in the Arctic has been revised in recent years. Earlier suggestions that the seed bank declined with increasing latitude (Johnson, 1975) have been refuted in a number of studies on tundra vegetation (Bennington *et al.*, 1991; Vavrek *et al.*, 1991). In these later studies, seeds of *Carex bigelowii* were found not only to remain viable for up to about 200 years but also to show clear post-emergence phenotypic and genotypic differences from modern populations. Seeds of *Luzula parviflora*, excavated from under a solifluction lobe at Eagle Creek, Alaska, have also been carbon dated from their seed coats and found to be capable of germinating even after 175 years of burial (Bennington *et al.*, 1991). Whole populations can lie dormant for decades, or even centuries, buried in the soil seed bank. A recent study in

Spitsbergen found that 71 of the 161 species indigenous to Spitsbergen had the capability of persisting in the soil seed bank (Cooper *et al.*, 2004).

Similarly, it has recently been found that even in the Antarctic the only two flowering plants that exist there naturally, *Colobanthus quitensis* and *Deschampsia antarctica*, have considerable seed banks (107–1648 seeds m^{-2}) which are comparable in size to those of arctic and alpine species (McGraw & Day, 1997). The age of the seed is unknown but as the buried seed density was not correlated with the local above-ground abundance where both species were present, it might be assumed that under the slow growing conditions of the Antarctic that the buried seed was of considerable antiquity. Not only do these studies show the potential for long-term viability in high-latitude seed banks and the variability of the seed populations, they have also demonstrated the latent capacity of Arctic and Antarctic ecosystems to respond to environmental change in that seeds taken from different soil strata were genetically differentiated. Seed banks, through their contribution to biodiversity, have therefore important implications for plant population responses to climatic change at high latitudes.

4.10.2 Warm desert seed banks

Semi-deserts are well known to bloom with fast growing plants from buried seed after periods of rain. How much of this seed is derived from recent seed fall and how much is buried in the soil for several years is, however, not clearly established. In the Mohave Desert there is a very high level of viable seed in the upper layers of the soil (Price & Joyner, 1997). Simultaneous monitoring of the seed bank and 'seed rain' over a 19-month period has demonstrated that the total seed bank averaged approximately 106 000 seeds m^{-2}, which is much higher than values reported for other North American desert sites. These numbers correspond roughly to the seed production of a single year, since daily seed rain averaged 262 seeds m^{-2}. However, as input from the seed rain did not accumulate in the soil and the seed bank decreased by a daily average of 114 seeds m^{-2} it suggests that virtually all seeds either germinate, die, or are harvested by granivores soon after being dispersed. However, in South American and Australian deserts, lower rates of granivory are found compared with their North American counterparts. Measurement of seed reserves

in two habitats of the central Monte Desert (Argentina) showed maximum seed standing crops of 16 000 and 23 000 seeds m^{-2} in shrublands and open forests, respectively. Seed banks in other South American semi-arid areas showed similar values. Total grass seeds as well as those presumably preferred by ants also seem to be similar in both continents. Hence, in South America and Australia granivory is lower than in North America in spite of the great similarity of seed bank sizes (Marone & Horno, 1997).

4.11 BIASED SEX RATIOS

Polar and warm deserts are marginal areas that are notable for a marked presence of dioecious species. In the Arctic, although the number of dioecious species is not significantly greater than the global average of 2–3%, the extent to which they provide a large degree of ground cover is very noticeable due to the ecological success of the various species of dwarf willow that grow at high latitudes. Observations of sex ratios in dioecious species often record a bias. A preponderance of males is commoner than a bias towards females (Lloyd, 1974). Nevertheless, female-biased sex ratios although less common are very frequently observed in arctic willows (Crawford & Balfour, 1990; Fig. 4.33). The advantages of dioecy are most frequently discussed in terms of the advantages of avoiding inbreeding. When closely related pairs of species are compared in which one member of each pair is dioecious then a greater allelic diversity and heterozygosity can be demonstrated in the dioecious species (Costich & Meagher, 1992). However, dioecy also allows the possibility of morphological and physiological specialization between pollen and seed bearing parents. These differences need not be confined to sexual organs, as separation of the sexes automatically allows for the differentiation of secondary sexual features in plants as well as in animals.

Studies of the form and function of non-floral parts of dioecious species have drawn attention to the role of physiological properties such as growth rate, source–sink relationships (Laporte & Soule, 1996), leaf demography, and gas exchange in influencing the relative extent of sex ratio bias (Dawson & Bliss, 1993; Marshall *et al.*, 1993). In many cases these physiological differences can be interpreted as possible cause for differential mortality between the sexes and thus account for the bias in their

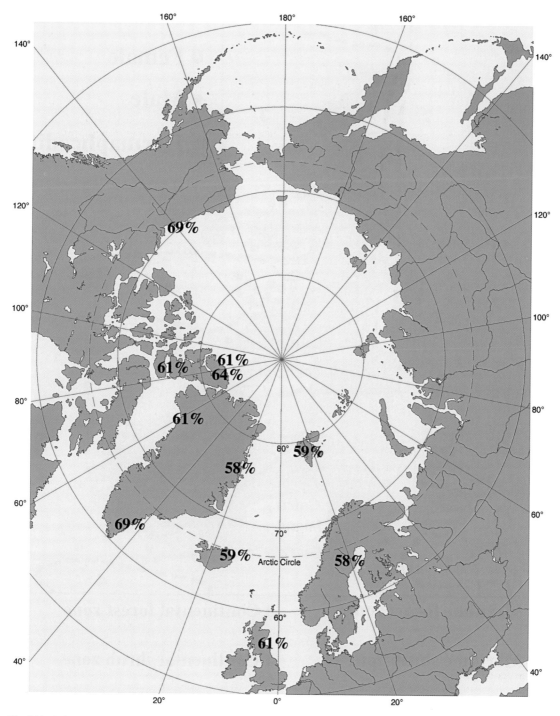

Fig. 4.33 Distribution of some female-biased sex ratios as observed in a number of arctic and subarctic species of willow *Salix polaris, S. arctica, S. glauca, S. myrsisinifolia, S. repens* (Crawford & Balfour, 1990).

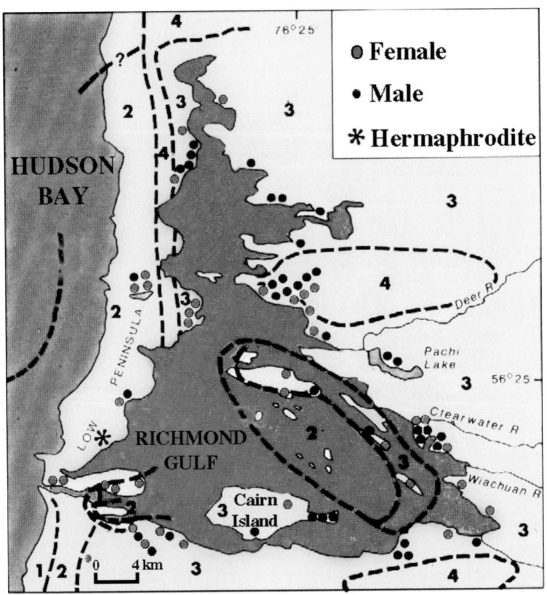

1 Maritime forest zone **3 Continental forest zone**

2 Maritime shrub zone **4 Continental shrub zone**

Fig. 4.34 Sex distribution in stands of balsam poplar (*Populus balsamifera*) at Richmond Gulf in subarctic northern Québec. (Reproduced with permission from Comtois *et al.*, 1986.)

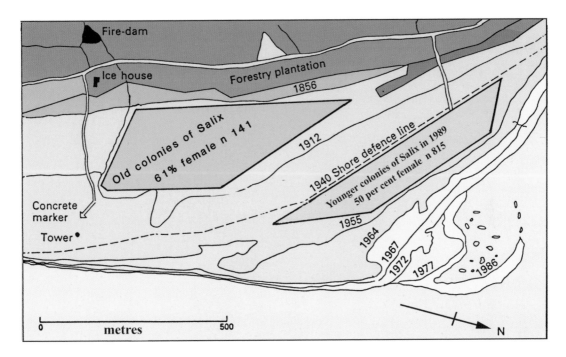

Fig. 4.35   Distribution of sex ratios in young and old colonies of creeping willow (*Salix repens*) in an actively accreting dune and slack system at Tentsmuir National Nature Reserve (Fife, Scotland) as observed in 1989. The dates refer to coastline changes. Note that the younger colonies of creeping willow did not begin to establish until the late 1960s. Re-examination of the younger colony in 1999 found a sex ratio similar to that in the old colonies. (Reproduced from Crawford & Balfour, 1990.)

sex ratios. In marginal areas such as semi–deserts and in the Arctic and northern limits of the boreal forest examples can be found of distinct microhabitat prefer- ences between the sexes of dioecious plants. In northern Québec, the balsam poplar (*Populus balsamifera*) grows as small stands which are normally unisexual and either poly or monoclonal (Comtois *et al.*, 1986). Although there was an overall equal male–female sex ratio there was a non-random distribution of the sexes in relation to habitat conditions (Fig. 4.34) with more female stands predominating in maritime regions. Conversely, eleva- tion gradients in adult sex ratios with a predominance of male clones at high elevations have been observed in *Populus tremuloides* (Grant & Mitton, 1979). In arctic willows there is frequently a bias towards females (Fig. 4.33) irrespective of species or geographic location (Crawford & Balfour, 1990). Contrary to intuitive expectations (Harper, 1977), the burden of bearing seed does not impair the growth of the female sex in arctic willows. On the contrary, in those cases where growth has been measured in terms of annual growth rings, the

female plants have shown themselves to be superior to male plants A differentiation between the sexes in either form or physiology raises the question as to whether the different sexes are opposed in their use of resources. In subsequent studies of arctic willows it was shown that in these female-biased populations the female plants were especially common in wet, higher nutrient, cold habitats, while the male plants predominated in the more xeric, low nutrient habitats with higher soil temperatures and greater drought exposure (Dawson & Bliss, 1989a,b). Similarly, in a study of five diverse dioecious wild- pollinated species in Utah it was found that male plants were more predominant in the drier microsites (Freeman *et al.*, 1976). The species included salt grass (*Distichlis spicata*), meadow rue (*Thalictrum fendleri*), box elder (*Acer negundo*), the gymnosperm desert shrub Mormon tea (*Ephedera viridis*), and saltbush (*Atriplex confertifolia*).

The causal mechanisms for these common trends with other examples of changes in sex ratio with habitat are still not clear. It cannot be substantiated that the

Fig. 4.36   Sea buckthorn (*Hippophae rhamnoides*) in fruit in autumn on a dune system in east Scotland. This is one of the few species outside the family Fabaceae which has nitrogen-fixing root nodules. The species is dioecious and as colonies become mature the female plants shade out the males and fruit production ceases. The species also spreads vegetatively with underground stolons (see Fig. 7.46). Inset shows details of fruits and foliage.

individuals in the separate sexes are themselves more fit in their respectively preferred microsites. In some cases the sex of the plants may be labile causing one sex to be more common in one habitat type than the other. Labile sex appears to be more common among dioe-cious and monoecious flowering plants than among hermaphrodites. The majority of plants with labile sex expression are perennials, which indicates that flexibility in sex is more important for species with long life cycles. Environmental stress, caused by less than optimal light, nutrition deficiencies, adverse weather or water conditions, often favours the development of male organs (Korpelainen, 1998). Alternatively, there can be a differential use of resources resulting in differing survival probabilities for the two sexes in the different microsites. Irrespective of the proximate cause of the differences in sex ratio it has been argued that for the marginal conditions of the sites in Utah, the male plants would be in a better position to disseminate their pollen from dry windy ridges while the reproductive effort of female plants would be favoured by the longer favourable moisture regime of depression sites. For creeping willows (*Salix repens*) it has been noted on a sand dune site in eastern Scotland (Tentsmuir National Nature Reserve) that on newly accreted land which had been colonized within the last 20 years (Fig. 4.35) the willows had a 50:50 sex ratio whereas the older colonies had a preponderance of females (Crawford & Balfour, 1990). It would appear that in this case at least, the females had a superior competitive power than males. A similar phenomenon is seen in sea buckthorn (*Hippophae rhamnoides*; Fig. 4.36) where the female plants are able to shade out the males with the result that mature colonies consist entirely of female plants isolated from a pollen source and where the lack of regeneration results in the ultimate demise of the colony.

4.12 CLONAL GROWTH AND REPRODUCTION IN MARGINAL HABITATS

Clonal reproduction is particularly suitable for plants in marginal areas and there are many examples of relict species where their continued survival appears to be due largely to being able to reproduce through clonal spread and fragmentation. It is therefore not surprising that marginal areas, whether they be on mountains, or at high latitudes, or in more widespread temperate habitats such as riverbanks and lakesides, are frequently rich in clonal species. Clones enable plants to resist environmental change and survive prolonged periods during which sexual reproduction does not take place. Asexual reproduction, as it is less hazardous and costs less in terms of resources, can aid survival in resource-limited situations. Clonal growth forms also permit the redistribution of resources from senescent to actively growing parts of the plant.

4.12.1 Asexual reproduction

Arctic and alpine habitats are among the most hostile environments for sexual reproduction for flowering plants. Growing seasons that are short and unpredictable can interfere with flowering and also hinder insect pollination. The long arctic winter and the constant risk of soil disturbance from erosion and cryoperturbation also makes seedling establishment hazardous. Thus, despite the fact that the Arctic in summer can be a botanist's delight in the variety and profusion of plants in flower, the amount of viable seed and established seedlings is often low. The fact that a large number of arctic species possess a capacity for vegetative reproduction has led some ecologists to assume that asexual reproduction is more important than sexual reproduction in arctic vegetation (Billings & Mooney, 1968; Grime, 2001). However, the use of molecular methods in population studies now reveals ample evidence of gene flow in the Arctic and therefore sexual reproduction can no longer be dismissed as of only minor importance. Marginal areas, from the deserts to arctic and alpine regions, have many examples of long-lived plants and therefore the contribution of sexual reproduction to the ecological success of such species cannot be assessed over short time spans.

As with sexual reproduction, asexual plants exhibit specialized mechanisms with regard to the manner and timing of asexual reproduction which usually reflect adaptations to particular types of low temperature habitats. The degree of reliance on asexual as compared with sexual reproduction also varies with the nature of the habitat and usually neither form of reproduction is completely excluded. The longevity of many clones makes it difficult to deduce from direct observation as to whether or not sexual reproduction can take place at spaced out and irregular intervals. A study of the alpine sedge *Carex curvula* in which clonal structure was analyzed from randomly amplified DNA (RAPDs) showed that one particularly large clone with more than 7000 tillers was estimated from annual growth rates to be around 2000 years old (Steinger *et al.*, 1996). Recruitment in this species from sexually produced seeds is extremely rare and the age of this plant demonstrates an ability for survival on one particular spot through a period of diverse warm and cold climates from mild periods in the early Middle Ages throughout a series of climatic changes including the Little Ice Age.

There are nevertheless many examples of arctic species that are very dependent on asexual reproduction. One of the most remarkable is creeping salt marsh grass (*Puccinellia phryganodes*) which dominates salt marshes over wide areas and has a circumpolar distribution, yet throughout its entire range the species is almost always completely sterile (Lid & Lid, 1994). The alpine bistort (*Polygonum vivipara*) has often been considered as a species in which asexual reproduction by bulbils is so predominant that ecologically sexual reproduction by seed can be ignored. However, recent research on high arctic, subarctic, and alpine populations of *Polygonum vivipara* has detected intermediate to high levels of genetic diversity within these various populations as well as showing physiological differentiation between locations. Plants from Scandinavia require considerably longer photoperiods for floral induction than plants from lower latitudes (Alps). Although *Polygonum vivipara* appears to reproduce almost exclusively asexually by bulbils, seed development can occasionally be observed even in arctic and alpine populations and may be sufficient to account for the differentiation of ecotypes and medium to high levels of genetic diversity found with the arctic and alpine populations (Bauert, 1996).

A similar situation is found in two circumpolar arctic species, *Saxifraga cernua* (Fig. 4.37), and *S. foliosa*, where it is sometimes assumed that both species have abandoned sexual reproduction. However, there are records of cases where seed is set irregularly although at low rates in both species. Field studies in northern Sweden have found that both species have an androdioecious mating system (a species which has both male and hermaphrodite individuals), a phenomenon which is extremely rare in natural plant populations (Molau, 1992). The two gender classes within each species are maintained by self-incompatibility in hermaphrodites, and a higher rate of vegetative propagation in female-sterile plants. Hermaphrodites are rare, especially in *S. cernua*, and most populations consist exclusively of female-sterile plants. The sparsity of fruiting material in herbaria may be a result of low frequency of collecting, due to the inconspicuous nature of the plants at this stage.

The tendency for asexual plants to have a more northern range than closely related sexual plants has been shown to be true for 76% of examined species (Bierzychudek, 1985) and has been described as

Fig. 4.37 *Saxifraga cernua*, an arctic species that flowers regularly but seldom produces seed and relies mainly on asexually produced bulbils for reproduction.

geographic parthenogenesis. A computer model designed to examine the causes of asexual plants being favoured in the north and at higher altitudes and in marginal resource-poor environments has suggested that as individuals move from areas from where they are well adapted to where they are poorly adapted then sexual reproduction may reduce fitness (Peck *et al.*, 1998). The mal-adaptation of sexually reproducing migrants due to lower population densities when they move to marginal areas can lead to a loss of fitness. Asexual populations will be protected from this tendency and this may account for their success in marginal habitats in the northern hemisphere and elsewhere and may be the ultimate cause for geographic parthenogenesis.

Fig. 4.38   The pseudo-viviparous alpine meadow grass (*Poa alpina*) bearing vegetatively produced plantlets.

A very successful circumpolar species is *Poa alpina*, commonly known as alpine meadow grass or alpine blue grass (North America). The species occurs in two forms: *P. alpina* var. *alpina* which reproduces sexually, and *P. alpina* var. *vivipara* which produces pseudo-viviparous vegetative spikelets (Fig. 4.38) that are then shed at the end of the growing period and root themselves in the ground. The spikelets are very hardy and can survive over winter frozen into snow and ice before rooting in the soil in the spring. Their ability to endure prolonged ice encasement, which deprives their tissues of access to oxygen, has already been described in Chapter 3 (Fig. 3.26). This tolerance of oxygen deprivation helps to explain their tolerance of prolonged freezing. The species is overall very adaptable in its choice of habitat, having low nutrient requirements as well as being tolerant of drought.

4.13   LONGEVITY AND PERSISTENCE IN MARGINAL HABITATS

Examples of long-lived trees can often be seen at treelines. The two species of bristlecone pines (*Pinus longaeva* and *P. aristata*) are probably the most famous of ancient trees that have survived for millennia on high ridges in hostile montane environments. *Pinus longaeva* is the best known from the Methuselah tree in California's White Mountains which has an age of over 4700 years (Fig. 4.39). The oldest known living Great Basin bristlecone pine had 4862 countable annual rings when it was cut down in 1974 (Currey, 1965).

Only 260 km away in the southern Rocky Mountains in Wyoming and extending through Colorado into New Mexico can be found the Rocky Mountain bristlecone pines (*P. aristata*). Some individuals of this species growing under arid environmental conditions achieve ages in excess of 2000 years, and one living specimen has been found which is at least 2435 years old (Brunstein & Yamaguchi, 1992). A study designed to determine if bristlecone pines show any signs of ageing examined live trees of *P. longaeva* ranging in age from 23 to 4713 years. No differences were found in annual shoot growth increments. The hypothesis that ageing results from an accumulation of deleterious mutations was examined by comparing pollen viability, seed weight, seed germinability, seedling biomass accumulation, and frequency of putative mutations, in trees of varying ages. None of these parameters had a statistically significant relationship to tree age. It would appear that the great longevity attained by some Great Basin bristlecone pines is unaccompanied by deterioration of meristem function in embryos, seedlings, or mature trees, and it might be concluded that the concept of senescence did not apply to these trees (Lanner & Connor, 2001). Despite the continued virility of the younger tree growth, older parts of the trees will inevitably exhibit deteriorating growth forms and heart rot. The bristlecone pines are not alone in their capacity to survive as trees to a great age in marginal sites. The monkey puzzle (*Araucaria araucana*) is an ancient species now restricted in the wild to a small area in central Chile and to the drought zone that lies between northern Argentina and Brazil at 37° to 40° S (Hueck, 1966). Many trees reach an age of about 1300 years and an age of 2000 years has been recorded (Lewington & Parker, 1999). Due to its thick fire-protecting bark, adult *Araucaria araucana* is better adapted to disturbances by volcanic activities and fire than other potentially competing species. Also in the southern hemisphere the Antarctic beech (*Nothofagus moorei*) can attain ages between 1000 and 3000 years old. Again this is a tree of marginal areas inhabiting mountain tops and ridges in eastern Australia.

The longevity of trees in marginal areas can give rise to what can be described as a *perched treeline*, where the upper limit of trees exceeds the capacity to reproduce without shelter from mature trees. An example of this is seen in another southern hemisphere beech, the lenga (*Nothofagus pumilio*). During warm

Fig. 4.39 *Pinus longaeva* in the Schulman Grove, White Mountains, California. Only part of this very ancient tree is still alive. (Photo Dr A. Gerlach.)

periods the upper limit of forest advances up the mountains in Chile and Argentina and then maintains its position during a subsequent period of climatic decline due to the longevity of the established trees. Old growth stands dominated by deciduous *Nothofagus pumilio* occupy more stable substrates, and probably represent the last stage of post-glacial succession. This long-lived tree species has recorded ages of over 200 years with a canopy which appears to be a mosaic of even-aged, old growth patches (Armesto *et al.*, 1992). *Nothofagus pumilio* therefore forms an abrupt alpine timberline (Fig. 4.4) and seedling emergence and density decreases with altitude within the forest so that the overall probability of adult establishment decreases with increasing altitude and becomes very low once the protection by the tree canopy is no longer available.

Persistence is therefore an advantageous property in marginal areas as it permits peripheral populations to wait for a suitable climatic window when they can again reproduce. In Orkney (Scottish Northern Isles) aspens

(*Populus tremula*), which are currently climatically limited from setting seed, appear to have persisted for centuries, possibly throughout the Little Ice Age due entirely to the vegetative renewal of their clones (Fig. 4.40).

The value of clonal proliferation in arctic and subarctic habitats has been closely studied in arctic species (Callaghan *et al.*, 1992). The polar willow (*Salix polaris*) has been found on extensive patches on small islands in the Spitsbergen archipelago with only one sex present, which suggests the presence of extensive clonal colonization (Crawford & Balfour, 1983).

Although clonal plants in general possess less genetic variation than those that reproduce sexually this does not mean that they are devoid of variation. In aquatic habitats large stands of the clonal common reed (*Phragmites australis*) have been shown to contain more than one genetic unit (Köppitz, 1999). Amphibious plants, although morphologically and physiologically well-adapted to surviving in habitats with fluctuating

Fig. 4.40   General view and detail of aspens (*Populus tremula*) growing in Britain's most northerly natural woodland at Berriedale on Hoy (Orkney). These trees are currently unable to set seed and may have persisted for centuries through vegetative renewal of their clones.

water tables, frequently require a respite from inundation in order that seeds can germinate successfully and establish young plants able to withstand flooding. Thus many amphibious species, e.g. *Glyceria maxima*, *Acorus calamus* (in Europe), survive almost entirely by vegetative reproduction. Apomictic species are also highly successful in marginal areas. Here again recent research is beginning to show that this highly specialized form of asexual reproduction can also contain within certain species groupings a significant degree of sexual variation (Chapman & Bicknell, 2000).

4.14 CONCLUSIONS

Marginal areas present a challenge to successful reproduction. It is particularly noticeable that at their limits of geographical distribution flowering plants can be particularly diverse in the strategies that are employed to aid reproduction and perennation, and a wide range of different strategies have evolved for overcoming the limitations of short growing seasons, disturbance through natural causes such as flooding and erosion, and other aspects of environmental uncertainty. The overall result of these evolutionary responses has been to maintain biodiversity despite the frequent physical constraints on reproduction. Hybridization, phenological and ecotypic specialization, sexual and asexual reproduction all combine in flowering plants to overcome the environmental and biotic limitations to reproduction. Depending on the nature of the environmental stress, flowering and fruiting can be accelerated sufficiently to accomplish seed production. The dispersal of propagules is vital in marginal areas and many diverse solutions are found from plants at river margins that employ fish, to high mountain forests where seed-hoarding birds serve to distribute seeds both up and below the treeline. Vegetative layering is also common in the ultra-short environmental window that is available for reproduction in semi-deserts as well as in the short growing seasons of the High Arctic. In polar regions, reliance on slow maturing perennials as opposed to quick growing annuals allows the development of floral structures to take place over several seasons. Asexual reproduction is particularly noticeable both in woody and non-woody species at high latitudes, and like sexual reproduction can take many forms ranging from apomictic seed production to pseudo-vivipary, bulbil formation and vegetative spread through stolons and tillers. Despite the reliance by many species in marginal areas on vegetative reproduction, genetic diversity is generally well maintained (see Chapter 2) and clonal longevity coupled with sporadic sexual production appears to be sufficient in many cases for the avoidance of loss of species variation.

Part III
Marginal habitats – selected case histories

Fig. 5.1 The tundra–taiga interface at Åbisko (68° N), Swedish
Lapland. Mountain birch (*Betula pubescens* ssp. *czerepanovii*) is in
the foreground and middle distance, and in the far distance is the
Lap Gate mountain pass, the traditional entry point for the
Lap reindeer herds to their summer pastures on the tundra.

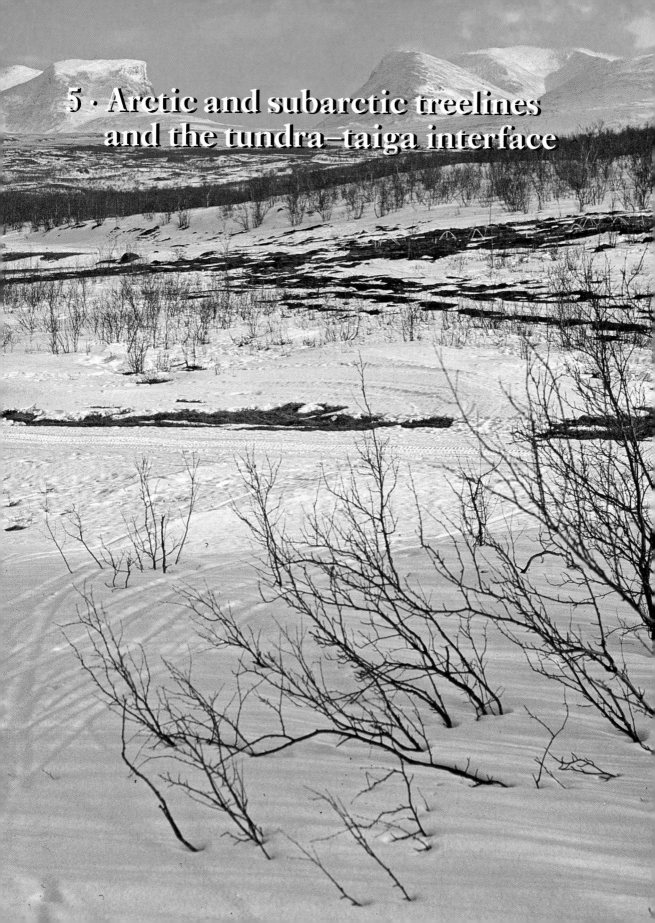

5 · Arctic and subarctic treelines and the tundra–taiga interface

5.1 THE TUNDRA–TAIGA INTERFACE

W ill climatic warming allow the boreal forest to advance onto the treeless tundra? This is one of the most tantalizing questions that can be asked in any discussion in relation to vegetation margins. The zone between the northern limit of the boreal forest (taiga) and the southern extent of the arctic tundra is the world's only circumpolar vegetation boundary and stretches for 13 400 km around the northern hemisphere and across three continents (Figs. 5.1–5.3). It is probably more exact to refer to this boundary as a zone, for in many places, and particularly in Russia, there is a well-developed interface region. This variable interface ecotone is best developed in Siberia where it is referred to by Russian ecologists as *lesotundra* (Russian *les*, forest). The term forest–tundra has also been adopted in describing similar ecotones or zones in eastern Canada created over the last 3000 years by deforestation as a result of the combined action of forest fires and climatic cooling (Asselin & Payette, 2005).

Such is the length of this boundary that it cannot be expected to have the same appearance or behave uniformly throughout its entire length. In some places it can be seen as a relatively abrupt change from forest to open tundra while in others, and particularly in Siberia, there is a lesotundra transition zone that can be several hundred kilometres deep. The forest element can be either evergreen or deciduous and the tundra can vary from ancient tundra–steppe communities to fellfield and bogs (see Table 5.1 for definitions).

The taiga–tundra interface has two unifying properties. The first is its recent origin. Throughout the Tertiary Period the lands bordering the Arctic Ocean were largely forested and little or no tundra existed. The second property is that the Holocene development of the circumpolar boreal forest has created a new major vegetation conflict zone. When temperatures rise it is possible for trees to advance over the tundra, changing not only the vegetation but also the microclimate and soil conditions. As conditions cool, trees can be suppressed by rising levels of permafrost, or possibly by being engulfed by bogs (paludification). This relatively new zone of vegetation conflict has never been static throughout its brief 10 000-year Holocene history, migrating often slowly as climates have changed. The question for this chapter is how it may move in response

to ongoing and future climatic change. Will trees be able to advance everywhere in the Arctic or are there regions, or 'no go areas', such as bogs and deep permafrost zones where trees will not readily advance?

5.1.1 Migrational history of the tundra–taiga interface

To understand the nature of the tundra–taiga interface we have to examine its past history and its present ecological condition. By the mid-Pliocene, the Cenozoic temperature decline had begun and the warmth-demanding trees of the earlier Tertiary period had been replaced in the Arctic with genera that are typical of the present boreal forest. Nevertheless, at 83° N in Peary Land (north-east Greenland), despite this dramatic cooling, a heterogeneous forest–tundra with trees, heathlands, and well-vegetated lakes, along with some present-day arctic species, is believed to have been present up to 2.0–2.5 million years ago at the Pliocene–Pleistocene transition (Bennike & Böcher, 1990). Forests then largely disappeared from these high latitudes and throughout the Pleistocene remained far to the south. However, an analysis of temporal–spatial patterns in pollen records suggests that some of the refugia that are known to have existed and supported herbs and shrubs in ice-free regions at high latitudes during the Pleistocene may have also allowed pockets of trees to survive during the Last Glacial Maximum (Brubaker *et al.*, 2005). The subsequent Holocene tree readvance, even during the warmest periods of the Holocene, never restored forest as far north as it formerly existed during the Pliocene, with the result that for the past 10 000 years a tundra–taiga interface or forest–tundra has existed at varying distances from the North Pole.

The Pleistocene ice sheets were never ubiquitous throughout the Arctic (see Chapter 2) and, particularly in regions bordering the North Atlantic and in Beringia, a number of ice-free refugia have been identified geologically (Section 6.4). Extinctions therefore will not have been uniform and reimmigration will have been influenced by the existence of refugia, the alignment of mountain chains and availability of ice-free coastal habitats. A fall in sea level will have exposed a greater area of the Arctic shelf, and provided an extensive polar desert which would have served both as a glacial refugium and as a pathway for subsequent plant migration. During the Last Glacial Maximum much of the Siberian

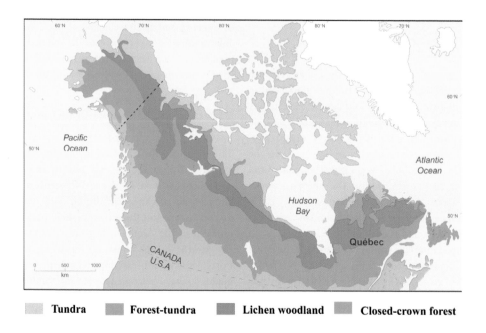

Tundra   **Forest-tundra**   **Lichen woodland**   **Closed-crown forest**

Fig. 5.2   Position of the North American forest tundra in relation to the other major subdivisions of the boreal forest. (Reproduced with permission from Payette *et al.*, 2001.)

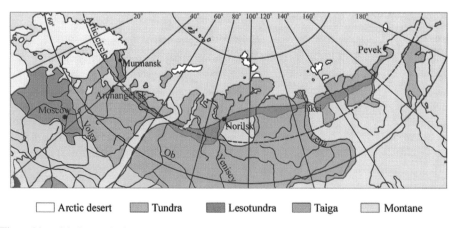

☐ Arctic desert   ▨ Tundra   ■ Lesotundra   ▨ Taiga   ☐ Montane

Fig. 5.3   The position of the lesotundra in Russia in relation to the major divisions of the boreal forest and principal cities of northern Russia. (Reproduced with permission from Vlassova, 2002.)

arctic coast was free of ice due to a lack of precipitation. Thus, the Taymyr Peninsula, like most of Siberia, lay bare of ice in a precipitation shadow at a time when there was still large-scale ice coverage in Scandinavia and north-western Europe.

Just as the severity of the Pleistocene cold period varied, so the Holocene has also altered climatically over the past 10 000 years. Early in the Holocene, a period of

rapid warming reached its maximum approximately 7000 years BP. The culmination of this warm period is variously referred to as the *Hypsithermal, Xerothermic* or *Climatic Optimum* (the warmest post-glacial period when temperatures rose to as much as 2 °C above present). The date and duration of the Hypsithermal also differed from one region to another. In south-western Saskatchewan there was a warm dry Hypsithermal

Table 5.1. *Terms, and abbreviations used in relation to arctic and subarctic treelines*

Arctic and subarctic treelines	The arctic treeline defines the northerly limit to where trees grow and beyond which there is only tundra (barren or treeless land). The subarctic treeline marks the upper altitudinal limits of boreal forest at high latitudes although isolated and small groups of trees can be found further north especially at lower altitudes.
Bowen ratio	The ratio of energy fluxes from one medium to another by sensible and latent heating respectively. For plants sensible heat loss and evaporative heat loss are the most important processes in the regulation of leaf temperature. The ratio of the two provides the Bowen ratio as used in plant ecology.
Cenozoic temperature decline	The cooling in global temperatures, particularly at high latitudes, over the past 60 million years which has probably been caused by tectonic uplift and the removal from the atmosphere of carbon dioxide by newly exposed rocks.
Cryoturbation	The physical mixing of soil materials by the alternation of thawing and freezing.
Fell field	An area of tundra with frost-shattered stony debris interspersed with finer rock particles supporting only sparse vegetation with much open ground and subject to freeze-thawing activity in the soil.
Holocene	The geological epoch within the Quaternary period, from *c.* 10 000 BP to the present.
Hypsithermal	The post-glacial period when temperatures were warmer than at present. The time interval varies with location (see text) but lies approximately within the range 9000–2500 years BP.
Köppen's Rule	A long-established rule of thumb often used to relate climatic tolerance of forest with temperature in North America and Europe, which states that trees do not survive when the mean temperature of the warmest month fails to rise above 10 °C.
Krummholz	(German *krumm*, crooked, bent, twisted; *Holz*, wood) is used to describe trees with distorted and prostrate or stunted forms that are frequently found at or just beyond the treeline either on mountain slopes or at the tundra–taiga interface. Sometimes this growth form is genetically determined while in other cases it is due entirely to phenotypic plasticity or death of terminal buds.
LAI (Leaf Area Index)	The ratio of total upper leaf surface of a crop, a natural stand of vegetation, or an individual plant, divided by the surface area of the land on which the plant or plants grow.
Last Glacial Maximum	The time of maximum extent of the ice sheets during the last glaciation. The timing of this event varied with location (see text).
Lesotundra or forest tundra or tundra–taiga interface	An interface region sometimes several hundreds of kilometres wide where pockets of tundra and taiga intermingle. This variable interface ecotone is best developed in Siberia where it is referred to by Russian ecologists as *lesotundra* (Russian *les*, forest). The term forest–tundra is also used in describing similar ecotones or zones in North America.
Lichen woodland	Open forest with the open ground dominated by lichens. Frequently these are areas with shallow permafrost which limits the root zone available to trees and shrubs.

Table 5.1. (cont.)

Little Ice Age	A period of cooling lasting approximately from the fourteenth to the mid nineteenth centuries. There is no generally agreed start or end date. Some authors confine the period to 1550–1850. This cooler period occurred after a warmer era known as the *Medieval Climatic Optimum*. There were three minima, beginning about 1650, 1770, and 1850, each separated by slightly warmer intervals.
Mono- and polycormic stems	Refers to woody plants depending on whether they have a single or many basal stems.
Paludification	The growth of bogs.

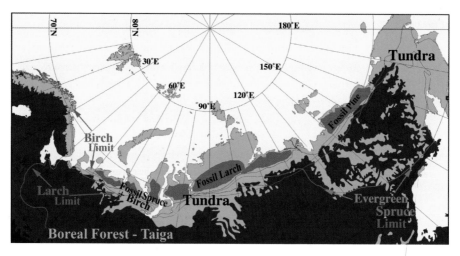

Fig. 5.4   Comparison of present and past northern limits for tree survival in northern Siberia. Present-day distribution of the boreal forest (brown) is based on the vegetation map produced by Grid Arendal and published by the World Wide Fund for Nature. Mid Holocene limits to forest trees are regional generalizations from locations of fossil remains (green is evergreen – pine and/or spruce spp.; red is tree birch; purple is larch) based on Kremenetski *et al.*, 1998. Modern limits for the northern survival of individual tree species (colours as above) are also taken from Kremenetski *et al.*, 1998, as drawn by Callaghan *et al.*, 2002.

period between 6400 and 4500 BP (Porter *et al.*, 1999), while in north-west Montana the equivalent warm dry period lasted from 10 850 to 4750 BP (Gerloff *et al.*, 1995). In western Norway the Hypsithermal is dated to between 8000 and 6000 BP (Nesje & Kvamme, 1991) when hazel (*Corylus avellana*) reached its most northerly expansion at the Norwegian Nordkapp (71° N). In Russia the early Holocene also saw a spread of spruce, larch, pine and tree birch to northern latitudes where they are no longer found (Fig. 5.4). In Siberia between 8000 and 4500/4300 BP, *Picea abies* ssp. *obovata* was farther north than at present and various *Larix* spp. were further north than now between 10 000 and 5000/4500 BP. Tree birches (*Betula pubescens*) reached the present-day shoreline of the Barents Sea in the Bolshezemelskaya Tundra (68° N) and in the Taymyr Peninsula (72° N) between 8000 and 9000 BP, and in the Yamal Peninsula (Fig. 5.4) the tree birch limit was near 70° N by 8000 BP (Kremenetski *et al.*, 1998).

5.2   CLIMATIC LIMITS OF THE BOREAL FOREST

5.2.1   Relating distribution to temperature

Throughout North America and Eurasia trees reach a northern distribution limit, which can be compared with a number of thermal indicators. However, this does not mean that in all these areas the same natural processes are limiting the northern extension of boreal forests. Traditionally, the northern boundary of the boreal forest (the tundra–taiga interface) has tended to be considered as a purely thermal phenomenon. In North America this has been related to the median July position of the polar front which matches approximately the 10 °C July isotherm (Bryson, 1966). Köppen's Rule, a long-established rule of thumb often used to relate climatic tolerance to temperature in North America and Europe, states that trees do not survive when the mean temperature of the warmest month fails to rise above 10 °C (Köppen, 1931).

A more accurate correlation can be obtained by using measurements that reflect the mean temperature of the entire growing season as detected in mean soil temperatures. A worldwide study of high-altitude tree-line temperatures found that, disregarding taxa, land use, or fire-driven tree limits, high-altitude climatic treelines can be associated with a seasonal mean ground temperature of 6.7 °C ± 0.8 SD, and a 2.2 K amplitude of means for different climatic zones, which is a surprisingly narrow range. Temperatures were higher (7–8 °C) in the temperate and Mediterranean zone treelines, and are lower in equatorial treelines and in the subarctic and boreal zone (Körner & Paulsen, 2004). In all cases this is significantly lower than the 10 °C isotherm that was the basis of Köppen's Rule. However, in any discussion between cause and effect it is necessary

to remember that mean temperatures, either of warmest months or entire growing seasons, do not exist in nature and therefore should be considered only as indicators and not causal factors (Holtmeier, 2003).

Despite attempts to find a common causal basis for the position of the arctic treeline (see Table 5.1) in North America and Europe, a generalized solution has proved elusive. This is not surprising given that tree-lines are frequently only approximate limits to the altitudinal or latitudinal distribution of scattered but upright trees (Sveinbjörnsson, 2000). The degree of scatter can vary, and in parts of Russia the development of a mosaic of vegetation at the tundra-taiga interface is particularly noticeable.

The difficulty that forest ecologists have in agreeing on the position of the arctic treeline can be seen in satellite images for North America and Siberia recording the distribution of evergreen vegetation by using the Normalized Difference Vegetation Index (NDVI) as seen in May 1998 (Figs. 5.5–5.6).

The NDVI is a simple, unitless index calculated from satellite-measured reflectance in the red and near-infrared regions of the electromagnetic spectrum (Tucker, 1979). The basis for the NDVI index is the strong reflectance of healthy green-leafed vegetation in the infrared region and strong absorption in the red region. The values range between values of −1 and +1, in the sense that it is mathematically impossible

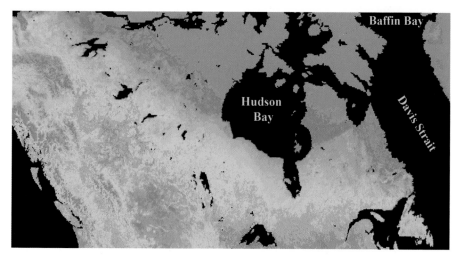

Fig. 5.5   Northern limits to the North American boreal forest: Normalized Difference Vegetation Index (NDVI) image recorded in May 1998 (8 km resolution). Colour scale: blue = 0; dark-green = 0.48–0.64; yellow-orange = 0.76–0.88. (Image prepared from 8 km resolution Pathfinder data set – US Geological Survey, EROS Data Center, Sioux Falls, South Dakota.)

to have a value outside this range. Normally there is a strong correlation between the leaf area index (LAI) and the NDVI. But this does not imply that when the green LAI is zero the NDVI must also be zero. Snow and ice give NDVI values close to zero, while water bodies can give negative values. In the 8 km resolution images recorded in May 1998 (Figs. 5.5–5.6) there is a clear discontinuity between the blue area (NDVI = 0) which represents the tundra still covered in snow and ice, while the already photosynthetically active evergreen forest with the trees emerging through the snow is shown in dark and bright green (NDVI > 0.25).

In North America a southern displacement of the evergreen coniferous forest can be observed in eastern oceanic regions, which are mainly east and west of the Hudson Bay as well as in Labrador and Québec (Fig. 5.5). Similarly in Eurasia, the West Siberian Lowlands show a marked southward displacement of the boreal forest between the rivers Ob' and Lena due to the influence of the Arctic Ocean (Fig. 5.6).

The NDVI images also illustrate the scepticism with which many Russian ecologists view the concept of a boreal forest treeline and the possibility of mapping a treeline with any degree of precision (Kriuchkov, 1971). As shown in Fig. 5.4 the northern limit for boreal forest proper generally lies to the south of the northern limit for individual stands of trees. This extensive transition zone (forest ecotone) between intact forest and completely treeless tundra is particularly noticeable in the West Siberian Lowlands. A high resolution NDVI image from this transition zone

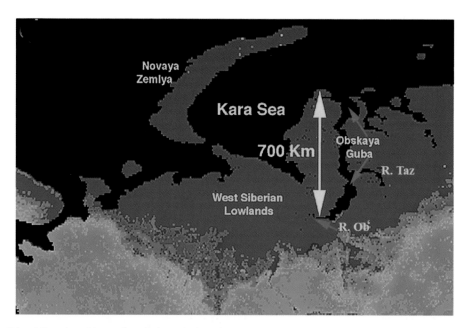

Fig. 5.6   (Above) Location of the northern limits to the boreal tundra forest and West Siberian Lowlands, as recorded by a Normalized Difference Vegetation Index (NDVI) satellite image recorded in May 1998 while the tundra was covered in snow. The snow-covered region does not register on the NDVI scale as there is no photosynthetically active visible vegetation. The evergreen forest emerging above the snow gives positive readings, shown by various shades of green with the more intense colours indicating more positive values. Note the southern depression of the evergreen treeline in the region of the Siberian Lowlands where forest has been replaced by the development over the past 6000 years by a very extensive bog, creating probably the largest example in the world of paludification of former forest. (Opposite) Detail of the transition zone between forest and tundra as seen in the normalized vegetation index recorded in May 1998 (1 km resolution). Colour scale: blue = 0; blue-green = 0.11–0.25; dark-green 0.26–0.40; bright green = 0.41–0.64. It is considered (see text) that this mosaic represents a self-renewing cyclic process taking place over hundreds of years as patches of forest develop on land that dries out after being raised by frost-heave and then reverts again to bog as tree cover cools the underlying ground. (Images prepared from 8 km resolution Pathfinder data set – US Geological Survey, EROS Data Center, Sioux Falls, South Dakota.)

Fig. 5.6   (cont.)

ciently to allow the establishment of trees. The trees persist only for a period until they shade the ground and cause the permafrost to rise and favour once again the growth of mosses as opposed to trees. Under these conditions determining the possible movements of tree limits in relation to climatic oscillations is almost a cartographic impossibility in a transition zone that can be as much as 600 km wide (Fig. 5.6).

The many and varied changes that take place in vegetation at the tundra–taiga interface therefore make it difficult to formulate pan-arctic generalizations as to the probable extent of treeline movement at present or in the foreseeable future, as the climatic history of the different regions in relation to trees is highly varied (Skre *et al.*, 2002). A similar phenomenon is seen on a smaller scale with the east Siberian dwarf pine (*Pinus pumila*) which frequently shows a very patched distribution (Figs. 5.7–5.9).

The extent of the permafrost zone is much more extensive in Eurasia than in North America (Fig. 5.10). The large expanse of Siberia that is underlain by a permanently frozen subsoil can be related to a number of ecological features, including the widespread growth of bogs during the Holocene and the presence of patchy woody scrub vegetation over large areas of drier ground. When making comparisons between North America and Eurasia in relation to treeline movements it is important to note that it is customary in North America to distinguish between the arctic and the subarctic treeline (Fig. 5.11). This is due in part to the extent of the Precambrian Shelf which covers a large part of northern Canada with a hilly terrain of sufficient elevation to produce subarctic treelines. It is these subarctic treelines that show the greatest tendency to respond to climatic change, while the more northerly arctic treeline has remained relatively stable over the past 3000 years (Payette *et al.*, 2001).

5.2.2   Krummholz and treeline advance

The term *krummholz* (German *krumm*, crooked, bent, twisted; *Holz*, wood) is used to describe trees with distorted and prostrate or stunted forms that are frequently found at or just beyond the treeline either on mountain slopes or at the tundra–taiga interface. The dwarfed or prostrate tree forms often have the appearance of being severely pruned by wind and

shown in Fig. 5.6 (detail) reveals a mosaic of tree and bog cover that extends north–south at this location for approximately 600 km from the south-eastern shore of Obskaya Guba (Russian *guba*, bay or inlet) to the east–west flowing section of the River Ob'. This mosaic is considered by Russian ecologists to be the result of a self-renewing cyclic process taking place over hundreds of years. Cryoturbation causes the soil surface in localized areas to rise above the general level of the bog (Chernov & Matveyeva, 1997). In some regions this permits the *active layer* (the layer of peat in a bog that dries out during the growing season) to dry out suffi-

Fig. 5.7   Aerial view of patches of dwarf Siberian pine (*Pinus pumila*) growing in northern Kamchatka. The lighter green trees are *Betula ermanii* growing on the south-facing slopes while tall forbs meadows cover the narrow valleys, where snow cover still exists at the beginning of July and returns again in October. (Photo Dr P. Krestov.)

Fig. 5.8   Close-up detail of bush of *Pinus pumila*. (Photo Dr P. Krestov.)

ice blast. Sometimes this growth form is genetically determined while in other cases it is due entirely to phenotypic plasticity.

The dwarfing of woody species in marginal habitats is a worldwide phenomenon and can be observed from the fringes of the boreal forest to the montane forests of the tropics. Typical American species that can be found as dwarf forms at the treeline are sub-alpine fir (*Abies lasiocarpa*), Engelmann spruce (*Picea engelmannii*), limber pine (*Pinus flexilis*), lodgepole pine (*P. contorta*) as well as the famous long-lived bristlecone pines of the Rocky Mountains (*P. aristata*, *P. longaeva* and *P. balfouriana*).

In North America black spruce (*Picea mariana*) is the main *krummholz*-forming species and shows a greater plasticity in form than white spruce (*P. glauca*). Black spruce is also better adapted to poorly drained acidic soils and high permafrost tables. Due to its particular krummholz habit of prostrate growth it is able to survive the dangers of wind throw in shallow soils and is more able to propagate vegetatively by layering than white spruce (Gamache & Payette, 2005). In Europe the mountain pine (*Pinus mugo*) is common throughout the Alps and the Apennines while the dwarf Siberian pine (*P. pumila*) is extensive in the more eastern regions of Eurasia. The mountain birch (*B. pubescens* ssp. *czerepanovii*) is the dominant treeline species for much of northern Europe from Scandinavia to Iceland. It is not strictly a *krummholz* species as its many limbs normally

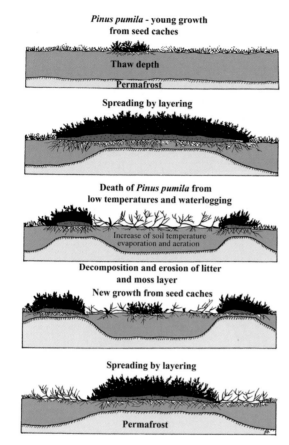

Pinus pumila - young growth from seed caches

Thaw depth

Permafrost

Spreading by layering

Death of *Pinus pumila* from low temperatures and waterlogging

Increase of soil temperature evaporation and aeration

Decomposition and erosion of litter and moss layer

New growth from seed caches

Spreading by layering

Permafrost

Fig. 5.9   Interaction between dwarf Siberian pine (*Pinus pumila*) and site conditions in the lesotundra (tundra–taiga interface) in northern Siberia. (Reproduced with permission from Holtmeier, 2003.)

maintain the persistence of the leading shoots. In exposed conditions these leading shoots can adopt a prostrate growth form (see also Figs. 9.13–9.14). Scots pine (*Pinus sylvestris*) and Norway spruce (*Picea abies*) can also exist in the *krummholz* form which they commonly adopt as a phenotypic response to exposure. In terms of latitudinal extent across the Eurasian Arctic, the dwarf Siberian pine is outstanding for the amount of terrain that it occupies and for being the most northerly pine species in the world, reaching 72° N in Yakutia (Fig. 5.12).

A Canadian study of regeneration patterns of black spruce (*Picea mariana*) along a 300-km latitudinal transect across forest–tundra in northern Québec compared forest–tundra interfaces at different latitudes; one from

the southern forest–tundra and two regions from the northern forest–tundra, including the arctic treeline (Gamache & Payette, 2005). The results showed that during the twentieth century the southern forest–tundra treelines rose slightly through establishment of seed-origin spruce, while some treelines in the northern forest–tundra increased the height of stunted spruce (*krummholz*) already established on the tundra. Despite these improvements in spruce reproductive success in the twentieth century in the southern forest-tundra, there was little evidence that recruitment of seed–origin spruce was controlled by 5- to 20-year regional climatic fluctuations, apart from the effects of winter precipitation. It was therefore concluded that local topographic factors rather than climate have influenced the recent rise in treelines. In particular, the effect of black spruce's semi-serotinous cones (retaining seeds in protective structures) and the difficulty of establishment on exposed, drought-prone tundra vegetation appeared to explain the scarcity of significant short-term correlations between tree establishment and climatic variables. Furthermore, the age data of the trees indicated that the development of spruce seedlings into forest was being retarded by the harsh wind-exposure conditions.

In Scandinavia, the upper treeline is formed mainly by mountain birch, which is not a well-defined taxon due to the high degree of polymorphism and introgressive hybridization which has created many widely different morphological types of *krummholz*. The Nordic mountain birches (Fig. 9.13) are therefore composed of many varieties, hybrids and subspecies of *Betula pubescens* and sometimes also *B. pendula* and the dwarf birch *B. nana* (Wielgolaski & Sonesson, 2001). All these combinations are usually referred to as the subspecies *B. pubescens* spp. *czerepanovii* (formerly spp. *tortuosa*). Within this taxon a variety of growth forms can be found (Väre, 2001; Holtmeier, 2003). These can be either one-stemmed (monocormic) or many stemmed (polycormic). In the former case the trunk sometimes extends itself close to the ground in a twisted form, often developing semi-upright knees, and is sometimes considered as the variety *appressa* (see Section 9.4). This type is probably genetically fixed. Polycormic forms of mountain birch are found typically in areas with nutrient deficiencies, or where there has been disturbance or attack from the autumn moths *Epirrata autumnalis* and *Operophtera brumata*, which

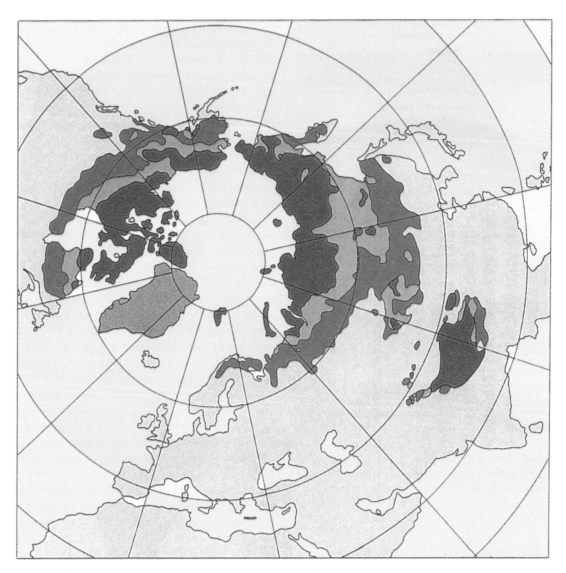

Fig. 5.10   Contemporary distribution of permafrost in the northern hemisphere classified into three categories: (1) sporadic is green; widespread discontinuous is blue; continuous is purple. (Reproduced with permission from Brown, 1997.) Light grey is no permafrost. (Reproduced with permission from Brown, 1997.)

feed on the leaves of the mountain birch and have population peaks approximately every 10 years. In the polycormic forms improved growth can be achieved with nutrient application and it has therefore to be assumed that here the *krummholz* form is phenotypic.

Irrespective of how the *krummholz* arises it has two principal advantages for tree survival at the timberline. First, there is the advantage in having foliage near the ground as this decouples low stature plants from air temperatures which are cooler than temperatures near the ground (Grace *et al.*, 2002). Secondly, the ratio of photosynthetically productive tissues to non-productive tissues is increased, especially in the polycormic (many stemmed) form of *krummholz*, as is also the dependence for survival on one upright tree trunk.

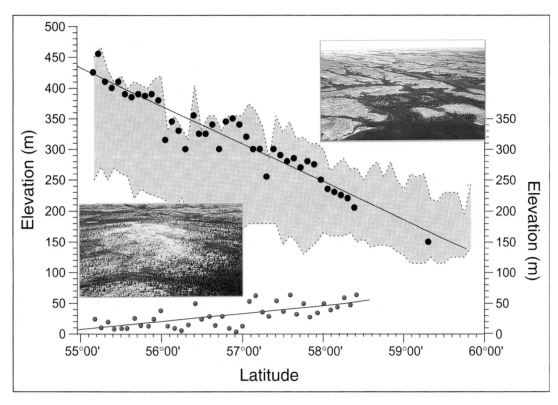

Fig. 5.11   Arctic and subarctic treeline elevation as seen in the maximum elevation for the occurrence of spruce stands (forest and *krummholz* black circles) according to latitude from the continuous forest limit to the arctic treeline and the species limit in the tundra. The amplitude of the lowering of the subarctic treeline (red circles) is shown according to latitude. The shaded area corresponds to the topographic corridor of maximum and minimum ground altitudes according to latitude from 55° to 60° N. (Reproduced with permission from Payette *et al.*, 2001 with additional photographic inserts by courtesy of Professor S. Payette.)

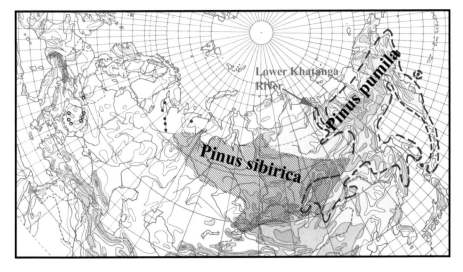

Fig. 5.12   Distribution of Siberian dwarf pine (*Pinus pumila*) and the east Siberian pine (*P. sibirica*). *Pinus pumila* has the most northerly distribution reaching 72° 32′ N in the Lower Khatanga River. (Reproduced with permission from Hultén & Fries, 1986.)

5.3   CLIMATIC CHANGE AND FOREST MIGRATION

5.3.1   Boreal migrational history

In the early Holocene, around 9000 BP, tree migration rates of 1 to 2 km yr^{-1}, have been reported both in Europe and Canada, causing a substantial and rapid reduction in the area of the tundra biome (Huntley & Birks, 1983; Ritchie, 1987; Huntley & Webb, 1988). However, since the passing of the Hypsithermal period, cooler weather has prevailed over the past 6000 years and this has been reversed only recently due to the current global climatic warming trend. Since the Hypsithermal temperature optimum, the extensions and contractions of the boreal forest in response to climatic oscillations have been geographically diverse. In some areas, small reductions in temperature have been followed by marked vegetation changes due to a parallel onset of more oceanic conditions. Across northern Russia (including Siberia), boreal forest advanced to or near the current arctic coastline between 9000 and 7000 BP and retreated to its present position between 4000 and 3000 BP (MacDonald *et al.*, 2000). The length of the cool period and the consequent substantial forest retreat that has taken place in some areas has allowed the long-term development at high latitudes of a variety of treeless plant communities including vast expanses of bog (Figs. 5.13–5.15). Consequently, there is now a high-latitude landscape that differs from that which existed in the early Holocene and which may prove resistant in certain areas to forest recolonization.

Fig. 5.14   Aerial photograph taken flying north over the West Siberian Lowlands at 75° E between 64° and 66° N showing an advanced state of paludification with lichen-rich open forest surrounded by bogs. (Photo S. Kirpotin.)

Fig. 5.15   Aerial photograph taken flying north over the West Siberian Lowlands at 75° E between 64° and 66° N showing the ultimate state of paludification. (Photo S. Kirpotin). (Figs. 5.13–5.15 reproduced with permission from Skre *et al.*, 2002.)

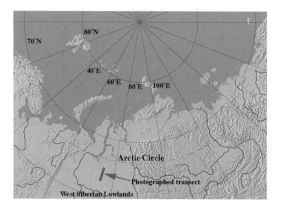

Fig. 5.13   Map of northern Russia showing the West Siberian Lowlands (between approximately 60° and 90° E) with the location of the area shown in Figs. 5.14–5.15.

This is a fundamental habitat change between what existed in the early Holocene and the communities that are now in place. The modern wetland vegetation in particular is well established and through its homeostatic properties will present a considerable barrier to forest migration. The northward movement of the rainbelts as in Russia, the rise in winter temperatures and the greater input of nitrogen to high latitude regions will all serve to encourage bog growth and thus hinder the establishment of forest.

It is therefore not possible to base predictions of the potential responses to current climatic warming based on the early Holocene rapid advance of trees that took place when the ice sheets melted after the last glaciation. In particular, the large expanses of arctic bogs that have developed in Siberia may prove to be an unsurmountable barrier to a poleward readvance of the boreal forest. Even in northern temperate zones this same trend to increased paludification can be found. In Shetland (Scotland's Northern Isles) a similar contemporaneous change took place and has been considered as accelerating a natural tendency for an autogenic succession from forest to bog under oceanic conditions rather than as a direct response to a decrease in temperature (Bennett *et al.*, 1992).

Plant migration in a modern world is therefore likely to differ from the recolonization that took place after the Pleistocene *tabula rasa*. Similarly, in northern Sweden and in the Kola Peninsula there was a regression from a boreal birch and pine forest to subarctic birch woodland tundra *c.* 3000 BP (Gervais *et al.*, 2002) which has existed to the present day. The change appears to be due not so much to a change in temperature, but to the onset of a more oceanic climate with greater soil leaching and erosion and the establishment of heath vegetation in place of forest (Holmgren & Tjus, 1996). Such phenomena indicate that sustaining forest cover depends not just on adequate thermal conditions, but also on water relations, with excessive humidity being just as detrimental to tree survival as drought.

These various forest retreats, due either to increasing oceanicity, or decreasing temperatures or both, were all gradual processes. During cooling periods treeline-altitude decrease has generally lagged behind changes in solar activity levels by 400 to 500 years as in the Sierra Nevada (Scuderi, 1994, and for up to 650 years in the mixed forests of southern Ontario (Campbell & McAndrews, 1993). Similarly, there can be expected a lag in the advance of forests with climatic warming. In Scandinavia, responses to temperature increase in birch forest are less than would be predicted from the meteorological record (Holmgren & Tjus, 1996; Dalen & Hofgaard, 2005). Visual reports of the establishment of tree seedlings above existing treelines in Scandinavia (Kullman, 2002b), might suggest that current climatic warming is already favouring an upward and polewards migration of the treeline. Nevertheless, close examination usually shows that the extent of the change falls short of what might be expected given the degree of climatic warming that has taken place over the past 100 years. Often the density of the saplings above the treeline is low and the fact that they have been recorded is possibly due to more

thorough searching than before climatic warming and plant migration became active research issues. Whether the seedlings will survive long enough to form a close forest canopy still remains an open question. Many cases can be seen where in the past seedlings have advanced up mountains (Fig. 5.16) in periods of favourable weather only to be killed at a later date by the return of adverse weather or the increased exposure they encounter as they rose above the shelter of the surrounding vegetation (Holtmeier, 1995).

In north-western Canada and Alaska the arctic treeline limits are now at higher latitudes than at any previous time during the Holocene. These favourable conditions for tree survival at high latitudes are evident in a general stability at the treeline coupled in many cases with a tendency to migrate northwards or upwards. In north-western Canada there has been increased establishment of white spruce (*Picea glauca*) resulting in increasing density, but no substantial increase in the altitude of the treeline during the past 150 years (Szeicz & MacDonald, 1995). These findings match a similar study along the coast of the Hudson Bay (Payette & Filion, 1985). Observations in the Seward Peninsula in western Alaska indicate a treeline advance with an increase of approximately 2.3% in the area of tundra that has become covered with trees over the past 50 years (Chapin *et al.*, 2005). A continuation of these trends in shrub and tree expansion, it is suggested, could further amplify the effects of atmospheric heating.

Moving from general to particular trends, Alaska provides examples of how factors other than temperature are likely to affect the eventual outcome of climatic warming on forest expansion at high altitudes. Sampling of 1558 white spruce at 13 treeline sites in the Brooks and Alaska Ranges has found both positive and negative growth responses in the same broad geographical area to climate warming, an observation which challenges the widespread assumption that arctic treeline trees grow better with warming climates. High mean temperatures in July decreased the growth of 40% of white spruce at treeline areas in Alaska, whereas warm springs enhanced the growth of an additional 36% of the trees while 24% showed no significant correlation with climate (Wilmking *et al.*, 2004). Further investigations on the effect of drought stress on radial growth of white spruce in the western part of the study area showed that a high number of trees

Fig. 5.16   Rowan (*Sorbus aucuparia*) maintaining the tree form above a *krummholz* scrub (*Pinus mugo*) at an alpine treeline in Bavaria photographed in 1976. Note the Norway spruce (*Picea abies*) in the distance which have colonized the treeline scrub zone during an earlier period and are now showing a high rate of mortality.

responded to recent warming with an increase in growth; while in the eastern part, trees responded predominantly with decreases in growth. These patterns coincided with precipitation decreases from west to east and local water availability gradients, therefore pointing to drought stress as the controlling factor for the distribution of trees (Wilmking & Juday, 2005).

Trembling aspen (*Populus tremuloides*) is the most important deciduous tree in the North American boreal forest and this species is also showing signs of suffering from increasing drought. A study in western Canadian aspen forests showed that during 1951–2000 the aspen forests underwent several cycles of reduced growth, notably between 1976 and 1981, when mean stand basal area increment decreased by about 50%. Most of the growth variation was explained by interannual variation in the climate moisture index in combination with insect defoliation. The results of the analysis indicate that a major collapse during the years 2001–3 in aspen productivity is likely to have occurred during the severe drought that affected much of the region at that time (Hogg *et al.*, 2005).

Similar situations have been reported using sequential aerial photography in examining the distribution of the subalpine fir (*Abies lasiocarpa*) at the Glacier National Park (Montana, USA) where over a 46-year period no altitudinal changes in the location of alpine treeline ecotone were observed (Klasner & Fagre, 2002). There was nevertheless, over this same period an increase in the area covered by stunted subalpine fir (*krummholz, sensu lato*) and an increase in tree density. In a similar study, at and near the subarctic and arctic treeline in three regions in Alaska it was found, contrary to expectations, that after 1950 warmer temperatures were associated with decreased tree growth in all but the wettest region of the Alaska Range (for location see Fig. 5.17). Growth declines were most common in the warmer and drier sites, and thus give further support to the hypothesis that drought stress may accompany increased warming in the boreal forest (Lloyd & Fastie, 2002; Holtmeier & Broll, 2005).

In Alaska, a comparison of treeline movements based on palaeoecological studies has provided a regional analysis of patterns of recent treeline advance, and given an estimate of how much lag exists between recruitment onset and forest development beyond the Alaskan treeline (Lloyd, 2005). Treeline advance has been ubiquitous throughout the region, but asynchron-

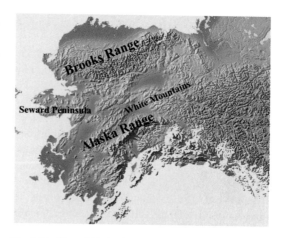

Fig. 5.17   Map of Alaska showing the relative positions of geographical features referred to in the text.

ous in time, occurring significantly earlier in the White Mountains in interior Alaska than in western Alaska or the Alaska Range. The mean lag between initiation of recruitment and forest development was estimated at approximately 200 years, similar to that predicted from contemporary modelling studies. It appears that although a continued advance of white spruce forests is the most likely outcome of future change, variability in the rate of forest response to warming will be modified by problems with seed dispersal and tree establishment which may be further hindered by the slow warming in highly permafrost-affected sites. There appears therefore to be considerable resistance to both expansion and retraction of boreal forest in response to climatic warming, both at present and in the past.

Treelines, it would appear both from this and other studies, represent ecological boundaries of great complexity that develop over long periods of plant–environment interaction. Consequently, it seems doubtful that forest migration will respond either directly or in direct proportion to any degree of short-term climatic change.

It has also been observed that there is substantial variability in response to climate variation according to the distance from the treeline. Inverse growth responses to temperature were more common at sites below the forest margin than at sites at the forest margin. Together, these results were taken to suggest that inverse responses to temperature are widespread, affecting even the coldest parts of the boreal forest. As growth declines, as is common in the warmer and drier sites, it would appear that increased warming if accompanied by increased drought stress will nullify any positive responses to temperature amelioration from climatic warming (Lloyd, 2005).

In northern Québec, a study of logs buried in peat underlying stunted clonal spruce enabled an assessment to be made of the stability of the forest limit during both warm and cold periods of the late Holocene. This study strongly suggests that the forest limit has remained stable during the last 2000–3000 years BP (Lavoie & Payette, 1996). The stability of the forest limit during warm periods (c. 2000 BP, early medieval times, and this century) and cold periods (c. 3000 and 1300 BP, and the Little Ice Age) of the late Holocene demonstrate that mechanisms allowing forest limit advance or retreat are not easily triggered by climatic change. A summary of expected changes to the boreal treeline in response to present warming trends has been given in the Arctic Climatic Assessment Report (ACIA, 2005). In North America where black and white spruce are the principal tree species at the tundra–taiga interface it was concluded that it is possible that the northern white spruce tree limit in Alaska and adjacent Canada would not advance readily (Juday, 2005). For black spruce, in Alaska at least, it would appear that warm years would result in strongly reduced growth and that it is unlikely that black spruce would survive in some types of sites.

5.4 FIRE, AND PALUDIFICATION AT THE TUNDRA–TAIGA INTERFACE

Many studies have questioned whether or not climate alone is the main controlling factor for the northern limit of forests and have suggested alternative scenarios. One notable argument is that the effect of natural fires may be particularly severe at the tundra–taiga interface (Fig. 5.18). At high latitudes, the interactions between vegetation and disturbances such as forest fire are particularly important, as the changes induced in the vegetation may produce significant alterations in local climatic conditions. Natural forest fires may be so severe as to prevent the vegetation from returning to its original forested state. Furthermore, the frequency of natural forest fires makes it essential that trees retain a capacity for subsequent regeneration (Laberge et al., 2000). In this connection it is relevant to note that in

Siberia along the tundra–taiga boundary there is an especially high frequency of thunderstorms, suggesting an increased fire probability due to the warmer conditions of the forest increasing evapotranspiration from the adjacent bog (Valentini et al., 2000). The removal of trees, whether by fire or any other agency (e.g. insect attack) will lead to rising water tables and hence will promote paludification and thus hinder regeneration. Fire can also aid bog formation as particles of ash and carbon deposited into the soil profile can reduce drainage and therefore help to initiate peat growth (Mallik et al., 1984).

5.4.1 Post-fire habitat degradation

Tree growth on peatland is more successful in regions with cold climates than in oceanic areas with warm winters. When the winter is cold the roots become fully dormant and do not suffer the adverse effects of prolonged exposure to anaerobic conditions as their metabolic processes are much reduced. Tree colonization of peatlands in cold climates is therefore widespread but frequently regeneration presents problems and deteriorating environmental conditions can cause a disappearance of trees from formerly forested bogs. In a study covering the Little Ice Age period in northern Québec (Arseneault & Payette, 1997) it was found that black spruce continued to colonize the peatlands as long as the adjacent well-drained sites were occupied by seed-producing forest.

The need for neighbouring forest is two fold. In addition to the need for a seed source, there is also a requirement for shelter to reduce the damaging effects of winter snow-drifting conditions. When neighbouring areas succumb to periodic fires then both the seed source and shelter are removed. Under these conditions the surviving trees suffer dieback of supranival stems. The trees then finally succumb to drowning in permafrost-induced ponds. The post-fire degradation and disappearance of the conifer stands from the peatlands represents the ultimate stage of a positive feedback process triggered by a modification of the snow regime at the landscape scale. The region is dominated by cold ocean currents, with the result that in summer frosts are frequent and cause a stunting of the growth of trees regenerating in open spaces, hence the frequent occurrence of krummholz (S. Payette, pers. com.). The same sequence of events will probably not

Fig. 5.18   Remains after forest fire in black spruce (*Picea mariana*) forest, Québec.

occur in non-oceanic areas with lower levels of precipitation. This scenario appears to explain the southern position of the treeline in Québec as well as the anomalous occurrence there of lichen–spruce forest at a latitude of only 47° N (Fig. 5.19).

In more continental subarctic Finland the after-effects of fire on local climate and its implications for forest regeneration were studied after a widespread forest fire in the Tuntsa area of Finnish Lapland in 1960 (Vajda & Venalainen, 2005). Here it was found that fire-induced deforestation increased the wind velocity by 60%, which changed the soil thermal

regime through a 20–30 cm reduction in snow cover, lowered the evapotranspiration and diminished the Bowen ratio to 0.4. In this case the absence of forest recovery after fire appears to have been due to the imposition of an increase in climatic adversity for the trees in an area which was already at the boundary for tree survival.

5.4.2   Treelines and paludification

Studies of treeline position and movement in the northern hemisphere are most commonly carried out

Fig. 5.19   View of an open frost-prone site at the southernmost occurrence of lichen spruce forest in Quebéc at 47° N. The arrow indicates a *krummholz* tree several decades old that has been regularly stunted in growth by frost damage, as shown in the progressively enlarged insets. Note also the high degree of lichen diversity.

in regions where an uninterrupted boundary between tundra and taiga can be found extending over many hundreds of kilometres. This natural bias towards areas with extensive treelines tends to emphasize continental areas and neglects the influence of oceanic conditions on the position of the treeline. Proximity to the ocean can have both positive and negative effects on the growth of trees. Freedom from drought and reduction in exposure to frost enables active forestry and agriculture to be carried out at many locations north of the Arctic Circle. However, there are also many aspects of the oceanic environment which have a negative influence on long-term tree survival and can depress the position of the treeline and impoverish soil conditions. Oceanic cooling of the growing season and its negative effect on long-term tree survival has been noted in a number of sites including the western shores of the Hudson Bay, Fennoscandinavia, and the Siberian lowlands (Crawford, 2000; Crawford *et al.*, 2003).

5.4.3   History of paludification

The end of the twentieth century witnessed a pronounced increase in the intensity of maritime conditions which varies with the periodic behaviour of the North Atlantic Oscillation (Fig. 5.20). This change in climate has also had a noticeable ecological impact at high latitudes, altering the seasonal spatial dynamics of plant growth and the seasonal spatial dynamics and social organization of musk oxen (Forchhammer *et al.*, 2005). Over the last three decades, the phase of the NAO (Fig. 5.21) has been shifting from mostly negative to mostly positive index values, and an increase in oceanicity is also evident (Fig. 5.20), as measured by a reduction in Conrad's Index of Continentality (the converse of oceanicity).

The recent trend to increased oceanicity in the Siberian Arctic can be viewed as a continuation of a long-term trend from continentality to oceanicity. Examination of peat stratigraphy and carbon dating of buried tree stumps (as mentioned above) has shown that there has been a marked southward movement of the Siberian treeline over the past 6000 years, probably as a consequence of the mid-Holocene sea level rise and the flooding of the Arctic Ocean onto the Siberian lowlands and the consequent imposition of cool, moist summers (Kremenetski *et al.*, 1998).

The early to mid Holocene expansion of arctic waters altered the delicate balance between temperature and air humidity so that in both arctic and subarctic regions bogs began to replace forest over a wide area in northern Québec (Payette, 1984), Scandinavia, Scotland (Tipping, 1994), and in the Siberian lowlands (Kremenetski *et al.*, 1998). A marked climatic deterioration, commonly termed the 8200 BP event (probably due to freshwater fluxes in the final deglaciation of the Laurentide ice sheet), appears to have been accompanied by a reduction in tree cover throughout Europe (Klitgaard-Kristensen *et al.*, 1998).

A trend to more oceanic conditions with cool moist summers appears to have continued in many areas probably as a result of these early Holocene rises in sea level. The Hudson Bay and Arctic Ocean would therefore have begun to exert a greater maritime influence on climate in Québec and northern Siberia respectively. Examination of the changing abundance of wetland forests located at the arctic treeline (northern Québec, Canada) during the last 1500 years (Payette & Delwaide, 2004), has shown a remarkable coincidence of events such as the mass mortality of wetland spruce and the post-fire failure of forest regeneration during the late 1500s, suggesting that this was connected with local flooding, probably attributable to greater snow transportation and accumulation after fire disturbance. This same study demonstrated a marked increase in paludification with the highest water levels recorded so far (nineteenth and twentieth centuries), causing lake and peatland expansion. The conclusion that any future moisture increases in these subarctic latitudes in oceanic eastern Canada will result in important spatial rearrangements of wetland ecosystems.

The late Holocene retreat of the Eurasian treeline coincides also with declining summer insolation (McFadden *et al.*, 2005). Although this decline in insolation would have been global, the effect appears to have been greater in arctic Siberia due to the proximity of the Arctic Ocean. The increase in the extent of arctic water as a result of sea level rise would have created at northern latitudes a more oceanic climate with cool moist summers, which would allow permafrost levels to rise and facilitate the development of bogs.

By *c.* 5000/4500 BP the northern limit of tree birch in Russia had retreated from its Hypsithermal maximum to more or less its present limit and this was mirrored in the southward retreat of other tree species

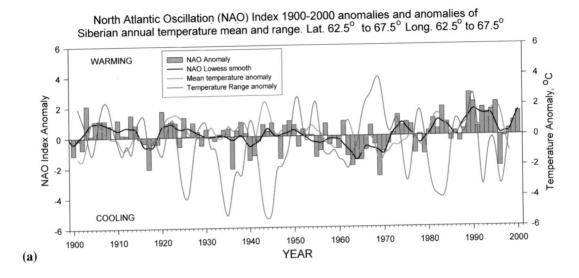

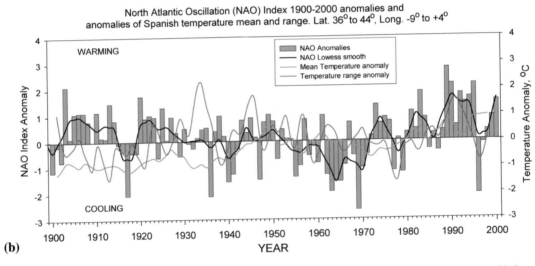

Fig. 5.20   North Atlantic Oscillation (NAO) anomalies for 1900 to 2000 (grey columns with superimposed lowest line in black: data from Hurrell & VanLoon, 1997) compared with anomalies of annual mean temperature °C (green) and annual temperature range °C (red) for (a) a restricted part of the West Siberian Lowlands, lat. 62.5° to 67.5° N, long. 62.5° to 67.5° E, showing a significant relationship between NAO and the annual mean temperature. The temperature range anomaly is typically of opposite sign to the mean anomaly (b) Spain, lat. 36° to 44° N showing weak relationships between NAO index and mean temperature but a stronger relationship between NAO and annual temperature range. (Reproduced with permission from Crawford et al., 2003.)

(Fig. 5.22). This treeline retreat was accompanied by development of dwarf shrub tundra and extensive bog growth. The considerable antiquity of the peat in the upper layers of bogs in the region of the Pur-Taz rivers of western Siberia (for location see Fig. 5.6) has been taken to suggest that much of the peat growth took place at an early date (Peteet et al., 1998). However, comparison of methane emissions from the Holocene Optimum (5500–6000 BP) with modern levels from forests of northern Eurasia, suggests that the area of

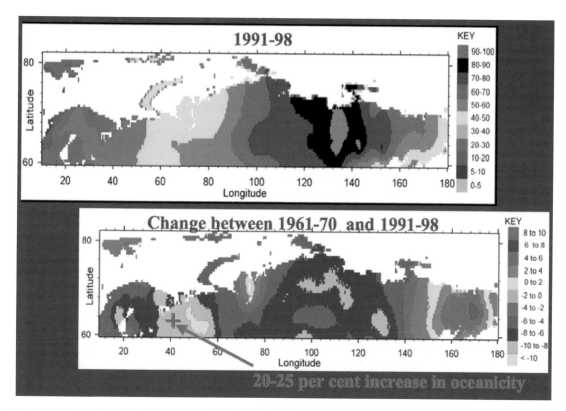

Fig. 5.21 Changes in continentality across arctic Siberia from the 1960s to the 1990s as calculated by Conrad's Index (see below). There is a notable increase in oceanicity in western Siberia which gradually changes to an increase in continentality in the far east. (Map prepared by Dr C. E. Jeffree using temperature data from the Climatic Research Unit 0.5° gridded 1901–1995 Global Climate Dataset). Conrad's Index of Continentality $K = (1.7A/\sin (\o +10))-14$ where K is the index of continentality; A the average annual temperature range; and ø is latitude. The scale provides values approximating to nearly zero for Thorshavn (Faroe: 62° 2′ N, 6° 4′ W) and almost 100 for Verkoyansk (Russia: 67° 33′ E, 133° 24′ E).

tundra and the proportion of wetlands within the boreal forest zone is probably greater now than at any time in the past (Velichko *et al.*, 1998). Examination of the recent growth of bogs suggests that bogs appear to engulf the forest both in the maritime regions of Canada and northern Siberia as well as in other locations, particularly near the present tundra–taiga interface. The situation is complex, as at this interface between forest and bog there is (as mentioned above) a cyclical alternation between tree cover and wetland vegetation due to the rise and fall of permafrost levels (Fig. 5.9). The predictions therefore that are made in relation to the movement of northern vegetation types in relation to climatic warming probably do not take fully into account the homeostatic properties of bogs in retarding the northward migration of the boreal forest.

5.4.4 Bog versus forest at the tundra–taiga interface

A lengthy debate has taken place, particularly among Russian ecologists, as to whether in certain locations climatic warming, instead of causing an advance of the treeline, will result in a continuation of the retreat that began after the Hypsithermal (Callaghan *et al.*, 2002). Those who believe that climatic warming causes an advance of the treeline northwards subscribe to the eventual overriding influence of climatic factors. There is, however, in the north Siberian lowlands a powerful argument that climatic warming can cause a retreat of the treeline, based on the belief that the presence or absence of trees is due to edaphic factors influenced by the proximity of the Arctic Ocean and the long-term

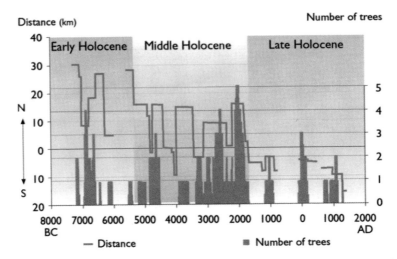

Fig. 5.22   A proposed division of the Holocene into three stages based on climatic data in relation to the dynamics of the polar treeline on the Yamal Peninsula in Siberia. The graph illustrates the relative distance of sampled tree remains from the present position of the most northerly open stands of larch in river valleys and the number of tree samples for each radiocarbon date. (Diagram based on data from Hantemirov & Shiatov, 1999, and reproduced with permission from Juday, 2005.)

Fig. 5.23   Tree survival along river courses at the tundra–taiga boundary in the western Siberian Lowlands. (Photo I. A. Brown, reproduced with permission from Callaghan *et al.*, 2002.)

persistence of permafrost. The open cold water in summer in the coastal areas of northern Siberia creates over the region a cold air with high humidity that dominates the present lowland Siberian tundra, with its excessively wet and treeless terrain supporting only an impoverished flora and fauna (Sher, 1996). In such an environment any advance of tree vegetation is limited to better-drained sites along river banks (Fig. 5.23).

A common presumption in assessing the direction of future change in high latitude biomes as a result of climatic warming is that forest is the natural climax vegetation and this will move north. However, in areas subject to maritime climatic influences, and these can extend inland considerable distances from the sea, there is always the possibility that the natural climax vegetation may be bog rather than forest (Klinger, 1996). In oceanic environments mineral soil impoverishment is frequently aggravated, as soil leaching accelerates nutrient removal and results in iron pan formation. Waterlogging then follows and impedes mineralization and nitrogen fixation, leading to the process of paludification and establishment of bogs.

Paludification, however, is not just limited to the coastal fringe areas of western Europe. Extensive regions of Newfoundland, north-west Europe and northern Siberia have become covered in extensive bogs during the past 6000 years (Crawford, 2000). Whether the tundra–taiga interface is moving north or south or remains stationary at present is uncertain and requires further investigation, particularly in areas where cool moist summers may accelerate the rate of peat formation as the growing season extends its length and facilitates bog growth.

Oceanicity is a complicating factor in many studies on climatic warming. In the past, some of the major influences of warm temperatures observed on plant distribution have been due to oceanic influences as manifested through increased precipitation and milder winters. Already, examples are available where milder conditions at high latitudes are producing more variable winters. One such consequence of climatic warming has been ice encasement. Instead of a continuous covering of snow, thaw periods with rain falling on frozen ground have resulted in vegetation ice encasement which can completely destroy large areas of improved pasture with disastrous results for agriculture in the north (Gudleifsson, 1997) as well as causing high mortality rates from winter starvation in reindeer. The natural grasslands at these latitudes do not appear to be damaged. It appears to be reseeding with grass-seed mixtures from more southern provenances that has made these once hardy northern pastures now sensitive to ice-encasement damage which is lethal and leaves a soil surface in spring that is entirely devoid of vegetation.

The relationship, or rather conflict between bogs and forests for dominance of the landscape, is complex. Trees survive on bogs in many northern areas and when they do so it is usually in regions where the soil is frozen to a considerable depth throughout the winter period and root metabolic activity will therefore be minimal during periods of prolonged oxygen deprivation. The black spruce (*Picea mariana*) and tamarack (*Larix laricina*) are examples of trees that are notable for their ability to grow on wet peatlands in the continental cold-winter regions of North America. When spring does arrive there is a rapid transition from frozen soils to a warm growing season. Consequently, during the summer growing season the active layer of the soil horizon will normally be aerated. However, when late spring flooding occurs on these forested peatlands it can be detrimental to growth for both *P. mariana* and *L. laricina* (Roy *et al.*, 1999; Girardin *et al.*, 2001).

The Siberian forests are particularly vulnerable due to the proximity of the Arctic Ocean coupled with the greater extent of the permafrost zone. Extensive boreal forest could exist under present temperature conditions within much of the southern zone of the subarctic tundra were it not for widespread bog development. In the Eurasian zone there is a region where the treeline is depressed southwards for over 200 km. In many of these locations colonization by shallow rooting native trees (*Larix sibirica*, *L. gmelinii*) would be expected on the basis of latitude, as these species are able to survive with a shallow active layer above the permafrost. Their absence may therefore be due to the counter-argument that climatic warming leads to a retreat of the forests due to bog and swamp formation.

In the West Siberian Lowlands over the past 6000 years there has been a 300–400 km retreat of forest, which in some areas used to extend almost to the shores of the Arctic Ocean (Kremenetski *et al.*, 1998). Figure 5.4 shows the changes in the northern limits of the boreal forest proper and the approximate modern limits for the sporadic occurrence of individual tree species north of the limit of closed forest. Coloured zones to the north of the current treelines indicate the areas where Holocene macrofossils have been located and carbon dated. A pronounced southerly depression of the northern limit of the boreal forest in the region of the West Siberian Lowlands is due to the presence of enormous areas of bog which now cover terrain that was once covered with forest (MacDonald *et al.*, 2000).

5.5 HOMEOSTASIS AND TREELINE STABILITY

Trees die slowly. The ability to postpone the moment of death enables many trees to survive long enough to

Fig. 5.24   Norway spruce (*Picea abies*) clonal group at an altitude of *c*. 450 m in Yllästunturi Fjell, Finnish Lapland (67° N).
All the lower lateral branches contacting the ground and embedded in the litter layer and moss cover have produced
adventitious roots. Seedlings on this exposed site are rare but occur at high density in the closed forest below.
(Photo Prof. F.-K. Holtmeier.)

outlive the duration of adverse climatic oscillations. Slow senescence is apparently a vital process for trees in marginal situations and a lag phase can therefore be expected before a period of climatic adversity removes a marginal tree population. As growth, development and senescence all take place slowly at high altitudes and latitudes it should be expected that tree stands in these regions should be of considerable age. At the treeline there can often be found individual trees that are thousands of years old. However, in the northern regions of the boreal forests ages of 200–500 years are more common. In both situations vegetative reproduction by layering can maintain populations for centuries without sexual reproduction (Fig. 5.24).

Equally evident is the lag phase between seedling advance and the establishment of a forest. In addition there are physical obstacles to the advance of the treeline in the form of wetlands and mountains. From time to time outliers can become established, in the form of isolated or small cluster of seedlings and saplings that manage to survive beyond the limits of their parent trees as a result of periods of favourable climatic conditions (Fig. 5.16). However, these advancing colonists may not always be able to establish the extent of tree cover that would be sufficient to qualify as an advance of the treeline. It is therefore not surprising that the re-establishment of tree cover with an intact forest canopy is a vegetation change that is not readily accomplished.

Once established, forests with a well-developed canopy layer create within themselves a climatically buffered environment not only in the understorey but also in relation to soil temperature and nutrient status. The eventual disappearance of trees will take place only when conditions eventually deteriorate to such an extent as to make accretion of biomass untenable. The location where this balance is achieved varies with climate, but catastrophic events and human interference make correlation with climate tenuous and future predictions difficult (Sveinbjörnsson, 2000). The establishment of trees can create both positive and negative feedbacks through the action of tree density on tree growth. Increased tree density provides shelter and leads to increased air temperatures and decreased wind damage. It also has negative aspects by lowering soil temperature, reducing nutrient availability, and increasing nutrient competition (Sveinbjornsson *et al.*, 2002). There is

even the extreme condition, already mentioned above, where forest advancing over descending permafrost eventually cools the soil so that the permafrost becomes re-established near the surface and kills the trees (Holtmeier, 2003).

5.6   BOREAL FOREST PRODUCTIVITY AT HIGH LATITUDES

The general question of whether or not trees are responding positively to climatic warming in boreal regions can elicit different answers depending exactly on whether the response is measured as a northward expansion of forest or an improvement in tree growth. In terms of forest expansion there is considerable inertia for movement in much of the tundra–taiga interface in response to climatic warming. The most noticeable advances appear to be a spread north in sheltered river valleys as seen in Alaska where the continued advance of white spruce (*Picea glauca*) forests is the most likely scenario of future change. There are, however, limiting factors due to restrictions on spruce establishment in highly permafrost-affected sites, together with problems in seed dispersal and early establishment which may cause spruce populations to exhibit non-linear responses to future warming. It follows therefore that an uncritical extrapolation from recent trends may be unwarranted (Lloyd, 2005).

In terms of growth rates there has been a noticeable improvement in boreal tree growth in recent years throughout the northern world, from Mongolia (Jacoby *et al.*, 1996) to Siberia (Briffa *et al.*, 1995) and from central Europe (Spiecker *et al.*, 1996) to North America (Lavoie & Payette, 1994). Although this improved growth is usually coincident with increased temperatures, there are also cases which are associated with a demonstrable influence of increased precipitation (Graumlich, 1987; Whitlock, 1993) as well as higher levels of atmospheric carbon dioxide. Growth trends of Scots pine (*Pinus sylvestris*) at its northernmost extent in the northwest Kola Peninsula may be an indicator of changes in the carbon cycle of terrestrial forest ecosystems. Using a method which removed age trends from the data, a time-series analysis of annual radial increments in wood over the last few decades compared with an earlier period of registered warming (maximum around 1920–40), revealed elevated growth, particularly for younger trees. In this northerly site the higher

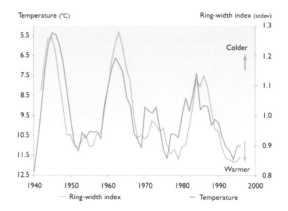

Fig. 5.25   Relationship of radial growth of black spruce at Caribou-Poker Creeks Research Watershed to the mean of April (growth year) and February (two years prior) temperature, smoothed with a 5-year running mean. N.B. The temperature axis is inverted. (Reproduced with permission from Juday, 2005.)

global level of atmospheric carbon dioxide is the most probable reason for the marked recent increase in radial increment growth of the younger populations of Scots pine (Alekseev & Soroka, 2002).

It is important to note that although seasonal changes in temperature can have a strong influence on tree growth they can, depending on species and site conditions, show very different responses. In a study of radial stem growth of Alaskan black spruce (*Picea mariana*) at four different sites in Alaska three showed a negative response to warming and one a positive response. At one of the sites (Caribou-Poker Creeks Research Watershed) where the growth response was negative, warmer temperatures could promote the onset of photosynthetic activity in early spring when the ground was still frozen, causing desiccation and damage to the needles early in the growing season (Fig. 5.25). By contrast at another site at Fort Wainwright the growth of the trees was positively correlated with winter temperatures (Fig. 5.26). It was also noted that this was one of the few species and site types in Alaska where empirically calibrated growth rates can be inferred to improve under projected higher temperatures (Juday, 2005).

A further complication in assessing whether or not the tundra–taiga interface is on the move or not arises from the changes in land use that have taken place in

marginal areas over recent decades. The long history of extensive grazing by domestic animals is now changing with the abandonment of many northern and upland pastures. In many places in northern Europe, including Scandinavia, trees can be seen spreading over former summer pastures.

A comparison of birch forests in different regions of Scandinavia (Dalen & Hofgaard, 2005) has shown that growth rates varied through time and between regions, with an apparent decrease in the north since the 1940s. Although Scandinavian treelines are expected to advance in response to climate warming, this was not evident as a general pattern for all regions. Seasonally different climate patterns, browsing, and abrasion are all involved. These regionally different patterns have to be taken into account in predictions of future responses to avoid overestimation of ecosystem responses to climatic change. Nevertheless, there still appears to be a noticeable limit to the growth and establishment of birch forests (*Betula pubescens*) at higher altitudes in Norway.

Examination of the birch treeline position showed that both the number of trees and their basal areas decreased continuously with increasing altitude from 300 m below the treeline (Hofgaard, 1997). The number of birch saplings also decreased from about 150 m below the treeline towards higher altitudes. Viable but browsed populations of birch were present along the whole length of all transects, irrespective of aspect and geological substrate, with saplings present up to summit positions at 420 m above the treeline. Due to browsing by sheep, mean height of saplings established above the treeline was only 0.2 m. In situations such as this it has to be concluded that future vegetation responses to diminished grazing pressure are likely to override responses forced by changing climate (Holtmeier, 2003).

5.6.1   Physiological limits for tree survival at the tundra–taiga interface

Research into the limitations to tree growth in recent years have centred on the acquisition and utilization of resources for growth. Consequently, attempts to make generalizations as to the limit of tree growth have been mainly related to aspects of the growing season as they affect the performance of the whole tree. The

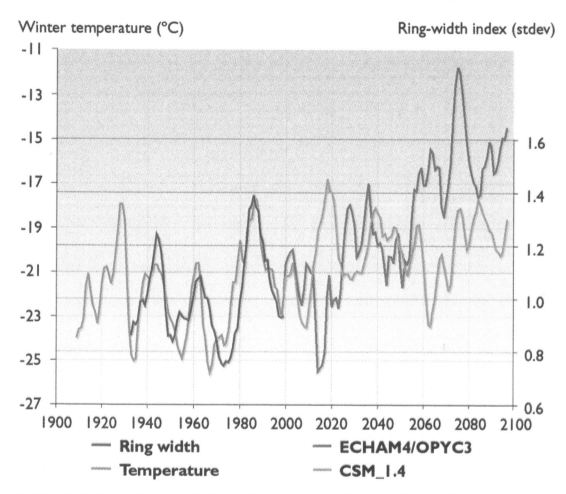

Fig. 5.26 Relationship of radial growth of black spruce (*Picea mariana*) at Fort Wainwright, Alaska (*n* = 20 trees), to a 4-month climate index (mean of monthly temperature at Fairbanks, Alaska, in January of growth year and January, February, and December of previous year). Scenario lines show projections of the 4-month climate index using two models (ECHAM4/OPYC3 projecting strong winter warming for the Fairbanks grid cell and CSM_1.4 a lesser degree of winter warming. (Data from Juday & Barbour, 2005, and reproduced with permission from Juday *et al.*, 2005).

hypotheses that have been put forward as to the major causes for the absence of trees at high latitudes and altitudes have therefore revolved around the question of whether limitations in available thermal time affect the ability of the trees to make a net gain in carbon, or whether it is the time that is necessary for growth and development that is crucial to survival in the marginal conditions of boreal and alpine treelines. Both these hypotheses seek to find generalizations that can be used to make global predictions as to whether or not any particular climatic regime will be either favourable or disfavourable for tree survival.

Despite the emphasis that carbon balance studies give to the holistic reaction of trees to their environment, there are other more specific aspects of tree physiology that are affected by climate and thus have consequences for survival and distribution limits. Certain specific tissues or parts of the tree can suffer physiological and metabolic damage by adverse climatic conditions. If the susceptible tissue is part of a vital component (organ) even just intermittent stress exposure can reduce the viability of the whole tree. Examples of this type of localized injury, with widespread consequences for long-term survival, are found in apical buds, stem cambium, and root meristems. All these tissues can be fatally damaged by unfavourable conditions such as drought, flooding, frost, and attacks from pathogenic organisms which can lead to the eventual demise of the entire tree.

5.6.2 Carbon balance

The concept of carbon balance has a logical appeal when predictions are required for forest productivity in relation to changing climatic conditions. Increasing temperatures and higher levels of atmospheric carbon dioxide are climatic variables that demand an assessment of their likely effects on growth, photosynthesis, respiration, litter production and quality, soil respiration and other aspects of carbon sequestration in forests. As conditions become more severe with increasing altitude or latitude it is also logical to enquire if there is a limit above which trees fail to make a net gain in carbon over the year. As already discussed (Section 1.3.2), numerous studies have attempted to ascertain whether or not trees are limited in their growth by a carbon deficiency, particularly at high altitudes. There are numerous difficulties in obtaining an integral estimate of total photosynthetic and respiratory activity throughout the year in the severe conditions that exist at either the tundra–taiga interface or the montane timberline. Nevertheless, some studies have succeeded in estimating the extent to which increased atmospheric carbon dioxide levels in the atmosphere will benefit trees at the upper limits of their distribution. The alpine treeline where these experiments were carried out is not the tundra–taiga interface. However, the eventual significance of increased levels of carbon dioxide for tree growth at the northern limits of the boreal forest, like that in the Alps, is likely to be influenced by the interplay between biotic and abiotic processes, and the species of tree.

Experimentally it has been possible to provide carbon dioxide enrichment and combine this with foliage removal to test the effect of altered source–sink relationships on tree growth and leaf level responses in mature larch (*Larix decidua*) and mountain pine (*Pinus*

uncinata) *in situ* near the treeline in the Swiss central Alps at 2180 m above sea level (Handa *et al.*, 2005). A three-year long study found that elevated carbon dioxide levels enhanced photosynthesis and increased non-structural carbohydrate concentrations in the needles of both species. While the deciduous larch trees showed longer needles and a stimulation of shoot growth over all three seasons when grown *in situ* under elevated carbon dioxide, pine trees showed no such responses. After three years, the results suggested that for the deciduous larch carbon is limiting at the treeline, while for the evergreen pine it is not a limiting factor. It might be expected therefore that depending on the functional types of the treeline species increasing carbon dioxide availability may improve the carbon balance in some but not all cases. In complex cases it is sometimes useful to use modelling to simulate the likely outcome of changes in environmental conditions on some feature which is difficult to quantify in the field. A comparison of dwarf trees (stunted trees that retain an arboreal growth form) and *krummholz* trees (assumed in this study to have a mat-like growth form) has shown that these differences in canopy types have significant effects on simulated carbon balance. Dwarf tree canopies appear in computer simulations to have higher carbon balances for all leaf area index values above 4, while carbon balances for both canopy types decrease with increases in LAI. The results demonstrate the importance of canopy type in relation to survival at the alpine treeline (Cairns, 2005). Unfortunately, models tend not to include other important factors such as the distribution of carbohydrates, which varies with age in trees, and estimations based on one particular age of tree or tree seedling may not be representative of the forest as a whole. A resolution of this problem is obtained when an actual assessment is made in the field where it can be seen that there is usually no problem for trees at the timberline to maintain a surplus of non-structural carbohydrate throughout the year. Such investigations reveal that although temperature is related to the position of the timberline it does not reflect a negative carbon balance but is due instead to a shortening of the thermal time available for developmental processes (Körner, 2003). An explanation of just what processes are involved and how they are limited by temperature nevertheless remains elusive even though it has long been observed that arborescent growth is restricted by direct climatically caused damage before reaching an altitudinal limit set by an insufficient carbon balance (Holtmeier, 2003).

5.6.3 Carbon balance versus tissue vulnerability at the treeline

While carbon balance has much practical significance for forestry and other productivity studies it is not necessarily a deciding factor in relation to the survival of natural plant populations. Plants being autotrophic have an inherent capacity for matching carbon utilization with availability. As the growing season shortens or temperatures fall, growth can be reduced and an overall carbon deficit may be avoided. Woody plants become progressively smaller at high latitudes. Trees are replaced by scrub, and scrub is then replaced by woody plants with prostrate stems. The most northerly willows in the world, the polar willows (*Salix polaris*, *S. reticulata*) of Spitsbergen, are sometimes described as the arctic forest. They give no indication of suffering a carbon deficit and produce copious quantities of seed even at their most northern locations.

What sometimes is unavoidable, however, is a carbon deficit in certain vulnerable tissues. Root meristems that are depleted of carbohydrate by anaerobic respiration when flooded will also likely be deficient in antioxidants, making them vulnerable to post-anoxic injury when water tables drop. Buds that are depleted of sugars by intermittent warm periods in winter become sensitive to late frosts (see Section 9.8.2). It follows therefore that under natural conditions it is likely that the sensitivity of certain vulnerable tissues will determine the long-term viability rather than the carbon balance of the entire tree. However, once the essential tissues are damaged a reduction in the carbon balance of the whole tree will follow.

5.6.4 Winter desiccation injury

The length of the growing season is undoubtedly critical for the survival of trees and early research first carried out in Germany (Michaelis, 1934) considered that the length of the growing season and the development of the cuticle and scale leaves on overwintering buds was essential for surviving the dangers of winter frost desiccation (Tranquillini, 1979). Observations at the treeline in oceanic Scotland have not found evidence of excessive dehydration damage due to inadequate cuticle development (Grace *et al.*, 2002). However, in more continental climates, significant water losses from trees that do not survive well at the timberline are frequently reported. Winter desiccation-induced foliage loss incurred by *krummholz* growth forms of subalpine fir (*Abies lasiocarpa*) at treeline locations in National Glacier Park, Montana, USA, for the winter of 1998/1999 affected an average 8.68% of the *krummholz* canopy (Cairns, 2001. Winter desiccation, however, was not found to be related to any single environmental variable. Nevertheless, when outliers were removed, winter desiccation showed a strong correlation with elevation ($r = 0.97$) and was highly predictable in relation to various habitat characteristics, e.g. elevation, slope and aspect. In general, injury increased with elevation and on more south-westerly facing hill slopes, while sheltered locations showed decreased winter desiccation. Within patches, most winter desiccation was found at the windward edge of patches. This trend was attributed to the presence of leading shoots above the mean canopy surface of the *krummholz* patch. In these more extreme environments high winds and ice particles can cause severe abrasion of cuticular surfaces (Hadley & Smith, 1989).

A further risk to the maintenance of the water supply to tree stems at montane treelines comes from the incidence of embolisms in the water conducting tissues. A study of the two dominant species of the European central Alps timberline, namely Norway spruce (*Picea abies*), and stone pine (*Pinus cembra*), compared the seasonal courses of embolism and water potential at 1700 and 2100 m during two winter seasons. Stone pine is the hardier tree and usually reaches higher altitudes than Norway spruce. Embolism was observed only at the timberline and only in Norway spruce. Both species showed a significant drop in hydraulic conductivity, but in stone pine critical levels in water potential were avoided. It would appear that water losses in Norway spruce make winter embolisms a relevant factor in defining the treeline position for this species (Mayr *et al.*, 2003).

5.6.5 Overwintering photosynthetic activity

Studies on *Picea abies* and *Pinus cembra* near the alpine timberline have been able to explain the depression in photosynthetic activity that occurs during winter as

(a)

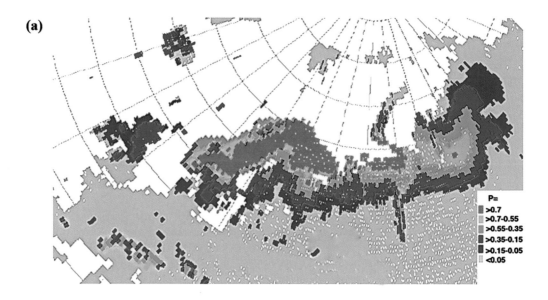

(b)

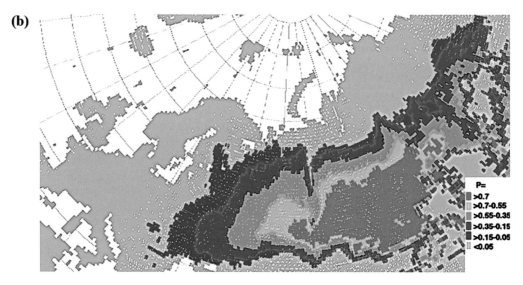

Fig. 5.27   Comparison of possible contrasting changes in distribution of *Pinus sylvestris* with varying climate conditions as compared with the temperature regime from 1961–1990. (a) Winter 4 °C colder, summer 4 °C warmer. (b) Winter 4 °C warmer, summer 4 °C colder. Note the retreat from western Europe with the imposition of warmer winters. The colour key relates to the probability of occurrence of pine. (For details see Crawford & Jeffree, 2007.)

the result of oxidative stress due to the coincidental effects of high light and cold temperatures. Potential efficiency of photosystem II (F-V/F-M), light and CO_2-saturated rates of photosynthetic oxygen evolu- tion (P-max) and contents of the antioxidants ascor- bate and glutathione in the reduced and oxidized state were measured in sun-exposed and shaded needles at the timberline. It was found that needles with the

Fig. 5.28   The Siberian alder (*Alnus fruticosa*) on Paramushir Island, northern Kurils. In this oceanic environment the alder and willow scrub is found where dwarf Siberian pine (*Pinus pumila*) would be found in more continental regions. (Photo by courtesy of Dr P. Krestov.)

lowest photosynthetic activity contained high levels of antioxidants in the reduced state. It was therefore concluded that the reduction of F-V/F-M during winter is not the result of oxidative stress, but is instead an injury-preventive down-regulation of PS II (Stecher *et al.*, 1999).

5.7   FUTURE TRENDS AT THE TUNDRA–TAIGA INTERFACE

Climatic warming in northern latitudes will undoubtedly create new ecological opportunities for vegetation advance as ice sheets retreat and permanent snow cover is reduced and the climate extremes are reduced with a

great degree of oceanicity (Fig. 5.27). For some species, however, particularly those that show a preference for continental climates there may be disadvantages from warmer winters, which are a marked feature of the current warming trend. Examination of potential changes in distribution for *Pinus sylvestris* (Fig. 5.27) has shown that winter warming is likely to cause a retreat from western oceanic areas. Similar trends have been found in woody scrub species, for example *Cassiope hypnoides*, *Vaccinium myrtillus*, *Calluna vulgaris* and *Salix polaris* (Crawford & Jeffree, 2007). Such a trend mirrors the existing tendency for certain species to retreat from maritime habitats in response to increases in oceanicity. Other explanations are also possible. Patho–

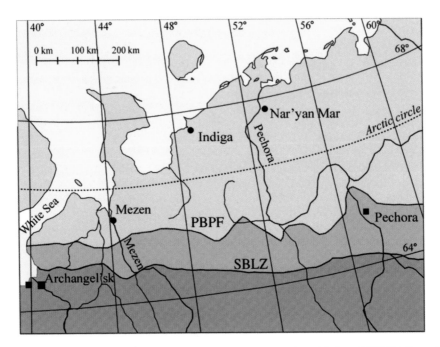

Fig. 5.29   Displacement of the southern border of the Russian lesotundra in the region of Archangel. PBPF indicates the Protection Belt of Pre-tundra Forest that was designated in 1959. SBLZ indicates the Southern Border of the lesotundra in 2002. (Reproduced with permission from Vlassova, 2002.)

genic organisms may also play a role. It is frequently observed that Scots pine can be damaged by snow blight, e.g. *Phacidium infestans*, due to long-lasting cover of wet snow with temperatures close to zero in late winter as has been observed in the Alps (Senn, 1999) and also Lapland (Holtmeier, 2003). How different winter and summer warming scenarios will affect the various woody plants that inhabit the tundra–taiga interface is far from certain. The ability of species such as *Pinus pumila* to reach a latitude of 71° N (Fig. 5.28) in one of the coldest and most continental regions of Eurasia shows that they are highly adapted to cold winters and short growing seasons which would suggest that warmer winters may eventually prove less favourable to this species. In the more oceanic areas *Alnus fruticosa* replaces *Pinus pumila* as the dominant woody species (Fig. 5.28). Woody plants have come and gone from the Arctic during the Holocene as can be seen in examinations of the pollen record in the Taymyr Peninsula. A shrubby tundra with dwarf birch species and willow scrub together with alder (*Alnus fruticosa*) persisted from the Late Pleistocene/Holocene transition, *c.* 10 300–10 000 BP, and then disappeared by

3500–3000 BP (Andreev *et al.*, 2003). It is probable therefore that these shrubby communities which are already adapted to more oceanic conditions will advance northwards more readily if winters are warmer than the more continental forest species (Fig. 5.27). In North America, the black spruce (*Picea mariana*), with its outstanding morphological plasticity, appears to be the tree species that is best adapted to expand its range and replace the tundra. Whether or not the forest tundra (Russian *lesotundra*) will also develop into a forest vegetation at high latitudes will depend on the degree of disturbance that will be inflicted on these regions in the years ahead. In many regions the tundra–taiga zone is not only a natural ecotone, but also a unique fringe zone with particular socioeconomic properties. In Russia, there is at present much evidence of deforestation and ecosystem degradation in different regions of the forest–tundra zone. Industrial activity has accelerated and despite a protection declared in 1959 for the lesotundra, there are now regions where the northern limit of the zone has been displaced southwards by 40–100 km (Fig. 5.29). In regions such as this it cannot be predicted with any certainty that climatic warming will lead to a northward movement of the boreal forest treeline (Vlassova, 2002). The manner of destruction with large-scale pollution is not only destroying the vegetation but is also causing a drastic reduction in the population of the indigenous peoples as well as reducing the life expectancy of the immigrant industrial population. The very considerable natural resources of the Arctic worldwide for timber, pulp and paper industries, as well as mining, oil and gas extraction, will have to be managed with great care if the environmental health of the region as a whole is to be preserved.

Fig. 6.1   Habitat heterogeneity in the Arctic at Hornsund,
Spitsbergen (77° N). The close juxtaposition of contrasting sites
(early and late, wet and dry), creates habitats with plants that will
differ in their reactions to climatic warming. (Little auk, *Alle alle*,
on boulder in foreground.)

6 · Plant survival in a warmer Arctic

6.1 DEFINING THE ARCTIC

The Arctic is a large and heterogeneous area (Figs. 6.1–6.2) and it cannot be expected that all regions will respond equally to global warming. It is therefore necessary to define the Arctic and its major ecological regions. These divisions can be geographical, climatic, or ecological. Climatically, the Arctic can be considered as that region of the Earth's surface that is underlain by permafrost, or permanently frozen ground. This is not entirely satisfactory as permafrost underlies approximately 20% of the Earth's terrestrial surface and occurs not only at high latitudes but also in some non-arctic locations at high elevations (Brown *et al.*, 1995). Permafrost is, however, one of the main restricting factors limiting the northward expansion of trees; the absence of trees is the key ecological feature that distinguishes the tundra from the taiga or boreal forest. Geographically, the Arctic can be simply defined as the portion of the Earth's surface that lies north of the Arctic Circle (66° 33′ N), but this contains a great diversity of habitats including portions of the boreal forest (taiga). In Chapter 5 the tundra–taiga interface was considered as the boundary

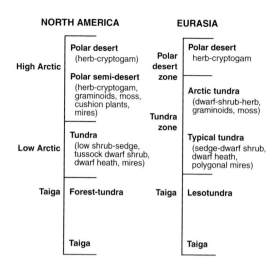

Fig. 6.3   Classification schemes for arctic vegetation in North America and Eurasia. (After Bliss & Matveyeva, 1992; see also Calow, 1998.)

zone between the arctic tundra and the boreal forest and therefore an ecological approach will also be used here and the southern extension of the Arctic will be defined by the northern limit to tree growth. Approximately 5.5% of the Earth's land surface comes under this heading. Many different types of plant communities exist within this zone and each has a northern limit. It follows therefore that there are many marginal areas within the Arctic.

A first approximation in distinguishing the various limits to plant distribution in the Arctic is the recognition of two major zones, namely the Low Arctic and the High Arctic. The former is dominated by woody shrubs (e.g. *Alnus*, *Betula*, *Salix*). The latter, the High Arctic, describes the region where much of the land is covered by permanent snow and ice and the vegetation is limited to a thin discontinuous cover of diminutive flowering plants, mosses and lichens. Even within the Low and High Arctic there are many recognizable margins which differ between North America and Eurasia and ecologists in these different regions have adopted terminologies that reflect these differences (Fig. 6.3).

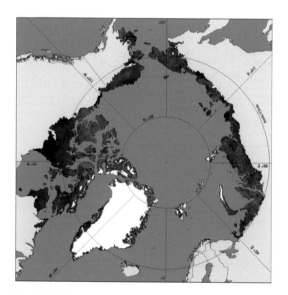

Fig. 6.2   False-colour satellite infrared image of the circumpolar Arctic. Red areas represent greater amounts of green vegetation; light and dark green represent sparse vegetation; black areas represent fresh water, and white areas represent ice. (Walker *et al.*, 2002, with permission of Taylor and Francis Ltd.)

6.2   SIGNS OF CHANGE

Global warming, already a noticeable phenomenon at high latitudes (Figs. 6.4–6.5), raises a number of

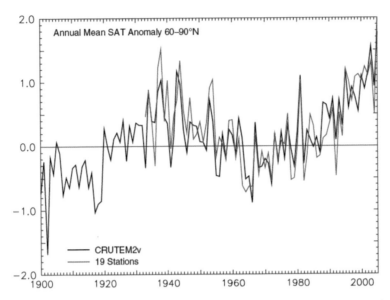

Fig. 6.4   Arctic-wide and annual averaged surface air temperature anomalies (60°–90° N) over land for the twentieth century based on the CRUTEM2v monthly data set. (Reproduced with permission from Richter-Menge *et al.*, 2006.)

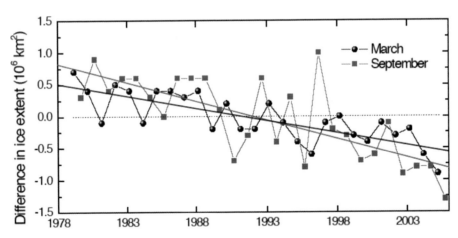

Fig. 6.5   Time series of the difference in ice extent in March (maximum) and September (minimum) from the mean values for 1979–2005. Based on a least-squares linear regression, the rates of decrease in March and September were 2% per decade and 7% per decade, respectively. Recent data from March 2006 are also shown and represent a new record minimum for the period of observation. (Reproduced with permission from Richter-Menge *et al.*, 2006.)

questions concerning the long-term survival of arctic vegetation. Will climatic warming lead to a new flourishing of arctic plants, or will the unaccustomed heat cause extensive destruction and loss of biodiversity in an ancient cold-adapted flora?

Much of the High Arctic, particularly in Eurasia, can be described as coastal (Fig. 6.2) and consequently the timing of the summer retreat of the sea ice has a strong influence on plant development. Recent years have seen a distinct trend for an earlier retreat of sea ice (Fig. 6.5) and one particular response noted at Hornsund in Spitsbergen has been remarkable coastal floral displays by the early flowering purple saxifrage (*Saxifraga oppositifolia*; Fig. 6.6).

Fig. 6.6   A vegetation response to climatic warming at high latitudes. Vigorous early summer flowering display of coastal populations of the purple saxifrage (*Saxifraga oppositifolia*) in recent years at Hornsund, Spitsbergen (77° N) appears to be related to an early retreat of sea ice.

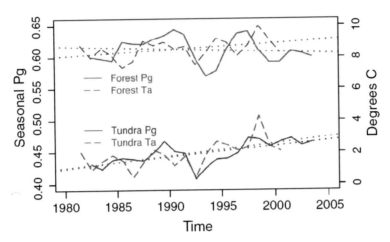

Fig. 6.7　Trends of gross primary production (Pg) as interpreted from NDVI satellite measurements (see Section 5.2.1) in the tundra regions versus boreal forest areas. The dotted lines represent linear trends of the plotted parameters. Both areas have shown a general increase in air temperature (Ta), but with a drop in temperature following the Pinatubo eruption in 1992. Although both areas have warmed since the eruption and Pg initially recovered in both areas, Pg has tended to increase in the tundra since 1997, whereas Pg declined in the forested areas. (Reproduced with permission from Goetz *et al.*, 2005.)

A time series analyses of a 22-year record of satellite observations in which the Arctic is defined somewhat widely in terms of latitude (60–90° N) and not by vegetation (as in this discussion) has shown that only about 15% of the extended region displays significant positive trends, of which just over half involved temperature-related increases in growing season length and photosynthetic intensity, mostly in the tundra (Fig. 6.7). Trees growing north of 60° N in this study (Goetz *et al.*, 2005) and described as arctic forest, in areas not affected by fire were found to have suffered a decline in photosynthetic activity possibly due to drought as there was no noticeable change in growing season length.

In relation to the true tundra (the treeless Arctic) these satellite observations confirm some of the effects already noticed by arctic dwellers particularly in the Low Arctic. Willows and alders have been noted as growing taller, with thicker stem diameters and producing more branches, particularly along shorelines. Indigenous human communities have also reported increases in vegetation, particularly grasses and shrubs – stating that there is grass growing in places where there used to be only gravel. Further north, on Banks Island, in the western Canadian Arctic, it has been observed that the musk oxen are staying in one place for longer periods of time, which is taken as additional evidence that vegetation is richer. Arctic sorrel (*Oxyria digyna*) is described

as coming out earlier in the spring, with noticeably bigger, fresher, and greener leaves (Callaghan, *et al.* 2005). However, negative effects are also reported. In northern Finland, marshy areas are said to be drying out. Sami reindeer herders in Utsjoki (northern Finland 70° N) have observed that berries such as the bog whortleberry (*Vaccinium uliginosum*) have almost disappeared in some areas. Other berry-bearing species such as cloudberry (*Rubus chamaemorus*) and cowberry or lingonberry (*Vaccinium vitis-idaea*) have been noted to be suffering adverse effects as a result of high temperatures early in the year followed by inadequate moisture. Declining cloudberry production has been noted over the last 30 years (Callaghan *et al*, 2005).

In arctic Alaska, air temperatures have warmed 0.5 °C per decade for the past 30 years and over this same period shrub abundance has increased. It has been suggested (Sturm *et al.*, 2005) that winter biological processes are contributing to this conversion through a positive feedback that involves the snow-holding capacity of shrubs, and the insulating properties of snow. It is suggested that hardy microbes profit from higher winter soil temperatures, due to the snow trapping of the shrubs providing great soil insulation. The resulting increase in microbial activity increases plant-available nitrogen which then stimulates more shrub growth and yet more snow trapping.

Physically adverse effects can arise from high temperatures which bring about the melting of surface snow and ice thus exposing unstable slopes and screes. Climatically, the rain belts in Russia are moving north increasing the flow of Asian rivers into the Russian Arctic Ocean (Richter-Menge *et al.*, 2006); this together with melting of the permafrost is in danger of destroying large areas of tundra. With the melting of the permafrost layer, the risks of flooding and erosion from increased river flow in spring will cause large-scale erosion in the Russian Arctic, and in areas where there has been nuclear weapon experimentation and related activities in Soviet times there is a real danger of releasing of radioactive isotopes into the Arctic Ocean (for review see Crawford, 1997b).

Fig. 6.8   Juxtaposition of contrasting habitats on the Brøgger Peninsula, Spitsbergen (79° N). Note the dry beach ridges and gravel screes that alternate with the dark-coloured mires and greener patches of bog growing in the effluent from the Mørebreen bird cliffs.

The question therefore remains: is the Arctic a fragile ecosystem, which will suffer major perturbation should climatic warming continue or has it a hidden biological resilience to change and disturbance that is not yet fully understood? The lack of uniformity of response is not surprising. The tundra contains a multitude of habitats both at large- and small-scale levels. In the latter, different microhabitats, which are likely to vary in their response to climatic warming, are often found in close juxtaposition to each other. Dry gravel ridges may suffer from drought while only a few metres away cold wet shores that harbour a large part of the arctic flora may in fact be growing better and flowering more profusely as a result of the earlier retreat of the sea ice (Figs. 6.6, 6.8).

6.3 THE ARCTIC AS A MARGINAL AREA

The Arctic is possibly unique in being the northern hemisphere's ultimate peripheral habitat; for it is here that continental convergence brings together the northern limits of the flora of three continents.

The observation that in many polar areas some populations of plants are not fully self-sustaining has prompted the assertion that the Arctic as a whole should be considered as a marginal area (Svoboda & Henry, 1987). Such a concept implies that polar deserts and semi-deserts are marginal for the establishment of vascular plants (*sensu strictu*) as frequently they do not produce viable seed and therefore depend on propagules dispersed from more favourable habitats. It is also argued that these areas are marginal as the vegetation succession rarely progresses beyond the initial invasion stage of succession. The second phase rarely occurs, when it might be expected that under more favourable conditions stand formation and a build-up of a critical standing crop would lead to habitat improvement and finally to the replacement of the pioneer species.

Coastal sites being close to the sea frequently have their growing seasons limited by the proximity and duration of the sea ice. In some cases the growing season here can be so short that even though the plants may flower, there is not sufficient time for seed production and such sites are dependent on adjacent areas for an input of seeds. This is the case with the colonization of the lower shore by *Saxifraga oppositifolia* in western Spitsbergen, as the creeping ecotype of this species which survives on the cold wet shore has in the past rarely produced seed and regeneration has largely been dependent on seed produced by neighbouring populations on the beach ridge (see Fig. 2.24).

It is sometimes taken for granted that areas of extremely sparse vegetation which are often observed in polar deserts are due to the prevailing unfavourable climatic conditions inhibiting plant establishment, growth and survival. In a study of such areas in arctic Canada it has been claimed (Lévesque & Svoboda, 1999) that the antagonistic factor is in fact the result of historical episodic adverse climatic anomalies. Such a case was considered to have taken place in Alexandra Fiord Lowland on Ellesmere Island (see map Fig. 6.20) in the Canadian High Arctic as a result of the recent Little Ice Age cooling which caused a dieback and even large-scale extinction of High Arctic plant communities that had taken centuries to develop. The Little Ice Age brought about new glacial advances, expansion of permanent snow cover and ice crusts over entire landscapes. It was argued that the newly formed ice (and snow) killed the underlying vegetation, thus creating what is referred to in the geological literature as 'lichen-kill zones'. In these zones the current plant diversity and abundance are exceedingly low and the plants are all relatively young and even-aged, as they are all part of a recolonization process that has been occurring only over the past 100–150 years. This vegetation, it is argued, has not yet reached equilibrium with the present prevailing climate and is still in an initial stage of succession (Lévesque & Svoboda, 1999).

6.3.1 Mapping arctic margins

There are many boundary zones within the Arctic. Low and High Arctic and various subdivisions have already been mentioned (Fig. 6.3). There are many other lines that can be drawn in relation to the zonation of arctic plant communities. Despite the general recognition of these zones (Fig. 6.3), vegetation maps of the Arctic have proved very difficult to standardize on a circumpolar basis. The circumpolar map of Walker *et al.* (2005) in an attempt to reconcile North American, Scandinavian and Russian concepts of boundary zones includes 23 different floristic zones. The simplest circumpolar classification that was produced consisted of identifying bioclimatic types as defined by Elvebakk (1999), rather than mapping actual vegetation units. This simplified

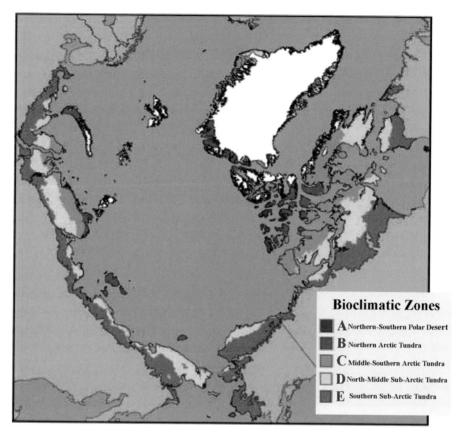

Bioclimatic Zones

◼ **A** Northern-Southern Polar Desert
◼ **B** Northern Arctic Tundra
◻ **C** Middle-Southern Arctic Tundra
◻ **D** North-Middle Sub-Arctic Tundra
◼ **E** Southern Sub-Arctic Tundra

Fig. 6.9   A bioclimatic zonation map of the Arctic. (After Elvebakk, 1999.) For climate and vegetation details of zones see Table 6.1. (Reproduced with permission from Walker *et al.*, 2005.)

approach reduces the number of circumpolar zones from 23 to five (Fig. 6.9 and Table 6.1).

The possibility of identifying at least 23 vegetation types highlights the extensive variation that is found between different regions of the Arctic. The process is made even more difficult because the actual geographical limits to the different zones are not abrupt boundaries but are instead a series of merging zonations (*limes divergens* or *ecoclines*; see Section 1.1).

Exceptions to the pattern of gradual change can be found in areas where there is an abrupt alteration in edaphic factors due to differences in substrate pH. Such a case is found at a sharp pH boundary along the northern front of the Arctic Foothills in Alaska which separates the non-acidic (pH>6.5) ecosystems to the north and the predominantly acidic (pH 5.5) ecosystems to the south. The edaphic boundary also marks abrupt changes to ecosystem processes. Comparison of

two sites on either side of the boundary but only 7 km apart (sites 3 and 4, Fig. 6.10b) showed that the moist acidic vegetation had twice the gross photosynthetic activity and three times the respiration rate of the non-acidic site just to the north. Moist non-acidic tundra has greater heat flux, deeper summer thaw (active layer), is also less of a carbon sink, and is a smaller source of methane than the more southern moist acidic tundra (Walker *et al.*, 1998).

6.4   PLEISTOCENE HISTORY OF THE ARCTIC FLORA

6.4.1   Reassessment of ice cover in polar regions

Past climatic histories mark out the Arctic as an area that has probably endured an unending series of

Table 6.1. *Climate and vegetation characteristics of bioclimatic arctic vegetation zones*

Subzone	Mean July temp (°C)	Summer warmth index[a]	Vertical structure of plant cover
A	1–3	<6	Mostly barren. In favourable microsites 1 layer of lichen or moss < 2 cm tall, very scattered vascular plants hardly exceeding the moss layer
B	4–5	6–9	2 layers, moss layer 1–3 cm thick and herbaceous layer 5–10 cm tall, prostrate dwarf shrubs < 5 cm tall
C	6–7	9–12	2 layers, moss layer 3–5 cm thick and herbaceous layer 5–10 cm tall, hemi-prostrate dwarf shrubs < 5 cm tall
D	8–9	12–20	2 layers, moss layer 5–10 cm thick and herbaceous and dwarf shrub layer 10–40 cm tall
E	10–12	20–35	2–3 layers, moss layer 5–10 cm thick, herbaceous /dwarf shrub layer 20–50 cm tall, sometimes with low shrub layer to 80 cm.

[a] Sum of monthly mean temperatures above 0 °C.
Source: Based on Elvebakk (1999).

extinctions, migrations and invasions. However, even at the height of the Pleistocene glaciations the Arctic was never entirely covered in ice. The extent of the Pleistocene glaciations in subarctic regions in western Europe (Fig. 6.11) can lead to an erroneous assumption that all the land north of the Arctic Circle must have also been subjected to a prolonged and deep ice cover, with the inevitable consequence that terrestrial plant life would have been extinguished – the *tabula rasa* effect. It would follow from this assumption that the entire present-day arctic flora must be the result of recent immigration from more southerly latitudes, as has more probably been the case for many insect species (Buckland & Dugmore, 1991).

The evidence that has been used in earlier discussions of whether hardy refugial populations of arctic plants survived in ice-free areas and nunataks during the Last Glacial Maximum has rested mainly on indirect arguments based on phytogeographical evidence. On this question of arctic refugia it is necessary to distinguish carefully between inland nunataks on high mountain chains as in Scandinavia, where serious doubts have been raised as to their role as refugia (Nordal, 1987), and coastal low altitude nunataks at the edge of ice sheets. In coastal areas, mountains to seaward of stable ice sheets need be of only moderate

height to rise above terminal ice cliffs. There are many such areas where these terminal ice cliffs, or semi-nunataks (Figs. 6.12–6.13) as they are sometimes called, could have provided Pleistocene refugia in North America as well as in south Iceland and the islands of the Arctic Ocean.

Earlier estimates of the past extent of polar ice were mostly based on oceanic sedimentation studies (Mangerud *et al.*, 1998) which suggested that from Spitsbergen to northern Norway there was extensive sea ice cover adjacent to the coast. However, research using rock-exposure dating techniques based on cosmic-ray-produced isotopes has now provided a reassessment of the extent of Pleistocene ice cover which has downgraded earlier estimations of both the depth of the ice and its seaward extension. Consequently, there are various sites where the lower ice levels indicate the existence of nunataks in western Scotland, Norway (Ballantyne *et al.*, 1998) and also in Spitsbergen (Landvik *et al.*, 2003).

In northern Norway the island of Andøya has long been the subject of geological and botanical investigations as its situation close to the continental shelf has prompted the suggestion that it supported a possible Late Weichselian unglaciated enclave. Early deglaciation of the island has now been clearly recognized, as

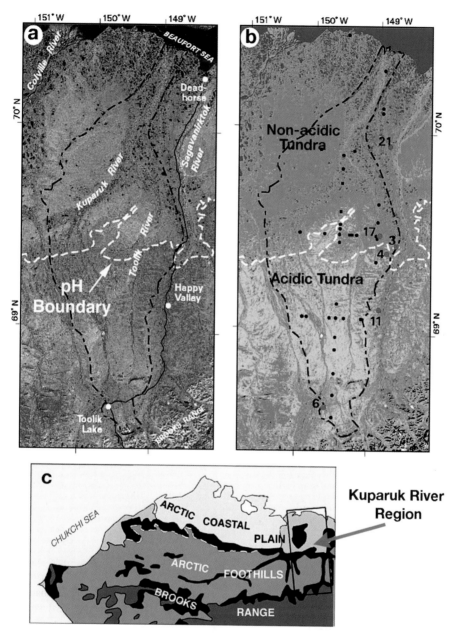

Fig. 6.10   Maps illustrating a sharp pH boundary along the northern front of the arctic foothills in Alaska which separates the non-acidic (pH>6.5) ecosystems to the north from the predominantly acidic (pH 5.5) ecosystems to the south. (a) Landsat false-colour infrared mosaic of the Kuparuk River Basin, northern Alaska. The pH boundary (dashed line) separates the redder tones to the south of the acidic soils from the greyer tones to the north of the non-acidic soils. (b) Generalized distribution of acidic and non-acidic vegetation types in northern Alaska. (c) The location of the pH boundary west of the Colville River (white dashed line) is less distinct and has therefore attracted less investigation. (Reproduced with permission from Walker et al., 1998.)

has also the existence on Andøya of ice-free areas with possible nunataks on the nearby islands of Senja and Grytøya. Geomorphological mapping suggests that

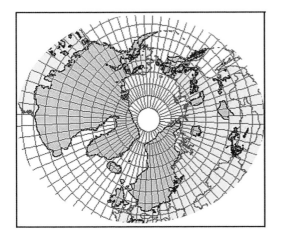

Fig. 6.11   Map for the northern hemisphere at the Last Glacial Maximum (Weichselian, Valdaian, Würmian, Devensian, Wisconsinan, MIS 2) compiled by Jürgen Ehlers from data and maps assembled and published in *Quaternary Glaciations* (Ehlers & Gibbard, 2004).

the upper surface of the Late Weichselian ice sheet at the Last Glacial Maximum rose eastwards from sea level on north-west Andøya to *c.* 1200 m in the longitude of Narvik (17° E), and that numerous mountain summits remained above the ice as nunataks (Vorren *et al.*, 1988).

Spitsbergen provides another well-studied example of ice-free terrain at high latitudes (Fig. 6.12). Using the isotope dating technique of the exposure of rock surfaces to cosmic radiation, it has been shown that the upper levels of late Weichselian ice sheet on the north-west islands of Amsterdamøya and Danskøya (79° 45′ N) have had ice-free nunataks since the Last Glacial Maximum. More extensive ice-free areas have also been postulated along the west coast of Spitsbergen during the last glaciation by studies of radiocarbon dates in mollusc shells and whalebone (Landvik *et al.*, 2003). A typical west coast Spitsbergen nunatak is shown in Fig. 6.13.

The present-day arctic flora is thought to have originated during the late Tertiary period approximately 3 million years ago, partially from some species of the previous arctic forest biome, but also from mountain species in subarctic regions which then migrated north as

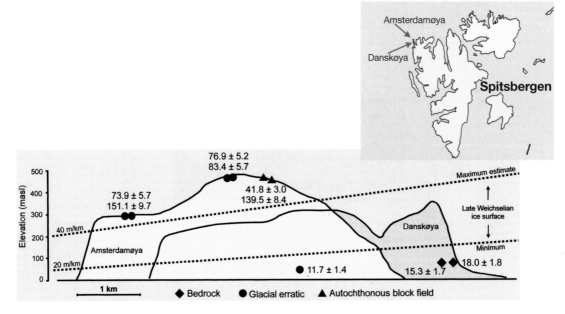

Fig. 6.12   Topographic profiles of Amsterdamøya and Danskøya (see location in inset) with dates for the exposure ages of sampled rocks in ka. The dotted lines are estimates of the minimum and maximum elevations of the Weichselian ice sheet. (Reproduced with permission from Landvik *et al.*, 2003.)

Fig. 6.13   A probable former semi-nunatak (red arrow) on the Casimir Périerkammen in Krossfjorden (north-west Spitsbergen). This is one of a number of regions in north-west Svalbard considered by a number of authors to have been free of ice at various times during the Weichselian glaciation. Such areas could have been glacial refugia sites (see Ingólfson & Forman, 1997). Note also the remains in the foreground of a former Russian (Pomor) hunting camp now colonized by mosses.

the climate became cooler, forests retreated from the Arctic and the present tundra conditions developed.

The dates now available for the retreat of the ice sheets at high latitudes show that timing and extent of the ice cover at the Last Glacial Maximum varied considerably with location. Estimates of the extent of ice cover during the Last Glacial Maximum range from those who consider that there was only limited glacial extension in arctic Canada and northern Eurasia (Pavlidis et al., 1997), to others who maintain that the cover was more widespread (Grosswald, 1998). The minimalist view also asserts that glacial maxima were not synchronous throughout the northern hemisphere; for example, the Novaya Zemlya ice sheet reached its maximum extent c. 39 000 BP, approximately 20 000 years earlier than the maximum extent of the contiguous Scandinavian ice sheet. Even during the coldest periods of the Pleistocene, glaciation did not extend to all areas of arctic Russia.

The rain shadow area of north-eastern Siberia has been relatively ice free throughout the Last Glacial Maximum, and traces have been found there of where Palaeolithic man was hunting mammoth 40 000 years ago. Plant macrofossils found on the Bykovsky Peninsula (west Beringia) indicated the existence of productive meadow and steppe communities during the late Pleistocene and could have served as a food resource for large populations of herbivores (Kienast et al., 2005). Other areas further west where the climate was more oceanic became free of ice only much later.

However, even here there were areas which are now known to have been free of ice at the Last Glacial Maximum. Consequently, when discussing the arctic flora it is essential to remember that the evolutionary history of the vegetation varies greatly from one region to another.

The Taymyr Peninsula, like much of north-eastern Siberia came into a precipitation shadow and was consequently ice free during times when there was large-scale ice coverage in Scandinavia and north-western Europe (Möller et al., 1999). A case can also be made for ice-free refugia in the south-west region of the South Island of Novaya Zemlya on the basis of its proximity to the edge of the Last Glacial Maximum ice boundaries (Tveranger et al., 1999) and the demonstration that pollen and radiocarbon data on the western coast of Novaya Zemlya are sufficient to disprove earlier theories of extensive glaciation during the Upper Pleistocene (Serebryanny & Malyasova, 1998). In the lands to the east, both western and central Siberia and the Chukotsky Peninsula were all marked by limited glaciation during the Last Glacial Maximum (Velichko et al., 1997). The dry, cold climate of Asia during the Pleistocene was associated with a specific tundra–steppe biome which was floristically, historically, and ecologically distinct from the polar deserts of the more high latitude sites.

Figure 6.14 shows some of the sites where there is now a corpus of mainly geological evidence for ice-free areas in the Arctic at the time of the Last Glacial Maximum which makes it probable that there were

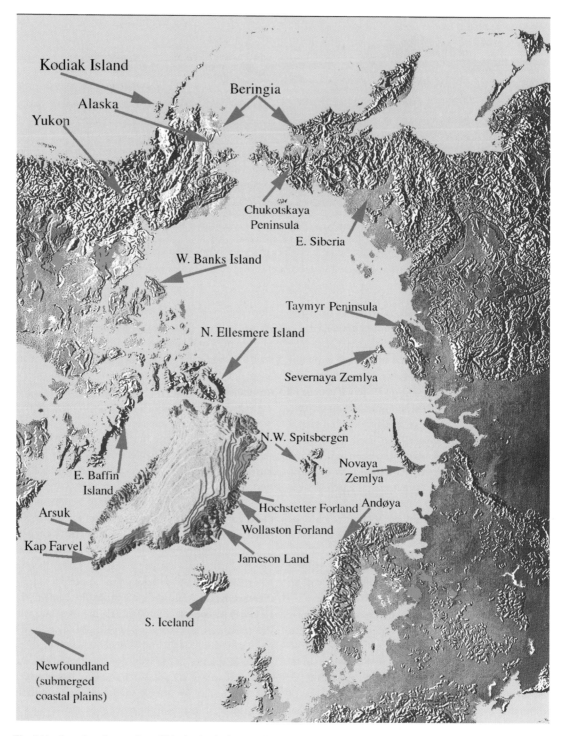

Fig. 6.14 Location of some sites within the Arctic that were ice-free and may have served as refugia for flowering plants during at least part of the Last Glacial Maximum. All examples are based on geological or palaeological evidence with the exception of south Greenland (Arsuk and Kap Farvel) and south Iceland where the evidence is phytogeographical. (For sources of information see Crawford, 2004.)

numerous outlying populations of flowering plants surviving, at this time north of the major ice sheets.

6.4.2 Molecular evidence for the existence of glacial refugia at high latitudes

To investigate the hypothesis that plants survived north of the major ice sheets a major investigation was undertaken of the widespread polymorphic and highly successful arctic colonizer the opposite-leaved purple saxifrage (*Saxifraga oppositifolia*). The species was sampled for its chloroplast DNA variation at a number of sites throughout its circumpolar distribution (Fig. 6.15) with the aim of establishing how this species colonized

the Arctic during the late Tertiary or early Pleistocene and whether or not refugia for this and other species occurred in the Arctic as well as at more southern latitudes during the Pleistocene ice ages. Fourteen different haplotypes (A–N) of chloroplast DNA (cpDNA) were identified. Phylogenetic analysis showed that *S. oppositifolia* is composed of two major DNA lineages: a 'Eurasian lineage' distributed westward from the Taymyr Peninsula in north-central Siberia, through Europe and Greenland to Newfoundland and Baffin Island and a 'North American' lineage, distributed eastward from the Taymyr Peninsula, through north-east Siberia and North America to North Greenland. Haplotypes basal to each lineage co-occur only in north-central

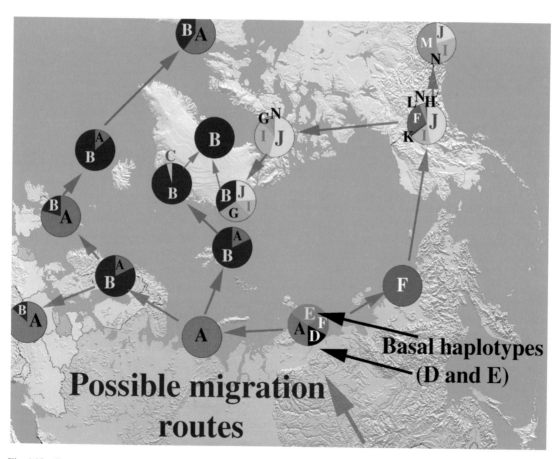

Fig. 6.15 Summary of results from a survey of chloroplast DNA variation in purple saxifrage (*Saxifraga oppositifolia*). This study was conducted to reconstruct the evolutionary history of this circumpolar arctic plant and to establish how this species colonized the Arctic during the late Tertiary/early Pleistocene, and whether refugia for this and other arctic plants occurred in the Arctic as well as at more southern latitudes during Pleistocene ice ages – for detail see text. (Adapted with permission from Abbott & Brochmann, 2003.)

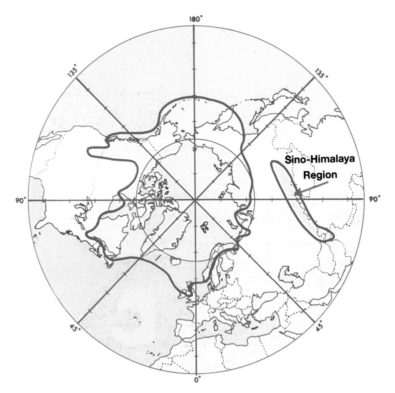

Fig. 6.16   Circumpolar distribution of *Saxifraga oppositifolia* and the location of the Sino–Himalyan region where other opposite-leaved species of the sect. *Porphyrion* subsection *oppositifoliae* also occur. (Map adapted with permission from Webb & Gornall, 1989.)

Siberia (Taymyr). A possible explanation of this phylo-geographic phenomenon is that *S. oppositifolia* first occurred in the Arctic in western Beringia in the Sino–Himalayan region (Fig. 6.16) during the late Tertiary before migrating east and west to obtain complete cir-cumpolar distribution (Abbott *et al.*, 2000) and could have migrated northwards along the Altai Mountains to reach the shores of the Arctic Ocean.

Saxifraga oppositifolia has long been regarded as one of the ancient species of the arctic flora (Tolmatchev, 1966), largely maintaining the diploid state over a wide geographical range together with much morphological variation. Tetraploid cytotypes are found, but the rela-tive distribution of the two cytotypes is not known. In Spitsbergen it is claimed that only the diploid type is present (Brysting *et al.*, 1996).

Other species of opposite-leaved saxifrages can be found in the Altai Mountains. It would therefore appear from the pattern of cpDNA distribution that *S. oppositifolia* is probably derived from ancestral stock native to high mountains in central Asia. From these

ancestral populations the species migrated north during the late Tertiary to northern Siberia along mountain ranges that connect these two regions, and then spread around the shores of the Arctic Ocean.

The high cpDNA diversity present in Alaskan populations of *S. oppositifolia* (Fig. 6.15) supports fossil evidence that a major refugium for arctic plants was present in eastern Beringia during the last full glacial period. The possibility that a northern refugium also existed in unglaciated parts of the Canadian Arctic and north Greenland is not excluded but requires fossil evi-dence for confirmation. Similarly, it appears equally feasible that an arctic refugium occurred in parts of the Taymyr Peninsula which were never glaciated. Low cpDNA diversity is evident throughout the distribution of the mainly Eurasian lineage, which occupies mostly areas that were heavily glaciated during the last ice age. It is likely that migrants from the south, west and east of the main ice sheets colonized much of Europe, north-west Russia, Iceland, Greenland and north-east America during the post-glacial period (Abbott *et al.*, 2000).

6.4.3 Evidence for an ancient (autochthonous) arctic flora

The improved knowledge of past migration routes and locations of important Pleistocene refugia for arctic plants adds substance to earlier phytogeographical speculations concerning the history of the arctic flora. Taxonomists who have paid particular attention to the arctic flora and its relationship to its Eurasian elements have long argued that not all species of arctic plants are of the same age. In areas where there has been little glaciation, as in eastern Siberia, Beringia and Alaska, there are species that have had a presence in the Arctic throughout the Pleistocene. Species with widespread distribution and chromosome counts that are simple diploids with no evidence of allo-polyploidization are likely to be ancient and indigenous members of an arcto-alpine flora that probably evolved on mountain chains outside the Arctic. The circumpolar species *Saxifraga oppositifolia* ($2n = 26$) is included in Tolmatchev's list (Table 6.2) of widespread arcto-alpine species; it has varying ecological forms but is probably not of arctic origin (Tolmatchev, 1966).

The cpDNA migrational history for *S. oppositifolia* described above provides substantial molecular evidence for the apparent long-term presence in the High Arctic of this particular species. Further research is needed to determine if the same molecular evidence for long-term residence in the Arctic will be found in the other species listed by Tolmatchev as autochthonous. With no total purging (the Pleistocene *tabula rasa* effect), it is to be expected that the Arctic flora as a whole will have had diverse origins, with some species surviving north of the ice sheets at the Last Glacial Maximum (e.g. *Saxifraga oppositifolia*) and then extending southwards as the ice retreated. The Arctic flora although lacking the species numbers found in warmer climates is nevertheless highly heterogeneous, not only in its genetic structure but also in its biogeographical history.

6.5 HABITAT PREFERENCES IN HIGH ARCTIC PLANT COMMUNITIES

The flora of the High Arctic is made up predominantly of a series of coastal plant communities. There are of course plants on mountains. There are also many coastal nunataks (mountain tops that are not perman-

Table 6.2. *Examples of ancient, indigenous (autochthonous) species widespread in the Arctic*

Saxifraga oppositifolia	($2n = 26$)
S. hyperborea	($2n = 26$)
S. tenuis	($2n = 20$)
Ranunculus pygmaeus	($2n = 16$)
Dryas octopetala	($2n = 18$)
Loiseleuria procumbens	($2n = 24$)
Cassiope tetragona	($2n = 26$)
Diapensia lapponica	($2n = 12$)

Note the low ploidy level of the chromosome numbers.
Source: Tolmatchev (1996). Chromosome numbers taken from Lid & Lid (1994).

ently covered by snow and ice), which are usually more accurately described as semi-nunataks as they frequently expose bare rock to seaward while their landward side is covered in snow and ice. Such locations support a few hardy pioneer species that can often be found in microsites that are particularly sheltered and escape the cooling influence that arctic sea ice exerts at lower altitudes. In recent years sea ice has noticeably retreated (Fig. 6.6), which is extending the growing season for the shore vegetation.

As long as the tilt of the Earth's rotational axis does not disappear, the High Arctic will continue to have a long polar night; consequently short growing seasons and long winters will remain ever-present features that all arctic-inhabiting organisms have to tolerate if they are to survive. In a sense cold tolerance is the entrance qualification for joining the Arctic Community. As already stressed, the Arctic is extremely heterogeneous, both within localities and in different areas around the North Pole. As cold is ever present at these high latitudes it is less likely to be a factor that will discriminate directly between species or populations in relation to their particular regional or microhabitat preferences. If we are seeking explanations of the ecological and physiological adjustment to alterations in climate or habitat within the Arctic it is more profitable to examine other potential causes of limits to plant survival at high altitudes. These are likely to include stresses that are not ubiquitous or permanent features of the environment. The occurrence of such stresses

Table 6.3. *Summary of range of variations found in two contrasting ecotypes (tufted and prostrate; see Chapter 2) of* Saxifraga oppositifolia *populations at high latitudes contributing to long-term fitness and which may pre-adapt the species to climatic change*

Drought tolerance	Arctic tufted form, conserves water – intolerant of flooding – slow growth rate	(Crawford & Smith, 1997; Teeri, 1973)
Flooding tolerance	Prostrate form, flood-tolerant, rapid growth rate	(Crawford & Smith, 1997; Teeri, 1973)
Morphological variation	Local, *in situ* continuous ecoclinal differentiation with only one ploidal level	(Brysting *et al.*, 1996)
Epicuticular wax variation	Leaf waxes of plants from snow-free, wind-swept microsites had significantly higher abundance of n-alkanes than in those plants growing in adjacent areas where snow accumulates in winter	(Rieley *et al.*, 1995)
Phenology	Late-flowering genets found to have accelerated phenology	(Stenström *et al.*, 1997)

will vary depending on aspect, topography and exposure, as well physical and biotic disturbance.

Within any one area in the Arctic, despite the uniform imposition of low temperatures and short growing seasons, a great deal of variety can be found between microsites, not only in specialist species that are present, but in the growth and form of the more widespread species. It would appear that the ability of plants to withstand these different stresses is highly variable and is reflected in the many specific habitat preferences that are found within the arctic landscape. The fact that even within such unproductive environments as the High Arctic, species and populations still exhibit well-defined habitat preferences demonstrates that there must be both positive and negative consequences in the possession of specific adaptations.

6.5.1   Incompatible survival strategies

It has been claimed that *adaptation is the first step on the road to extinction* (Crawford, 1989). Adaptation increases habitat specialization and therefore makes species vulnerable to environmental change.

As the Arctic has one of the most variable climatic histories in the entire world there is therefore a risk that some adaptations, although increasing immediate fitness, may prove disadvantageous in the long term. An unfortunate aspect of both morphological and physiological adaptations is the incompatibility of many adaptations to opposing conditions. Visual inspection of plants is frequently sufficient to be able to determine the difference between a drought-resistant and a non-drought-resistant plant or between sun and shade plants. In some cases it is immediately obvious from form alone that adaptations for contrasting conditions cannot exist in the same individual plant.

This same argument can also be applied to unseen characteristics such as the physiological properties that determine whether a plant is a calcifuge or a calcicole, a halophyte or a glycophyte, or even tolerant or not of anoxia. Within one species, e.g. *Saxifraga oppositifolia* and *Dryas octopetala*, it is possible to detect population differences that are found in contrasting ecotypes, which could not exist in one individual (Table 6.3).

6.5.2   Ice encasement and the prolonged imposition of anoxia

The probability of the imposition of anoxia on over-wintering vegetation is probably higher in the Arctic than anywhere else on Earth. When snow melts and refreezes, or late rains fall on frozen ground, a thick covering of ice can result which can last from November until June. Temperatures are low under ice and metabolism will therefore be reduced. However, this does not mean that low temperature anoxia is not

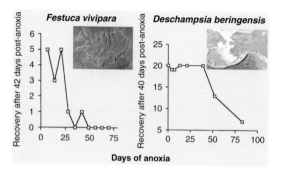

Fig. 6.17   Tolerance of anoxia in high-latitude grasses using (Left) a subarctic Icelandic population of pseudo-viviparous plantlets of *Festuca vivipara* and (Right) seedlings from an arctic population of *Deschampsia beringensis*. The plants were kept under total anoxia for different lengths of time at 5 °C and then allowed to recover in a cold room at 10 °C. With *F. vivipara* 6 plants per treatment were used and with *D. beringensis* 25 per individual time treatment.

dangerous. Throughout the period under ice there is a total deprivation of oxygen and all aerobic metabolism ceases. After months of encasement in ice, the arrival of spring and snowmelt exposes plants to the full oxygen concentration of the air, often within a matter of hours. Plants that have survived months of oxygen deprivation then face the additional hazard of post-anoxic injury. This is exactly the same situation that is faced by temperate plants that have been subjected to flooding, where plant survival requires an ability to overcome the double stress of first oxygen deprivation through flooding followed by the dangers of post-anoxic injury when tissues that have endured anoxia for a prolonged period are returned suddenly to air (Fig. 6.17; see also Chapter 8).

Tolerance of anoxia is energetically expensive, particularly in the carbohydrate reserves that have to be conserved to support overwintering anaerobic metabolism and provide the antioxidants for defence in spring against post-anoxic injury. In an area with limited resources as in the Arctic, it would be advantageous for those species that inhabit areas such as slopes and ridges where ice encasement is less likely to occur not to use their limited overwintering reserves for protection against ice encasement, and its associated anoxic and post-anoxic stresses.

Examples such as these demonstrate the evolutionary responses of plants to the less immediately

obvious aspects of habitat diversity within a landscape that seems at first sight to offer minimal potential for habitat differentiation. Although the Arctic may not be rich in species it nevertheless possesses considerable diversity within populations. Furthermore the juxtaposition of contrasting habitats, wet and dry, cold and warm, is very immediate at high latitudes. The minute stature and sparse nature of the vegetation does not allow for the buffering of the climate elements and the creation of mesic habitats. Consequently, populations inhabiting contrasting environments may be only metres from each other as the transition from one type of habitat to another is often very abrupt.

6.6   MUTUALISM IN ARCTIC SUBSPECIES

Subspecies variation has been repeatedly noted as present in many arctic species (Table 6.2). A common explanation advanced for ecotypic variation is based on the need for plants to optimize their use of resources in response to competition. There is, however, an alternative interpretation for the frequency of ecotypic variation, which does not depend on competition and which has a particular aptness for the High Arctic situation where competition is minimal, and that is the increase in long-term fitness that comes from *mutualism*.

Selection acting on individuals gives rise to population variations, which as a result of competition are associated with specialized habitats. This increases the habitat range occupied by the species and can be considered as increasing immediate fitness. In the Arctic these ecotypic variants are frequently associated with warm or cold, wet or dry, early or late sites. Thus they can also be considered as increasing long-term fitness by providing a reservoir of genetic variability which has pre-adapted species to climate change. This argument for the mutualistic advantages of ecotypic variation for species survival in fluctuating environments becomes apparent in those regions where the major constraints on survival are physical and not biotic. Competition does exist in the Arctic in favoured sites, but over large expanses of polar desert and semi-desert as well as tundra, there is much open ground where competition is minimal. In such situations a series of polymorphic metapopulations (see Table 1.1) can act as a genetic reservoir.

For the ancient (autochthonous) arctic species which are largely diploid and of low ploidy level the advantages of gene flow can be readily expressed due to the lack of masking genes.

At high latitudes *Saxifraga oppositifolia* exists with distinct ecotypes adapted to differences in growing season length. In areas with late snow-lie and cold, wet soils, increased metabolic rates and rapid shoot production compensate for ultra-short growing seasons (see Section 3.2.3) but do not conserve carbohydrate or water for adverse periods. An opposing strategy is evident in ecotypes living in sites with an earlier resumption of growth, where soils are warmer and drier and the growing season longer. Here metabolic rates are lower and result in a greater ability to conserve both carbohydrate and water. The existence of opposing strategies for survival in warm and cold habitats suggests that even in the minimal thermal conditions of the High Arctic a high degree of population diversity gives the species as a whole a wider ecological amplitude. This degree of diversity not only increases the range of sites in which the species can survive but confers an ability to adapt to climate change by altering ecotype frequencies to accommodate climatic fluctuations. Such a facility may have contributed to the survival of these polymorphic populations of this and other species in the High Arctic during the Last Glacial Maximum. Table 6.3 lists a number of incompatible characters, both morphological and physiological, which have the capacity to adapt *Saxifraga oppositifolia* ecotypes to different environmental conditions.

In a review of regional and local vascular plant diversity in the Arctic (Murray, 1997), attention was drawn to the observation that in the Arctic it is only common species that exhibit a broad ecological amplitude, which may be due to phenotypic plasticity of the individuals, or else is the result of the species in question being an aggregation of a series of ecotypes. It is this latter possibility that is advanced here as accounting for the outstanding capacity for survival in the Arctic and sub-Arctic over a very long time of *Saxifraga oppositifolia* and other common and variable arctic species. The phenomenon, particularly noticeable at high latitudes, by which a species maintains a range of interfertile ecotypes, instead of evolving breeding barriers as a response to competition, has been described as *suspended speciation* (Murray, 1997) and can be considered as one of the consequences of mutualism.

6.7 POLYPLOIDY AT HIGH LATITUDES

The distinction between long-term and short-term residence in the Arctic makes it possible to re-examine the properties that allow certain plant species to have a long history of survival at high latitudes. One characteristic which is particularly notable in arctic species is polyploidy.

The Arctic is one the Earth's most polyploid-rich areas; it is also noted for a high incidence of recently evolved polyploids (Brochmann *et al.*, 2004). The frequency and level of polyploidy increases markedly on moving northwards within the Arctic. A detailed examination of the levels of polyploidy throughout the Arctic could not detect any clear-cut association between polyploidy and the degree of glaciation for the arctic flora as a whole due to the number of widespread species. However, for 'arctic specialist' taxa with restricted distributions, it was claimed that the frequency of diploids was high in the Beringian area (Fig. 6.18), which remained largely unglaciated during the last ice age. Such observations support the hypothesis that polyploids are more successful than diploids in colonizing after deglaciation (Brochmann *et al.*, 2004). It could therefore be concluded that the post-glacial evolutionary success of polyploids in the Arctic may be due to their fixed-heterozygous genomes, which buffer against inbreeding and genetic drift through periods of dramatic climate change (Brochmann *et al.*, 2004).

Based on these arguments it is frequently suggested that polyploid species are better adapted physiologically to cold climates than diploid species. However, as noted above the exceptions to this habit are the autochthonous (ancient and indigenous) species which are usually diploid or occasionally tetraploid (Table 6.2). Even in the relatively recently deglaciated areas of Svalbard, 20% of the flowering plant species are diploid. Consequently it can be argued that a high level of polyploidy, although common at high latitudes, is not necessarily an adaptation for long-term survival in the fluctuating arctic environment. More probably, the presence of so many polyploid species in the Arctic is merely the consequence of plant migrations caused by periods of cooling and warming, with the climatic

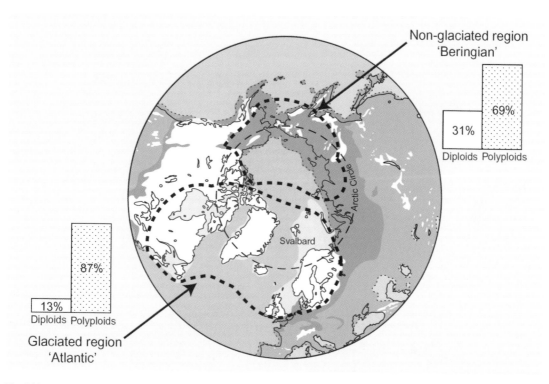

Fig. 6.18   Frequency of polyploidy among 'arctic specialist taxa' that are either restricted to the region that was heavily glaciated during the last ice age ('Atlantic') or restricted to the region that remained mainly unglaciated ('Beringian'). The maximum extent of the late Weichselian/Wisconsian ice sheets (white) and tundra (dark grey) are shown, modified after Abbott & Brochmann (2003) and Brochmann *et al.* (2003). (Reproduced with permission from Brochmann *et al.*, 2004.)

disturbance bringing species into proximity in unusual combinations (Stebbins, 1971).

As in many other marginal areas, e.g. aquatic habitats, so in the Arctic there are some very successful sterile polyploid species. In terms of distribution of species and genetic exchange between populations it has to be remembered that much of the arctic flora is coastal and that coastal species in general enjoy a relative freedom from physical barriers to dispersal of ramets and other aids to vegetative reproduction. The sterile and triploid creeping salt marsh grass (*Puccinellia phryganodes*) is outstanding for the extent of its distribution throughout the Arctic as a major component of salt marsh vegetation (Jefferies & Gottlieb, 1983; Fig. 6.19). Accounts even exist of stolons of this plant, found at sea frozen into sea ice, that have been rescued, brought ashore and successfully grown (R. L. Jefferies, pers. com.). Dispersal throughout the Arctic of plant fragments frozen into pack ice must therefore be considered

as an effective method of migration. In Beringia the species is found as the fertile diploid ($2n = 14$), while over Canada, Greenland, northern Norway, Spitsbergen and Fennoscandinavia the sterile $2n = 21$ occurs, while in Finnmark, north Norway and Svalbard triploid $2n = 21$, tetraploid $2n = 28$, and hexaploid $2n = 42$ individuals have all been recorded (Aiken *et al.*, 2001).

As discussed above, most adaptations when examined closely have both positive and negative aspects in relation to survival potential. The question therefore has to be asked if there are similar distinctions between polyploid and diploid species in relation to survival fitness in the Arctic. It has never been demonstrated that polyploid species are more cold hardy or better adapted to short growing seasons than diploid species. The genetic advantages of polyploidy include restoring fertility to hybrids, and the ability to form new species within the home range of their parents. Polyploidy also has obvious advantages in

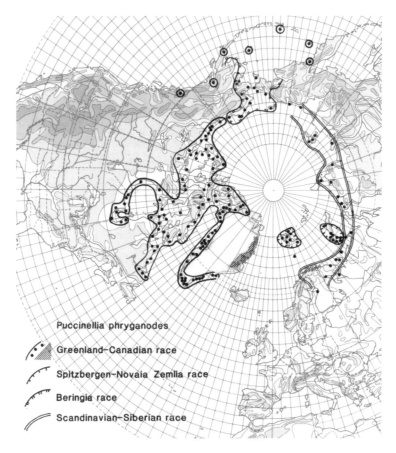

Fig. 6.19 Circumpolar distribution of creeping salt marsh grass (*Puccinellia phryganodes*). (Reproduced with permission from Hultén & Fries, 1986.)

masking the effects of harmful alleles. More questionable, however, is the assertion that polyploids benefit from having a greater store of genetic variability merely because there are more alleles at any one locus. If harmful alleles are masked by polyploidy, then the same must also be true for helpful alleles. In polyploid species mutations are less likely to be expressed and phenotypic variability is likely to be low. In diploid species neighbouring populations with different environmental preferences will have the advantage of being able to profit more readily than polyploids from gene flow as interchange of alleles from one population to another is less likely to be masked. It is therefore perhaps no coincidence that the ancient (autochthonous) arctic species (see Table 6.2) are mainly diploid (or else have only a low level of polyploidy) as they can adapt readily to climatic

change by exchanging genes between different ecotypes of the same species which are readily expressed and not masked by other alleles. The manner in which mutualism (see above) rather than competition between ecotypes can contribute to species survival will function only if genetic exchange leads to readily expressed adaptations. This is more likely to take place between adjacent diploid than polyploid populations.

Conflicts between adaptations at the physiological, biochemical and morphological level inevitably limit the choice of adaptations that can be expressed by any individual plant. No matter how much genetic information is present within the genome of any individual it is of little use if it cannot be expressed. It follows therefore that polyploid populations because of their genetic stability are less likely to be able to adapt to changing circumstances by exchanging small amounts of genetic material. In short, the adaptive significance of polyploidy, in contributing to physiological fitness in general, or in the Arctic in particular, has never been proven.

6.8 ARCTIC OASES

In the Arctic it is possible to detect sites which in comparison to adjacent areas are relatively rich either in species or genetic diversity or both. Genera that show high species diversity at high latitudes in such areas include *Salix, Saxifraga* and *Draba*. Arctic species-rich hotspots, sometimes referred to as polar desert oases, include Peary Land, Devon and Ellesmere Islands as well as Bathurst Inlet (Figs. 6.20–6.24).

The first polar oasis to be studied in detail was Truelove Lowland at 75° N in Devon Island (Bliss, 1977). Here gently sloping coastal lowlands, protected from the harsher upland climate, enjoy a warmer environment with a locally longer growing season together with availability of water throughout the growing season.

Further north two other outstanding polar oases have been extensively studied on Ellesmere Island. The more southerly, at Alexandra Fiord at 78° 53′ N, 75° 55′ W (Fig. 6.20) has an extremely short growing season yet supports a vascular plant flora comprising over 90 species (Svoboda & Freedman, 1992). The Sverdrup Pass which lies to the north west is an 80-km-long deglaciated valley that separates the two Ellesmere Island ice fields. The vegetation in the pass is reported as richer at the east end and when studied in the period between 1986 and 1994 by Professor Svoboda and colleagues sustained a resident population of 45–60 musk oxen (Fig. 6.23).

The other notable polar oasis lies further north at Lake Hazen (81° 50′ N, 70° 00′ W) which in spite of being at a higher latitude has greater biodiversity than Alexandra Fiord. Lake Hazen area is a large open basin with a longer growing season than Alexandra Fiord due to a higher number of 24-hour days and earlier snowmelt compared with Alexandra Fiord. About 117 vascular species were found there compared with 92 at Alexandra Fiord. At this latter more southern location, the lowland is north-facing, losing some radiation energy early in the spring which can delay snowmelt. However, in mid-season the noon temperatures during the sunny, calm days are higher than at Lake Hazen, making Alexandra Fiord a true thermal oasis (J. Svoboda, pers. com.).

Examination of the flora of these polar hotspots reveals that the plants are predominantly aggregates of neoendemic or ancient relict species, which suggests that their existence is a consequence of local moderation of environmental extremes persisting through shifting climatic periods, permitting populations of unique species to survive in these places. There are parallels that can be drawn between the biological oases in Devon and Ellesmere Islands, which have probably had a long history of local environmental amelioration due to local topographic effects on climate, with concentrations of endemic species detected elsewhere. In Central Africa it has been shown that species hotspots are characterized by a local reduction in ecoclimatic variability, or by being located on the boundary to a stable region (Fjeldsa *et al.*, 1997). This is the same situation that exists in the South African Cape flora. In all these cases the floristic hotspot possesses a longer history of occupation by the current vegetation type than is found in neighbouring areas with more variable climatic histories.

At high latitudes, climatic oases are exceptional phenomena and throughout the Arctic the more typical situation is for extensive areas which are poor in species (barrens) to be interspersed with sites where vegetation is both more plentiful and varied. This is most strikingly seen along river courses where sometimes trees can survive many tens of kilometres further north of the main boreal forest. Recent studies using remote sensing have investigated the vegetation of the Hood River region in the central Canadian Arctic. Using satellite imagery (Fig. 6.24) and examining the 'normalized difference of vegetation index' (NDVI) it has been possible to detect sites where high NDVI values corresponded with species richness as estimated from ground surveys. Ground-based sampling showed species richness to vary between 69 and 109 vascular plant species per 0.5 km^2 of sample area (Gould, 2000). Sites with the lowest species richness were in the upper reaches of the river, with species numbers generally increasing downstream. Variation in richness along the river was correlated with increasing topographical and microclimatic heterogeneity and reflected changes in the range of site-level variation due to alteration in substrate type and texture, topography, moisture, and

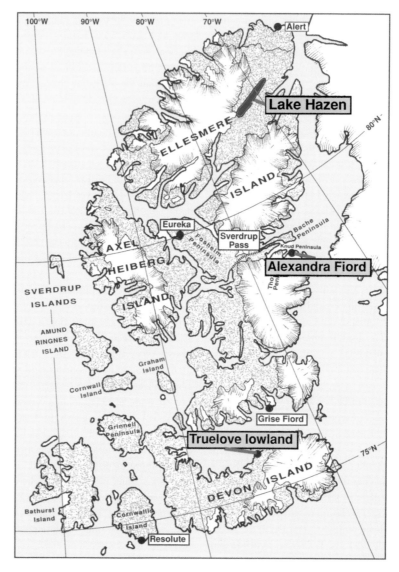

Fig. 6.20   Location of regions studied for botanical diversity in the Canadian Arctic. Note in particular the location of the first arctic biodiversity hotspot to be studied in the Canadian Arctic at Truelove Lowland on Devon Island and also the more northerly locations of subsequent studies at Alexandra Fiord and Lake Hazen in Ellesmere Island. (Reproduced with permission from Svoboda & Freedman, 1992.)

soil pH. The most significant component of the index was an increase in the range of soil pH. Soil pH tends to increase downstream, due to the presence of uplifted marine sediments and tills (Gould & Walker, 1997). The structure and diversity of the vegetation along the arctic river showed also that variation in species richness along the corridor is structured in relation to increasing landscape heterogeneity, with increases in

the floristic community distinctiveness (beta diversity) together with species richness within communities.

The facility to use remote sensing to detect species-rich pockets even in distant and uninhabited regions of the Earth should prove a useful conservation tool in providing a definitive global database of areas which could have particular ecological value for their biodiversity.

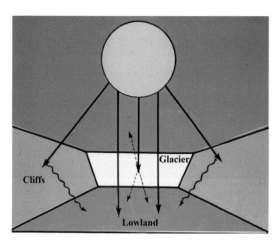

Fig. 6.21 Diagrammatic representation of the enhanced thermal environment associated with polar oases. A combination of 24-hour days together with reflection of light from the sea, glaciers and re-radiation of heat from adjacent cliff faces enhances the temperature regime of the adjacent lowland. (Diagram prepared by Professor J. Svoboda.)

6.9 PHENOLOGICAL RESPONSES TO INCREASED TEMPERATURES

This account of arctic vegetation so far has been orientated towards a discussion of the likely long-term effects of climate change in the belief that for an ancient and heterogeneous arctic flora the main impacts of climatic change will be found at a population or species level. There is no doubt that phenotypic responses to increasing temperatures are already taking place. In the Arctic these have already been mentioned in relation to the profuse flowering by the purple saxifrage that can now be found on cold shores (Fig. 6.6). Whether or not these phenological responses will have a long-term effect on the future status of high-latitude vegetation is still a difficult question. In the case of *Saxifraga oppositifolia*, which everywhere in the Arctic can be found as a highly successful pioneer species, increased flowering, if it leads to increased seed production, should facilitate its spread to newly deglaciated terrain. This species has already demonstrated its ability to persist once glacial conditions have retreated by its well-documented occurrence in areas from which the ice has long vanished. One particularly striking location is the lowland Scottish site at the Falls of Clyde. For other species the question still remains: will

unaccustomed heat cause extensive destruction and loss of biodiversity in an ancient cold-adapted flora?

Many experimental studies have been carried out on the short-term phenological responses of vegetation to climatic change by placing plastic shelters in various habitats around the Arctic and making detailed observations of the effects of the artificially induced higher temperatures. In some cases this has also included the addition of nutrients. Most notable among these studies has been the International Tundra Experiment (ITEX), a collaborative experiment using a common temperature manipulation to examine variability in species reactions to increased temperatures across a wide variety of tundra sites. The data recorded the vegetation responses in terms of plant phenology, growth, and reproduction. Details of the results as reported after the first four years of these observations revealed many phenological differences between sites and species. A general conclusion in the report (Arft *et al.*, 1999) indicated that key phenological events such as leaf bud burst and flowering occurred earlier in warmed plots throughout the study period, but that there was little impact on growth cessation at the end of the season. A shift away from vegetative growth and towards reproductive effort and success in the fourth treatment year was taken to suggest a shift from the initial response to a secondary response. The change in vegetative response may be due to depletion of stored plant reserves, whereas the lag in reproductive response may be due to the formation of flower buds one to several seasons prior to flowering.

Warmer, low arctic sites produced the strongest growth responses, but colder sites produced a greater reproductive response. It was suggested that greater resource investment in vegetative growth might be a conservative strategy in the Low Arctic, where there is more competition for light, nutrients, or water, and less opportunity for successful germination or seedling development. By contrast, in the High Arctic, it was speculated that heavy investment in producing seed under a higher temperature scenario may provide an opportunity for species to colonize patches of unvegetated ground, as appears to be illustrated by the flowering success already mentioned for the purple saxifrage.

The main interest of this research lies in the diversity of responses that can be found between different life-forms and different species, depending on

Fig. 6.22   Aerial view looking to Lake Hazen from above the thermal oasis. Although further north than Alexandra Fiord this oasis supports a flowering plant flora of 117 species. (Photo Professor J. Svoboda.)

whether they are in the Low Arctic or the High Arctic. However, in terms of predicting the long-term consequences of climatic change on arctic vegetation these experiments still leave considerable uncertainty, and consequently reliable predictions are difficult to make.

When experiments are pursued over longer periods of time some of these initial phenological responses can disappear altogether.

In a series of studies carried out manipulating light, temperature, and nutrients in moist tussock tundra

Fig. 6.23   Oblique aerial view showing the eastern portion of Sverdrup Pass. This deglaciated 80-km-long pass studied by Professor Svoboda and colleagues between 1986 and 1994 separates two of Ellesmere Island's major ice fields. The vegetation in the pass is reported as richer at the east end and sustained a resident population of 45–60 musk oxen. (Photo Professor J. Svoboda.)

near Toolik Lake, Alaska, it was found that short-term (3-year) responses were poor predictors of longer term (9-year) changes in community composition (Chapin *et al.*, 1995). Instead the longer-term responses showed closer correspondence to patterns of vegetation distribution along environmental gradients that were evident in changes in the availability of soil nutrients. In particular, nitrogen and phosphorus availability tended to increase in response to elevated temperature, reflecting increased mineralization. The major effect of elevated temperature was to accelerate plant responses to changes in soil resources and, in the long term (9 years), to increase nutrient availability through changes in nitrogen

mineralization. It has been suggested that the lag in response is due to the time needed for litter fall to alter the nutrient status of the underlying soil.

In Alaska there is increasing evidence of climatic warming on woody plants. Shrubs trigger several feedback loops that influence their expansion rate (Chapin *et al.*, 2005). The responses of vegetation to increased temperature, as seen in Alaska with its extensive moist tussock-tundra, cannot be taken as indicative of what will happen elsewhere in the Arctic. Detailed discussion on the whole question of the availability of resources in marginal areas including the Arctic is given in Chapter 3 and the diverse effects of

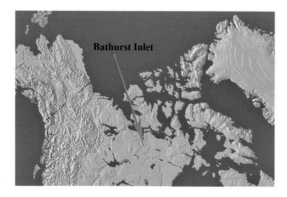
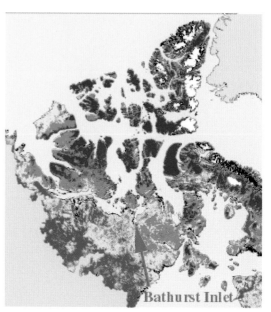

Fig. 6.24   Location of an arctic biodiversity hotspot along a river course. (Left) Location of Bathurst Inlet and the Hood River. (Right) Satellite NDVI image of Bathurst Inlet and the surrounding area in July. The NDVI image provides a relative comparison of photosynthetic activity between the surrounding barrens (blue) and the river course (green). The more continental location of this relatively southern polar hotspot (66° 49′ N – just north of the Arctic Circle) results in a later start to the growing season than either of the two northern sites described above at Lake Hazen and Alexandra Fiord. (Satellite image by courtesy of Dr W. Gould.)

climatic warming on the tundra–taiga interface have already been reviewed in Chapter 5.

6.10   CONCLUSIONS

An increasing corpus of knowledge on the history of the Arctic flora is showing that plant life at high latitudes has a long history and that therefore many of the species that inhabit these regions will have endured extensive and prolonged climatic fluctuations over the past 2–3 million years. The chloroplast DNA studies on *Saxifraga oppositifolia* and other studies suggest that the founding stocks of the arctic flora appeared at the end of the Pliocene (approximately 3 Ma BP) and the genetic imprint of this migration can still be traced around the shores of the Arctic Ocean (Abbott & Brochmann, 2003). The presence of many subspecies and local ecotypes indicates the existence of considerable genetic variability despite the peripheral nature of these high-latitude sites.

The long-term effect of warming in the Arctic may therefore be expected to operate at a genetic level as probably it has done in the past. The polar night will continue to impose a long dormant period in the High Arctic for which the present flora is well adapted. The prospect of the liberation of large areas of land from permanent ice cover should provide a wealth of terrain for the continued existence and probable new flourishing of this remarkable flora. Climatic warming may cause the polar bear to disappear, but this is only a species that has evolved over the past 100 000 years, and cannot compare with plants in long-term occupation of the Arctic. Botanists in the future may look forward confidently to relaxed exploration of a diverse and plentiful flora as far north as land exists without the inconveniences and risks of ursine disturbance.

7 · Land plants at coastal margins

Fig. 7.1 Machair-forming sand dunes at Luskentyre (Gaelic, *Losgaintir*), Harris, Outer Hebrides. As these dunes erode the sand is blown landwards and contributes to the renewal and maintenance of species diversity in the coastal machair (*machair* from Scots Gaelic – a low-lying plain).

7.1 CHALLENGES OF THE MARITIME ENVIRONMENT

Coastal regions present both opportunities and challenges to plants. On one hand, reduction of temperature extremes, combined in many instances with freedom from frost and drought, extends the potential distribution for species that are intolerant of climatic extremes. On the other hand, the constant threat of habitat destruction, coupled with the physical stresses of wind exposure, salt drenching, burial and flooding, renders coastal regions marginal areas for many species (Fig. 7.1).

Climatic warming and its consequences for sea-level rise are already having an increasing impact on coastal habitats. Coastal vegetation has always had to adjust to changes in sea level. Since the peak of the last ice age about 18 000 years ago the sea has risen more than 120 metres. The greater part of this change took place over 6000 years ago. Over the past 3000 years the sea level has been largely constant, rising by about 0.1 to 0.2 mm per year. Over the last 50 years, however, the average rate of sea-level change obtained from tidal gauges has risen to $+1.8 \pm 0.3$ mm yr^{-1}. The rate now seems to be rising more rapidly. Since 1992 satellite altimetry measurements have shown an average rise of $+3.1 \pm 0.4$ mm yr^{-1} (Nerem *et al.*, 2006). It is now probable that this recent acceleration (Fig. 7.2) represents the first signs of the effect of global warming on sea level (Houghton *et al.*, 2001).

The potential impact on vegetation of the expected large rises in sea level will vary depending on the geological nature of the coastline and its vegetation cover. Coasts can be divided into soft coasts, such as sand dunes and salt marshes, and hard coasts where the interface between land and sea is made up of hard rocky shores or cliffs. It is in the soft coasts that the stability of the shoreline is determined by the resilience of the maritime plant communities to the onslaught of oceanic tides and winds. Soft shores have the greatest ability to accrete and respond to rising sea levels provided there is a sufficient supply of sediment. However, they are also the most vulnerable when physical erosion outstrips their ability to consolidate fresh sediment.

When examined in detail, the vegetation of coastal regions reveals not one but a series of margins depending on the relative tolerance of different plant species in relation to proximity to the sea. As discussed in Chapter 1, aerial photographs, coupled with ground inspection on cliff tops, sand dunes and salt marshes, frequently show a gradient of marginality in the zonation of plant communities in relation to distance from the sea (Figs. 1.5, 7.12).

On a worldwide basis, plant forms that are able to survive near the sea range from diminutive herbs to substantial forest trees (Fig. 7.3). A paradox of the maritime environment is that as well as being hazardous for plant survival it also provides ecological opportunities for species that are not found in more densely vegetated and competitive inland communities. It is the coastal environment that now provides the only refuge for some of the coniferous trees that once had a worldwide distribution. The coastal Californian redwood (*Sequoia sempervirens*) in common with several species of *Chamaecyparis* and *Taxodium* all had transcontinental distributions during the Tertiary period but now are found only in very restricted coastal habitats (Fig. 7.4). Apparently, the oceanic niche with its diminution of climatic extremes, together with a reduction in competition from more recently evolved tree species, provides many relict tree species with an environment in which they are still viable (Laderman, 1998a).

The coastal habitat, although often marginal in terms of physiological stress, is not usually limiting in terms of geographic distribution. Shorelines provide an unequalled highway for plant migration. In this modern age where human impact on the landscape has reduced many inland plant communities to isolated pockets or

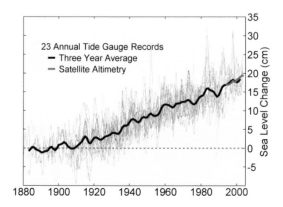

Fig. 7.2 Recent sea level rise as recorded from 23 annual tide gauge records and satellite altimetry. (Reproduced with permission from www.globalwarmingart.)

Fig. 7.3   Shingle beach showing the diversity of plant forms that can be found growing close to the sea. Scots pine (*Pinus sylvestris*) and bearberry (*Arctostaphylos uva-ursi*) at Gothemshammar on the island of Gotland. The Baltic Sea is noted for its low salinity, which may account for the survival of bearberry so close to the shore.

discrete patches, the constant physical onslaught of the oceans has preserved the open shore as a long-distance migration pathway. The ocean itself aids this dispersal not just for species with floating, salt-tolerant seeds, but for many others that can be transported by rafting with driftwood and other maritime flotsam, or by birds using coastal migration routes (Fig. 7.5). Even the Arctic Ocean may have been a significant dispersal route during the ice ages. The movement of drift ice and driftwood dendrochronologically dated from the Late Weichselian or early Holocene suggests that the Arctic Ocean has long been a possible dispersal route for diaspores from eastern Siberia and north-west Russia to parts of the North Atlantic region. The extremely disjunct

distribution of some vascular plants in northern Scandinavia and East Greenland which are also found in eastern Siberia (e.g. *Draba sibirica*, *Oxytropis deflexa* ssp. *norvegica*, *Potentilla stipularis* and *Trisetum subalpestre*) may be the result of this type of long-distance dispersal (Johansen & Hytteborn, 2001).

7.1.1   The concept of oceanicity

Ecologically, the term oceanicity is used to describe the climatic factors which modify the environment and ecology of a region within reach of maritime weather systems. In common with other terms used to denote environmental syndromes, 'oceanicity' describes a multi

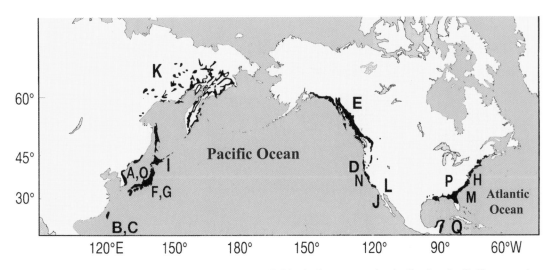

Fig. 7.4   Distribution of coastal forests shown in black and identified by dominant tree species. A, *Alnus japonica*; B, *Chamaecyparis formosensis*; C, *C. taiwanensis*; D, *C. lawsoniana*; E, *C. nootkatensis*; F, *C. obtusa*; G, *C. pisifera*; H, *C. thyoides*; I, *Picea glehnii*; J, *Pinus muricata*; K, *P. pumila*; L, *P. radiata*; M, *P. serotina*; N, *Sequoia sempervirens*; O, *Thuja orientalis*; P, *Taxodium distichum/Nyssa aquatica*; Q, Yucatan species. (Map adapted with permission from Laderman, 1998b.)

faceted situation. Species found in proximity to the ocean may live there for a variety of ecological, physiological, and even historical reasons. Equally, species absent from areas near the sea, and therefore responding negatively to oceanicity, are also likely to have a diverse range of primary and secondary causes for this restriction in their distribution (Figs. 7.6–7.8). A lack of summer warmth may be important in some cases, but secondary causes, such as the unavailability of their preferred habitat and exposure to salt, may be sufficient to ensure their exclusion rather than any direct climatic factor. Quantification of 'oceanicity' in all its various manifestations has therefore to be pursued with caution, as it is merely a human perception of what may be involved in determining why proximity to the sea creates a particular environment which is suitable for certain species of plants but excludes others.

The commonest meteorological assessment of oceanicity is mean annual temperature range. In some cases this can be used directly; in other cases it can be adjusted for latitude and related to defined extremes of 'oceanicity', or its converse 'continentality', as with Conrad's Index of Continentality (Conrad, 1946).

A direct consequence of oceanicity is a decrease in potential water deficit due in part to the reduced evaporative power of oceanic climates and partly to

increased precipitation in areas with high relief on their windward shores, as in western Scotland and Norway. Consequently, increasing oceanicity is likely to be associated with greater water saturation of the soil profile for lengthy periods of the year. The ecological consequences of prolonged soil saturation for bog growth need little elaboration. Bogs are important sources of information for the reconstruction of climatic history and as their surface topography is highly sensitive to changes in moisture they also serve as indicators of how oceanicity has varied with time. Many British bogs have profiles that show changes in bog surface vegetation from wet lawn to pool and hummock topography as the hydrological element of the oceanic environment fluctuates through the centuries (Barber, 1981; Chambers *et al.*, 1997; Mauquoy & Barber, 1999; Fig. 7.9).

The differences between maritime and continental climates are most commonly described by modification of temperature extremes and as such can be meteorologically quantified and mapped (Fig. 6.7). This simplification of the concept of oceanicity to an expression of annual temperature amplitude, although it is unambiguous and meets many human environmental needs, nevertheless neglects many of the ways in which the ocean alters the terrestrial environment.

Fig. 7.5   Hoary cress (*Lepidium draba*) growing on a boulder beach site at its most northern UK distribution limit on the west mainland of Orkney (see inset). This species has a mainly southern distribution in the British Isles but is now becoming more common in coastal habitats possibly as a result of climatic warming.

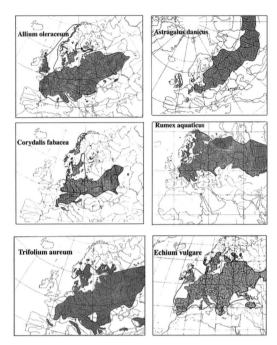

Fig. 7.7   Distribution of six species of flowering plants which are less able to survive in areas adjacent to the sea. Note that there is a tendency for a north–south orientation in the western distribution limit which matches the orientation of zones of similar oceanicity in Fig. 7.6. (Maps reproduced from Meusel & Jäger, 1992.)

Fig. 7.6   Distribution of six species of flowering plants limited in distribution to areas adjacent to the sea. (Maps reproduced from Meusel & Jäger, 1992.)

Plants are highly sensitive to many aspects of the maritime environment and respond to the influence of the ocean, even in regions that are a considerable distance from the sea, as a result of changes in temperature, rainfall, and variability and length of the growing season. Paludification can also be a consequence of oceanicity. The growth of Atlantic bogs in Scotland is one example, and another is the replacement of tundra by bog due to the proximity of the Arctic Ocean to the West Siberian Plain (see Section 5.4.3).

7.1.2   Physical versus biological fragility

In examining survival in any habitat it is essential to distinguish between physical and biological fragility. A shoreline, or dune system or even a mangrove forest may be physically fragile and suffer loss of terrain through storm erosion or excessive grazing and trampling, yet despite this material damage, there may be no loss in biological diversity. The plants that grow in these physically disturbed sites are usually adequately adapted to their surroundings and even require a certain degree of disturbance to provide opportunities for regeneration, which also aids habitat renewal and the preservation of species diversity.

Regeneration in many such communities is dependent on disturbance. For conservation activities this can present a dilemma. Large areas of valuable terrain can be physically destroyed by erosion or biologically impoverished by herbivores. Sheep, deer, cattle and goats have all been known to damage coastal vegetation. Hurricanes have always caused periodic widespread destruction to mangrove forests from which they recover. Removal of herbivores can restore an initial lushness to vegetation with improved flowering. However, this can then lead to an increasing dominance of scrub and graminoid species with loss of diversity in the herbaceous flora. Similarly, coastal

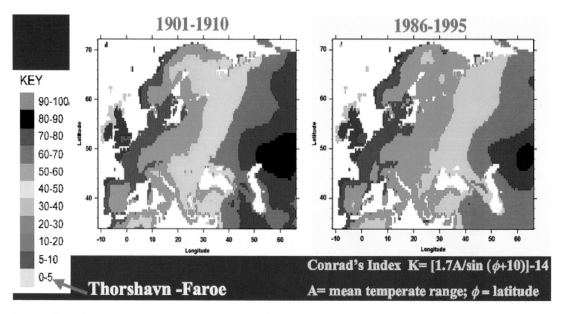

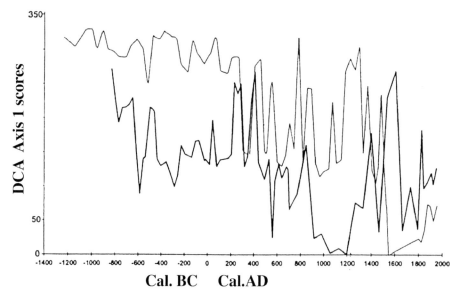

Fig. 7.8   Reduction in continentality as measured by Conrad's Index (see text) in central Europe between the beginning and the end of the twentieth century penetrating into central Europe. Note the gradients in oceanicity (the inverse of continentality run parallel to the eastern distribution limits of the oceanic species shown above. (Map prepared by Dr C. E. Jeffree using temperature data from the Climatic Research Unit, 0.5° gridded 1901–1995 Global Climate Dataset; New *et al.*, 1999, 2000.)

Fig. 7.9   Detrended correspondence analysis of eigenvalue scores for plant fossil data in two northern England bogs. The position of each sample relative to the *y*-axis reflects the degree of soil surface wetness. (Reproduced with permission from Mauquoy & Barber, 1999.)

Fig. 7.10   An eroding salt marsh and falling shore level threatens to remove part of the historic St Andrews Old Course. Over the last 50 years many remedial measures in the forms of walls and groynes have been tried to arrest the erosion but the shore level continues to fall and erosion remains a serious threat to the integrity of the golf course.

heaths can be invaded by trees, which if left undisturbed can dominate the landscape and reduce the diversity of the natural coastal plant communities. Coastal nature reserve management plans therefore usually attempt some form of robust protection policy. Similarly, when erosion encroaches on golf courses and other areas where territorial protection is paramount, concrete walls and groynes are frequently employed

(Fig. 7.10). Unfortunately, this form of protection frequently results in the loss of much of the upper shore vegetation causing the shore level to fall, eventually undermining the coastal defences which then require continual repair.

Finding a balance between protection and still allowing the periodic disturbance that is necessary for species regeneration is becoming an ever more serious problem in coastal habitats. Rising sea levels increase the demands on land for housing, industry, and recreation, and these multiple onslaughts tend to reduce coastal plant communities to vestigial ribbons along the edges of golf courses. Where once series of dunes, slacks and heathlands provided a broad band of varied communities that have been resilient over centuries to fire, floods, grazing and erosion, all that remains is a single dune front with some marram (*Ammophila arenaria*) and a few other grasses (Figs. 7.11–7.12). In the past the foreshore communities, although sometimes physically damaged during periods of stormy weather, could nevertheless recover as they were backed up by a natural coastal hinterland with a reserve of biodiversity. Today this ecological space is absent and the disturbance so frequent that recovery is greatly hindered.

Not all shores are equally susceptible to disturbance. The classical hard shores where the terrain consists of hard rock or cliffs may be subject to attack from the sea; nevertheless, they retain their status as hard shores and cliffs. Chalk cliffs may erode but the biological habitat is renewed rather than destroyed.

Fig. 7.11   The eroding West Sands of St Andrews (east Scotland) where removal of strand-line litter by regular raking in order to be awarded the prestigious European Blue Flag for beach cleanliness is resulting in the destruction of the last remains of a once vigorous and regenerating dune system.

Fig. 7.12   Actively growing line of yellow dunes at Tentsmuir National Nature Reserve at a time of rapid expansion in the 1970s due to the dominance of a lyme-grass (*Leymus arenarius*) creating low level dunes. Later colonization by marram (*Ammophila arenaria*, far distance) raises the height of the dunes and replaces the lyme-grass. This naturally accreting dune system is on the same region of the east coast of Scotland only 8 km to the north of the eroding dunes shown in Fig. 7.11.

It is the soft shores that are most likely to suffer from disturbance and erosion.

Soft shores vary in their plant communities and can therefore be expected to differ in their responses to erosion and disturbance. Some examples from different parts of the world are therefore discussed below not only in an attempt to highlight the ecological richness of maritime margins but also to draw attention to the range of adaptations that permit coastal plants to survive in such physically hazardous environments. If we wish to minimize the damage that they are likely to suffer in the years ahead it will be particularly important to give attention to the preservation of marginal coastal plant communities.

7.2   NORTHERN HEMISPHERE COASTAL VEGETATION

7.2.1   Foreshore plant communities

The foreshore flora, although at the mercy of the sea, nevertheless derives from this perilous situation the advantage of oceanic seed dispersal. Consequently, beaches on either side of the North Atlantic from

Québec, Newfoundland and Labrador, to the north-western shores of Europe have very similar strand-line plant communities dominated in summer by annual species such as orache (*Atriplex* spp.), sea rocket (*Cakile maritima*), and sea sandwort (*Honckenya peploides*) as a result of transatlantic seed dispersal (Figs. 7.13–7.14). Some coastal species are notable for their widespread distribution. Scots lovage (*Ligusticum scoticum*) has a disjunct distribution and can be seen on the cliffs from Norway's Nordkapp (71° N) to southern Scotland (55° N) and from western Greenland (67° N) to Newfoundland (Fig. 7.14), Nova Scotia and New England (42° N).

Seedling establishment on the upper beach is extremely hazardous. With the ever-present dangers of erosion, burial, overheating and desiccation, as well as inundation by seawater, it is not surprising that much of the foreshore plant cover in summer is composed of annual species. Colonization of this zone is highly dependent on the shelter and nutrition that can be provided by the flotsam and jetsam that washes up on the strand line. Shorelines with a good supply of detritus can develop a luxuriant growth of annual plants in summer (Fig. 7.12). Seaweed is one of the best

Fig. 7.13   Strand vegetation flourishing on a shore in Holandsfjorden, Arctic Norway, with copious deposits of seaweed, mainly bladder wrack (*Fucus vesiculosus*). The large green patches are sea sandwort (*Honckenya peploides*). (Inset) Detail from strand vegetation shown above with orache (*Atriplex* sp.) flourishing in a bed of washed-up bladder wrack.

ameliorators of environmental extremes for young foreshore plants as it provides not only shelter and anchorage but also nutrition.

The strand line has always been a natural repository for seaweed and timber even before human activity contributed to the rubbish that is now found on most shores. Modern detritus may be unsightly but some litter either natural or artificial is essential for diminishing sand movement, reducing excessive temperature fluctuations, conserving moisture, and fostering the development of the all-important embryo dunes. Flotsam and jetsam also provide a habitat for the

Fig. 7.14 *Ligusticum scoticum* growing on the south-west shore of Newfoundland: an example of an amphi-Atlantic coastal species. In Québec this species is commonly called the '*Livèche Écossaise*'. (Inset) North-Atlantic distribution of *Ligusticum scoticum*. (Reproduced with permission from Hultén & Fries, 1986.)

invertebrate fauna that is sought by the many different bird species that feed along the strand line (Llewellyn & Shackley, 1996). All too often a short-sighted environmental policy leads certain local government authorities, anxious to promote the touristic attraction of their beaches, to remove this detritus with regular raking. Although the desire to remove some of the more objectionable items of rubbish is understandable, it is unfortunate that regular beach raking is often so thorough that it removes completely the first line of plant colonization and an important bird feeding area. Sadly, the economic desire to gain prestigious awards

for clean beaches, such as the European Blue Flag, can result in the disappearance of the foreshore vegetation and expose the front dunes to erosion by wind and tide (Fig. 7.11).

The annual vegetation cover on the foreshore is only ephemeral, but during its brief summer existence it reduces the steepness of the shore profile with accreted sand (Figs. 7.15–7.16). Once the summer growing season is over the annual plants die, the upper shore profile again steepens, and the perennial rhizomatous species are left to hold the foredunes against the winter storms. The buried rhizomatous network most commonly

Fig. 7.15   Sloping foredune in midsummer on the island of Sanday, Orkney. The band of annual species colonizing the foreshore immediately in front of the dune reduces the steepness of the dune profile. At the end of summer the gradient of the dune profile gradually increases in the face of the winter storms.

created by graminoid species can hold the young embryo dunes in place and through their carbohydrate reserves produce new shoots in spring even though they may be buried to considerable depths.

One of the commonest rhizomatous grasses holding in place the first embryo dunes in northern and western Europe is sand couch-grass (*Elytrigia juncea*; Fig. 7.17). Once an initial reserve of tillers is established the oblique growing stolons of this species can emerge from burial below 30 cm of sand. Tiller fragments are also easily spread, but the young plants in newly formed sand accretions produce little seed. Thus, seed for spreading the species and aiding the formation of new embryo dunes is dependent largely on the state of the yellow dunes to landward, which provide the main source of seed.

When the shore is denuded by erosion or biologically impoverished, as when concrete walls replace the yellow dunes, recolonization is hindered due to the lack of propagules of this important pioneer species. Once the embryo dune is destroyed, the main front dunes become vulnerable and inevitably erode (Fig. 7.18). There is therefore a mutualistic relationship between young foredunes and the first line of yellow or mobile dunes on their landward side, with the embryo dunes being dependent on the mature dunes for seed regeneration, and the mature dunes dependent on the younger dunes for protection against erosion.

7.2.2   Dune systems of the North Atlantic

In the northern hemisphere dune systems are mainly anchored by grass species. Grasses produce their leaves sequentially, which means a considerable portion of the growing season has passed before they develop their full photosynthetic potential. Consequently, as growing seasons shorten at higher latitudes, grasses tend to be disadvantaged due to their failure to maximize their full leaf-area potential early in a short growing season. Marram (*Ammophila arenaria*) is the principal European dune-forming grass, and reaches a northern limit at Faerna (63° N) on the Norwegian coast, only just north of the limit for sand couch-grass (*Elytrigia juncea*; Fig. 7.17) at 62° N (near Ålesund, Norway).

Along the east coast of North America, the St Lawrence and the Great Lakes, the vicarious

Fig. 7.16   Sea rocket (*Cakile maritima*) growing on the
Orkney sand dune in Fig. 7.15. This is a common and variable
species that colonizes drift-line communities on sand and
shingle. When various vicarious and subspecies are included it is
one of the most widespread species around the shores of the
North Atlantic and Mediterranean Sea. *Cakile islandica* was the
first species to be recorded when the island of Surtsey was
created by submarine volcanic eruptions between 1963
and 1967.

Ammophila species, American beach grass (*A. brevigulata*) is the dominant primary sand dune stabilizer, as far north as Newfoundland and Labrador. However, the American beach grass – like the vicarious European species (*A. arenaria*) – does not extend into the Arctic. Both *Ammophila* species are well adapted to sand burial, wind, lack of moisture and nutrients. In both species growth is even stimulated by burial with wind-blown sand, and new shoots can emerge from burial when covered to a depth of one metre. The leaves are also rich in silica, which enables them to withstand sand blasting.

In the more southern regions of the European Arctic, lyme grass (*Leymus arenarius*) is the major species as far north as Norway's North Cape (71° N). The vicarious North America dune-grass (*Leymus mollis*) is equally dominant on arctic shores from Alaska and northern Canada to Greenland. A study of this species carried out on the east coast of the Hudson Bay (northern Québec) found that populations of *L. mollis* have different phenotypic responses depending on whether they are growing on the low foreshore or on the stabilized dune. On the latter, *Leymus mollis* ramets tend to have a lower net carbon assimilation rate and water use efficiency, and a higher substomatal CO_2 concentration than on the foredune. However, under controlled conditions the differences observed in the field disappear, suggesting that these are not genetic but determined by environmental changes along the

Fig. 7.17   Sand couch grass (*Elytrigia juncea*) – one of the most important perennial species in forming embryo dunes which are essential for the protection and renewal of the main dune system.

Fig. 7.18   Eroding sand dune at Tres Ness Point in Sanday (Orkney, UK). The once thriving arable farm is now in danger of disappearing due to coastal erosion.

foredune-stabilized dune gradient. It has been suggested that the higher net carbon assimilation rate on the foredune might be related to higher sink strength in relation to the growth-stimulating effect of sand burial (Imbert & Houle, 2000).

7.2.3   Arctic shores

The most marginal coastal sites for plant colonization are the shoreline fringes of the ice-covered lands of Antarctica and the continents and islands bordering the

Arctic Ocean. Each summer the ice and snow retreat sufficiently for plants that inhabit these shores to have a growing season that can vary from as little as 6 weeks to up to 3 months. It is on the low shores nearest to the sea that the shortest polar growing seasons are found. In addition, the beach is reworked annually by the on shore movement of winter sea ice creating a gravel beach ridge (Figs. 7.19–7.20). This ice scouring leaves behind a low-lying flood-prone plain bordered to landward by the displaced material. The polar shores therefore have two distinct habitats: a low coastal plain and a beach ridge. In the low coastal plain, proximity to the cold sea causes snow and ice to persist longer in early summer than on the adjacent beach ridge. Even during the height of the growing season the soil and air temperatures down on the low foreshore are lower than on the beach ridge (Fig. 7.21). It is one of the botanical marvels of high-latitude ecology that there are so many species of flowering plants that succeed in living in these low-lying shores.

Arctic shorelines do not provide the quantities of sand that are needed to form dunes. The tidal amplitude is low, often only 0.2–0.3 m. Salinity is low in arctic coastal waters, which allows many non-halophytic species to colonize foreshore habitats. The plants that survive on polar shores are therefore a highly varied group of pioneer species that can be found in many other habitats throughout the Arctic.

In continental areas such as northern Canada and Greenland the upper shore is drought prone due to the lack of summer precipitation. Despite the absence of dunes there are nevertheless a number of characteristic arctic grass species that are widespread along arctic shorelines. Some of the most widespread grasses throughout the Arctic are the arctic salt marsh grasses (*Puccinellia* spp.), of which the most widespread species is the creeping salt marsh grass (*P. phryganodes*), which has a circumpolar distribution throughout the High Arctic with the exception of the extreme north of Greenland (see also Section 6.7).

The drier parts of the arctic shore are not very different physically from open land anywhere else at high latitudes. One of the more successful species on the low arctic shore, and also on exposed mountain slopes and screes, is the purple saxifrage (*Saxifraga oppositifolia*; Section 6.4.2).

The arctic shore does, however, differ due to the colder conditions and shorter growing seasons found here than on sun-trapping mountain slopes. In many years the melting of the residual snow banks, and therefore the onset of growing season on the shore, can be delayed. Although the purple saxifrage may flower and even produce pollen, the shortness of the growing season often results in little viable seed being produced (see Chapter 4). However, the proximity of the warmer and drier beach ridge provides an adjacent population (a different ecotype) that can be fertilized by the plants on the low shore and thus preserve the genetic characteristics of the shore ecotype. As already mentioned (Chapter 4) the warmer dry ridge and the cold shore support different ecotypes of the purple saxifrage and this can replenish the populations on the low shore with hybrid seed. In some areas the shore is now enjoying an earlier beginning to the growing season and producing remarkable flowering displays (Fig. 6.6).

Other species capable of surviving in this habitat include moss campion (*Silene acaulis*) and the alpine bistort (*Polygonum viviparum*). Moss campion being a cushion plant maximizes internal tissue temperatures with a minimal surface area in reaction to the body mass of the plant. The viviparous bistort (*Polygonum viviparum*) despite the fact that it reproduces mainly vegetatively with bulbils nevertheless retains considerable genetic diversity, as can be observed in different enzyme phenotypes and bulbil colours (Fig. 7.22). The differentiation of ecotypes and medium to high levels of genetic diversity in arctic and alpine populations is thought to be the result of occasional sexual reproduction (Bauert, 1996).

Notwithstanding the short growing season some polar shores are highly productive and are sought out as feeding places by various species of migrating geese. In the more sheltered bays and fiords salt marshes develop and the sedges and grasses of these coastal flats are grazed by barnacle geese (*Branta leucopsis*; Figs. 7.23–7.24) that fly from the Solway Firth in southern Scotland (55° N) as far north as Spitsbergen (79° N). In North America, the lesser snow goose (*Anser caerulescens*) undertakes an even longer migration. Birds that winter on the coast of the Gulf of Mexico (30° N) fly up the Mississippi and Missouri rivers and can reach the Hudson Bay coast (60° N) and further north into the eastern Arctic (70° N). The mutual nutritional interactions between goose grazing and arctic salt marsh ecology during the long growing

Fig. 7.19 Arctic shore and beach ridge at Ny Ålesund, Spitsbergen, as viewed from the air with the beach ridge and the drier land behind it to the left of the picture. The position of the beach ridge where it adjoins the low shore is indicated by a red arrow.

Figs. 7.20   Arctic shore and beach ridges at Ny Ålesund, Spitsbergen. Note the heterogeneous nature of the arctic shores with dry ridges, wet hollows, and late-lying snow patches.

days of the arctic summer have already been discussed in Section 2.4.1 where sugar-rich salt marsh grasses and sedges, fertilized by goose droppings, provide nutritious feeding (Fig. 7.24). The advantages to the geese of these polar pastures are demonstrated in the speed with which the goslings which are hatched in July grow and become sufficiently strong to begin their long migration south. In August, barnacle geese leave Spitsbergen

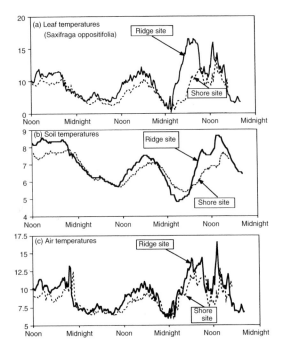

Fig. 7.21   Comparison of (a) leaf (*Saxifraga oppositifolia*), (b) soil and (c) air temperatures over a continuous 60 h period from 12 noon on 25 July 1994 on adjacent ridge and shore sites as shown in Fig. 7.19 at Ny Ålesund, Spitsbergen. (For further experimental detail see Crawford *et al.*, 1995.)

flying first to Bear Island (74° N), and then continue to the British Isles, arriving in the Solway Firth (55° N) between the end of September and early October.

Despite the extreme thermal limitations of the environment, the plant species that inhabit northern shores even into the High Arctic do not lack for genetic diversity (see Chapter 2). The reserves of genetic diversity found at these high latitudes probably reflect two important aspects of these unique shores. The first is the antiquity of the habitat. During the Pleistocene, when sea levels were as much as 140 m lower than at present, the shore area and polar desert hinterland north of the major ice sheets would have been more extensive and would therefore have provided a sufficiently open habitat for plant survival. Secondly, the arctic shores are not geographically isolated from one another. The circumpolar coastlines have never at any time been entirely encased in ice and plant migrations will have continued in different areas throughout most of the Pleistocene. The evidence for the possibility of this remarkable survival has been much debated.

Fig. 7.22   The alpine bistort (*Polygonum viviparum*), a widespread species in arctic and alpine habitats that produces mainly vegetatively with bulbils. Occasionally the flower sets fertile seeds which may account for the considerable genetic variability found in both arctic and alpine populations of this species (see text).

However, the study of chloroplast DNA in present-day populations of the purple saxifrage (*Saxifraga oppositifolia*) has substantiated this view of the antiquity of at least this species (Abbott & Comes, 2004) and by implication other high arctic species with similar distributions (see Section 6.4.2 for details).

Fig. 7.23   Barnacle geese (*Branta leucopsis*) with goslings grazing in late August on the shore at Ny Ålesund, Spitsbergen.

Fig. 7.24   A coastal marsh much visited by geese in Kongsfjord (Spitsbergen) with clumps of the anoxia-tolerant sedge *Carex misandra*, a favoured fodder for barnacle geese.

7.3   SOUTHERN HEMISPHERE SHORES

7.3.1   Antarctic shores

The ultimate marginal coastal environment for the survival of flowering plants is without doubt the Antarctic Peninsula and adjacent islands where the flora consists of 380 species of lichens, 130 species of bryophytes but only two species of flowering plants (Alberdi *et al.*, 2002). The Antarctic pearlwort (*Colobanthus quitensis*) and the Antarctic hairgrass (*Deschampsia antarctica*) are the only two angiosperm species to survive south of 56° S and occur in small clumps near the shore on the west coast of the Antarctic Peninsula (Fig. 7.25–7.27). It is in this area of maritime Antarctica, where mean air temperature tends to be above zero during the summer, that most of the Antarctic vegetation is found. Climatic warming in Antarctica has

prompted a number of studies in recent years into these two remarkable flowering plants. This low number of species, compared with the Arctic, may be due to a long history of low temperature and isolation from sources of propagules. The degree of isolation for plant populations in Antarctica is not just from other continents, but also between different regions of the Antarctic Peninsula. This can be seen in *Deschampsia antarctica* where two distinct populations, one from the maritime Antarctic, namely Signy Island in the South Orkney Islands, and the other from the Leonie Islands 1350 km further south (67° S) were found to be genetically distinct from each other with low levels of historical gene flow between them. Their genetic structure suggests that new populations of *D. antarctica* are founded by one or just a few individuals and that vegetative reproduction and selfing are therefore likely to have been key factors

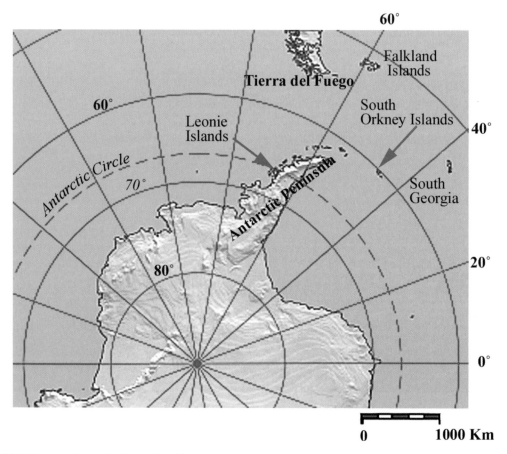

Fig. 7.25   Location of antarctic sites mentioned in text.

Fig. 7.26   A coastal colony on the Antarctic Peninsula with a plant community containing both *Deschampsia antarctica* and *Colobanthus quitensis*. (Photo Dr T. A. Day.)

Fig. 7.27   Details of the only two flowering plants native to Antarctica. (Left) The antarctic pearlwort (*Colobanthus quitensis*). (Right) The antarctic hair grass (*Deschamspsia antarctica*). (Photos Dr T. A. Day.)

in the establishment of *D. antarctica* at new sites in the Antarctic during recent years (Holderegger *et al.*, 2003).

It is of interest that despite the harsh conditions of the environment both *D. antarctica* and *Colobanthus quitensis* succeed in establishing significant seed banks of between 107 and 1648 seeds m^{-2}, which are comparable in size to arctic and alpine species (McGraw & Day, 1997). Physiologically, the existence of these plants in such a permanently harsh environment makes them of particular interest for the study of adaptations to cold environments and mechanisms of cold resistance in plants. Both species have a high resistance to freezing and can show a high photosynthetic capacity at low temperatures (Alberdi *et al.*, 2002). Despite the fact that these two species share closely the same habitat the nature of their cold resistance differs. Comparisons of the thermal properties of leaves and the lethal freezing temperatures (LT50 – the temperature required to induce a 50% mortality in the leaf tissues) have shown that the grass *D. antarctica* was able to tolerate freezing to a lower temperature than *C. quitensis*. Super cooling (cooling of a liquid below its freezing point without freezing) was found to be the main freezing resistance mechanism for *C. quitensis*. Thus, the grass *D. antarctica* is mainly a freezing-tolerant species, while *C. quitensis* avoids freezing (Bravo *et al.*, 2001). The species also differ in the nature of their foliage. *D. antarctica* is highly sclerophyllous while *C. quitensis* only has sclerenchymatic tissues and thin cuticles (Mantovani & Vieira, 2000). Thus, as well as their hardiness in sharing the same habitat, these species show that even in these minimal conditions it is possible for plants to differ in how they adapt to such a marginal situation.

The photosynthetic temperature response of the Antarctic vascular plants *Colobanthus quitensis* and *Deschampsia antarctica* have also been found to differ and to be more efficient at low than high temperatures (Xiong *et al.*, 1999). Measurements of whole-canopy CO_2 gas exchange and chlorophyll fluorescence of plants growing near Palmer Station along the Antarctic Peninsula have shown negligible midday net photosynthetic rates on warm, sunny, days (canopy air temperature > 20 °C), but nevertheless attained positive photosynthetic rates (Fig. 7.28) on cool days (< 10 °C). It was therefore concluded that although

continued warming along the Peninsula will increase the frequency of supra-optimal temperatures, the site averaged increase would be unlikely to exceed the temperature optima for photosynthesis for these species. It is therefore not surprising that the recent warming trend has already resulted in more seeds germinating and an increasing number of seedlings and plants. One report indicates a 25-fold increase in plants together with a southward extension of their range (see also Convey & Smith, 2006).

An additional modern hazard for flowering plants in Antarctica is exposure to high levels of UV-B radiation. Along the west coast of the Antarctic Peninsula springtime ozone depletion events can lead to a two fold increase in biologically effective UV-B radiation. Studies which have examined the influence of solar UV-B on the performance of the Antarctic vascular plants (*Colobanthus quitensis* and *Deschampsia antarctica*) have shown that leaf longevity decreased from the first growing season through to the fourth, suggesting that UV-B growth responses tended to be cumulative over successive years.

7.3.2   New Zealand

All scientific examinations require a control. This is just as true for field-based ecological studies as it is for laboratory-based experiments. Before any general conclusions can be made about the nature of the limitations on plant distribution in coastal habitats some attempt should be made to examine shores in a totally different situation. Ideally, an ocean shore on another life-supporting planet would be ideal. However, lacking this possibility at present, the coasts of New Zealand provide an ecologically informative example of a region where both the flora and the fauna have had a very different evolutionary history from that in the northern hemisphere. In New Zealand the native plants evolved without the threat of being grazed by mammals until the Maoris introduced rats. The date of this rodent invasion has been much disputed. However, the application of ^{14}C AMS (accelerator mass spectrometry) has now allowed this invasion to be dated confidently to the thirteenth century (Wilmshurst & Higham, 2004). This was followed later in the eighteenth and nineteenth centuries by other mammals from Europe and elsewhere. Before the arrival of European

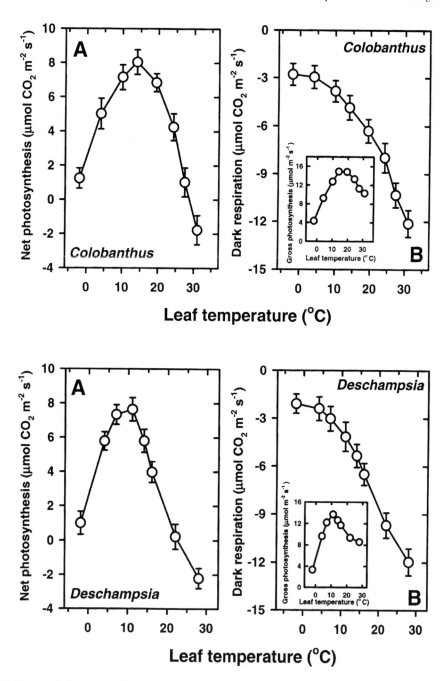

Fig. 7.28 (A) Photosynthetic responses (Pn) to temperature of the whole canopy as shown for *Colobanthus quitensis* (top) and *Deschampsia antarctica* (bottom). (B) Dark respiration (Rn) of the species in relation to temperature. The insets show the temperature dependence of Pg calculated as the sum of Pn and Pr assuming that respiration rates are similar in light and dark. (Reproduced with permission from Xiong *et al.*, 1999.)

settlers the principal graminoid native species that stabilized New Zealand's foredunes were the pingao (*Desmoschoenus spiralis*), a robust sedge used traditionally by the Maoris for making cloaks and matting, and spinifex grass (*Spinifex sericeus*) which is common on sand dunes along the coasts of Australia and New Caledonia. Pingao (*Desmoschoenus spiralis*; Fig. 4.18) has thick rope-like stems forming a vegetative entanglement inside the dune which provides robust anchoring against the strong winds of the South Pacific. Unfortunately, many New Zealand plants lack protection against small grazing mammals as they evolved in an environment devoid of mammals and where herbivory risks were confined to the pecking of shoots by grazing birds, such as the long-extinct giant moa. Consequently, pingao, although physically robust, lacks adequate protection against rabbits.

The introduced marram grass has high silica deposits in its leaves and shoots, like the common reed (*Phragmites australis*) and the sea club-rush (*Bolboschoenus maritimus*), and is more resistant to rabbits; as a result the native pingao has been brought near to extinction (see Section 4.6.1, Fig. 4.18). Fortunately, there is now a renewed interest in preserving the pingao due to the revival of Maori weaving that uses its leaves, and attempts are being made to protect and propagate the species (Wardle, 1991). Spinifex has also been greatly reduced by grazing but with protection is proving capable of recovering. Prostrate woody species with a capacity for layering have always been successful in establishing themselves on foredunes in both the northern and southern hemispheres. Coprosma (*Coprosma acerosa*) is a notable species in New Zealand that can be compared to northern hemisphere creeping willow (*Salix repens*) which although more commonly a dune-slack species will grow on sand dunes in areas with sufficient rainfall.

The New Zealand foreshore has now lost much of its native vegetation to alien invaders. In addition to *Ammophila arenaria* from Europe, now *Leymus racemosus* from China, *Carpobrotus edulis* from South Africa, and tree lupins (*Lupinus arboreus*) from California provide a cosmopolitan grazing-tolerant beach flora. This inter-hemisphere comparison is therefore a pertinent reminder of the powerful effects of herbivory even in a very marginal environment where it might be thought that physical constraints on plant survival were paramount and biotic factors would be minimal.

7.4   GLOBAL SHORE COMMUNITIES

7.4.1   Salt marshes and mudflats

Salt marshes and mudflats are soft shores and therefore vulnerable to both natural and human disturbance. Given a slow and steady rise in sea levels and an adequate input of sediment, it might be expected that mudflats and salt marshes will be able to grow apace with rising sea levels and provide a natural means of increasing coastal defences. If this were to happen it might offer some redress for the long-term reduction of salt marshes due to so-called 'land reclamation' and infill from dumping that has taken place worldwide on mudflats and salt marshes over many centuries. However, this would require a cessation of the ever-increasing construction of coastal defences, extensions to promenades, car parks, and other amenities which continue to destroy or else interfere with the natural growth of this particular type of soft shore.

Unfortunately, it is far from certain that salt marshes will be able to respond positively to rising sea levels. Throughout the British Isles the tendency during the past four decades has been for a reduction and fragmentation (Hughes *et al.*, 2000). It is therefore probable that many already fragmented salt marshes will be even further reduced and may perhaps disappear entirely as sea levels rise (Fig. 7.29). Whether or not there is one specific cause for this habitat loss is at present far from certain. For some salt marshes in south-east England it has been suggested that the failure of the marshes to regenerate, even when encouraged by the provision of sediment, is due to the grazing activities of the invertebrate fauna, particularly the polychaetes *Nereis diversicolor* (the ragworm), and the mud shrimp *Corophium volutator*. Organic pollution and agricultural run-off have also been noted for a number of years as causing ecological problems for the natural flora and fauna of low-lying coastal habitats. In Scotland the most detailed study has been on the River Ythan where the total oxidized nitrogen input has increased four fold over the past 30 years (Gillibrand & Balls, 1998) causing blooms of opportunistic green macro-algae (Raffaelli, 2000).

In south-east England, studies on the disappearance of salt marshes began in the 1930s as a result of the loss of inter tidal eelgrass (*Zostera marina*). Recent studies have recorded rising levels of organic matter in the estuarine sediments which have been accompanied

Fig. 7.29   Fractionated salt marsh at Taobh Tuath (Northton, Harris, Outer Hebrides) which may be liable to further erosion as sea levels rise.

by increasing abundance of a number of polychaete and copepod species (Hughes, 1999). In transplant experiments, ragworms have been noted to reduce the survival of *Zostera noltii* by grasping the leaves and pulling them into their burrows. From this and other studies it has been suggested that two conditions exist on the mudflats. In one state, the mudflats are dominated by plants, including algal mats and *Zostera* spp., which prevent colonization by the burrowing invertebrate fauna, as in the Ythan Estuary. The other possibility is that the invertebrate fauna inhibit plant colonization and therefore the estuary does not develop a significant plant cover. Consequently, attempts to restore mudflat and salt marsh communities will inevitably involve careful management of this contest (Hughes *et al.*, 2000). Laboratory-based experiments have also demonstrated significant negative effects of *N. diversicolor* abundance on the survival of *Spartina anglica* seeds

(Emmerson, 2000). The extent of the area of unvegetated shore is also crucial to the regeneration of salt marshes. Where areas of reclaimed agricultural land have become reflooded due to sea-wall failure, smaller, sheltered sites have been noted as re-establishing salt marsh more readily, while larger sites more commonly revert to unvegetated tidal flats (French *et al.*, 2000).

7.4.2   Rising sea levels and mudflats

The environmental problems that surround mudflats are numerous. Rising sea levels, which in the past might have been generally beneficial in extending this habitat in sheltered estuaries and inlets, are now more likely to be disadvantageous as plant colonization will probably be hindered by pollution, human disturbance and interference with natural sedimentation processes. Remedial action, including planting with species that might raise the shore level, e.g. *Spartina anglica*, is unlikely to take place due to the belief that such artificial intervention could reduce the access of wading birds to mudflats. This is not necessarily the case for all mudflats. In Scotland the planting of *Spartina* has been discouraged even though the late flowering of the species makes it unlikely that it will prove as dangerous in this respect as further south (Fig. 7.30). Anxieties concerning erosion and landscape preservation demand instant action, which sadly usually means that public opinion has more faith in civil engineering enterprises than the power of vegetation to hold the landscape in place.

In North America there is considerable concern about the expansion of the common reed into areas which in the past supported salt marshes. The common reed (*Phragmites australis*), although it can be regarded positively in terms of its ability to stabilize estuary shores in brackish waters and provide shelter for birds and other animals, tends to produce monospecific stands and therefore reduce plant biodiversity. In recent years in the coastal marshes of the north-eastern United States *P. australis* appears to have acquired a competitive advantage over a broad range of habitats, from tidal salt marshes to freshwater wetlands. The change in the ecology of the common reed may be due to a variety of causes which it is suggested are all connected to some degree with human disturbance. It has been claimed that the North American population has been infiltrated with aggressive European genotypes (Burdick & Konisky, 2003). Any human disturbance that lowers salinity in brackish waters is likely to aid the

clonal spread of *Phragmites australis* as small rhizome fragments are established more readily under conditions of reduced salinity. Experimental studies (Chambers *et al.*, 2003) have shown that salinity, sulphide, and prolonged flooding combine to constrain the invasion and spread of *Phragmites* in tidal wetlands through their physiological effects on ionic and carbon balance coupled with oxygen availability. Consequently, invasion takes place more readily in marshes occupying lower-salinity regions of estuaries as well in marshes that have been hydrologically altered. Climatically, it is noteworthy that periods of increased rainfall aid the growth of *Phragmites australis*. A study carried out during the 1997–98 El Niño event showed that soil pore-water salinities were negatively related to precipitation during the three years of the study, and that the growing season during the El Niño year was one of the wettest of the past century. These changes were associated with a 30% increase in shoot density, stems which were 25% taller, and an order of magnitude increase in inflorescences (Minchinton, 2002). It is therefore possible that should there be an increase in the frequency of El Niño years a spread of less salt-tolerant invasive species throughout brackish habitats such as salt marshes is to be expected.

7.5   HARD SHORES

7.5.1   Cliffs and caves

Irrespective of their height, it is inaccessibility that makes cliffs probably the least disturbed of all habitats. Provided cliffs are of sufficient size they are generally free from grazing, cutting, burning and detailed scientific investigation. Some exceptions are notable. Pickled rock samphire (*Crithmum maritimum*; Fig. 7.31) was once so popular and valuable that people risked their lives to collect it from precipitous rock faces. Shakespeare describes the collection of this much sought-after herb: 'How fearful … half-way down hangs one that gathers samphire; dreadful trade!' (Shakespeare, *King Lear*). Apart from such exceptions and the occasional visits from adventurous goats and sheep, the cliff vegetation is one of the best examples of totally natural and understudied plant communities that remain in proximity to human habitation. In early summer the display of floral colour on the exposed tops and faces of coastal cliffs is spectacular. Cliff

Fig. 7.30   Cord grass (*Spartina anglica*) in the estuary of the River Eden (Fife, east Scotland) where it was planted in 1948. The colony was successfully raising the level of the shore and would have provided protection against rising sea levels had it not been systematically eradicated during the 1990s in the probably mistaken belief that it might have interfered with the access of wading birds to the mudflats.

Fig. 7.31   Rock samphire (*Crithmum maritimum*). The leaves were formerly much sought after for pickling. Common on cliffs and rocks in England and Wales but rare in Scotland.

vegetation can be highly variable. Such are the range of cliff environments in relation to geology, soil, exposure and moisture that they can provide suitable habitats for species associations with contrasting ecological requirements. Thus depending on location and climate, cliffs can shelter arctic-montane, maritime, calcicole and calcifuge plants and even woodland understorey species with and without an upper canopy of trees.

Cliffs can be subdivided into two contrasting types, namely hard cliffs and soft cliffs (Figs. 7.32–7.33). In the latter, variation in vegetation is maintained due to constant physical disturbance which prevents certain species becoming dominant and thus allows a variety of species to coexist (Cooper, 1997). Hard cliff faces, which are almost vertical or exposed to extreme

conditions, usually lack any significant vegetation cover. Hard and soft cliffs denote merely contrasting ends of a range of possibilities. Depending on the hardness of the rock the degree of erosion will vary and the contrasting effects of instability destroying vegetation as opposed to erosion providing fresh sites and nutrients will have a decisive role in the development of the vegetation. Further variation in the nature of the cliff face vegetation comes from the relative deposition of sea spray, freedom from desiccation on north-facing sites, and access to water through seepage and variation in rock types. One feature in common with all cliffs is that as vegetation builds up with time, it eventually becomes unstable and erodes, thus renewing the succession cycle and preserving biodiversity.

Fig. 7.32   A soft Old Red Sandstone cliff on the Island of Hoy, Orkney. The vegetation clinging to the cliffs is dominated by the greater wood rush (*Luzula sylvatica*).

7.5.2   North Atlantic cliffs

Even in the cold and stormy climate of North Atlantic shores the vegetation clinging to cliffs can vary from dense forest-like stands of trees to tundra-like assemblages of dwarf-herbaceous and woody plants. Starting at the base of the cliffs and working upward, a succession of plant communities can often be observed. Lichens in particular the bright orange *Xanthoria parietina*, are found at the base of most sea cliffs. Above this most disturbed zone, crevices and ledges provide a habitat for halophytic species such as sea plantain (*Plantago maritima*), the buck's-horn plantain (*Plantago coronopus*), rose root (*Sedum rosea*) and sea aster (*Aster tripolium*).

Fig. 7.33   A hard Stromness flagstone cliff at Marwick, Orkney, where vegetation cover is minimal.

Cliff faces are the preferred habit for species that can tolerate sea spray and summer drought but would not survive inundation or competition. Thrift (*Armeria maritima*), and sea plantain (*Plantago maritima*) have an even wider distribution extending throughout the northern hemisphere, including survival on mountain tops. These species would have been widespread at the end of the last ice age but are now restricted to altitudinally disjunct sites where they find habitats with minimal competition. Towards the upper areas of cliffs and extending onto the cliff top where there is no grazing, as on sea stacks, there can be found tall herb communities with wild angelica (*Angelica sylvestris*), red campion (*Silene dioica*), the common primrose (*Primula vulgaris*), foxglove (*Digitalis purpurea*) and sorrel (*Rumex acetosa*). Later in the summer there can also be found grass of Parnassus (*Parnassia palustris*).

Although coastal cliff plant communities escape the predations of grazing mammals they are subject to colonization by sea birds especially during the breeding season. Dense bird colonies cause major changes in cliff vegetation. Too much disturbance and excess guano can eliminate flowering plants entirely. However, the more normal consequence of large nesting colonies of auks and fulmars is to favour those species which respond to large inputs of nitrogen. Most striking among these is scurvy grass (*Cochlearia officinalis*). In areas at the cliff top, where colonies of greater black-backed gulls congregate, the habitat is marked by luxurious growth of grasses such as Yorkshire fog (*Holcus lanatus*) along with sorrel (*Rumex acetosa*) and related docks (*R. obtusifolius*, *R. crispus*) and often stinging nettle (*Urtica dioica*). Puffin colonies are often associated with colonies of mayweed (*Triplospermum maritimum*) and orache (*Atriplex* spp.).

Caves provide an opportunity to investigate the shade tolerance of cliff-inhabiting species. The depth of the Smoo Cave on the north coast of Scotland provides just such an opportunity. The last flowering plant to be found in proceeding to the rear of the cave (Figs. 7.34–7.35) is the opposite-leaved golden saxifrage

Fig. 7.34   The Smoo Cave, north coast of Scotland (Sutherland), the largest opening of a limestone coastline cave in Britain. The depth of this cave makes it possible to observe the relative shade tolerance of cliff-inhabiting plants, the most tolerant flowering plant being the opposite-leaved golden saxifrage (*Chrysosplenium oppositifolium*; see Fig. 7.35).

(*Chrysosplenium oppositifolium*). As plants adapt to shade, the leaves generally become thinner and prone to sugar leakage which can bring about fatal fungal infections. In the case of *C. oppositifolium*, the principal soluble sugar is sedoheptulose, which being a 7-carbon sugar is not readily metabolized by fungi, which may be advantageous in extremely shaded habitats where solute leakage encourages fungal attack. A type of heath community specific to cliff tops is maritime sedge heath (Fig. 7.36). This usually develops a short distance back from the edge of the cliff where salt drenching is reduced, and consists of a community that is rich in various species of *Carex* together with dwarf heather (*Calluna vulgaris*) and crowberry (*Empetrum nigrum*).

7.6   TREES BY THE SEA

Worldwide there are many areas where trees or even forests survive due to the combination of climatic amelioration, reduced competition and habitat protection from grazing and human disturbance that is found in many coastal habitats. Even in the exposed islands of the North Atlantic, sheltered gullies, protected from excessive sea spray, as on the upper areas of the cliffs, harbour a number of species that are more typical of woodlands, with large stands of the greater woodrush (*Luzula sylvatica*) and honeysuckle (*Lonicera periclymenum*). Sometimes, as in pinnacles and cliffs, there can be found relict populations from former woodlands, with rowan (*Sorbus aucuparia*) and aspen (*Populus tremula*; Fig. 7.37). On drier cliffs, as on sandstone block cliffs, a heath-type vegetation can be found with heather (*Calluna vulgaris*), bell heather (*Erica cinerea*), crowberry (*Empetrum nigrum*), wood sage (*Teucrium scorodonia*) and golden rod (*Solidago virgaurea*). On shores of Scottish sea lochs and the steep sides of Norwegian fiords tree cover can be extensive. The mild wet conditions of the Atlantic coastline of Scotland if free from too much disturbance allow the development of extensive oak woods and associated species.

7.6.1   Mangrove swamps

In cool temperate climates tree establishment on salt marshes is an ecological impossibility. Yet throughout the tropics and subtropics, vigorous and productive salt- and flood-tolerant mangrove forests survive on tidal mudflats and are a source of wonder for this

Fig. 7.35   The opposite-leaved golden saxifrage (*Chrysosplenium oppositifolium*) together with the hart's tinge fern (*Phyllitis scolopendrium*) a highly shade tolerant fern inhabiting the rear wall of the Smoo Cave – for details of metabolic adaptation see text.

Fig. 7.36   Cliff-top maritime heath vegetation with a mixture of sedge and heath species on North Hill, Papa Westray, Orkney, after a period of wet and stormy weather.

remarkable physiological endurance of both flooding and high salt levels.

The term mangrove is commonly applied to the habitat as well as to approximately 40 species of trees that dominate these inter tidal tropical forests which lie mostly within 25° north and south of the Equator. However, to avoid confusion some authors use the Portuguese term *Mangal* to refer to just the habitat. The change in salt marsh vegetation that takes place when trees replace grasses as the dominant vegetation on salt marshes is a worldwide ecological boundary of considerable ecological significance. This change from herbaceous to woody vegetation on salt marshes becomes possible when frost is rare. Due to the

movement of unusually warm waters in certain locations, mangroves can be found outside the 25° north and south general limits. The grey mangrove (*Avicennia marina*) can be found as far north as 27° N in the Red Sea (Fig. 7.38), while *A. resinifera* occurs at 38° S in New Zealand, where it succeeds in occupying tidal muds to a lower level than any other angiosperm other than *Zostera* and *Spartina* (Wardle, 1991). In North America, mangroves are found from the southern tip of Florida along the Gulf Coast to Texas.

Mangrove forests are usually in estuaries at the interface between salt and fresh water where impenetrable stands of woody vegetation develop on the mudflats. They are, however, very susceptible to frost.

Fig. 7.37   A clone of aspen (*Populus tremula*) clinging to a cliff at Waulkmill Bay, Orkney. In the Orkney Islands the only remnants of natural woodland are to be found in deep gullies or on cliff faces. Aspen has never been observed to produce seedlings in Orkney and DNA tests have shown that remnant stands in their various refuges consist of a limited number of clones.

Fig. 7.38   Mangroves approaching their northern limits. A colony of dwarf mangrove *Avicennia marina* at Bahrain (26° N) in the Persian Gulf.

As little as three to four nights of light frost can be sufficient to kill most mangrove species. Consequently mangroves are not found north of 32° N and 40° S (Fig. 7.38). Why only frost-sensitive woody plants are capable of surviving on salt marshes has long been a mystery.

A highly significant observation that the increasing water deficit caused frost-induced xylem failure in mangroves led to the discovery that frost-induced xylem embolism appears to set a latitudinal limit to the distribution of mangroves (Stuart *et al.*, 2007). Further study has also revealed that the susceptibility of the mangroves to freeze-induced xylem embolism (blocking of a xylem vessel by an air bubble) is related to xylem diameter, the species with small vessel diameters being able to survive at higher latitudes than those with larger diameters (Table 7.1). A significant relationship was also found between vessel diameter and the frost-induced percentage loss in hydraulic conductivity (Fig. 7.39).

It would be simplistic to suggest that cold temperatures alone are sufficient to exclude plants from saline habitats given the extensive growth of herbaceous halophytes even north of the Arctic Circle. However, herbaceous plants do not have to maintain woody stems with meristematic tissues that are liable to desiccation injury during the cold season. It has been suggested that the above-ground tissues in areas of

mangrove–salt marsh transition should therefore be likened to those in other areas that are treeless due to disturbance by fire or grazing.

The alternative question also arises as to why mangroves have to have such large vessels that they expose themselves to the risk of developing embolisms. Living as they do in the interface between saline and fresh water it may be that the shoot tissues require a very large supply of water to avoid osmotic injury and that having large vessels is the optimal solution, provided there is no risk of freezing. Although mangroves can reduce their transpiration rate when salinity levels are high they nevertheless under favourable conditions have transpiration rates that match those of lowland dipterocarp and tropical heath forests with a similar climate in north Borneo (Becker *et al.*, 1997). Other factors that are important for the survival of mangrove forests are an adequate supply of silt and nutrients and a high tidal range. It is this latter factor that has probably the greatest importance for their survival as it reduces the increasing salinity coming from high evapotranspiration rates, supplies the silt necessary for physical support, as well as the nutrients, and the intervals of exposure to air which provide a diurnal relief from the dangers of anoxia imposed by flooding at high temperatures. Mangroves with their silt and peat accumulations protect tropical coastlines against hurricanes and storm surges. They are, however, susceptible to mass

Table 7.1. *Vessel diameters for five mangrove species from the northern and southern limits of the worldwide mangrove distribution*

species	D (μm)[a]	D_h (μm)[b]	Latitude[c]
Aegiceras corniculatum	17.1 ± 0.68	24.2 ± 1.22	35° 42″30″S
Avicennia marina	19.1 ± 0.35	29.0 ± 0.66	35° 42″30″S
Avicennia germinans	31.0 ± 1.01	40.9 ± 1.41	29° 40″08″N
Rhizophora mangle	30.8 ± 2.52	38.3 ± 2.94	29° 4″35″N
Rhizophora stylosa	22.2 ± 1.17	35.2 ± 0.90	27° 46″41″S

[a] Mean vessel diameter (D) ± SE.
[b] Hydraulically weighted mean vessel diameters, calculated as $D_h = \Sigma D^5/\Sigma D^4$.
[c] Collection latitude.
Reproduced with permission from Stuart *et al.* (2007).

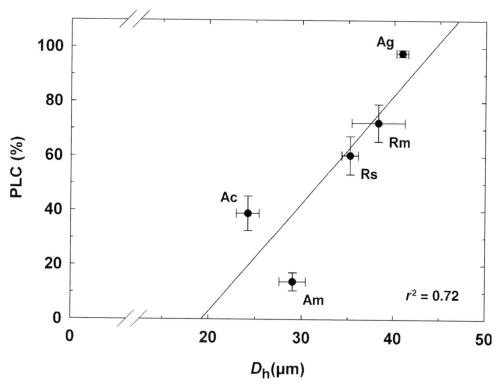

Fig. 7.39   Freeze-induced percentage loss in stem hydraulic conductivity (PLC) at native xylem tensions as a function of hydraulically weighted vessel diameter (D_h) in five mangrove species. Hydraulically weighted vessel diameters ($D_h = \Sigma D^5/\Sigma D^4$) account for the disproportionate contribution of larger vessels to conductivity. Regression line is $y = 3.961x - 76.08$, with $r^2 = 0.72$. Bars denote standard error of the mean, with $n = 5$–13 stems for PLC and $n = 5$ stems for vessel diameters. Two-letter abbreviations indicate genus and species: Ac, *Aegiceras corniculatum*; Ag, *Avicennia germinans*; Am, *Avicennia marina*; Rm, *Rhizophora mangle*; Rs, *Rhizophora stylosa*. (Reproduced with permission from Stuart *et al.*, 2007.)

mortality as a result of storms, excessive salinity and the deposition of large amounts of silt that exceed their ability to grow and raise their anchoring organs to match the rate of salt deposition. Mangrove development is principally cyclical, with older stands dying either through erosion or decay and new colonies advancing onto fresh silt deposits. It is the generally accepted view that mangroves are not land builders gradually encroaching on the sea but that they follow silt deposits and play a passive role in sediment accumulation. It can, however, be seen, as in the mangroves in the Bay of Bengal, that the pneumatophores that thrust upwards from the shallow rooting system are frequently laterally flattened with the broader side facing the oncoming current. Such an orientation implies an active role in slowing down the current and accumulating silt.

The largest mangrove forests are found in the Bay of Bengal (Fig. 7.40). Here in the world's largest delta formed by the rivers Ganges, Brahmaputra and Meghna between India and Bangladesh are the Sunderbans, a stretch of impenetrable mangrove forest of

Fig. 7.40   The Sunderban marshes in the Bay of Bengal. This is a densely forested region with at least 26 species of mangroves.

great size and biodiversity and a UNESCO World Heritage Site. The lower delta plain west of the modern river mouths has been accumulated in three phases over the past 5000 years (Allison *et al.*, 2003). Their present configuration is related to natural factors, such as eastward tilting of the delta, rapid sediment accumulation (to 0.7 cm yr^{-1}), marked land subsidence (to 0.5 cm yr^{-1}), and increasing anthropogenic influences, including large-scale land reclamation and decreased river flow influx (Stanley & Hait, 2000). A close network of rivers, channels and creeks intersect the entire Sunderbans forest area, which comprises hundreds of islands which get either partially or fully inundated during the diurnal high tides. The existing large rivers running from north to south are the remnants of the old courses of the Ganges.

The main current of the Ganges has gradually shifted eastwards over the last few centuries. A major tectonic movement in the sixteenth century appears to have caused a lifting of the upper crust towards the west, thus forcing the Ganges to drain mostly through Bangladesh, causing the sources of all the rivers in the western part of the Sunderbans to be progressively silted up, thus disconnecting the inflow of fresh water into the mangrove delta. This has increased the salinity of the river waters as well as making them shallower. Consequently, during the ebb tides the receding water level causes scouring and creates an innumerable number of small creeks, which normally originate from the centres of the islands. The receding water, while draining into the Bay of Bengal, carries away large volumes of silt. This silt is deposited along the banks of the rivers and creeks during high tide, increasing the height of the banks as compared with the interior of the islands. As a result, high tide cannot normally reach the interior of the islands (Blasco, 1977).

This environmental change from west to east in the Sunderbans highlights the conditions that mangrove forest requires for survival in relation to the quantity of silt together with optimal salinity and depth of the floodwaters. As already noted, there has been much discussion as to whether mangroves follow silting or cause silting. The answer is most probably a question of the degree in relation to the rapidity of the silting. In recent times the silt load in the great rivers of India has increased with the extensive erosion that is taking place in the Himalaya. This is likely to become even more severe as the mountain glaciers retreat and the glacial outwash is transported downstream. More sediment may enhance the cyclical processes in mangrove development in some places by providing fresh opportunities for colonization, while in others too much sediment may bury the mangroves at a rate at which they cannot survive.

7.7 PHYSIOLOGICAL ADAPTATIONS IN COASTAL VEGETATION

7.7.1 Drought tolerance

Tolerance of drought and the ability to regenerate in disturbed sites are two of the outstanding features of sand dune vegetation. Many of the moss and lichen species, which hold the surface of the sand dunes, have the remarkable property of being able to allow their tissues to dry out without losing viability. After prolonged periods of desiccation (e.g. lying on a herbarium sheet for 70 years!) some moss species begin to resume metabolic activity within 30 minutes of gaining access to water. One of the dangers of desiccation injury, especially when plants are exposed to sunlight, is the generation of highly destructive oxygen free radicals by transfer of energy from excited chlorophyll to oxygen. The dune moss *Tortula ruraliformis* when desiccated in the light has high concentrations of the antioxidants α-tocopherol and glutathione, which may contribute to its remarkable desiccation tolerance (Seel *et al.*, 1992).

Little thought is given to the needs of sand dune plants for water, probably because most perennial dune flowering plants are either economical with water supplies, as with the sand dune grasses, or else restrict growth to seasons when water stress is not a serious problem. Dunes have a characteristic flora of winter annuals with species such as common whitlow grass (a diminutive member of the cabbage family, *Erophila verna*) and spring vetch (*Vicia lathyroides*), which survive the heat and drought of summer as seeds, then germinate in the autumn, grow over winter and flower and seed in spring and early summer. Marram grass (*Ammophila arenaria*) of dune tops with its hard, inrolled leaves, appears to be the embodiment of drought resistance. Not only is transpiration reduced to a minimum, but also the deep root systems are able to access water from the lower moist layers in the dunes.

However, water is still an essential resource for dune vegetation. The transpiration needs of dune vegetation can exhaust the rainwater held in the rooting zone in a typical sand dune in four days (Salisbury, 1952). As discussed in Section 3.7.3 'Dew' the elevation of water by capillary action moves water no more than 40 cm above the water table. The average rooting depth of marram grass is in the order of 1–2 metres and the depth of the water table can be 6 metres or more below the surface of a high dune. The dune grasses are dependent therefore for their water supply in dry periods on the upward movement of water vapour, which takes place through internal condensation at night from the relatively warmer water table to the colder upper regions of the dune. Deeper-rooted plants also contribute to the water supply of the upper layers of the sand dune by a phenomenon described as hydraulic lift (Section 3.7.3).

7.7.2 Nitrogen fixation

The sand dune grasses *Ammophila arenaria* and *Leymus mollis* in common with several tropical grasses have been shown to harbour symbiotic nitrogen-fixing bacteria within their stem and rhizome tissues that may contribute to the nitrogen nutrition of the host plant. Cultivation of these sand dune grasses using surface-sterilized stem and rhizome tissue showed that these species possessed a capacity for acetylene reduction, which indicates an ability to fix atmospheric nitrogen. The stem and rhizome tissues also contained large bacteria populations which when cultured on N-free media revealed the presences of endophytic, diazotrophic bacteria (e.g. *Burkholderia* sp.). Evidence for a similar nitrogen-fixing association has also been detected in sea oats (*Uniola paniculata*) and the American beach grass (*Ammophila brevigulata*). It therefore seems probable that the success of these grasses on nutrient-poor sand is due at least in part to being able to compensate for the lack of nitrogen in sand dunes by fixation of atmospheric nitrogen (Dalton *et al.*, 2004).

A most intriguing aspect of sand dune ecology is the decline in vigour of the main dune-building grasses as the dunes mature. Marram grass (*Ammophila arenaria*), although it can withstand and thrive with vertical sand accretions rates of up to one metre per annum, nevertheless loses vigour as the sand level stabilizes, just when it might be reasonably assumed that having survived the risks and stresses of colonizing the unstable seaward side of the dunes it might flourish

on the more sheltered leeward slopes. This paradoxical loss in vigour in the seemingly more favourable environment has long intrigued ecologists (Moore, 1996) and is currently attracting particular attention in the north and mid-Atlantic coast of the United States where the American marram species (*Ammophila brevigulata*) is showing extensive dieback. Fresh burial of the stems of this plant have always appeared necessary for its continued vigour and a variety of explanations, including the need for new rhizome bud development and the adverse effects of soil compaction, have been discussed at various times. Both European and American marram grass have been shown to benefit from burial by sterile sand as an escape from pathogens and in particular nematode attack. Recently, however, it has been found that it is not the sterile soil per se which provides the escape from nematode attack, but the facility this sand provides for the development of fungal associations (mycorrhizal connections) which give the plants the vigour necessary to combat nematode infections (Little & Maun, 1996). Consequently, periodic dune destruction may be necessary in order to provide the fresh sand that is essential for the maintenance of vigour in marram grass.

7.7.3 Surviving burial

Burial is a constant danger for any plant that lives on a mobile soil. One of the most extreme examples is dune migration, which engulfs and annihilates whatever vegetation over which the dunes may pass. Coastal pine forests are engulfed and destroyed by large migrating dunes (Fig. 7.41). They also show a remarkable capacity to recolonize the dunes after the migration has moved on (Fig. 7.42).

More commonly burial is an insidious and less dramatic process; it is not always fatal and is sometimes beneficial for the stability of the ecosystem. The ability of sand dune grasses to re-emerge after a short period of burial is an important factor for dune growth and stability. The different responses of the major grasses has already been referred to (Section 7.2.2) with *Ammophila* spp. being able to emerge vertically through a metre of sand while *Leymus arenaria* tends to grow out laterally, and sand couch grass (*Elytrigia juncea*) emerges obliquely. On the western coasts of the British Isles burial continues to be a risk in the plains that lie inland from the major

Fig. 7.41   Forest of maritime pine (*Pinus pinaster*) being buried by a migrating dune at Arachon, France, Dune of Pilat. (Photo Dr A. Gerlach.)

Fig. 7.42   Migrating dune being recolonized by umbrella pine (*Pinus pinea*) in the Donaña, Spain. (Photo Dr A. Gerlach.)

dune systems. Where the sand is rich in shell content, a typical herb-rich pasture develops which can be found from Braunton Burrows in North Devon (Willis, 1985) to the Luskentyre Banks (Fig. 7.1) in the Outer Hebrides. Where dunes are rich in shell sand the slacks usually erode to form flat, herb-rich plains sometimes referred to as *machair* (Scottish Gaelic, a low-lying plain; Fig. 7.43). Since pre-historic times the plant communities that develop on the machair have provided fertile grazing and croplands along the Atlantic seaboard of the British Isles. Among the most well known of ancient machair farmers must be St Columba (*c.* AD 521–597) who with his 12 disciples landed on the Hebridean island of Iona in AD 563; they founded a new monastery and farmed the machair plain, which they referred to as the *Campulus occidentalis* and where they both pastured their animals and sowed crops (Anderson & Anderson, 1991).

The attraction of these peripheral storm-exposed areas on the western side of the island, as opposed to the more stable and sheltered areas on the east, was due to the combined benefits for field manure of fresh wind-blown shell sand and plentiful supplies of kelp and tangle (*Laminaria digitata* and *L. hyperborea*) washed up on the beaches exposed to the Atlantic storms. The securing of seaware for manuring the light but easily cultivated soils of the machairs appears to have been long practised in the Hebrides. Prayers and rituals have been traditionally carried out at the edge of the sea to ensure a bountiful harvest of kelp for manuring the machair.

Fig. 7.43   Machair in full bloom in May on the island of Tiree, Inner Hebrides, Scotland.

The fertility of the machair and the extent of its herb-rich pasture depend on fresh depositions of calcareous sand that is constantly blown across these level plains. Although such deposits are beneficial they also carry the risk of being excessive. For the plants growing on the machair, as on other sand dune systems, burial has both risks and disadvantages particularly for species that do not have substantial reserves in rhizomes and other perennating organs. Many machair species can recover from sand burial if it is not too deep. However, there are frequently periods of active sand deposition that buries plants beyond their limits of recovery, as in the stratification that is often seen in the buried soils horizons when the machair erodes. Experimental investigations of the physiological effects of burial on machair vegetation have shown that they possess an ability to maintain a potential for photosynthesis. They are therefore able to resume photosynthetic activity rapidly on emergence, which is an important adaptation as survival is dependent on being able to replenish carbohydrate reserves before the next burial event (Kent et al., 2005).

Typical machair develops through a growth, stabilization and degeneration cycle over a long period. Continued growth with fresh calcareous sand, blown inland from the dunes, continues until the surface rises so high above the water table that drought becomes a stress, and degeneration ensues. When the soil surface falls far enough for the water table to be available, colonization resumes and the growth cycle starts again (Dickinson & Randall, 1979). This slow renewal cycle has maintained the fertility of this type of pasture for thousands of years. Unfortunately much of the biodiversity of these coastal dune pastures is being lost due to the subsidized improvement of coastal grazings; the natural cycle of erosion and renewal is being suppressed due to the application of fertilizers, herbicides, and electric fencing. Even though in many cases the machair may not have received direct applications of fertilizer, it is sufficient for cattle that have been previously fed on nutrient-rich herbage to be allowed on to the area to transport enough minerals to stimulate the growth of a few rapidly growing species to the eventual exclusion of the less competitive plants.

Machair and dune slacks are highly dependent on reserves of fresh water below the soil surface. Fresh water flows over the denser salt water to provide the dunes and slacks in summer with a buried freshwater resource that unlike water added by sprinklers is protected from surface evaporation. However, all too frequently conflicting interests such as forestry and agriculture, which are intolerant of winter flooding in neighbouring areas, insist on drainage measures, which reduce the water reserves for the summer. When the water table falls to more than a metre below the surface in summer, distinct ecological changes take place in the slacks, with many of the moisture-demanding herbaceous species being replaced by grasses. Thus, in common with the dune system, removal of water from areas to landward can be deleterious, as reduced winter flooding and increased summer drought both contribute to a loss in biodiversity.

7.7.4 Flooding

In an ecologically ideal situation, dunes are not just a line of sandy, grass-covered hillocks between the coastal road or golf course and the sea, but occur as a series of ridges interspersed by level plains usually described as dune slacks (Fig. 7.44). The word slack (cf. slake, to allay thirst) implies a tendency for these plains to be flooded, particularly when the water table rises in spring. In well-developed coastal systems there can even be a successional series of dune-slack communities. These begin with slacks still exposed to intermittent sea flooding which have a characteristic salt-loving flora with sea milkwort (Glaux maritima; Fig. 7.45) and sea plantain (Plantago maritima). These are replaced by a freshwater, nutrient-poor slack where creeping willow (Salix repens) is usually a dominant feature. As dunes generally block the flow of water to the sea, the slacks furthest from the sea have higher water tables and longer periods of flooding than those near the sea. To landward, the slacks gradually develop into marshes where flooding is of longer duration and a greater supply of nutrients becomes available in the floodwater. In addition to the varied habitats of the different dunes and slacks, there can also develop long bands of alder from seeds that float to the edge of the flood line. Similar bands of birch may also intersperse the dunes and colonize some of the drier slacks.

Despite the fact that all these communities are supported by a uniform oligotrophic (nutrient poor)

Fig. 7.44   Winter flooding with fresh water in a dune-slack at Tentsmuir National Nature Reserve. The water table rises most in slacks furthest from the sea due to the impedance of water flow to the sea by the accreted sand dunes.

sandy soil, it is quite remarkable that different patterns of inundation, desiccation and exposure are sufficient to create an extraordinarily diverse series of habitats which contribute collectively to the overall resilience of the area to disturbance and erosion. In the unmanaged state, trees come and go as natural flooding of slacks together with areas of windblow and fire denude the tree cover. Unfortunately, these boundary zones are frequently neglected, as management plans for nature reserves tend to focus on preserving areas as one particular type of habitat unless the reserve is very large.

Dune slacks are remarkable for the diversity of the plant communities. Unfortunately, in recent years the features which maintained this biodiversity, the oligotrophic soil and the seasonally varying flooding regime, with the summer water table usually little more than a metre below the surface, are disappearing from many dune slacks. Eutrophication of the freshwater input to the slacks is promoting the advance of dominant grasses, and local drainage activities and boreholes for watering golf courses, together with climatic warming, are making summer drought more frequent.

Fig. 7.45   Sea milkwort (*Glaux maritima*), a salt-loving plant which is widespread in salt marshes and in the spray-zone on higher ground.

7.8   CONSERVATION VERSUS CYCLICAL DESTRUCTION AND REGENERATION IN COASTAL HABITATS

Coastal vegetation faces disturbance from rising sea levels and human disturbance on unprecedented levels. Under natural conditions the rich variety of coastal communities and the easy dispersal of propagules by sea has in the past enabled the plant communities of maritime habitats to recover from sea level changes and other natural disturbances.

Probably the greatest danger to maritime vegetation will be the measures employed by human populations to save themselves from sea inundations, including sea walls, groynes and other artificial barriers. If allowed to find their own equilibrium with changes in sea level, plants can be effective protectors of coastal margins. Cliffs have a rigidity of their own. Sand dune and slack systems are biologically rich, with foredunes, yellow dunes, grey dunes and slacks. Collectively, this constitutes a well-adapted coastal defence system. Provided the detritus of the sea, seaweed and other flotsam, is not removed and there is an adequate supply of sand, most sandy beaches can repair storm damage. However, resistant as they are to frontal attack from the ocean, these natural coastal defences are nevertheless vulnerable to an attack from the rear through windblow and sometimes river erosion as well as other unexpected alterations in their environment. Although sand dune vegetation is highly drought resistant the water table must be accessible. Falling water tables, removal of expected resources such as seaweed, truncation of their natural development by roads and golf courses, all contribute to weakening the resilience of dune systems to withstand physical disturbance, whether from human interference or natural disasters. The conservation of biodiversity in dune systems therefore requires more than just ensuring the physical preservation of the front line of dunes. Equally, over protection can also result in dominance of a few aggressive species with the result that diversity is lost. For long-term preservation of dune systems and their biota a careful balance has to be struck between ensuring adequate physical protection and enough disturbance to enable the natural cycles of denudation and recolonization to take place without causing a loss of species diversity.

Coastal systems are inherently unstable, for as soon as they are consolidated by the sand-retaining grasses these species suffer a decline in vigour and the dunes become liable to blow-outs and other forms of erosion. This raises a management problem for coastal conservation of whether or not to encourage the spread of woody species across dune systems (e.g. sea buckthorn *Hippophae rhamnoides*; Fig. 7.46). It was specifically on this point that his Scottish Majesty King James VI (King James I of England) forbade 'for

hereafter' in 1695 the removal from dunes of 'brent, broom, or juniper', all woody species (Gimingham, 1964). Where grazing from domestic livestock or rabbits is limited, dunes are readily colonized by pine and birch, with willow and alder in the wetter intervening slacks. This may increase stability by reducing wind erosion; however, it may also reduce species diversity and although the site may be preserved physically, species diversity is not necessarily similarly conserved.

Fig. 7.46   Sea buckthorn (*Hippophae rhamnoides*) spreading vegetatively by runners across a dry slack at Buddon Links, Angus, Scotland.

Fig. 7.47   Goats supposedly put into the service of conservation by Scottish Natural Heritage in an attempt to reduce birch colonization of dried-out slacks at the Tentsmuir National Nature Reserve.

The answer to whether trees have a role in sand dune conservation lies in our long-term concept of what constitutes a viable dune system. Sadly, we are too used to considering sand dunes as a fringe of grassy hillocks bordering the sea. Less than 0.2% of the land surface of the British Isles is occupied by sand dunes and there remain few natural dune systems where development is not truncated to landward by roads, car parks or golf courses. It is to the shores of the Baltic, in Estonia and Latvia, that we have to turn to see this vegetation at its best, largely due to the draconian methods used in Soviet times to prevent illicit human migration to Finland. Some degree of disturbance, however, is necessary, either from grazing or recreational use, to keep the forest cover sufficiently open and encourage the plants of the undergrowth. Management has to be carefully balanced. There have been instances where misguided conservation measures such as introducing goats to the dune slack system to restrict the spread of woody plants have not only destroyed the biodiversity of the slacks but also devastated the lichen flora of the grey dunes (Fig. 7.47).

7.9 CONCLUSIONS

Coastal habitats throughout the world offer opportunities for plant survival that do not exist in the more competitive sites of inland areas. In the polar regions coasts in areas that were not permanently covered by snow and ice during the Pleistocene provided refugia for plant survival north of the ice sheets. It is the coastal areas today that still provide a viable habitat for relict species from the circumterrestrial Tertiary forests. Shore habitats are almost the last remaining habitats that allow plants to migrate without impediment from human interference. Littoral plant communities, due to their ability to migrate between similar habitats, also provide some of the best sites for monitoring the effects of climatic change, particularly when they have a north–south geographic orientation. Coastal sites may be marginal, and the dangers of physical destruction may be great, but the resilience of the plants to accommodate themselves to change has ensured their survival in the past and hopefully will continue to do so in the future.

Fig. 8.1 Plants inhabiting the water's edge at varying depths of inundation in a shallow loch in north-west Sutherland, Scotland. Common reed (*Phragmites australis*) in the shallows and white water lilies (*Nymphaea alba*) in the deeper water.

8 · Survival at the water's edge

8.1 FLOODING ENDURANCE

From the poles to the tropics water table levels and their fluctuations are powerful discriminators in plant distribution. Any body of water – a lake, a river or just a small stream – strongly influences the zonation of neighbouring plant communities (Figs. 8.1–8.3). The fringes of willow, alder, reeds, rushes and sedges that flourish at the water's edge provide striking evidence of the ability of certain plant species to survive in these marginal situations with ever-fluctuating water tables.

The flowering plants that are able to inhabit the border between dry and wetland can be thought of as the botanical equivalent of the amphibious animals as they possess an ability to live in both aerobic and anaerobic environments. Flooding can take many forms and in doing so imposes different types of stress. It can be seasonal, varying between both winter and summer. Flooding can also be of short or long duration. It may be confined to just the rooting zone or it can be a total immersion of shoot and root. Consequently, surviving under such a range of diverse inundations requires differing adaptations. Particularly noticeable are the various strategies that have evolved for surviving long-term waterlogging in winter as opposed to short-term inundation in summer. During the growing season, flooding is generally brief and escape mechanisms based on growth and development are predominant. In winter, however, water-saturated soils have a high probability of persisting for months and, consequently,

Fig. 8.2 Willows (*Salix alba* and *S. fragilis*) surviving high water on the Danube in early summer as it flows through the Hungarian Plain north of Budapest.

275

Fig. 8.3   Scots pine (*Pinus sylvestris*) in Glen Affric National Nature Reserve, one of Scotland's largest remaining ancient pinewoods. Note the poor growth and pale needle colour in trees on the flood-prone land at the edge of Coire Loch.

survival frequently depends on physiological tolerance of low-oxygen regimes.

Tolerance of flooding as measured by tolerance of oxygen deprivation (anoxia tolerance) can vary both quantitatively, in terms of duration, and qualitatively in relation to depth of flooding and whether the water is stagnant or flowing. In addition the temperature of the water (seasonal), and whether or not it contains potentially toxic ions can also have an influence on plant survival. Anoxia tolerance can be assessed experimentally by the length of time either whole plants or just certain tissues can survive in an anaerobic chamber (Table 8.1).

There are many ways in which flooding can vary ecologically and which impact on various aspects of plant physiology and development and ultimately influence the capacity of the plant to survive. The multifaceted nature of these qualitative effects is illustrated in Figs. 8.4–8.5 and Table 8.2. It is therefore not surprising that the varied characteristics of flooding in relation to duration, temperature, substrate, and season of the year create an environmental diversity in wetlands which is reflected in the variety of amphibious plant communities that manage to inhabit land by the water's edge. From the mangrove forests of the tropics, to the salt marshes, swamps and bogs of the temperate zones and the Arctic, the interface between dry land and water appears to provide both a challenge and an opportunity for niche specialization that has been exploited by a wide range of plant species.

Location also alters the nature of flooding stress. The shores of lakes with fluctuating water levels are the exclusive sites for truly aquatic and amphibious species. Some species, as in the case of the shoreweed (*Littorella uniflora*), normally live vegetatively while

Table 8.1. *Maximum anoxia tolerance of underground stems and rhizomes as observed in an anaerobic incubator at 20–25 °C. Survival was measured in having an ability to grow new shoots after the period of anaerobic incubation*

Group 1. Species with minimal tolerance of anoxia, surviving only 1–4 days

Solanum tuberosum
Eriophorum vaginatum
Saxifraga hieracifolia
Saxifraga cernua
Oxyria digyna

Group 2. Species with moderate tolerance of anoxia, surviving 4–21 days

Carex rostrata
Mentha aquatica
Juncus effusus
Juncus conglomeratus
Eriophorum angustifolium
Saxifraga oppositifolia
Saxifraga caespitosa
Ranunculus repens
Iris germanica
Eleocharis palustris
Carex papyrus
Filipendula ulmaria
Tussilago farfara
Phalaris arundinacea
Glyceria maxima
Festuca vivipara (Iceland)

Group 3. Species with a high tolerance of anoxia, surviving 1–3 months

Iris pseudacorus
Spartina anglica
Phragmites australis
Typha latifolia
Schoenoplectus lacustris
Bolboschoenus maritimus
Deschampsia beringensis (N. Alaska)
Acorus calamus

Source: Crawford & Braendle (1996).

totally submerged but flower when lake levels recede and the flowering shoots can emerge into air (see Fig. 8.12). By contrast, bogs provide more predictable conditions in relation to water table fluctuations.

8.1.1 Life-form and flooding tolerance

When wetlands are viewed on a worldwide basis, there is no major plant life-form that does not have some species that are adapted to the amphibious environment. The water's edge appears under varying circumstances to offer an ecological niche that is suitable for annual and perennial herbaceous species as well as bushes, and even trees. There are, however, marked regional differences in the relative success of plant forms in relation to wetland colonization. Marshlands in tropical and subtropical climates appear to be more conducive to tree establishment than those in cooler oceanic regions. In the tropics and warm temperate regions, extensive stands of timber grow in wet bottom-land forests, while on sheltered tropical coastlines mangroves flourish in a regime of daily tidal seawater inundation. In the southern United States, swamps are typically wooded, while in Europe they are commonly treeless mineral mires. Forested wetlands in the British Isles are usually restricted to patches of willow and alder carr. In the hyperoceanic conditions of Ireland bog has replaced much of the forest that once flourished across the entire island. Forested wetlands that do remain are to be found in areas with only periodic flooding, as along the banks of the River Shannon and neighbouring waterways where alluvial woodlands prone to winter flooding support willow, alder, downy birch, and oak (Kelly & Iremonger, 1997).

The sensitivity of trees to changes in soil hydration is clearly seen around pools in forests and in raised bogs that are fringed with trees (Fig. 8.3). At the edge of raised bogs (the rand), the depth of the upper layer of aerated peat (the acrotelm) increases and facilitates the penetration of the peat by tree roots. Below the acrotelm lies unaerated peat (the catotelm) into which there is very limited root penetration, and where the water content is constant and conditions are permanently anaerobic, with microbial activity much reduced (Ingram, 1978). Throughout the Holocene, warm dry periods at various times have favoured the advance of trees onto the surface of both raised and Atlantic blanket bogs, often to such an extent that one or more layers of birch or pine stumps can be located at specific depths in the peat.

The widespread occurrence of these sub-fossil tree layers has been cited as evidence for large-scale periods of climatic change during the Boreal and sub-Boreal periods. With a return to wetter conditions, peat

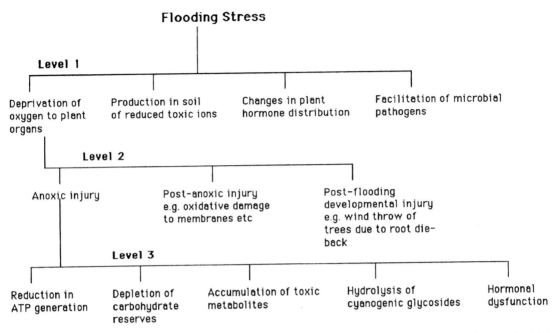

Fig. 8.4   An attempt at segregating the diversity of effects brought about by flooding on higher plants. Level 1 lists four ways in which flooding stress can impinge on plant growth and survival. In level 2 just one of these possibilities is subdivided, namely oxygen deprivation. Five different aspects of anoxic injury are then listed in level 3. (Reproduced from Crawford & Braendle, 1996.)

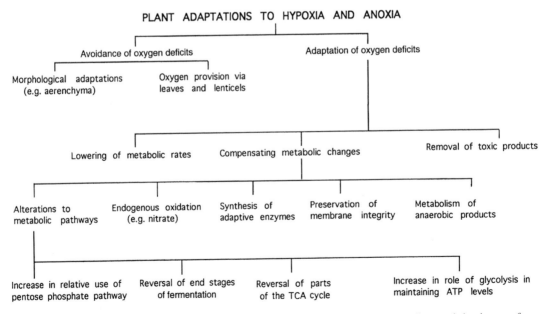

Fig. 8.5   Diagrammatic representation of the diversity of adaptations in higher plants that can contribute to their tolerance of flooding. (Reproduced from Crawford & Braendle, 1996.)

Table 8.2. *Summary of aspects of anaerobic physiology that impact on flooding tolerance and that can vary in their effect at different seasons of the year in wetland and dryland species*

Event	Wetland species	Dryland species	References
Survival under anoxia	Perennating organs of amphibious species (e.g. rhizomes, stolons and turions) have prolonged tolerance of anoxia	Rhizomes when flooded become unviable and are readily killed by short periods of anoxia (less than 1 week)	Crawford & Braendle, 1996
Ethanol accumulation	In intact amphibious species some accumulation on initial flooding then stabilization due to ethanol dispersal mechanisms	Continued accumulation to high levels in roots and rhizomes of species with only a short-term tolerance of anoxia	Schlüter & Crawford, 2001; Wuebker et al., 2001
Alcohol dehydrogenase induction	Less active induction in comparison with intolerant species	Normally rapid response with marked increases	Baxter-Burrell et al., 2002; Fukao et al., 2003; Tamura et al., 1996; Bertrand et al., 2001, 2003
Acetaldehyde generation and toxicity	Rice during early stages of germination is tolerant of acetaldehyde up to 200 mM	Highly damaging in *Avena sativa* and most other crop species	Kato-Noguchi, 2002; Boamfa et al., 2003
Carbohydrate utilization	Slow usage of carbohydrates and maintenance of high levels of free sugars	Rapid depletion of carbohydrate reserves including free sugars under low oxygen stress	Braendle, 1991; Hanhijärvi & Fagerstedt, 1995
Post-anoxic lipid destruction	Little damage in species tolerant of prolonged anoxia	Massive damage in anoxia-sensitive species	Hunter et al., 1983; Kolb et al., 2002
Seed reserves	Exclusively carbohydrates, fructans in grasses	Various	Bertrand et al., 2003; Crawford, 1992
Germination phenology	Inhibited by water; requiring oscillating temperatures	Rapid germination on imbibition for crop species	Baskin & Baskin, 1998; Jutila, 2001

Table 8.2. (cont.)

Event	Wetland species	Dryland species	References
Germination physiology	Elevated alcohol dehydrogenase activity; availability of free sugars under anoxia; ability to disperse ethanol and lactate	Seeds require prompt rupture of testa to ensure adequate access to oxygen	Kolb *et al.*, 2002; Crawford, 1992; Legesse & Powell, 1992, 1996
Seedling response to anoxia	Rice coleoptile and first leaf tolerant of anoxia; also tolerant of acetaldehyde. Weeds of rice crops have also evolved tolerance of anoxia over the germination period	Most seedlings are intolerant of anoxia	Biswas *et al.*, 2002; Kato-Noguchi, 2002; Crawford, 1989
Metabolic response of mature plant to anoxia	Down-regulation of metabolism during prolonged anoxia	Acceleration of glycolysis to overcome short-term oxygen shortages	Schlüter & Crawford, 2001, 2003; Crawford *et al.*, 1987
Ice encasement	Highest frequency of tolerance of ice encasement and anoxia tolerance found in plants of the High Arctic flora	Crop plants that are damaged by ice encasement have higher rates of winter anaerobic metabolism and glycolytic activity under anoxia than those that survive ice encasement	Crawford *et al.*, 1994; Bertrand *et al.*, 2003

growth is renewed, the acrotelm becomes shallower, and trees disappear from the surface of the bog. Raised bogs appear to be more sensitive to climatic change than Atlantic blanket bogs. There is a greater synchrony of the carbon dates for the buried layers of pine stumps in raised than in the Atlantic blanket bogs in Ireland, presumably due to the lower rainfall that is experienced by raised bogs (McNally & Doyle, 1984).

8.1.2 Seasonal responses to flooding

Flooding injury arises in winter and summer from the same basic cause, namely oxygen deprivation. Whether or not oxygen deprivation is more dangerous to plants in summer than in winter depends on the particular circumstances under which flooding occurs. In temperate regions with mild winters, summer water tables are generally lower than in winter, which makes flooding less frequent and of shorter duration. However, even though the summer flooding may be brief, the warmer conditions of the growing season will result in higher rates of metabolic oxygen demand.

Potentially acute oxygen shortages are usually avoided during the growing phase, as many species have available a range of adaptive escape mechanisms for overcoming or alleviating the dangers of hypoxia or anoxia. The principal adaptations are mainly based on phenotypic plasticity in growth responses to flooding which is possible in summer when plants are growing but is generally denied in winter. In summer, the renewed growth of adventitious roots and the speedy development of aerating tissues within the roots in response to flooding facilitate oxygen diffusion from the shoot to the root (Armstrong et al., 1996). Similarly, for species that can survive total inundation, extension of shoots and petioles enables leaves and flowers to be raised up to or above the water surface, and restores access to air for the submerged parts of the plant.

8.2 AERATION

8.2.1 Radial oxygen loss

Given the large quantities of air and oxygen that can circulate within the plant body during the growing season, it is not surprising that some is lost to the plant exterior. The extent of this loss varies with both the position and nature of the various parts of the plant. It can therefore happen, even though the vascular cylinder is hypoxic and the root meristems almost anoxic, that a *radial oxygen loss* from the root system takes place, as the diffusion resistance from the cortex of fine roots to the outside is often lower than that from the vascular cylinder.

Radial oxygen loss has both advantages and disadvantages for plant survival. Frequently, the loss of oxygen from the root to the soil environment in reducing soil conditions is high enough to oxidize the root environment up to a distance of several millimetres from the root surface. Laboratory investigations give some indication of the amount of oxygen that is released from the roots of wet plants into reduced soils. Experiments using individual plants of cat's tail (*Typha latifolia*) and soft rush (*Juncus effusus*) in hydroponic systems have shown that oxygen-release intensities vary between the species as well as depending on the redox state of the rhizosphere. *Typha latifolia* had the highest release rates with mean values of 1.1 mg h^{-1} plant while *J. effusus* recorded 0.5 mg h^{-1} plant at Eh values approximating to -200 mV for both species. The oxygen-release state was governed by the size of the above-ground biomass and intensification of illumination for *T. latifolia*. However, for *J. effusus* the intensity of illumination was less important (Wiessner et al., 2002).

The ability to raise the redox potential at the root surface oxidizes the reduced, and potentially toxic, soluble forms of iron and manganese, which then become insoluble and form a plaque on the root surface. The formation of this plaque is a property of living roots. Dying roots (e.g. roots affected by insects, fungi, mechanical damage, or senescence) are not able to increase the redox potential. The continued formation of plaque will, however, reduce the efficiency of roots as absorbing organs for water and nutrients.

Nevertheless, despite these adaptations some reduced soil ions pose an ever-present risk. In both the common reed (*Phragmites australis*) and the sweet flag (*Acorus calamus*) the formation of non-protein thiols indicate that the detoxification of sulphides can be detected more readily in rhizomes than in roots. The capacity for detoxification is limited, however, and there is always a possibility of sulphide injury (Fürtig et al., 1996). The plaques of iron and manganese deposited on their root surfaces have a high capacity for binding phosphorous. When water lobelia (*Lobelia dortmanna*; Fig. 8.6) was grown under such conditions for six months the formation of plaques on the lobelia roots restricted both phosphorous uptake and biomass production (Christensen *et al.*, 1998). A similar situation can be found in some typically aquatic submerged rosette plants with well-developed roots (often termed isoetids although taxonomically distinct) such as shoreweed (*Littorella uniflora*; Fig. 8.7) and quillwort (*Isoetes lacustris*). This is particularly likely when they are growing at depths of 2–5 m where Eh values are low

(<100 mV). Despite a high phosphate content in the sediment the isoetids showed low biomass and low phosphorus content (Christensen *et al.*, 1998). In both these cases the release of oxygen from the roots of the aquatic species, although preventing the uptake of potentially toxic concentration of iron and manganese, leads eventually through plaque formation to a reduction in growth due to phosphate deficiency.

Radial oxygen loss from the upper portions of the root can also limit the effectiveness of aerenchyma in facilitating oxygen diffusion downwards. The relative distribution of aerenchyma along the root has therefore a profound effect in enabling the more distal regions of the root to have an adequate oxygen supply. Assessments of the anatomy, porosity, and radial oxygen loss profiles from adventitious roots in the Poaceae and Cyperaceae have identified a combination of features characteristic of species that inhabit wetland environments. These included a strong barrier to radial oxygen loss in the basal regions through cell suberization of the adventitious roots and extensive aerenchyma formation when grown in both stagnant and aerated nutrient solutions (McDonald *et al.*, 2002).

8.2.2 Thermo-osmosis

A number of amphibious and aquatic species, e.g. aquatic species with floating leaves as well as the common reed (*Phragmites australis*), are able to use the heat energy of the sun during the growing season to pump air to their

Fig. 8.6   Water lobelia (*Lobelia dortmanna*) growing in a Scottish mountain loch. This species is confined to nutrient poor waters probably as a consequence of plaque formation from oxidized deposits on the roots inducing phosphate deficiency (see text).

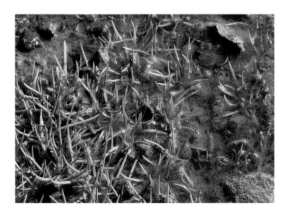 

Fig. 8.7   Shoreweed (*Littorella uniflora*) growing at the edge of a shallow loch in Orkney. (Left) While submerged this lake-margin species remains vegetative. (Right) Flowering when the lake level recedes. Photograph taken in late summer when the plant is seeding.

Fig. 8.8   White water lilies (*Nymphaea alba*), a species that uses solar energy via a process of thermo–osmosis to ventilate the submerged portions of the plant (see text).

submerged organs by thermo-osmosis (Figs. 8.8–8.9). A prerequisite for air transport by thermo–osmosis is a layer with very small pores or intercellular spaces ($<0.1\,\mu m$) and a high humidity inside the plant organ. Once air has passed through the small pores it is heated by the sunlight and mixed with the water vapour inside the organ. As a result of increasing temperature, the higher molecular motion of gas molecules leads to a longer critical path, which reduces the probability of escape from within the leaf. Pressure therefore increases as cool air continues to diffuse inwards. This increased internal gaseous pressure forces gas downwards which

develops into a flow-through system in rhizomatous species as air can exit through the underground organs into the older leaves where the pores (dilated intercellular spaces) are much bigger (Knudsen diffusion). It has been estimated that in the common reed (*Phragmites australis*) the flow through rate is sufficient to maintain the oxygen concentration of the submerged rhizomes at 90% of atmospheric concentration (Armstrong *et al.*, 1992).

The effectiveness of shoots in aerating underground rhizomes of emergent aquatic macrophytes was strikingly illustrated in a study of radial oxygen loss

Fig. 8.9   Colony of giant Amazonian water lilies (*Victoria regina*) growing typically in shallow nutrient rich water in a backwater (Varzea) of the River Amazon near Manaus (Brazil). A single leaf and petiole of this species can aerate the submerged organs by thermo-osmotic gas flow rates of up to 5000 ml h^{-1} (Grosse *et al.*, 1991).

from rhizome apices of *Phragmites australis* (Armstrong *et al.*, 2006). Radial oxygen loss from rhizome apices of *Phragmites* was increased by convective gas flow through the rhizome even when only the tips of the shoot emerged above the floodwater (Fig. 8.10). It was concluded that oxygen passes via internal gas-space connections between aerial shoot, rhizome and underground buds and into the phyllosphere regions via scale-leaf stomata and surfaces on the submerged rhizome buds. It was also suggested that the oxidized phyllospheres may protect rhizome apices against phytotoxins in waterlogged soils, just as oxidized rhizospheres protect roots.

8.3   RESPONSES TO LONG-TERM WINTER FLOODING

Less attention has been given to understanding the consequences of high water tables in winter than during the growing season, possibly because the study of flooding tolerance has been strongly influenced by the need for research on annual crop plants. The relative neglect of winter studies into flooding tolerance is also possibly due to the erroneous assumption that marsh and bog plants owe their ability to survive entirely to their capacity to aerate their submerged organs by the downward movement of oxygen from emergent shoots.

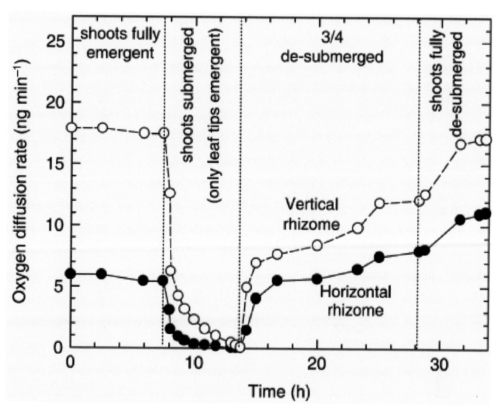

Fig. 8.10   Efficacy of shoot aeration of underground organs in the common reed *Phragmites australis*. Note the effects of shoot submergence and de-submergence on oxygen diffusion rate from submerged rhizome apices to phyllospheres. Closed circles, apex of horizontal rhizome; open circles, apex of vertical rhizome. Diffusion rate in agar remote from rhizome apices and roots was close to zero throughout the experiment. (Reproduced with permission from Armstrong *et al.*, 2006.)

However, those ecophysiological studies that have been carried out on wetland plants reveal considerable physiological and biochemical variation between species in their ability to withstand prolonged oxygen deprivation as a result of winter flooding.

8.3.1   Surviving long-term oxygen deprivation

Plants that over winter in flooded habitats have to adapt to more than a transitory deficiency in oxygen availability to their submerged organs. In many herbaceous species the perennating organs are buried for months in a water-saturated soil with no access to atmospheric oxygen. The underground organs (rhizomes, tubers, etc.) of many of these plants are remarkable for their tolerance of extended periods of waterlogging and oxygen deprivation. Experimental studies in which the

overwintering organs, instead of being merely submerged in water, were incubated in completely anoxic environments have shown that in many amphibious species perennating organs can survive months of total anoxia (see also Section 3.6.2). Some of these species are able to resume shoot growth in spring from a depth in anaerobic mud without the need for an air supply, while others wait for the floodwaters to subside before resuming growth. Species that emerge from a depth of flooded soil or water under total anoxia (e.g. *Schoenoplectus lacustris*, *Typha latifolia*) are capable of sustaining a viable growing shoot under total anoxia for several weeks. This tolerance of prolonged anoxia is frequently accompanied by an ability to down-regulate metabolism and thus reduce the risk of exhausting their carbohydrate reserves and accumulating toxic metabolites (see Section 3.6.3).

In most species the general dieback of herbaceous vegetation in winter renders aeration adaptations inoperative. However, a number of wetland herbaceous species do not die down completely in winter and still retain some green basal shoots. Both reed sweet grass (*Glyceria maxima*) and a number of rushes (e.g. *Juncus effusus*) retain some basal green leaves and can be expected therefore to have a source of at least some oxygen to alleviate the effects of winter flooding on their roots.

In the common reed (*Phragmites australis*) and common club-rush (*Schoenoplectus lacustris*) the dead stalks that survive above the level of winter flooding can act as a ventilating system for the submerged rhizomes. When wind blows across the tops of broken reeds it acts in a manner similar to air being blown over the holes of a flute, creating a pressure differential across tall dead culms, sucking air into the underground system (Venturi convection). It has been demonstrated (Armstrong *et al.*, 1992) that wind, by increasing *Venturi-induced convection*, can raise the oxygen concentrations in the rhizome system, thereby causing substantial fluxes of oxygen into both the root and rhizosphere.

8.4   FLOODING AND UNFLOODING

Those species that live by the water's edge or in seasonally flooded communities are exposed to a double jeopardy in relation to the level of the water table. The first is flooding, and the second is unflooding. The physiological problems that can arise from the first stress include cellular dysfunction due to a reduction in metabolic energy (ATP) and an increase of cytoplasm acidity due to proton pump failure (cytoplasmic acidosis), which can lead to rapid cell death.

Oxygen deprivation exposes all tissues, irrespective of whether they are plant or animal, to the dangers of anaerobic cytoplasmic acidosis. In the absence of oxygen it is impossible to maintain the ATPase activity of the tonoplast and the functioning of proton pumps. Consequently, there is no proton influx into the vacuoles from the cytoplasm, which results in the immediate onset of cytoplasmic acidosis (Ratcliffe, 1997). In plants the role of cytoplasmic pH in triggering a rapid switch to ethanol production within an hour of the onset of anoxia has been convincingly demonstrated in maize using 31P nuclear magnetic resonance (Fox *et al.*, 1995). In anoxia-tolerant plants basic amino acids (GABA and arginine)

and polyamines are formed to counteract pH decrease (Reggiani *et al.*, 1990). A more gradual loss of viability comes from an increased consumption of carbohydrate reserves, accumulation of toxic waste products of anaerobiosis, and leakage of metabolites leading to pathogenic infections. If the plant has no means of aerating its underground organs, and this is especially the case for large plants such as trees and deep-rooted perennial herbs, then a lack of oxygen can bring all these adverse factors into play.

8.4.1   Unflooding – the post-anoxic experience

The second stressful consequence of flooding is *unflooding*. With the coming of spring plants that have endured a long winter of oxygen deprivation through flooding have to face the additional hazard of post-anoxic injury when the soil profile once again becomes aerated or emerging shoots (or both) restore a pathway for oxygen diffusion to underground organs. Prolonged anaerobiosis, although dangerous in itself creates for the period of unflooding a potentially more dangerous state due to the rapid and often irreversible damage that can be caused when air is re-admitted to tissues that have been deprived of oxygen for a lengthy period. This type of damage, *post-anoxic injury*, as it is termed in plants, is similar to *post-ischaemic injury* in animal and human organs which arises when they are suddenly re-exposed to air after interruption of the blood supply, as after a heart attack or a transplant operation.

The return to air generates reactive oxygen species to which the tissues of many plant species become vulnerable after a prolonged absence of oxygen and cessation of aerobic metabolism (see below). Many amphibious plant species, and also those arctic species that face the risk of oxygen deprivation due to prolonged encasement in ice, show a remarkable capacity to survive post-anoxic injury. During flooding, the lack of oxygen causes the transition metals involved in aerobic metabolism (the iron in cytochromes etc.) to become reduced to the ferrous state. The turnover rate of the enzymes needed for aerobic metabolism is also much diminished. Consequently, sudden re-emergence into air as water tables subside presents an aerobic shock to unprepared tissues. Under such conditions many plants are liable to suffer post-anoxic injury. Tissues prone to this type of injury become soft and spongy on return to air, and rapidly lose their cell constituents. In these cases, oxygen exerts a definite toxic effect and membranes are destroyed irreversibly. This type of injury arises from the post-anoxic generation of reactive oxygen species (ROS), typically oxygen (O_2^-) and hydroxy radicals (*HO_2).

Both O_2^- and *HO_2 can undergo spontaneous dismutation to produce H_2O_2:

$$H^+ + O_2^- \rightarrow {}^*HO_2$$

$${}^*HO_2 + {}^*HO_2^- \rightarrow H_2O_2 + O_2.$$

In the post-anoxic state when cytochromes are reduced, the Fenton reaction can be particularly active. This is the iron-salt-dependent decomposition of dihydrogen peroxide, generating the highly reactive hydroxyl radicals:

$$Fe^{2+} \text{ complex } + H_2O_2 \rightarrow Fe^{3+} \text{ complex}$$
$$+ {}^*OH^* + OH^-.$$

Other enzymatic sources include the action of xanthine oxidase on dioxygen:

Xanthine oxidase

$$O_2 \rightarrow O_2^-$$

Superoxide dismutase

$$O_2^{*-} + O_2^{*-} + 2H^+ \rightarrow H_2O_2.$$

Ethanol that has accumulated under anoxia is also rapidly oxidized to acetaldehyde by the presence of H_2O_2 and catalase. The oxidation product, namely acetaldehyde, is highly damaging to membranes and therefore probably contributes more to membrane damage at the post-anoxic stage than ethanol during anoxia:

Catalase

$$C_2H_5OH + H_2O_2 \rightarrow CH_3OH + 2H_2O.$$

Other sources of ROS come from the reduced electron chains. In non-green cells the mitochondrial electron transport can give rise to ROS in living cells at any time. However, they are normally detoxified by enzymatic and non-enzymatic systems (SOD, catalase, ascorbate reductase, peroxidase, vitamins C and E, glutathione, and many others).

In stressed cells when these protective mechanisms are not fully developed or no longer equilibrated, ROS become effective agents of cell death. Although these active radicals are capable of reacting with many types of macromolecules, in plants their most deleterious action is in their reaction with lipids (e.g. lipid peroxidation) by initiating membrane damage leading to the destruction of cell organelles. Peroxidation products, such as ethane and malonedialdehyde, appear rapidly in the organs and tissues of non-tolerant plants (Braendle & Crawford, 1999). The inner membrane system of mitochondria seems to be particularly sensitive. The production of ROS is a common feature also for other stresses, including drought, salt injury, and pollution damage. Plants that are able to survive prolonged oxygen deprivation followed by a return to air all have the capacity to withstand both prolonged anoxia and avoid post-anoxic injury to their cell membranes with a protective antioxidant system (Braendle & Crawford, 1999). The highly anoxia-tolerant *Acorus calamus* (see below) maintains high concentrations of antioxidants, including ascorbate, phenolics and glutathione, throughout the period in which the plant is under anoxia, as well as the enzymes ascorbate reductase, peroxidase and catalase. *Iris pseudacorus* has a tolerance of anoxia that is only slightly inferior to that of *A. calamus*. This iris is also notable for the ability while under anoxia to synthesize superoxide dismutase, a key enzyme for active radicle detoxification on return to air (Monk *et al.*, 1987).

The metabolic adaptations associated with surviving low oxygen availability can be summarized as:

(1) anaerobic mobilization of starch reserves
(2) prevention of cytoplasmic acidosis under anoxia
(3) dispersal and excretion of products that transfer hydrogen from anoxic or hypoxic tissues, either to the external environment, or to parts of the plant with access to oxygen
(4) an active antioxidant defence system dependent on (a) antioxidants and (b) enzymatic activity for antioxidant reduction and the destruction of active radicals (e.g. superoxide dismutase).

8.5 RESPONSES TO SHORT-TERM FLOODING DURING THE GROWING SEASON

Plants that are tolerant of short periods of flooding in summer tend to respond to oxygen deprivation in the opposite manner to that which can be seen in the over wintering organs of flood-tolerant species. Instead of down-regulating metabolism there is usually observed an acceleration of glycolysis. In the growing season this type of response can maintain tissues with an adequate supply of metabolic energy that gives the plant time to improve root aeration through the development of adventitious roots and aerenchyma. It is essential in growing plants that aeration be restored by improved root aeration otherwise a rapid depletion of carbohydrate reserves by a hypoxia-induced acceleration of glycolysis (the *Pasteur effect*) will lead rapidly to cell death. The Pasteur effect can be an effective remedy in the short term for restoring ATP levels and will aid short-term anoxia tolerance and prolong the life of the anoxic tissues for a few days. This is often sufficient for overcoming short periods of flooding during the growing season but will not serve as a long-term strategy for the survival of inundated organs during long periods due to the rapid consumption of carbohydrate reserves (Crawford, 2003).

Under field conditions it is often difficult to determine to what extent the roots of flood-tolerant plants are subjected to an oxygen deficiency. All root tips contain very little free oxygen due to high consumption rates and the density of the tissues and ethanol can always be detected in the dense tissues of the root meristems. Consequently, any reduction in the availability of oxygen, as well as leading to higher levels of ethanol, will extend the zone of oxygen deprivation upwards from the distal regions of the root to other parts of the root system and increase the amount of root that accumulates ethanol. Flooding may initially induce a shortage of oxygen, but in summer flood-tolerant species usually can alleviate this condition and restore an adequate supply in a matter of days, by the development or enhancement of existing aeration tissues (aerenchyma), and the growth of adventitious roots (Joly, 1994).

A series of studies carried over many years in the Netherlands on the plants that grow on the banks of the Rhine have provided much information on the close match that exists between flooding risk and the ability of species to respond morphologically so that submergence does not expose them to the dangers of anaerobiosis (Voesenek *et al.*, 2004). A comprehensive survey of the ethylene-induced elongation response in 22 plant species occurring in the Rhine flood plain showed that the capacity for stem and petiole elongation upon exposure to ethylene correlates positively with flooding duration and negatively with drought. Based on this analysis, it was concluded that the capacity to elongate is an important selective trait in field distribution patterns of plants in flood-prone environments. However, rapid shoot elongation under water appears to be a favourable trait only in environments with shallow and prolonged flooding events, as the costs associated with this response make this an unviable strategy in sites with deep floods, or in sites where the floodwaters recede rapidly (Voesenek *et al.*, 2004).

The genus *Rumex* has received particular attention, as it possesses a number of closely related species that coexist along the banks of the Rhine and exhibit varying capacities for petiole extension in relation to their location of the floodline. *Rumex palustris* which lives low down on the bank and closer to the water's edge than any of the other species shows rapid petiole elongation mediated by the integrated action of the plant hormones ethylene, auxin, gibberellin, and abscisic acid. By contrast, the closely related *Rumex acetosa* which lives in the upper zone of reduced flooding risk is unable to switch on petiole elongation when submerged. Mature plants, with their greater carbohydrate reserves, generally show greater tolerance of flooding than seedlings (Fig. 8.11).

Adaptations to submergence during the growing season are mostly based on restoring the potential for photosynthesis to inundated plants. When the species are submerged in the dark, death follows in most cases within 6–9 weeks irrespective of whether the plants live at high or low levels on the riverbank (Blom *et al.*, 1994). It is this last finding which demonstrates most clearly the importance to the submerged plant of being able to maintain some photosynthetic capacity and generate oxygen if it is to survive inundation during part of the growing season. A similar effect can be seen in flooded over wintering cranberry plants (*Vaccinium macrocarpon*). They remain evergreen under water and even under ice. However, when the ice is covered with snow and light is excluded oxygen deprivation injury ensues (Schlüter & Crawford, 2003).

Once contact with air is established the efficiency of the aeration process will be dependent on the internal anatomy of the flooded species. The amount of aerenchyma differs considerably between wetland species, but usually varies between 10% and 60% of the total volume of shoots, rhizomes and roots (Studer & Braendle, 1984). Obviously, long-term flooding resistance is intimately correlated with aerenchyma formation, as it is the basis whereby gas transport is facilitated from shoots to fine roots. Oxygen can be channelled down to support

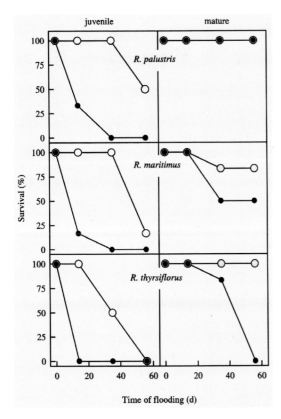

Fig. 8.11 Survival percentages ($n = 6$) of *Rumex* species after complete submergence in light (open circles) or dark (closed circles) after allowing 14 days for recovery and regrowth. Juvenile plants: 35–40 days old; mature plants: 85–110 days old. (Reproduced with permission from Nabben, 2001.)

respiration, and to remove waste products, e.g. CO_2 removed in the opposite direction. Analyses of the internal atmosphere have shown that the oxygen concentrations are high and in most cases match the needs for respiration. In the common club-rush (*Schoenoplectus lacustris*) the oxygen content in the submerged stalks and rhizomes is maintained at 13–15% (Braendle, 1991).

Underwater studies of the submergence-tolerant marsh dock (*Rumex palustris*) have demonstrated a remarkable ability to carry out photosynthesis under water (Mommer *et al.*, 2004; Mommer *et al.*, 2005) and maintain relatively high internal oxygen pressures under water, sufficient even to cause a release of oxygen via the roots into the sediment. This acclimation to submergence was found to involve both an increase in specific leaf area and a significant reduction of diffusion

resistance for gas exchange between leaves and the water column with the development of thinner cuticles (Fig. 8.12).

8.5.1   Disadvantages of flooding tolerance

As with all adaptations to specific habitats it is necessary to consider the disadvantages that can result in relation to specialization for one particular type of environment. The fact that wetland plants grow in wetland areas is so obvious that the question is seldom asked as to why they cannot also live in areas where flooding is unlikely. The question has rarely been investigated as to whether adaptation to drought is fundamentally incompatible with flooding tolerance. Most flood-tolerant species can be cultivated successfully in pots without flooding and the presence of other species, which suggests that under natural surroundings flood tolerance renders plants in some way uncompetitive in drier habitats. One possible disadvantage of the presence of aerenchyma in dry land species is that it reduces the capacity of the roots for nutrient uptake (Koncalova, 1990). Roots adapted to flooding with aerenchyma are usually thick, with little branching, and with a consequently reduced surface/volume ratio. An added adverse effect of this strategy is the diminution of the total surface areas for nutrient uptake. The upper regions of the roots are frequently suberized against premature oxygen leakage, which further hinders their absorptive powers.

Research on root growth in a selection of species with varying tolerances of flooding that are common on Dutch flood plains has shown that in general the more flood-tolerant species are less selective than the flood-sensitive ones in placing their roots in soil patches enriched with nutrients (Jansen *et al.*, 2005). Metabolically, long-term flooding tolerance which enables plants to over winter in flooded habitats is achieved only by incurring extra costs in terms of carbohydrate utilization. Despite the down-regulation of metabolism to conserve carbohydrate reserves against the dangers of accelerated glycolytic activity, the need for high levels of these reserves to resume growth from submersion or water-logged soils in spring is a cost that is imposed by anaerobic habitats. The provision of adequate antioxidants to counter post-anoxic injury in spring is also dependent on adequate carbohydrate reserves. The use of carbohydrate reserves to meet these costs will thus place adapted species at a competitive disadvantage

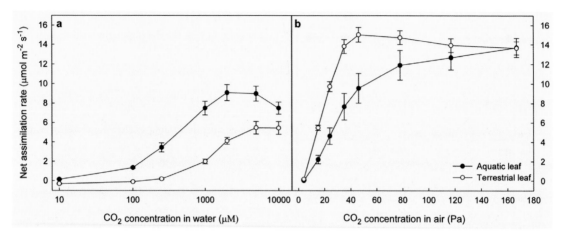

Fig. 8.12 Comparison of photosynthetic efficiency in terrestrial and aquatic leaves of *Rumex palustris* as measured with carbon dioxide response curves (a) under water from oxygen production and (b) in air from net carbon dioxide assimilation. Note that each leaf type achieves maximum effectiveness in its own particular environment. (Reproduced with permission from Mommer, 2005.)

when growing in unflooded habitats (Crawford, 2003). There are therefore a number of attributes, both morphological and physiological, associated with flooding tolerance that will disadvantage adapted species if they attempt to grow in natural communities that are flooded only rarely.

8.6 AMPHIBIOUS PLANT ADAPTATIONS

Most amphibious herbaceous species in temperate zones are either rhizomatous or tuberous geophytes (plants with their perennating buds below ground) and rely more on vegetative propagation than seed for regeneration. Seed germination is often very low in many amphibious species and successful sexual reproduction takes place only during periods of exceptional drought when water table levels have receded long enough for seeds to germinate and produce established juvenile plants. This restricted environmental window for seedling establishment explains the small number of annual species that are found in natural wetland habitats (see also Chapter 4).

As already mentioned, shoreweed (*Littorella uniflora*) is among the more successful perennial amphibious species of western Europe. It is a diminutive but abundantly stoloniferous plant that typically inhabits the drawdown zone in lakes and reservoirs where it

can descend to a depth of 3–4 m. It can also be found in winter-flooded depressions on cliff tops and heaths and in seasonally flooded temporary pools. In these areas shoreweed also has to endure prolonged periods in summer when these habitats dry out completely. In lakes it can form carpets, which survive under continuous immersion relying entirely on vegetative reproduction. Flowering can take place under shallow flooding (50 cm) but no seed production takes place. Successful flowering with seed production takes place only when the shoots emerge above the water level. Although there is only a limited production of viable seed this species is normally successful in having a large seed bank with the seeds remaining viable for decades (Preston, 1997).

8.6.1 Phenotypic plasticity in amphibious species

Leaf heterophylly, the phenomenon where floating and submerged leaves on the same plant are strikingly different in morphology, has long been recognized as a feature of emergent aquatic macrophytes, and appears to play a role at the margin between submergence and emergence. Many experiments have been carried out to elucidate the control mechanisms, which determine leaf form, and to assess the physiological significance of the change in terms of growth rate, productivity and survival of macrophytes under varying submergence and emergence regimes. Shoreweed (*Littorella uniflora*), the small perennial, amphibious rhizomatous plant that is common along lake margins (Fig. 8.7; see also above), functions more efficiently when lake levels fall and leaves are exposed to air, even though it survives as extensive widespread colonies in many permanently submerged locations. On emergence this small plant exhibits a very rapid phenotypic adjustment which includes the production of a new set of terrestrial leaves with reduced lacunal volume and increased stomatal density. Once these leaves are formed there is a rapid increase in leaf growth rate, leading to flowering within 3–4 weeks (Robe & Griffiths, 1998). It was also found that the terrestrial plants incorporated three- to fourfold more carbon and nitrogen into above-ground biomass than submerged plants. Shoreweed is a striking example of a species which has the possibility of using two divergent survival strategies: low growth rate and vegetative reproduction when submerged, and rapid growth with flowering and seed production when water levels recede. Together these two adaptations provide a flexibility that appears to account for the success of this plant in the amphibious niche.

The heterophyllous aquatic Loddon pondweed (*Potamogeton nodosus*) produces morphologically distinct leaves above and beneath the surface of the water. Application of abscisic acid (ABA) for 4 hours has been found to be effective in inducing entirely submerged plants to produce leaves of the form normally found at the water surface. There appears to exist 'a window of responsiveness' to ABA in that changes in leaf morphology were evident within 2 to 3 days after the treatment and continued for only 4 to 5 days thereafter (Gee & Anderson, 1998). Conversely, terrestrial shoots of the American aquatic *Ludwigia arcuata* formed narrower, submerged-type leaves when treated with ACC (1-aminocyclopropane-1-carboxylic acid), a precursor to ethylene, suggesting that ethylene might also be an endogenous factor responsible for change in leaf shape (Kuwabara *et al.*, 2001). There may also be some measure of nuclear differentiation between submerged and emergent leaves. In the heterophyllous macrophyte water chestnut (*Trapa natans*) cytological analyses have provided evidence for polymorphism between the genomic DNAs of floating and submerged leaves (Bitonti *et al.*, 1996).

8.6.2 Speciation and population zonation in relation to flooding

Some closely related amphibious species specialize in particular habitats through speciation and population zonation instead of responding phenotypically and morphologically with different types of leaves as discussed above. Such a species differentiation in habitat preferences can be seen on the closely related club-rushes. Adults of the emergent common club-rush (*Schoenoplectus lacustris*) and the closely related zebra rush (*B. tabernaemontani*) and the sea club-rush (*Bolboschoenus maritimus*) occur respectively along a gradient in water depth from deep to shallow water (see also Fig. 3.29). Seedlings of the two *Schoenoplectus* species showed their highest relative growth rate under terrestrial growth conditions, whereas *Bolboschoenus maritimus* grew best under submerged growth conditions. However, submerging seedlings that were established in unflooded conditions reduced growth in all three taxa. This effect became stronger with increasing age of the seedlings at the time of submergence. When transferred the other way round, from submerged to unflooded conditions, seedlings of *S. tabernaemontani* and *B. maritimus* adapted quickly to the terrestrial growth conditions, whereas the thin leaves of *S. lacustris* partly dried out. It was concluded that although seedling establishment of all three species will be most successful under terrestrial conditions, subsequent fluctuating water levels may act as a strong selective force which finally determines the distribution of these taxa along a gradient in water depth (Clevering *et al.*, 1996).

The genus *Eriophorum* (cotton grasses) also shows differences in anoxia tolerance. In common with many graminoid species, the northern temperate species, *E. vaginatum* and *E. angustifolium* are not tolerant of prolonged anoxia as tested experimentally in an anaerobic incubator at 20–22 °C (Fig. 8.13). Two days of oxygen deprivation can cause a total kill in *E. angustifolium*. However, an arctic provenance of *E. scheuchzeri* still had a 20% survival of its stolons after 12 days of total anoxia at 20–22 °C. At the much lower temperatures that prevail during arctic winter a higher survival rate can be expected which will be almost a necessity (Figs. 8.13–8.14) as in both Greenland and Spitsbergen the arctic populations of this species may have to endure many months of ice encasement which will deny any access to oxygen.

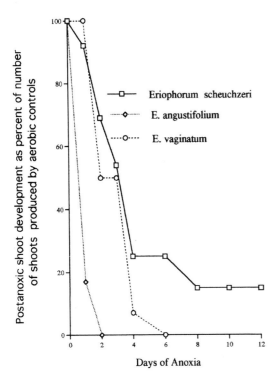

Fig. 8.13 Comparisons of anoxia tolerance in the genus *Eriophorum* based on the percentage survival of shoots compared with aerobic controls after being kept under strict anoxia in an anaerobic incubator (see Chapter 3). *Eriophorum scheuchzeri* was of arctic origin from Spitsbergen; the other two species, *E. angustifolium* and *E. vaginatum*, were collected in Scotland. For experimental details see Crawford *et al.* (1994).

Other amphibious species exist as distinct populations in relation to their ability to survive winter flooding. A striking example is found in the American speedwell (*Veronica peregrina*). This species is a winter annual that has been successful in inhabiting shallow vernal pools in California. Plants at the centre of pools have been observed to be less variable and produce larger seeds than those at the edge as well as showing some metabolic adaptations to flooding including increased accumulations of malic acid (Linhart & Baker, 1973).

A dissected-leaved form of the creeping buttercup (*Ranunculus repens*) occurs in temporary limestone lakes or turloughs (Irish *tuar loch*, a dry lake) that can be found on limestone pavements in the Burren region in western Ireland (Fig. 8.15). Turloughs fill with ground water for up to eight months of the year and then dry out in the summer. Comparisons of the effects of experimental flooding up to ground level revealed localized populations of plants of *R. repens* growing within the turlough that differed both physiologically and morphologically from populations of this same species growing on neighbouring ruderal unflooded sites. Comparisons of the effects of experimental flooding up to ground level on populations of plants of *R. repens* collected from the turloughs as compared with ruderal locations showed no differences. However, when the plants were submersed with their shoots also under water this led to direct tissue death in the ruderal population but had no effect on the turlough population. There was no detectable difference in the proportion of aerenchyma in drained, flooded and submerged roots of plants from either population. However, the dissected leaf form in the populations that grew in the flooded turloughs demonstrated a higher rate of aerial and submerged photosynthesis than populations of the more typical broad-leaved ruderal form. The turlough populations also had higher rates of stomatal conductance and exhibited a higher stomatal index on the upper leaf surface and a lower index on the lower leaf surface than the ruderal populations. It would therefore appear that the more dissected leaf shape of the turlough population may have a thinner boundary layer and thus enhance gas exchange in submerged conditions (Lynn & Waldren, 2002, 2003).

The Alpine krummholz pine (*Pinus mugo*) is a very variable species and several ecologically distinct subspecies have been described. In particular the ecotype that is found in alpine bogs (*P. mugo* ssp. *pumilio*) appears to be a form that is particularly adapted to living by the water's edge (Fig. 8.16).

8.7   AQUATIC GRAMINOIDS

Rushes (*Juncus* spp.), sedges (*Carex* spp.) and reeds (*Phragmites*, *Scirpus*, *Bolboschoenus*) are among the commonest groups of species to be found at the edges of streams and lakes. The rushes (*Juncus* spp.) are also prevalent in poorly drained pastures. Despite their common occurrence in wet soils, rushes (*Juncus* spp.) are generally not tolerant of prolonged anoxia and the rhizomes of all species of this genus so far tested die after a few days of being placed in an anaerobic incubator. A notable feature of overwintering rushes is the

Fig. 8.14   Scheuchzer's cotton grass (*Eriophorum scheuchzeri*) – an arctic and alpine wetland sedge with a disjunct distribution from the Alps to the Arctic. In this photograph it is growing in a bog at Mesters Vig, in north-east Greenland (73° N). Arctic populations of *E. scheuchzeri* have been found to be anoxia tolerant, a property which has not been detected in non-arctic populations of *E. angustifolium* and *E. vaginatum* (Crawford *et al.*, 1994). These relatively anoxia- and flood-tolerant plants will have their rhizomes encased in ice throughout the many months of the arctic winter and have therefore to endure a prolonged period of oxygen deprivation until they are suddenly re-exposed to oxygen when the ice melts in early summer.

persistence of green shoot bases throughout the winter. Aeration of the roots from the shoots, possibly assisted by some winter photosynthesis, can therefore play an important role in their survival. By contrast, the club-rushes and common reeds (*Schoenoplectus* spp. and *Phragmites australis*), which inhabit areas that are frequently subject to prolonged and deep winter flooding in oxygen-deficient muds, have rhizomes which are capable of surviving lengthy periods of anoxia (Braendle & Crawford, 1999).

Fig. 8.15   Population differences in leaf dissection in and near turloughs in creeping buttercup (*Ranunculus repens*). (Left) Leaves of *R. repens* cultivated for one year collected from the lowest elevation (left) and the highest elevation (right) from Hawkhill turlough, Co. Clare, Ireland. (Photo by courtesy of Dr D. Lynn.) (Right) Leaf dissection index of leaves derived from the field and cultivated plants in relation to elevation in the turlough. (Reproduced with permission from Lynn & Waldren, 2003.)

Fig. 8.16   An alpine bog extensively colonized by the flood-tolerant form of the alpine krummholz pine (*Pinus mugo* ssp. *pumilio*). The bog lies just below the summit of Snizka, the highest mountain in the Czech Republic with an altitude of 1602 m (5300 feet) in the Krkonose range. The mountain range is also known as the Giant Mountains (German, *Riesengebirge*). The Czech name is ancient, appearing in the name of a people listed in Ptolemy as the *Corconti*. The mountains stretch from north-west to south-east and form the border between Poland and the Czech Republic. (Photo Dr Tomas Kucera.)

Fig. 8.17   Reed canary grass (*Phalaris arundinacea*), a semi-aquatic marsh grass with rhizomes that are tolerant of long-term anoxia when overwintering. In the growing season, provided flooding is either shallow or well aerated as here in a moving stream, the species is capable of aerating its submerged organs.

Anoxia-tolerant species can therefore be further divided into two groups, namely those that can extend their shoots in the total absence of oxygen and those that are tolerant of prolonged anoxia but wait for the water table to fall and soil aeration to be restored before resuming growth in spring (see Table 8.1). There are many intermediates between these two extreme types of behaviour in relation to imposed oxygen deprivation. Plants that are tolerant of anoxia over winter can become susceptible to oxygen deprivation in early summer when their carbohydrate reserves have been reduced in the support of spring growth (see *Phalaris arundinacea*; Fig. 8.17). These variations illustrate the delicate ecological balance that exists between a phenology in which species maintain a prolonged tolerance of anoxia throughout the period of risk from flooding and a more opportunistic strategy which favours a rapid resumption of growth in spring, even if there is a risk of periodic growth checks due to late flooding.

8.7.1   *Glyceria maxima* versus *Filipendula ulmaria*

Such a case can be seen in the competitive relationship between queen of the meadow (*Filipendula ulmaria*) and reed sweet grass (*Glyceria maxima*; Fig. 8.18). Successive mapping of the distribution of these two species over an interval of 24 years in a series of dune and slack communities in Tentsmuir National Nature Reserve, Scotland, that had suffered a change in their drainage regime was able to record the ecological consequences of the change that had taken place (Figs. 8.19, 8.20). The alterations in the flooding regime were particularly noticeable in spring when flooding was shallower and disappeared sooner than in the past. The effect on the vegetation was for *G. maxima* to expand its territory into areas that had formerly been dominated by *F. ulmaria*. *Glyceria maxima* is less tolerant of anoxia than *F. ulmaria* but, like many *Juncus* species referred to above, maintains green basal leaves during the winter and, provided the flooding is not deep, will resume growth in spring before *F. ulmaria*. The latter species relies on its anoxia tolerance for survival and does not extend its new season's shoots until the water level has fallen below the surface of the soil. Thus, when flooding is shallow the less-tolerant species has the competitive advantage due to a more precocious phenology. When flooding is more prolonged, maintaining a depth of 10 cm or more, *Glyceria maxima* cannot compete with the more anoxia-tolerant *F. ulmaria* (Studer-Ehrensberger *et al.*, 1993; Figs. 8.21–8.22).

Overwintering in wetland habitats necessitates the ability to survive prolonged periods of inundation, which in some species may also include prolonged periods of anoxia. As is evident from much of the discussion above, it is not just the anoxia tolerance of the flooded part of the plant that is important for survival but its relationship to the plant as a whole. The presence or absence of ventilating stalks, or some residual activity in basal shoots, or even photosynthetic activity within the bark of wetland trees such as common alder, can all serve to mitigate the risks of oxygen deprivation in winter. The presence of potential competitors also has a bearing on whether or not flooding creates a marginal situation for the flood-tolerant species.

8.7.2   Sweet flag (*Acorus calamus*)

The sweet flag (*Acorus calamus*) is a non-fertile (n = 36, triploid) neophyte (Fig. 8.23), probably introduced from one plant that belonged in 1574 in Vienna to the famous herbalist Charles de L'Ecluse (Latinized as Clusius, 1525–1609; see Schröter, 1908). Nevertheless, despite this apparent lack of population heterozygosity this

Fig. 8.18   Reed sweet grass (*Glyceria maxima*), an example of a semi-aquatic grass with numerous vegetative shoots and stout rhizomes spreading over wide areas in marshes and banks of slow running rivers.

Fig. 8.19   Aerial view of dune and slack system at Tentsmuir, Fife. Rapid accretion of sand has caused this coastline to advance rapidly seawards with the development of a series of dune and slack communities which show distinct boundaries controlled by topography, nutrients and water supply. The yellow box indicates the location of the transect shown in Fig. 8.20. (Photo Cambridge University Collection of Air Photographs: copyright reserved.)

species is a competitive and aggressive invader of heterotrophic European lake edges. This invasive success appears to stem in part from the extreme tolerance of anoxia found in the rhizomes of this species. The rhizomes have high carbohydrate reserves throughout the year which are more than sufficient to sustain ethanolic fermentation for several months. Furthermore, ATP production is considerably greater than in potato (1.55–2.33 µmol ATP g^{-1}fr.wt.h^{-1}) and adenylate pools and energy charge levels remain stable and high throughout prolonged periods of anoxia (Sieber & Braendle, 1991). These figures indicate a state of equilibrium between ATP production and consumption coupled with metabolic rates that are high enough to allow an extended viability for tissues under anoxia. In addition, the porous nature of the rhizomes and their

position on the surface of lake muds allows excess ethanol to diffuse out into the lake water and avoids the dangers of post-anoxic conversion to acetaldehyde discussed above (Studer & Braendle, 1984). Messenger RNAs for glycolytic and glycolysis related enzymes are induced anaerobically under artificial anoxia as well as in the natural habitat during winter (Bucher *et al.*, 1996). The proteins formed under anoxia in the laboratory and the field have been shown to be highly active. Moreover, in contrast to non-tolerant species, many additional proteins other than those involved directly in anaerobic metabolism are also synthesized under field and laboratory anoxia and clearly indicate that oxygen depletion is less of a metabolic perturbation and less likely to lead to cellular dysfunction in this species than in potato (Armstrong *et al.*, 1994). In addition, rhizomes of *A. calamus* are able to store and detoxify nitrogen that has been taken up as ammonium by the roots by transfer into alanine. The main nitrogen storage compound, however, in winter rhizomes is the nitrogen-rich amino acid arginine (Haldemann & Braendle, 1986, 1988). Arginine is readily converted into transport amino acids in spring when growth starts (Weber & Braendle, 1994). Nitrogen recycling and continuous uptake favours growth and development and protein synthesis in this species in comparison with other marsh plants. A similar strategy is used for the detoxification and utilization of sulphide formed in anaerobic soils. It is stored in the rhizomes as glutathione (Weber & Braendle, 1996). Glutathione is used as a sulphur source, but can also serve in the antioxidative defence mechanisms, in addition to tocopherol and phenolics (Larson, 1988). The most outstanding strategy with regard to anoxia tolerance in this species is probably the stability of membrane lipids under anoxia and their protection against peroxidation damage when the tissues are re-aerated (Henzi & Brändle, 1993). After 70 days of anoxia, lipids show only minimal alterations to the saturation level, with the principal change being a shift in fatty acid saturation from 18:3 linolenic acid to 18:2 linoleic acid. Furthermore, free fatty acids in the tissues are minimal and there is little evidence of membrane breakdown.

By comparison, in the anoxia-tolerant common club-rush (*Schoenoplectus lacustris*) rhizomes begin to show some signs of injury only after 35 days of anoxia. Moreover, in the rhizomes there is only a minor production of the peroxidation products malonedialdehyde and ethane. The membranes of this species are clearly

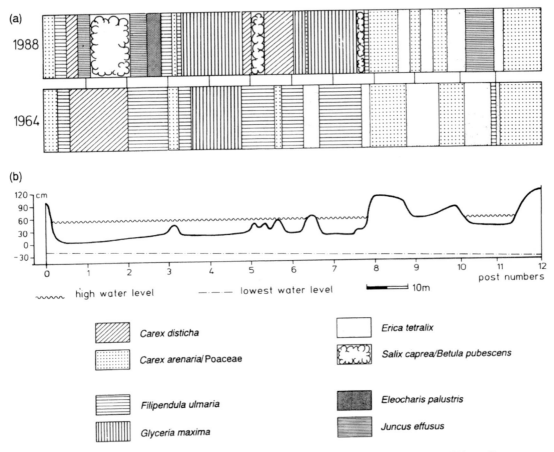

(a)

1988

1964

(b)

- ‍〜〜‍ high water level
- — ‍·‍— lowest water level
- ▬▬▭ 10m

Carex disticha	Erica tetralix
Carex arenaria/Poaceae	Salix caprea/Betula pubescens
Filipendula ulmaria	Eleocharis palustris
Glyceria maxima	Juncus effusus

Fig. 8.20 Relationship between vegetation and relief along a dune and slack transect at Tentsmuir National Nature Reserve as recorded over 24 years from (a) 1964 to (b) 1988. For approximate location see Fig. 8.19. The main change between 1964 and 1988 was the advance of the less anoxia-tolerant *Glyceria maxima* displacing the highly anoxia-tolerant *Filipendula ulmaria* as a result of a new drain reducing flooding levels. (Reproduced with permission from Studer-Ehrensberger *et al.*, 1993.)

well adapted to withstand prolonged periods of oxygen deprivation. Lipid metabolism under anoxia differs from that of proteins in that lipids are preserved while proteins can be synthesized *de novo*. This distinction is not unexpected given that lipids will require desaturases and molecular oxygen for their synthesis. *Acorus calamus* is therefore particularly well defended against the dangers of anoxia, both in terms of internal resistance to anoxia and resistance to externally generated anaerobic products in the soil solution, and can be considered better adapted than even such anoxia-tolerant species as *Phragmites australis* and *Spartina alterniflora* where high sulphide concentration can be damaging (see below).

8.7.3 Reed sweet grass (*Glyceria maxima*)

Among amphibious plant species, reed sweet grass (*Glyceria maxima*) represents an important example of an ecological strategy for wetland survival, namely seasonal tolerance of anoxia in early spring. Despite this seasonal variation in anoxia tolerance, *G. maxima* is able to outcompete the more anoxia-tolerant species such as *Filipendula ulmaria* (Studer-Ehrensberger *et al.*, 1993) due to its capacity for early spring growth enabling it to pre-empt the occupation of sites by other later developing species. In summer *G. maxima* rhizomes are not as tolerant of anoxia as some other wetland species and cannot survive prolonged deep flooding. When

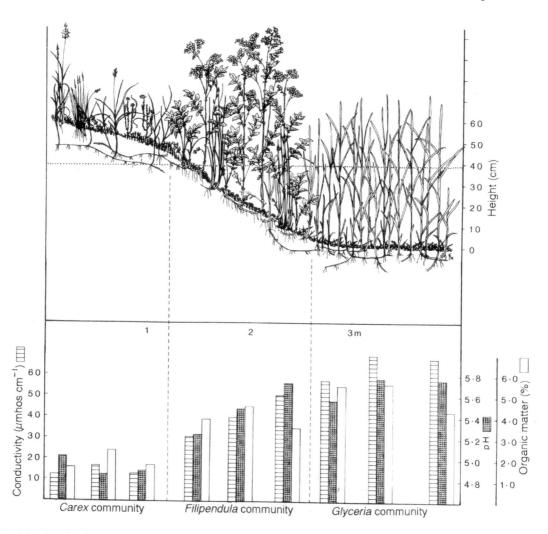

Fig. 8.21   Interface between retreating patch of *Filipendula ulmaria* and an advancing stand of *Glyceria maxima*. Note the gradual changes in soil pH, conductivity, and organic matter. Despite the gradual change in properties along this section of the transect there is nevertheless an abrupt boundary between the dominant plant species. Compare this with Fig. 8.22 (see text). (Reproduced with permission from Studer-Ehrensberger *et al.*, 1993.)

deprived of oxygen at high summer temperatures (22 °C) the rhizomes can lose 50% of their total non-structural carbohydrate reserves in 4 days (Barclay & Crawford, 1983). In this case, the underlying cause of anoxia intolerance is energy starvation.

By contrast overwintering rhizomes survive up to three weeks at 22 °C in the laboratory and probably survive even longer under field conditions. The physiological basis for this seasonal dependence of anoxia tolerance is not fully understood, but may be due in part

to high carbohydrate reserves in winter coupled with a less active metabolism of the overwintering organs. In sites where reed dieback is associated with abnormally high mineral nutrient concentrations, *Glyceria maxima* can be used as a successful replacement species. Planted into such seemingly phytotoxic sites, as seen from the decline of *Phragmites australis*, cuttings of *Glyceria maxima* are able to produce new roots within a few days. The survival strategy of *G. maxima* in wetland sites is that of a stress avoider rather than a stress tolerator. The

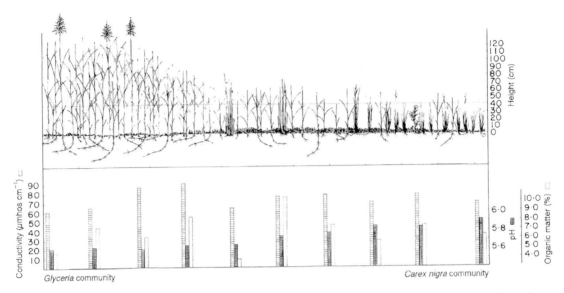

Fig. 8.22   Interface between *Glyceria maxima* and *Carex nigra*. Note the minimal changes in soil pH, conductivity, and organic matter along this section of the transect and the absence of an abrupt boundary between the dominant plant species and compare this with Fig. 8.21 (see text). (Reproduced with permission from Studer-Ehrensberger *et al.*, 1993.)

Fig. 8.23   The sweet flag (*Acorus calamus*) invading a lake margin in Switzerland where it has replaced former vigorous stands of the common reed (*Phragmites australis*).

species has a powerful capacity for oxygen transfer from shoots to roots and the well-developed aerenchyma is protected against accidental flooding by the presence of subdivisions of longitudinally arranged cells. This arrangement prevents inundation of the whole gas lacunae in event of any accidental physical damage and, therefore, reduces the risk of a sudden anoxic stress developing in the underground organs (Armstrong *et al.*, 1994). The capacity for flooding tolerance in this species, however, is strictly limited and the species is not a suitable replacement for *Phragmites* in areas with large fluctuations in the levels of the water table. The short and soft leaves of *Glyceria maxima* decay readily and thus fail to provide a snorkel for prolonged periods of submergence.

8.7.4   The common reed (*Phragmites australis*)

The common reed (*Phragmites australis*), in common with *Acorus calamus* and *Spartina alterniflora* (also wetland geophytes), combines all year round anoxia tolerance with a high capacity for oxygen transport from shoots to roots during the growing season. These species can therefore be considered as both avoiders and tolerators of the anoxic stress condition of wetland sites. Nevertheless, despite this two-pronged tolerance of inundation stress both *Phragmites australis* and *Spartina alterniflora* can suffer from elevated levels of sulphide (Bradley & Morris, 1990). High sulphide concentrations, above about 1 mM, occur frequently in reduced sediments of eutrophic lakes, polluted areas and in estuarine muds. An extensive study of sulphide tolerance in *Phragmites australis* showed that sulphide applied under hypoxic conditions (<0.6 ppm oxygen) severely affects root energy metabolism of reed plants with weakly developed shoots mainly by inactivation of metalo-enzymes. Normally sulphide has no direct access to rhizome tissues because of the thickened rhizome surface. However, grazing by the larvae of the reed beetle (*Donacia claviceps*) can create lesions which cause flooding of the rhizome air spaces with sulphide-rich water (Ostendorp, 1993). Sulphide can be translocated into the rhizome where it is partially detoxified by the formation of glutathione. The intermediate compounds are cysteine and glutamylcysteine with O-acetylserine acting as the sulphide acceptor. Rhizomes are less sensitive to sulphide poisoning than roots, but the detoxification capacity is limited and sulphide accumulates. The following phenomena have been observed in roots suffering from sulphide poisoning: (1) a decrease in adenylate energy charge and total adenylates, (2) a decrease of alcohol dehydrogenase activity, and (3) a decrease in post-hypoxic respiratory capacity (Fürtig *et al.*, 1996).

8.7.5 Amphibious trees

Trees with large trunks and deep anchoring roots represent the ultimate challenge in withstanding oxygen deprivation in wetland habitats. It is therefore surprising that in many parts of the world flooding is no detriment to tree growth. The forested bottomlands of the Mississippi Basin, the swamps of Louisiana and South Carolina, and the mangrove forests of tropical coastal regions, are all testimony to the ability of trees to grow, even where prolonged inundation is inevitable. In common with all other higher plant species the trees of swamp forests have an upper limit for the length of time that they can endure constant inundation, which is determined by the need for access to oxygen for tissue renewal.

The trees of bottomland forests in North America are also dependent on intermittent periods of low water levels for regeneration. For swamp cypress trees it can take four or more years of continuous inundation before many trees are killed (Ewel & Odum, 1984). In the past it was sufficient for this to take place once every 20–30 years. However, improved river level regulation has removed the amount of occasional

drawdown of river levels with the result that the regeneration of these forests in several areas such as the Mississippi Basin and the swamp forests of Louisiana is seriously threatened. Even the recruitment of such flood-tolerant trees as the swamp cypress (*Taxodium distichum*) and the water tupelo (*Nyssa sylvatica*) is prevented when there is no relief from constant flooding (Conner *et al.*, 1981). Similarly, it is predicted that higher flood levels on the Rhine will reduce the establishment of hardwood tree species in the low-lying sites in this river's flood-plain forests (Siebel *et al.*, 1998).

8.8 TROPICAL VERSUS TEMPERATE TREES IN WETLAND SITES

Outside the tropics most of these ultra-flood-tolerant swamp forests usually have either a relatively short winter or no winter, and trees do not have the stress of preserving an extensive root system in anaerobic conditions throughout long, non-productive winter periods. In the cold and cool-temperate regions of the world, particularly where the climate is oceanic, flooding is generally unfavourable for tree survival unless the ground is frozen during the period of potential inundation. The few woody species that survive in wetlands in cool oceanic regions are generally bushes and scrub of a limited number of species in which the genera *Salix* and *Alnus* are probably the most successful. The alder (*Alnus glutinosa*) is a typical species of riverbanks and lake edges. A common feature of alder stands is the formation of floodline communities due to the seeds washing up on banks during periods of flooding where they subsequently germinate. Alder seedlings do not establish readily under water and the river or lake edge can provide a suitable refuge from the dangers of flooding during the vulnerable period of establishment. Once established, alders are tolerant of flooding although they develop differently and show the many stemmed (polycormic) form in areas that are frequently flooded as compared with the pole (monocormic) form which is more usual in unflooded sites (see below).

In relation to tissue aeration in trees it should be noted that several species (e.g. birch, poplar, alder) have greenish tissues below the outer peridermal or rhytidomal layers. These chlorophyll-containing tissues within the stems are able to use the stem internal CO_2 and the light penetrating the bark for photosynthesis. When the photo-assimilation process is examined quantitatively

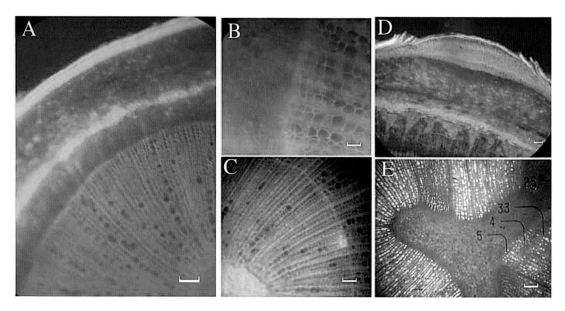

Fig. 8.24   (A) Transverse section of sector of a 1-year-old alder stem: red autofluorescence under blue light is due to chloroplasts in secondary cortex, secondary phloem, secondary xylem and medulla. Scale bar = 200 mm. (B) As (A) but at higher magnification showing individual chloroplasts fluorescing red in secondary phloem and in secondary xylem ray cells. Scale bar = 25 mm. (C) Sector of a 3-year-old alder stem showing red autofluorescence from chloroplasts in medulla and the inner two annual rings. Scale bar = 200 mm. (D) Transverse section of a 3-year-old alder stem through a lenticel. Some algal cells are fluorescing red on the outer surface of the flaking lenticel. Some red fluorescence is seen from chloroplasts in secondary cortex and secondary phloem. The pink colour in banded tissues of lenticel is not fluorescence but is due to anthocyanins. Scale bar = 100 mm. (E) Transverse section of a 1-year-old alder stem with transmitted white light. Chloroplasts (green) in secondary xylem rays and medulla. Scale bar = 200 mm. (Reproduced with permission from Armstrong & Armstrong, 2005.)

(Pfanz, 1999) net photosynthetic uptake of CO_2 is rarely found. Instead, internal re-fixation of CO_2 in young twigs and branches may compensate for 60–90% of the potential respiratory carbon loss. Corticular photosynthesis is thus thought to be an effective mechanism for recapturing respiratory carbon dioxide before it diffuses out of the stem. Furthermore, chloroplasts of the proper wood or pith fraction also take part in stem internal photosynthesis. Although there has been no strong experimental evidence until now, it has been suggested that the oxygen evolved during wood or pith photosynthesis may play a decisive role in reducing internal stem anaerobiosis. Whether or not this internally generated oxygen reaches the deeper roots in mature trees is still an open question. In some deciduous tree species (e.g. *Betula pubescens*, *Alnus glutinosa* and *Populus tremula*) the bark retains some photosynthetic activity in winter which, like the overwintering leaves in *Juncus* spp., will provide a source of oxygen.

Alders are particularly remarkable for the chloroplast content of their woody tissues (Fig. 8.24). A combination of anatomical and physiological studies of alder has shown that a significant oxygen flux occurs from stems in high light periods, indicating a net carbon gain by stem photosynthesis (Armstrong & Armstrong, 2005). Chloroplasts are abundant in the secondary cortex and secondary phloem, and occur throughout the secondary xylem rays and medulla of 3-year-old stems. Marked diurnal patterns of radial oxygen loss have been observed when light reaches submerged portions of the stem, and it has been suggested that this internally produced oxygen may improve root aeration, especially when temperatures are low (Fig. 8.25).

Investigation of the xylem flow in birch using microsensors has shown that the oxygen status of the sapwood was related to the mass flow of xylem sap but was reduced by oxygen depletion in the root space

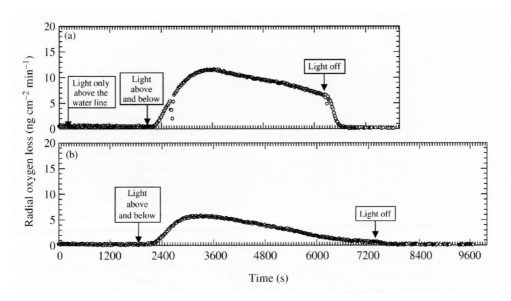

Fig. 8.25   Alder sapling showing the effects of light on ROL from a 13.5 cm long intact root. (a) ROL was unaffected by north light (50 mmol m^{-2} s^{-1}) or supplementary light (550 mmol m^{-2} s^{-1}) on an emergent shoot only but rose rapidly when emergent and submerged parts were exposed to a PAR of 550 mmol m^{-2} s^{-1} followed by a steady decline. ROL returned rapidly to background (approximately zero) when the light was switched off. (b) Continuation from (a) showing a similar pattern but lower ROL peak after exposure of emergent and submerged parts to 550 mmol m^{-2} s^{-1} PAR. (c) Continuation from (b) showing two further cycles of light application and a further diminished ROL peak despite some CO_2 enrichment of the bathing medium. (Reproduced with permission from Armstrong & Armstrong, 2005.)

(Gansert, 2003). Such a finding makes it problematical as to whether or not tree roots will receive any substantial benefit to their oxygen from winter stem photosynthesis.

Tree species that live in wetlands often show a polycormic growth form. In dry habitats, both willow and alder will grow as pole trees, but when flooded, the basal buds are stimulated to develop and the bush form predominates. *Alnus* species appear particularly successful in exploiting the polycormic bush growth form for surviving in wetlands. The polycormic growth form is so frequently a characteristic of wetland alder carr that the woods can appear as if they have been coppiced. The greater area of stem surface that the polycormic form produces close to the ground may facilitate aeration, and also maximize the potential benefits of stem photosynthesis in winter. It is possible therefore that the combined effects of a greater bark surface together with the presence of adventitious roots in willow, and negatively geotropic roots in nitrogen-fixing alders, could provide a supply of oxygen to roots growing in waterlogged soils.

Trees appear to be most at risk from flooding in oceanic climates where winters are long, wet and not particularly cold. In the more continental regions of the boreal forests of Canada and Siberia tree growth is not prevented by prolonged cold winters. In the bogs of Latvia and Estonia it is possible to find large stands of pine growing on raised bogs. Further north, in some of the coldest regions of the boreal forest *Picea mariana* and *Larix dahurica* can grow on bog surfaces even when permafrost is not far below the surface. The most northerly location of the boreal forest (Section 5.2.1) at 72° 34′ just to the west of the mouth of the Khatanga River in Siberia coincides with the greatest degree of continentality in the Siberian climate. However, in all these situations the soils are frozen for most of the winter, the roots are relatively shallow and oxygen demand is lower and supply less impeded (Crawford, 1992).

An entirely different winter situation is found in northern oceanic regions such as the British Isles and parts of western Norway where roots can be subjected to

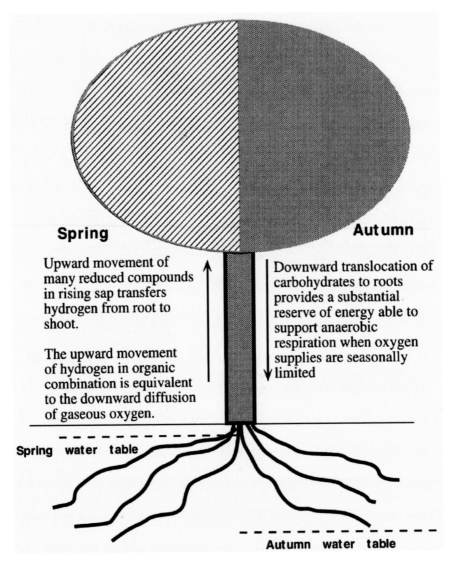

Spring

Upward movement of many reduced compounds in rising sap transfers hydrogen from root to shoot.

The upward movement of hydrogen in organic combination is equivalent to the downward diffusion of gaseous oxygen.

Spring water table

Autumn

Downward translocation of carbohydrates to roots provides a substantial reserve of energy able to support anaerobic respiration when oxygen supplies are seasonally limited

Autumn water table

Fig. 8.26   Diagrammatic representation of the seasonal cycles of the movement of carbohydrate in relation to the ability of birch trees to produce a strong upward flow of sap in spring even when soils are fully saturated or flooded and aeration is impeded (Braendle & Crawford, 1999).

winter flooding before they become dormant. In Sitka spruce (*Picea sitchensis*) flooding at temperatures above 6 °C leads to extensive death of root tips (Coutts & Philipson, 1987). Experimental studies of overwintering trees of Sitka spruce in which the flooding is carried out both at ambient temperature and ambient temperature plus 5 °C have shown that flooding under milder winter conditions causes a marked decrease in root carbohydrate reserves. This does not lead to the immediate death

of the root, but when carbohydrate-depleted roots are exposed once again to air when the water table is lowered, there is extensive dieback of the root system, presumably due to post-anoxic injury (Crawford, 1996a). Death of the whole tree will not be immediate, but reduced root development will make the larger trees unstable and therefore prone to windblow.

The long winter, and the need to conserve carbohydrate for the resumption of sap flow and bud burst in the spring, creates a need for adequate overwintering supplies of carbohydrate. In birches, the active upward movement of xylem sap in spring carries with it substantial quantities of soluble carbohydrates, ethanol, organic and amino acids. This flow has been suggested as the means whereby roots of some woody plants compensate for the lack of oxygen in the soil-atmosphere by exporting hydrogen to the shoot and thus maintaining the redox balance of the inundated root system (Crawford, 1996a). The upward movement of hydrogen in reduced compounds as a constituent of the xylem flow is a more efficient means of repaying the oxygen debt of submerged organs than the slow diffusion of gaseous oxygen through more than a metre of woody tissue (Fig. 8.26). This replacement of oxygen import by hydrogen export does, however, have a metabolic cost, namely the provision of a high carbohydrate store in the overwintering root. The flow of sap in birch trees should not be confused with that of the sugar maple (*Acer saccharin*). In the latter the sap flow is dependent on diurnal freeze–thaw cycles affecting the trunk and stem and is not a function of root pressure. In birch, however, root pressure drives the spring ascent of sap (Kozlowski & Pallardy, 1997). It is relevant to note that in the bottomland trees of the USA it has been shown that high pre-flood root-starch concentrations are an important characteristic allowing flood-tolerant species to survive inundation. The bottomland *Fraxinus pennsylvanica* (green ash) and flood-tolerant *Nyssa aquatica* (water tupelo) are able to store more carbohydrate in their roots and retain less in their leaves than the less flood-tolerant *Quercus alba*.

8.9 CONCLUSIONS – PLANTS WITH WET FEET

Ecophysiological studies of plants in marginal flood-prone habitats have demonstrated that there are differences between adaptations to oxygen deficiency as it operates in plants in winter and in summer. The winter adaptations may not be as active as those operating in summer. However, they operate under different conditions and are well suited to function in a prolonged and sustainable manner which will satisfy, at least in part, the lower demands for oxygen in overwintering plant organs.

Throughout the world plants are divided ecologically as to whether or not they can survive with wet feet. This does not mean that all plants that survive flooding endure similar stresses or adopt the same adaptive strategies. Flooding in the moving tidal zones of tropical mangrove swamps or even bottomland trees of South Carolina or Louisiana is a very different stress from the long wet winters and short summers encountered by plants that grow in bogs at higher latitudes. Even within the flora of any one particular bog there can be found a variety of adaptations in plants that grow in close proximity to one another. Surface rooting plants with evergreen shoot bases, as in the rushes (*Juncus* spp.), survive by ventilating their underground organs by diffusion supplemented by internal generation of oxygen from photosynthesis. Deeper-rooted plants endure these adverse conditions by ensuring that their perennating organs have the ability to tolerate total oxygen deprivation and resume growth once the winter water tables subside.

The tree habit appears to be particularly disadvantaged by flooded soils in areas where flooding in winter is prolonged but not cold enough to ensure complete dormancy of the roots. Flooding of non-dormant root systems early in the winter can induce severe carbohydrate depletion which is then associated with dieback of the root systems on re-exposure to air when water tables drop. Re-exposure to air is potentially hazardous for tissues that have been denied access to oxygen for a prolonged period. Most plant organs that can endure long-term tolerance anoxia also have to be resistant to post-anoxic injury and this incurs extra metabolic costs in ensuring adequate carbohydrate reserves.

Metabolically, the pathways that are available under anoxia to plants that can endure anaerobiosis are not substantially different from intolerant dry land species. The major differences appear to be in regulation, in accessing and translocating carbohydrate reserves under anoxia, together with the minimization and dispersal of the oxygen debt and avoidance of anoxia-induced cytoplasmic acidosis. In terms of biochemistry, the modifications may not appear to be particularly remarkable, yet they are probably sufficient to enable tolerant species to do just a little bit better than dry land species when deprived of oxygen. These minor biochemical differences, however, when translated into anoxia and hypoxia survival time limits, can make all the difference between life and death at the margins for plants with wet feet.

Fig. 9.1   Coastal heathland, Papa Stour, Shetland (UK) in August with dwarf heather (*Calluna vulgaris*) in flower.

9.1 WOODY PLANTS BEYOND THE TREELINE

Beyond the limits for the survival of forest, woody plants in shrub form exist as viable and even major components of ecosystems. Plants with lignified stems and branches can be found on mountains, moors, coastal heaths (Fig. 9.1), as well as across the tundra to the very north of Greenland and south to remote sub-Antarctic islands. The woody shrub or bush form is highly flexible and can be almost of tree stature as in *krummholz* pine or diminutive as in some arctic heather species (Figs. 9.2–9.3). Every possible variation exists in the extent of tissue lignification from bushes that are entirely woody, as in the dwarf willows, mountain alders and birches to plants where the shoots are only partly woody. The minimal woody stem condition, where only the roots and the base of the stem are lignified (defined botanically as *suffrutex*), is not discussed here, as the shoots of these plants are essentially herbaceous. The versatility of woody species in both size and habit probably contributes much to their success in marginal areas whether it be in hot or cold deserts or on wind-swept oceanic heaths.

The presence of woody plants, and the canopy they create, whatever their size, alters the temperature,

Fig. 9.3 A marginal woody plant of diminutive size. The clubmoss mountain heather (*Cassiope lycopodioides*), native to Alaska, Kamchatka and Japan. (Photo approx. 2×life size.)

light, and evapotranspiration regimes around the leaves in summer, and the degree of exposure to adverse climatic conditions in winter. Branches can be erect, giving the shrub a narrow outline (*fastigiate*), or they can be spreading at wide angles (*divaricate*). In between these extremes many forms exist. When the branches are procumbent, they can be stoloniferous and spread as a mat over the surface of the ground as in the arctic bearberry (*Arctostaphylos alpina*; Fig. 9.6),

The abandonment of the single pole form in favour of the many-stemmed bush habit also facilitates regeneration after damage, whether from climatic adversities such as drought or frost or from disturbance by grazing or fire. These properties serve to make woody shrubs long-lived and major components of a wide variety of plant communities.

Highly exposed sites at sea level can support very tenacious heath communities. Although heather (*Calluna vulgaris*) and crowberry (*Empetrum nigrum*) may suffer erosion as a result of strong winds they nevertheless manage to survive by slowly migrating in terrace formation before the wind (Fig. 9.4). In more sheltered sites behind islands the natural succession is to dune heath (Fig. 9.5) provided the sand is not rich in seashells and the sea salts are leached out by rain to leave a soil pH value of 6 or less.

Fig. 9.2 Versatility of woody shrub form habit. Krummholz pine (*Pinus sylvestris*) at the treeline at Creag Fhiaclach at 620 m a.s.l. in the Cairngorm Mountains, Scotland.

Fig. 9.4   Maritime heath on the west coast of Orkney. (Top) Wind terracing with heath ridges eroding on their downhill side and recolonizing the ground on the upper side. (Above left) Detail of advance of heath species onto bare ground. The crowberry (*Empetrum nigrum*) (bright green) is particularly successful in putting out new roots around edges of turf and spreading across the open ground in areas with high wind exposure. (Above right) Detail of erosion on windward side and ability of heath to recolonize on the downwind face of the eroding terrace.

Fig. 9.5 Lowland dune heath dominated by bell heather (*Erica cinerea* – see inset) in north-east Scotland. Active grazing by sheep and cattle control the further spread of Scots pine (*Pinus sylvestris*).

Woody shrubs can be aggressive invaders, as can be seen in the spread of *Rhododendron ponticum* in the British Isles and *Calluna vulgaris* in New Zealand. Asian tamarisks (saltcedars, *Tamarix* spp.) were first imported into the United States in the nineteenth century as ornamental plants and then used for erosion control. This has now led to the invasion of almost all watercourses and other wetland habitats throughout the south-west, taking over more than one million acres of wetland. In New Zealand heather (*Calluna vulgaris*) can invade burnt surfaces more rapidly than the native New Zealand species, with the result that it has displaced much of the lower and subalpine natural shrub and tussock vegetation (Wardle, 1991). In western Europe, the removal of much native forest has led to the downhill migration of the heather from its natural lower-alpine habitat to create extensive heathlands where once trees flourished. These many-stemmed woody invaders from the lower-alpine zone have survived 5000 years of fire and grazing, presumably because the conditions under which they evolved above the treeline produced a plant form that was pre-adapted to the burning, grazing and slashing regimes that were imposed on woody plants with the Neolithic development of agriculture.

The ecology of woody plants in relation to climate change can therefore be examined with examples ranging from the north of Greenland (83° N) to the southernmost outpost for woody plants on New Zealand's Southern Oceanic Auckland and Campbell Islands (*c.* 51° S).

9.2 WOODY PLANTS OF THE TUNDRA

The tundra is commonly described as treeless. Trees growing upwards with a clearly visible trunk are only found in isolated pockets near the tundra–taiga interface (see Chapter 5). However, this does not necessarily exclude the presence of woodlands in the tundra. Whether or not trees can be considered to be present in polar regions is merely the imposition of an arbitrary human judgment. Willows that grow upwards in the temperate zone survive better in the Arctic if their trunks lie flat along the ground, and this should be recognized as a natural development for woodland survival in this particular situation. The vast mats of polar willow that cover the tundra and reach as far north as 80° N in Spitsbergen can have horizontal trunks several metres long and over 200 years old. For this achievement they have been called Spitsbergen's 'forest'. At ground level the colourful autumnal display of this prostrate woodland is as striking as any North American maple forest.

The woody flora of the Arctic comprises relatively few taxonomic groups. Birches (*Betula* spp.), willows (*Salix* spp.), alders (*Alnus* spp.) are all examples of tree genera that can grow in the pole form in temperate climates but can exist either in the bush form or as prostrate mats in polar regions. The phenotypic and genotypic forms of *krummholz* of several coniferous tree species are also found in the southern fringes of the arctic tundra where it borders with the boreal forest (see Chapter 6). In the northern hemisphere, the Ericaceae is the predominant family in providing many widespread woody arctic-alpine and heath genera such as *Arctostaphylos* (Fig. 9.6), *Vaccinium*, *Erica*, *Calluna*, *Rhododendron* and *Kalmia*.

In terms of geographic ubiquity, first place belongs to the genus *Empetrum*. The northern hemisphere crowberry, *Empetrum nigrum*, together with its almost identical, vicarious South American counterpart *E. rubrum*, are in their combined ranges the most widespread of all the dwarf woody plants in having a distribution range that extends from the High Arctic to the heaths of the Tierra del Fuego and the Falkland Islands. In all cases they inhabit either acid peatlands, or cold coniferous forest, and acidic rocky slopes. Crowberry also colonizes calcium-depleted sand dunes and cliff tops. In Europe there exist two subspecies: *E. nigrum* subsp. *nigrum* which is dioecious (Fig. 9.7), and *E. nigrum* subsp. *hermaphroditum* which is hermaphrodite. The hermaphrodite subspecies has a more northern and montane distribution but can occur at sea level in highly oceanic regions such as the Shetland Islands. The hermaphrodite form is easily identified from the withered stamens that persist at the base of some of the berries. Both subspecies can be found in marginal areas and are particularly successful in putting out new roots around edges of turf and spreading across open ground in areas with high wind exposure.

The preference of *Empetrum nigrum* for cooler moister sites is seen more clearly at lower latitudes outside the Arctic. This is seen in the East Friesian island of Spiekeroog where crowberry is dominant only on the cooler and more humid north-facing slopes (Fig. 9.8).

Maintaining the woody form even in the polar deserts of the High Arctic is a remarkable phenomenon both in terms of being able to support non-photosynthetic tissues in habitats with short cold growing seasons but also in being able to survive the stresses of winter and dangers of herbivory in unproductive habitats. In northern regions woody plants are particularly at risk as they are a principal source of forage for overwintering mammals. Domesticated sheep and reindeer, and a wide range of wild herbivores, including caribou, reindeer, red deer, musk oxen, hares, lemmings and voles, use dwarf woody plants beyond the treeline as a source of fodder. There are many examples in marginal habitats in the subarctic regions of Europe where overgrazing of dwarf woody shrubs, including *Calluna vulgaris*, can lead to the destruction of this vital winter resource which in subarctic regions is often replaced by unproductive acid grasslands.

Woody, non-productive stems and branches might appear as a morphological luxury in terms of photosynthetic efficiency in marginal areas where the balance between carbon surplus and carbon deficit may be problematical. A study of net primary production (NPP) in the polar willow (*Salix polaris*) in Spitsbergen has shown that this species is well adapted under current conditions to maximize its use of the potential growing season. Maximum values for photosynthesis rates and stomatal conductance are reached within one week after leaf emergence, which takes place immediately on snowmelt and then gradually decreases. Depending on leaf age, photosynthetic rates were found to be saturated at a photosynthetically active photon flux density (PPFD) of 200–400 μmol m^{-2} s^{-1}, which is the light level usually available in this habitat. Optimum leaf temperature for photosynthesis was in the range 10–18 °C, while air temperature in the habitat varied between 8 and 20 °C. These light and temperature responses of photosynthesis permit efficient carbon gain in a natural habitat characterized by highly variable light and temperature conditions. Model-based predictions for one particular year gave values for a probable net primary productivity for the year of 26.1 g cm^{-2}. However, this model also predicted that rising temperatures would cause a reduction of NPP due to potentially large increases in respiration (Muraoka *et al.*, 2002).

When the respiratory activity of the below-ground parts (roots + below-ground stems) of three dominant arctic species (*Salix polaris*, *Saxifraga oppositifolia* and

Fig. 9.6   Arctic bearberry (*Arctostaphylos alpina*) providing a mat of vegetation in full late summer colour in August at Mesters Vig (73° N), East Greenland.

Fig. 9.7   Detail of fruits of crowberry (*Empetrum nigrum* ssp. *nigrum*) together with the vicarious species *E. rubrum* make crowberry one of the world's most widespread species inhabiting heathlands from the Arctic to the southern Andes, the Falkland Islands and Tristan da Cunha.

Luzula confusa), were determined under laboratory conditions it was found that both the respiratory activity and the Q_{10} value for respiration were higher in *S. polaris* than in the other two species. This suggests that the polar willow will be more likely to suffer a carbon deficit with rising soil temperatures than the other species examined in this study (Nakatsubo *et al.*, 1998).

The woody plants of the Arctic differ in their nutrient requirements and it might therefore be expected that the deposition of pollutants at high latitudes may affect the species composition of the tundra. A study in Spitsbergen comparing the arctic heather (*Cassiope tetragona*), mountain avens (*Dryas octopetala*) and the polar willow (*Salix polaris*) was able to demonstrate basic differences in the carbon and mineral nutrient economies of the shrubs, related to their growth form. This was seen in the ability of the shrubs to respond to nitrogen and phosphorus treatments (Fig. 9.11). *Cassiope* was conservative and there were no significant treatment effects. *Salix* was the most responsive, showing increases in leaf

nitrogen concentration, biomass and photosynthetic rate (Baddeley *et al.*, 1994).

9.3   MONTANE AND ARCTIC WILLOWS

As a genus, the willows are a widespread group containing approximately 400 species occurring mostly in the northern hemisphere. In the arctic and subarctic regions of Europe and North America there are approximately 28 boreal species with numerous subspecies and hybrids. However, only five of these species achieve circumpolar distribution, namely *S. arctica*, *S. glauca*, *S. lanata*, *S. phylicifolia*, and *S. reticulata* (Figs. 9.9–9.12; Hultén & Fries, 1986). The genus *Salix* also stands out for its unsurpassed ability to maintain the woody form higher up mountains and further north than any other genus. In the Arctic, the polar willow (*S. polaris*; Fig. 9.10) can be found as carpets of vegetation at Biskayerhuken (79° 50′ N) in Spitsbergen while its North American vicarious counterpart *S. polaris* subsp. *pseudopolaris* reaches 75° N in the Canadian arctic archipelago. In Greenland, *Salix arctica* reaches 83° N in Peary Land where along with *Papaver radicatum* and *Saxifraga oppositifolia* it belongs to the most northerly terrestrial plant community in the world. In terms of high-altitude survival, the least willow (*S. herbacea*) occurs at 2170 m in central Norway (Lid & Lid, 1994) while in the Alps it can be found at over 3000 m. In the hyperoceanic climate of Shetland (Scotland's Northern Isles – 60° N) it occurs at much lower altitudes from 150 m down to sea level (Beerling, 1998).

In Scotland montane willows are much depleted due to grazing. Species such as the woolly willow (*Salix lanata*; Fig. 9.12) are now restricted to steep slopes where they are more vulnerable to chance events such as erosion, rock falls and avalanches.

All willows share a number of common characteristics. They are dioecious, developing male and female flowers in different individuals. Flowering is commonly precocious with the catkins appearing before the leaves. However, in the most northern species this is less marked. In *S. polaris* and *S. herbacea* leaves and catkins appear almost simultaneously. The species are insect-pollinated, even in the Arctic where small flies are attracted by the ability of the willows to secrete nectar. Seeds are widely dispersed by wind. Vegetative reproduction is common. In larger willows rooting can take

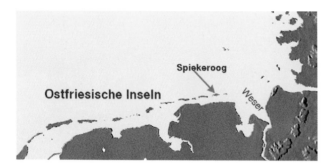

Fig. 9.8   Site preferences for crowberry (*Empetrum nigrum*) on sand dunes on the island of Spiekeroog. (Above left) Location of Spiekeroog in the East Friesian Islands (North Germany). (Above right) Aerial view of Spiekeroog. (Photo Dr A. Gerlach.) (Below) View of crowberry colonies (dark coloured areas of dune slopes). Although *Empetrum nigrum* is a very widespread species geographically it is restricted in these dunes to north-facing slopes.

Fig. 9.9   The net-leaved willow (*Salix reticulata*) growing in Möller Fjorden (79° N) in Spitsbergen.

place whenever the branches touch the ground. In the creeping arctic forms such as *S. herbacea*, *S. polaris*, *S. arctica*, and *S. reticulata* (Figs. 9.9–9.10) underground creeping stems ensure vegetative reproduction and lead to the establishment of extensive and long-lived clones.

The potential hazard of biomass loss by herbivory can be countered either by maximizing resource acquisition and compensatory growth, or by minimizing loss of resources by grazing deterrents. The latter strategy depends on investing in structural defence or synthesizing secondary metabolites that render the plant unpalatable, toxic or non-nutritious to the herbivore. A study of *Salix polaris*, which is regularly browsed by reindeer in Spitsbergen, examined the response one year after simulated browsing (by repeated clipping) in early, mid, and late summer. The simulated grazing greatly reduced the number of leaves as well as the total and individual biomass of leaves, and the number of catkins (Skarpe & van der Wal, 2002). There was no increase in phenolics but a tendency to an increase in nitrogen content in the leaves one year after the treatment. It appears that *S. polaris* responds to summer browsing the previous year by allocating resources to compensate for loss of biomass rather than stimulating the synthesis of secondary metabolite grazing deterrents.

The final result is a herbage that is reduced in quantity but may have some increase in quality.

In other Spitsbergen studies, the length of the growing season was manipulated by adding or removing snow so as to effect a two-week difference in snowmelt. This was equivalent to an approximately one-sixth alteration of the growing season between advanced (first to be snow-free) and delayed (last to be snow-free) treatments. The shift in 'phenological time', led to the Spitsbergen reindeer (*Rangifer tarandus platyrhynchus*) selecting the advanced, longer-exposed snow-free plots, presumably because of the greater biomass of *Salix polaris* and *Luzula confusa*, both major components of reindeer diet at the early part of the year. By contrast, plant quality, measured as nitrogen content and C:N ratio of leaves, was lowest in the preferred plots. Here again phenolic content did not differ among treatments. It would appear that in this northern arctic region, grazing preferences are for quantity rather than quality, which is the reverse of the usual tendency in temperate regions. This grazing habit of consuming large quantities of herbage is common to many lactating mammals in the Arctic and may be due to the need to secure a sufficient supply of minerals and in particular calcium for their lactation periods.

Fig. 9.10   The polar willow (*Salix polaris*) showing pairs of leaves (length 5–10 mm) emerging through moss from a buried rhizome. (Upper) Male plants. (Lower) Female plants – the predominant sex, 59% female; see Fig. 4.33). Photographed in Möller Fjorden, Spitsbergen (79° N).

A curious feature of willow populations that is very noticeable in the Arctic is the bias towards females which is found in a number of species and appears to be a circumpolar phenomenon. Various explanations have been postulated to explain the skewed sex ratio in arctic willows in favour of female plants first noted in *S. polaris* in Spitsbergen and *S. herbacea* in Iceland (Crawford & Balfour, 1983). On one hand it might be due simply to differences in growth rates with more aggressive females outcompeting male plants, as has been observed in sea buckthorn (*Hippophae rhamnoides*; Fig. 4.36). On the other hand, variation in growth rates might lead to differences in the relative investment

in metabolic grazing deterrents between the sexes bringing about a skewed sex ratio due to differential herbivory (see also Chapter 4). An experiment in Spitsbergen where there are no lemmings, and therefore where reindeer are the main grazers, found that excluding reindeer for three years increased the abundance of male flowers in one of two vegetation types investigated (Dormann & Skarpe, 2002). However, growth rates differed only slightly between the sexes, with females investing more in inflorescences. The concentration of chemical defence compounds (phenolics and condensed tannins) did not differ between the sexes. Consequently it was concluded that the

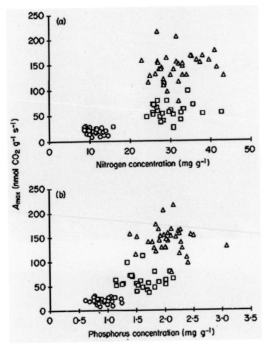

Fig. 9.11   Light saturated photosynthetic rates (A_{max} on a leaf-weight basis) in relation to leaf concentration of (a) nitrogen and (b) phosphorus. The differences in nutrient concentration have been induced by feeding sustained treatments. Open circles, *Cassiope tetragona*; squares, *Dryas octopetala*; triangles, *Salix polaris*. (Reproduced with permission from Baddeley *et al.*, 1994.)

Fig. 9.12   The woolly willow (*Salix lanata*), a common mountain willow in Scandinavia but now rare in Scotland and proving difficult to re-establish in former habitats. Photograph taken in Iceland.

hypothesis was not tenable that growth rate-dependent herbivory caused the unbalanced sex ratio in *S. polaris*. The universality of this skewed ratio in northern willows, irrespective of the degree of grazing, suggests that the bias in favour of females is an intrinsic property of montane and arctic willows. The skewed sex ratio may be connected in some as yet unknown manner to inherently different survival rates as young populations have sex ratios near equality, with the sex bias developing only as the populations age (Crawford & Balfour, 1990).

Arctic and montane willows do not appear to thrive in oceanic areas with warm winters and may therefore be forced to retreat from certain parts of their present distribution should climatic warming continue. Already in more oceanic subarctic regions, as in Scotland, great difficulty is being experienced in restoring the mountain willows that were formerly more widespread but are still such a characteristic feature of the colder mountain regions of Norway. In Scottish mountains the long-term survival of the montane shrubby willows (*S. lapponum*, *S. lanata*, *S. arbuscula*, *S. myrsinites* and *S. reticulata*) is giving particular cause for concern. Seedling establishment is low even when grazing is absent. The populations are small and fractionated and frequently the separate male and female plants are so distanced from each other that pollination is inefficient. Why these Scottish willows lack the vigour of the Scandinavian populations is not clear. There may be an adverse effect from the hyperoceanic Scottish environment. The possibility that woody species in general in northern habitats are disadvantaged by oceanic conditions is discussed further below.

9.4   MOUNTAIN BIRCHES

The term *mountain birches* encompasses a diverse taxonomic group which can be defined most easily in terms of growth habit and location. Mountain birches are distinguished from all other birches by their ability to live in the alpine or subarctic zone above the treeline and have the potential to adopt the many-stemmed growth (polycormic) habit as opposed to the pole form (monocormic). The silver birch (*Betula pendula*) grows only as the pole form, which in Scotland requires a minimum of 1100 day degrees for survival. Consequently, silver birch fails before the treeline is reached. By contrast the downy birch (*B. pubescens*), which has a capacity to grow also in the bush form, can exist in Scotland with as little 700 day degrees (Forbes & Kenworthy, 1973). In the Nordic regions of Europe mountain birches (Figs. 9.13–9.15) are treated as a subspecies of *Betula pubescens* ssp. *czerepanovii* (formerly ssp. *tortuosa*) which grows today from Iceland to the central Kola Peninsula (Wielgolaski & Nilsen, 2001). The species is highly polymorphic as it has probably arisen on more than one occasion from hybrids between downy and dwarf birch (*Betula nana*) which can then backcross with downy birch. The Iceland form has most probably arisen independently of the Nordic forms, as have also the Scottish populations of *B. pubescens* ssp. *carpatica* which is distinct in having the round odoriferous leaves characteristic of *B. nana* (Fig. 9.16).

In Iceland it has been shown that hybridization between diploid (2n = 28) dwarf birch *Betula nana* L. and the tetraploid (2n = 56) downy birch *B. pubescens* Ehrh. has occurred in natural populations (Anamthawat-Jonsson & Thorsson, 2003). About 10% of birch plants randomly collected for this study were triploid hybrids (2n = 42) as confirmed by ribosomal gene mapping. The triploid hybrids are morphologically distinct from the diploid and tetraploid parental plants with an intermediate morphology. It appears that the triploid hybrids have played an important role in driving bidirectional gene flow between these two species.

In Greenland *B. glandulosa* and *B. pubescens* form hybrids, as do *B. glandulosa* and *B. papyrifera* in Alaska. Mountain birches therefore are not a coherent taxon and show much clonal variation (Väre, 2001). For the purposes of this present discussion they will therefore be defined simply, as stated above, as those birches that can live in the alpine or subarctic zones above the treeline and possess the ability to adopt the many-stemmed (polycormic) as opposed to the pole (monocormic) form (Figs. 9.13–9.14).

The woody habit of the mountain birches is highly variable even within specific sites with mixtures of monocormic and polycormic individuals. A third type can also be found in some situations which has a semi-prostrate undulating main stem which extends in a general horizontal direction supporting vertical shoots (Fig. 9.14). The causes of polycormic growth in birch have been much discussed (Wielgolaski & Nilsen, 2001). Hybridization between *B. pubescens* and *B. nana* is one possible reason for polycormic growth and may explain why mountain birches have more basal shoots than the lowland birches. The hybrids normally also grow at somewhat lower temperatures than *B. pubescens*.

The prevalence of polycormic as to monocormic forms of mountain birch varies regionally and may be due to various causes including soil conditions, flooding and herbivory. Severe outbreaks of insect defoliation are often cited as one reason for the prevalence of the polycormic form in mountain birch. The most frequent defoliators are *Epirrita autumnata* and *Operophtera brumata*, the autumn and winter moths, which show marked cyclical activity particularly in coastal regions. In the more continental areas, the outbreaks are more irregular but when they do occur can have catastrophic results and leave a mark on the landscape that is visible for decades. In the Åbisko valley of northern Sweden (68° N), the mountain birches were completely defoliated by *Epirrita autumnata* caterpillars during an outbreak in 1954–55. The defoliation resulted in an 80–90% mortality of the leaf-carrying shoots, which then triggered extensive vegetative stand rejuvenation with very little regeneration from seed. The surviving plants regrew and increased production of long shoots from surviving shoots and basal sprouts, with the result that damaged stands developed a higher proportion of polycormic individuals than comparatively undamaged stands (Tenow & Bylund, 2000). The larvae of the geometrid moths also eat the foliage of the bilberry (*Vaccinium myrtillus*) and bog bilberry (*V. uliginosum*) to such an extent that these heath species can be replaced with wavy-hair grass (*Deschampsia flexuosa*). A similar outbreak in northern Finland in 1965 destroyed over 1000 km^2 of birch forest (Kallio & Lehtonen, 1973).

Other reasons for strong development of the polycormic form (sometimes referred to as coppicing), may be browsing by sheep or reindeer. Some field observations also indicate that mountain birch trees tend to develop many stems in poor and dry soils. This seems particularly to be the case in cold, wind-exposed habitats with poor snow protection. By contrast, in moister, nutrient-rich soils, monocormic forms are more frequent (Wielgolaski & Nilsen, 2001). There may also be a connection between local climatic conditions and insect attack as the topo-climatically favourable,

Fig. 9.13   Mountain birch (*Betula pubescens* spp. *czerepanovii*) growing with typical twisted stem *krummholz* form in the Lofoten Islands (68° N, Norway).

warmer, earlier areas provide better overwintering conditions for insect defoliators. In landscapes with marked relief, as in the areas around Åbisko where many studies on mountain birch have been made (Neuvonen *et al.*, 2001), there are differences in the location of tree forms, climate and insect attack. Winter moth (*Operophtera brumata*) outbreaks are more frequent and usually start on south-facing slopes, as

Fig. 9.14   Varying forms of the mountain birch (*Betula pubescens* spp. *czerepanovii*). (Left) Creeping *krummholz* form (with knees) growing at the treeline in central Norway. This is sometimes referred to as var. *appressa*. (Right) Many-stemmed form (polycormic) growing on the Dovrefjell, Norway.

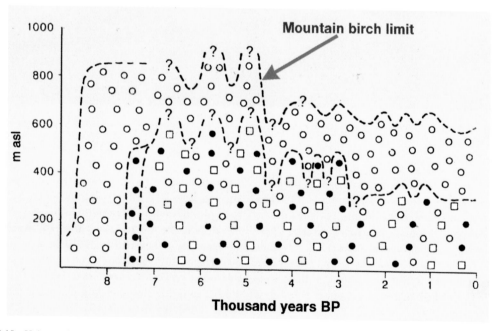

Fig. 9.15   Holocene forest line oscillations in central Troms, northern Norway. Open circles, forest dominated by mountain birch (*Betula pubescens* spp. *czerepanovii*); closed circles, dominated by pine (*Pinus sylvestris*); squares, dominated by grey alder (*Alnus incana*). (Reproduced with permission from Aas & Faarlund, 2001.)

Fig. 9.16   The dwarf birch (*Betula nana*) – note the rounder (orbicular) leaves, a character that this species brings to hybrid mountain birches.

this species is less cold-hardy than *Epirrita autumnata*. The periodicity and frequency of the attacks could be a significant long-term factor in the selection of the polycormic growth form in these stands. In areas where local climatic conditions ensure cold winters, the stands appear to be dominated by old monocormic trees. When warmer conditions lead to an outbreak in these areas many trees are killed and recovery is slower especially if grazed by reindeer (Tenow *et al.*, 2001).

9.4.1   Biogeographical history of mountain birch

Mountain birch has a long history in Scandinavia and neighbouring North Atlantic islands. In Swedish Lapland, megafossil wood remnants of *Betula pubescens* ssp. *czerepanovii* buried in peat at an altitude of 999 m a.s.l., provide evidence of an early Holocene birch belt at 68° 20′ N in the Åbisko-Lake Torneträsk area. Radiocarbon dating of these remains, which lie approximately 500 m higher than today's tree limit of *Pinus sylvestris* in this region, yielded values of *c.* 5400–4500 BP (Kullman, 1999). Figure 9.15 illustrates the extent of fluctuations in forest altitude limits as observed in northern Norway (Aas & Faarlund, 2001). In other more southern areas the birch belt can be traced to earlier dates. In the south-central Norwegian mountains megafossils of birch and pine from altitudes of 900–1370 m have been found with datings between 2900 and 8660 BP, and in the eastern Jotunheimen Mountains at 1288 m from 8000 BP and 1370 m from 6930 BP (Aas & Faarlund, 2001). The earliest macrofossil remains of birch so far reported in Scandinavia are from a study site at 1360 m a.s.l. close to the summit of Mt Åreskutan (63° 26′ N, 13° 06′ E). In the alpine region of the southern Swedish Scandes, fossil remains of mountain birch dated to 14 000 BP have been found 400–500 m above modern tree limits (Kullman, 2002a).

Elsewhere in the North Atlantic region, as mentioned above, mountain birch may have different historical and genetic origins. In Iceland, there is a suggestion that birch may have survived in glacial refugia or may have evolved more recently from parental immigrants of downy birch crossing with *Betula nana* that may have survived in situ or migrated in the early Holocene. In Greenland, the earliest record of birch forest is 4400 BP and was already in decline at the time of the Norse settlement in the late tenth century. Ruins of Norse farms have been found located at what appears to have been the forest altitudinal limit at the time of Norse settlement (Albrethsen & Keller, 1986). With the onset of the Little Ice Age both people and trees disappeared from these marginal sites.

9.4.2 Current migration

In modern times the possible effects of climatic warming and increased levels of atmospheric carbon dioxide have stimulated careful monitoring of the performance as well as the distribution of birch at altitudes above the main mountain forest belt. Among forest trees birch appears to be well adapted to take advantage of both higher levels of atmospheric carbon dioxide and soil nitrogen. Several reports have been published of seedlings being found well above the present birch treeline. Examination of the late-Holocene tree-limit history (*Betula pubescens* ssp. *czerepanovii*), on the east-facing slope of Mt Storsnasen in the southern Swedish Scandes (63° 13′ N, 12° 23′ E) has detected an upward migration of the tree limit by 75 m over the past century, which matches the meteorological records for summer warming (Kullman, 2003). The new tree colonies are, however, limited to small clumps of trees growing in minor depressions or else limited to young seedlings 5–20 cm high.

Treelines are not easily mapped and the presence of isolated seedlings above the main occurrence of trees has probably always taken place, particularly during warmer climatic intervals, with the plants failing to

survive as they grew above the immediate shelter of their surroundings. Such ephemeral migrations could have happened many times in the past and might not have been detected at a time when investigators were less aware of climatic warming phenomena. Definite confirmation of an advancing treeline still awaits a demographic demonstration that can show that what has so far been observed is not just a temporary advance of ephemeral seedlings.

9.5 DWARF BIRCHES *BETULA NANA* AND *B. GLANDULOSA*

The dwarf birches *Betula nana* and *B. glandulosa* share the same propensity for hybridization as the mountain birches. *Betula nana* is circumpolar in distribution and represented by two subspecies: ssp. *nana* in Europe and western Asia and ssp. *exilis* in North America and central and eastern Asia. *Betula glandulosa* is also a closely related shrub found across North America and Greenland, and where the two species overlap there is much hybridization as well as taxonomic confusion (DeGroot *et al.*, 1997; Fig. 9.16). Both of these dwarf birch species have chromosome counts of $2n = 28$ with triploids and tetraploids reported only in hybrids where both euploid and aneuploid chromosome counts are found ranging from 28 to 56 (Anamthawat-Jonsson & Thorsson, 2003).

The various dwarf birches differ in their climatic tolerances. *Betula nana* extends further north than *B. glandulosa*, reaching 79° N in Spitsbergen and 80° N in north-west Greenland where both subspecies of *B. nana* occur. By contrast, *B. glandulosa* is restricted to more oceanic conditions with greater winter snow cover, as in the maritime provinces of Canada where it is typically associated with bryophyte communities in areas of lichen–spruce (*Picea mariana*) forest (see Fig. 5.19). The most northerly location for *B. glandulosa* is in Baffin Island at approximately 68° N. The interior of southern Baffin Island between 64 and 68° N contains a locally rich and diverse vegetation, which is indicative of the low arctic bioclimatic zone and marks the present northern limit of that zone in the eastern Canadian Arctic.

9.5.1 Biogeographical history of dwarf birch

Palynological studies of lake sediments in the eastern Canadian Arctic have revealed a presence of *Betula* since 4750 BP. Elements of a low arctic vegetation association have been present in the area since this time, indicating a local bioclimatic system that has been relatively stable (Jacobs *et al.*, 1997). Given this long-term persistence of *B. glandulosa* at this locality it is of interest to note that at this northern limit less than 0.5% of the seeds (samaras) are viable and these marginal populations are maintained by asexual reproduction (Weis & Hermanutz, 1993). The most probable cause of this sterility, the failure of pollination, was discussed in Section 4.2.

Both species (*B. nana* and *B. glandulosa*) are successful at colonizing a wide range of habitats, although within any one region the ecological preferences of the species may be restricted. In the Colorado Rockies *B. glandulosa* occurs in the southern boreal region at altitudes up to about 3000 m a.s.l. in the subarctic and alpine tundra, while *B. nana*, although restricted to blanket peat in Britain, nevertheless survives on xeric and rocky sites in the Arctic. In Greenland where *B. nana* became a dominant plant in the early Holocene, a climatic change *c.* 5000 BP almost exterminated the species at more oceanic sites. It lost only little ground inland and has retained its position there since in many kinds of heath and other vegetation types (Fredskild, 1991). This is one more example of the widespread tendency for woody species to react adversely to the warmer wetter winter conditions that accompany increases in oceanicity (see below).

9.6 ECOLOGICAL SENSITIVITY OF WOODY PLANTS TO OCEANIC CONDITIONS

Sensitivity to oceanic versus continental conditions is found in many northern dwarf woody species (Crawford *et al.*, 2003; Crawford & Jeffree, 2007). Oceanic conditions have long been known to lower both the altitudinal and latitudinal position of the treeline. It appears that this is also the case for a number of shrubby, woody species. The blanket bogs of Scotland and Ireland stand out as highly oceanic habitats. The basis on which the blanket bog vegetation is distinguished from less oceanic mire associations in the British National Vegetation Classification is not from the plants that are present in this community, but by the species that are absent (Rodwell, 1991). In particular, it is the absence of woody shrub species such

as *Vaccinium vitis-idaea*, *V. uliginosum* and *Empetrum nigrum* ssp. *hermaphroditum* that typifies blanket bogs with bog myrtle (*Myrica gale*) as the most distinctive woody species.

In these blanket bogs, although bryophytes flourish, oceanicity is nevertheless predominantly a negative influence for many higher plants in that it gives rise to communities that are species poor, particularly in woody species, with the ericoids being represented mainly by the flood-tolerant *Erica tetralix* while the presence of *Calluna vulgaris* is much reduced. With *Betula nana* there is a marked difference in Scotland between the oceanic west and the continental east coast. In the warmer, wetter, west coast, *B. nana* is restricted to higher ground but descends to nearer sea level in the colder, drier east coast (McVean & Ratcliffe, 1962).

Neither *B. glandulosa* nor *B. nana* is tolerant of continuous flooding. They do, however, survive on blanket peat but tend to occupy elevated hummocks when the ground is prone to prolonged waterlogging (Rodwell, 1991). *Betula glandulosa* is similar to *B. nana* in its ability to become re-established after fire. Glasshouse experiments on the effects of fire of varying severity on plant populations from northern and western Canada have shown that burning significantly increased growth at all temperature treatments and that this effect was most apparent at the highest growth temperature. The warmer growing conditions of post-fire microsites appear to provide this shade-intolerant plant with a competitive advantage over other invading pioneer and resprouting species by enhancing the fire-stimulated, height-growth response. Due to its fire ecology, *Betula glandulosa* populations might therefore be expected to expand and thrive under a future warmer climate regime (DeGroot & Wein, 1999).

Another aspect of how *B. glandulosa* may profit from fire has been illustrated in a number of studies on the causes of spruce decline in northern Québec where the environmental changes associated with fire have induced a shift from old-growth lichen–spruce *krummholz* to lichen–tundra. Tree-ring measurements, together with growth-form patterns from black spruce (*Picea mariana*) remains lying on the ground in a lichen–tundra community, have enabled a reconstruction of the structure of a conifer stand at the time that a burn that took place at approximately AD 1750 (Arseneault & Payette, 1992). Before the 1750 fire event, the spruce *krummholz* was predominantly maintained by layering, and overwintered under a considerable snow cover due to the severe climatic conditions that have persisted in this region since the beginning of the Little Ice Age (approximately AD 1580). However, after the fire, the site remained deforested due to the limited regenerative potential of stunted spruce. The post-fire shrubs, mostly dwarf birch (*Betula glandulosa*), of the lichen–tundra community were unable to trap sufficient drifting snow to permit spruce regeneration and consequently lichen–heath and birch have remained the dominant vegetation for the past 250 years.

The future ecological success of the dwarf birches does not appear to be in doubt. There may be expansion in northern Canada if fire damage continues to increase. In oceanic climates, a tendency to wetter, warmer winters may cause the species to retreat both inland and upwards to regions in which the winter temperatures are lower.

Bog rosemary (*Andromeda polifolia*; Fig. 9.17) is a small heath species (Ericaceae) growing generally up to no more than 10–20 cm, and widespread across much of the northern hemisphere. Despite its circumpolar distribution bog rosemary appears nevertheless to have a marginal existence, as witnessed by its current decline in many of its former habitats. The circumpolar distribution is achieved by two subspecies: *Andromeda polifolia* var. *polifolia* in northern Europe and Asia and *A. polifolia* var. *glaucophylla* in north-eastern America. Bog rosemary is only found in cold climate bogs with accumulating peat. It is present throughout the whole of Scandinavia including Finland but, curiously, is absent in bogs from northern Scotland and is rare in northern Ireland, possibly due to peat cutting and drainage (Godwin, 1975). It is also absent in Faroe and Iceland. The species is showing a decline in many habitats, probably due to increased drainage and in some places to greater forest growth. By comparison, a two-year study of flowering which compared temperature responses in cloudberry (Fig. 9.18) and bog rosemary in Swedish Lapland found that both species are very responsive to climate change, with warmer

Fig. 9.17   Bog rosemary (*Andromeda polifolia*), an inhabitant of cold north temperate bogs which can store 75% of its fixed carbon below ground (see Chapter 3). Bog rosemary has also been found to have high antioxidant activity and high total phenolic content (Kahkonen *et al.*, 1999).

Fig. 9.18   Cloudberry (*Rubus chamaemorus*) in fruit. Cloudberries are responsive to rising temperatures and when ripe the fruits turn from red to yellow. The fruits ripen readily in Scandinavia and also in Labrador and Newfoundland where they are much prized. However, in Scotland's oceanic climate they do not fruit consistently and the berries only ripen beyond the red stage in rare favoured places. (Photo Professor R. M. Cormack.)

weather advancing flowering by two weeks. This response appears, however, to be dependent on specific climate events. In both species flower production was only stimulated by the spring-warming treatments, which suggests that more attention in high-latitude climate change experiments should be given to winter and spring events than has been the case so far (Aerts *et al.*, 2004). The fact that bog rosemary occurs at altitudes up to 1250 m in Eidfjord and Ullensvang (60° N) in Norway (Lid & Lid, 1994) makes it curious that it does not occur even on lowland bogs north of Perth (56° N) in Scotland.

9.7 JUNIPER

Juniper (*Juniperis communis*) is a highly polymorphic species with a widespread circumpolar distribution. Juniper, like willow, is dioecious. It is a wind-pollinated shrub or small tree and produces cones which develop into berry-like fruits which are readily dispersed by birds. Several subspecies are commonly recognized, which together with intermediate forms provide a source of extensive genetic diversity. As a genus, the junipers are slow-growing conifers of marginal areas as

they can survive in full sunlight on dry mineral soils and are capable of withstanding exposure and drought. Juniper is therefore capable of living on sand dunes and chalk grasslands, as well as in birch woods and heaths, and as scrub communities at high altitudes and latitudes. Two subspecies are regularly recognized. The typical lowland variety *Juniperus communis* ssp. *communis* (2n = 22) occurs as a low-growing tree or columnar (fastigiate) bushes or scrub up to and also above the treeline (Fig. 9.19). This subspecies is distinguished by having spreading leaves almost at right angles to the stem, which makes the species prickly to touch. The second form (*J. communis* ssp. *nana* syn. *alpina*, *J. sibirica*) is a procumbent variety with ascending or loosely appressed leaves which grows on rocks and moors and mountains (Fig. 9.20). It occurs up to 1730 m in central Norway as well as on exposed coastal sites in western Scotland and lowland bogs in western Ireland (Stace, 1997). The is subspecies (*J. communis* ssp. *nana*) attracts attention with its northern, circumpolar distribution extending to 70° N at Disko in West Greenland and to the north of Norway where it reaches an altitude of 700 m (Hultén, 1971). Intermediates also occur between these two forms and

Fig. 9.19   Juniper bushes (*Juniperus communis* ssp. *communis*) in the columnar form growing in a clearing of coastal forest in Estonia.

Fig. 9.20   A dry heath dominated by the prostrate form of juniper shrub (*Juniperus communis* ssp. *nana*) at the Slochd Summit (370 m a.s.l.), Grampian Mountains, Scotland.

other subspecies are recognized in different regions of Europe, Asia, and North America.

In terms of species success in marginal areas in the British Isles the dwarf subspecies favours areas with cool summers and relatively mild winters and a high rainfall (Rodwell, 1991). In the Scottish north-west Highlands *Calluna–Juniperus* heath is found typically between 300 and 600 m at the junction of the sub- and low-alpine zones except in areas where there has been frequent burning. So sensitive is juniper to fire that there are even places where it has apparently vanished after a single burning (McVean & Ratcliffe, 1962). Juniper also needs protection by the winter snow pack. However, young seedlings do not tolerate prolonged snow cover due to their susceptibility to the brown snow felt fungus *Herpotrichia* (Holtmeier & Broll, 2005).

Examination by DNA fingerprinting in 12 different populations of juniper from widely dispersed locations has shown that the species as a whole can be considered as two main groups: one from the western hemisphere, and one from the eastern hemisphere (including Greenland and Iceland). There is also a minor Kamchatka population that is dissimilar to any other population so far examined (Adams *et al.*, 2003). This study suggested that the current arctic populations are recent in origin as the present sites for this species in the north were probably covered with ice or otherwise inhospitable for juniper during the late Pleistocene (*c*. 12 000 BP). The genetic affinities of the differing populations appeared to indicate that the path of recolonization was northward in North America while Greenland appears to have been colonized from Icelandic plants, which in turn came from northern Europe. The Kamchatka population seemed likely to have come from Japan.

In Britain, juniper has two main centres of distribution: a highland zone in the north and west, in which populations of dwarf juniper (*J. communis* ssp. *nana*) are still extensive and sexually reproducing, and a southern zone on chalk downlands. In the latter populations of common juniper (*J. communis* ssp. *communis*) are small and fragmented and currently suffer from a decline in fertility. Vigorous stands of columnar juniper can be seen in coastal areas in the Baltic as well as in north German heathlands (Fig. 9.19).

It might have been expected that the large sexually viable populations in the north would possess high levels of within-population genetic variation, while the declining southern populations would be genetically depauperate. However, an analysis of amplified fragment length polymorphisms (AFLPs) has shown that all the populations studied had high levels of genetic variation (van der Merwe *et al.*, 2000). Juniper, if not disturbed by fire, can be long lived. On the upper slopes of Ben Eighe (north-west Scotland), many plants are over 100 years old and the oldest individual found in a recent survey was at least 202 years old (Dr Sarah Woodin, pers. com.) while in northern Finnish Lapland there can be found junipers that are almost 1000 years old (Kallio *et al.*, 1971). Thus, even though sexual reproduction may not be frequent longevity compensates for lack of new recruitment.

Geographical variation in seed production, predation and abortion has been analyzed for 31 populations of juniper in seven distinct regions throughout the species' distribution range in Europe, including both the northern and southern boundaries (Garcia *et al.*, 2000). Seed abortion shows a significant quadratic relationship with latitude, with higher values of abortion at either end of the gradient, especially at the southern limit. The production of filled seeds declined gradually towards both northern and southern distribution limits. In the Mediterranean mountains (southern limit), low seed production coincided with a marked limitation placed upon natural regeneration by summer drought, leading to a demographic bottleneck in populations. Although seed abortion levels were relatively high in the subarctic tundra (northern limit) populations, they were free from predispersal seed predators, which suggests that population viability here may be under less pressure (Garcia *et al.*, 2000).

There also exist equally peripheral lowland areas where common juniper is a characteristic feature of the vegetation provided burning is not frequent. On acid sand dunes in north-east Scotland, Denmark and the Baltic coasts of North Germany, Latvia and Estonia, juniper can form a scrub community along with the nitrogen-fixing sea buckthorn (*Hippophae rhamnoides*; Fig. 4.36).

In southern Europe, and around the Mediterranean, juniper scrub is a feature of high mountain vegetation, but here again it is vulnerable to human disturbance. Aerial images of the high summits of the Spanish Central Range reveal significant changes in vegetation over the period 1957 to 1991 in which high-mountain grassland communities, formerly dominated by *Festuca aragonensis*, have been recently replaced by shrub patches of *Juniperus communis* ssp. *alpina* along with *Cytisus oromediterraneus* from lower altitudes. Climatic data for this mountainous region indicate a shift towards warmer conditions since the 1940s, with the shift being particularly marked from 1960. Changes include significantly higher minimum and maximum temperatures, fewer days with snow cover and a redistribution of monthly rainfall. Total yearly precipitation showed no significant variation. There were no marked changes in land use during the period under examination, although there were minor changes in grazing practice in the ninetenth century. It may be that the advance of this drought- and heat-tolerant woody species into higher altitudes is related to climate change (Sanz-Elorza *et al.*, 2003).

9.8 HEATHLANDS

The concept of heaths is of ancient origin. The word itself can be traced back to Old Saxon *hetha* and Middle High German *Heide*, meaning wasteland. The people who lived on the heath were the heathens (Gothic *haithi*) in the same way that plants that grow on the heath. The genera that constitute the heaths all belong to the Ericaceae (e.g. *Erica*, *Calluna*, *Vaccinium*, *Empetrum* (Figs. 9.4–9.8), *Arctostaphylos*, *Cassiope*, *Andromeda*, *Ledum*, and *Loiseleuria*). All these genera of dwarf woody shrubs that grow on the oligotrophic soils are commonly referred to as heaths or heath species. As a vegetation formation, heaths are found worldwide from the Arctic to the Fynbos of South Africa (see Chapter 2) and the mountain ranges of the Himalaya (*Rhododendron* spp.) to the raised bogs and moors of Patagonia and the Falkland Islands (*Empetrum rubrum*, *Gaultheria* spp.) and the moorlands of the Atlantic coasts and islands of western and north-western Europe dominated by heather (*Calluna vulgaris*).

9.8.1 Relating heathlands to climate

Heathlands are typically communities that flourish in oceanic climates both at altitudes above the treeline and at sea level. Where trees fail or are removed by disturbance (fire, grazing, etc.) heaths can migrate downhill and flourish to an extent that is not seen in the high altitude sites. Any study of the factors which control the distribution of species in oceanic climates has to take into account the dual limitations of both summer and winter temperatures. Predicting plant distribution in terms of seasonal temperature regimes demands correlative observations from a range of variables. An objective manner of comparing species distribution with the interaction of winter and summer temperatures has been described and examined for a number of European species (Crawford *et al.*, 2003). This method enables the production of maps which compare the probability of the present occurrence of a species with that which would be expected under a range of specified winter and summer temperature combinations. Using this map system it is instructive to view the probable changes in the distribution in heath species in relation to varying degrees of summer and winter warming (Figs. 9.21–9.24).

(1) Heather (*Calluna vulgaris*) under present conditions is more oceanic in its distribution than *V. myrtillus* even though the temperature-based probability distribution suggests that its presence in Scotland is marginal. *Calluna vulgaris* also occurs on dry and hot sites above the forest limit in the relatively continental Central Alps (e.g. Ötztal, Upper Engadine). In northern Germany, *Calluna* has been favoured by human impact, probably the same factors as in Scotland. When grazing is removed, pine, birch, and also juniper invade *Calluna* heath as the light-demanding *Calluna* becomes rapidly outcompeted. In the model (Fig. 9.21) summer warmth appears from these predictions to be a limiting factor, as summer warming would expand the range, while winter warming would cause a further retreat from the west and advance eastwards. As might be expected, summer cooling by 4 °C below the present temperature levels would cause a south-western migration which would not take place if it were the winter temperatures that had been reduced.

(2) The bilberry (*Vaccinium myrtillus*) is a plant of northern Europe but with a much wider distribution southwards and eastwards than *Cassiope hypnoides*. In the Alps *Vaccinium myrtillus* occurs widely in pine and spruce forests on acidic soils. Above the forest limit in the Alps it is restricted to sites with a not too shallow snow pack and is not found on snow-free sites. The probability map for its present

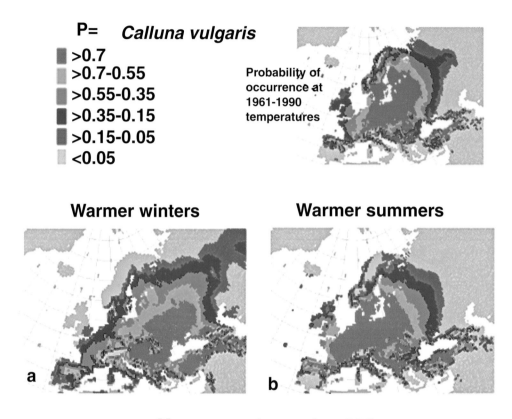

Fig. 9.21 *Calluna vulgaris* and possible responses to seasonal differences in relation to climatic warming. Probability density plots show the possible ranges in relation to changing summer and winter temperatures. The colours represent bands of increasing probability of the different combinations of winter and summer temperatures being suitable for the species. Red is most suitable (see inserted scale). In (a) the scenario is shown for a higher winter temperature of + 4 °C while the summer temperature is reduced by −4 °C. The reverse conditions pertain in (b) where the winter is −4 °C colder and the summer 4 °C warmer. In relation to annual mean temperature this represents no change from the 1961–1990 temperatures as given in the CRU Global Climate Dataset (New *et al.*, 1999, 2000). Note the prediction of a retreat for this species from the British Isles and Scandinavia with warmer winters and cooler summers. (For further details see Crawford & Jeffree, 2007). N.B. These maps are model predictions based on the present distribution of the species in terms of potential range in relation to temperature and not on present geographic distribution.

distribution based on temperature classifies its western extension in the British Isles, western France and northern Spain as marginal (Fig. 9.22). Summer warming would appear to lead to a major decline in this species while winter warming would result in a western retreat and an eastward expansion. Climatic cooling, if it took place largely through a reduction in summer temperatures, would cause a retreat from the northernmost habitats and an expansion southwards and eastwards. Winter cooling, however, would have a different effect and

would be likely to reduce the marginality of its presence in western Europe.

(3) The dwarf arctic heath (*Cassiope hypnoides*) is a species of transatlantic distribution. Under the present climatic regime it is found from eastern Canada where it is confined from Labrador northwards to 75° N (Fig. 9.23). It is widespread in Greenland to 75° N on the west coast and 80° N on the east coast. The species also occurs in central and northern Iceland and in the more northern montane regions of Scandinavia, the Kola Peninsula, and

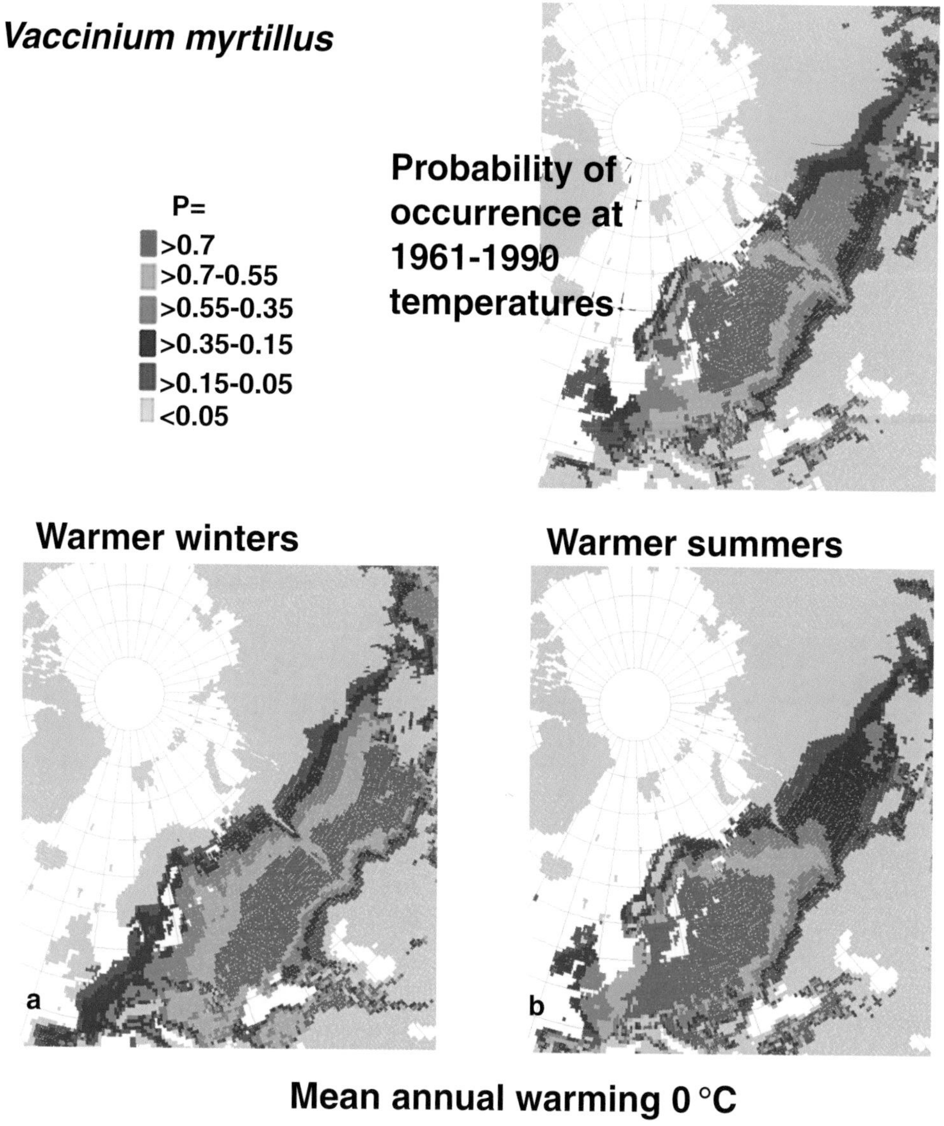

Vaccinium myrtillus

P=
■ >0.7
▨ >0.7-0.55
■ >0.55-0.35
■ >0.35-0.15
■ >0.15-0.05
▨ <0.05

**Probability of
occurrence at
1961-1990
temperatures**

Warmer winters

Warmer summers

a

b

Mean annual warming 0 °C

Fig. 9.22 *Vaccinium myrtillus* and possible responses to seasonal differences in relation to climatic warming. For explanation see Fig. 9.21.

northern Siberia, extending to the southern part of Novaya Zemlya. The most northerly present location is on the west coast of Spitsbergen at 79° N in Kongsfjord (Rønning, 1996). The present temperature conditions appear suitable for a wider extension in the region to the east of the Hudson Bay and in western Greenland. Increased summer warming would result in an increased presence in Greenland, Spitsbergen, and Novaya Zemlya but would reduce the occurrence of the species in Scandinavia and northern Siberia. The converse conditions with warmer winters would appear to favour an expansion westwards in North America and eastwards in Siberia. Winter cooling would be likely to favour a southern extension of the species in Europe and a retreat in eastern Siberia.

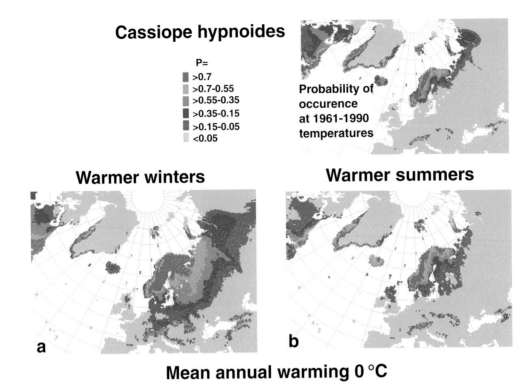

Cassiope hypnoides

P=
■ >0.7
▨ >0.7-0.55
▨ >0.55-0.35
■ >0.35-0.15
■ >0.15-0.05
░ <0.05

Probability of occurence at 1961-1990 temperatures

Warmer winters   **Warmer summers**

a   b

Mean annual warming 0 °C

Fig. 9.23 *Cassiope hypnoides* and possible responses to seasonal differences in relation to climatic warming. For explanation see Fig. 9.21.

(4) The polar willow (*Salix polaris*), shown here for comparison, also exhibits differential effects between summer and winter temperature changes, with winter warming being the most disadvantageous and winter cooling the most advantageous for its presence in north-western Europe (Fig. 9.24).

9.8.2 Possible migration behaviour

The maps in Figs. 9.21–9.24 offer insights into the possible effects of temperature on species distribution.

(1) Species migration is sensitive to existing temperature seasonality and the superimposed seasonality of temperature change.
(2) Migration cannot be predicted from annual mean temperature alone.
(3) For the same patterns of change, in different parts of their ranges, species could migrate in opposite directions.
(4) Seasonality gradients may present barriers to migration notwithstanding overall warming.

(5) The effects of changes in temperature and seasonality may affect species either directly or indirectly by changing the nature of competition with other cohabiting species.

Some physiological explanation is required to explain why milder oceanic conditions should be disadvantageous for woody species. The Norwegian plant ecologist Eilif Dahl was one of the first to distinguish between the positive and negative effects of oceanic conditions on alpine species. The relatively species-poor mountain flora of the Scottish Highlands and south-west Norway were considered by Dahl to be due to mild periods of winter weather that encouraged premature spring growth, causing severe dieback of non-hardy shoots. Dahl first drew attention to Norwegian montane species such as *Rhododendron lapponicum* and *Aconitum septentrionale* that are absent from more oceanic mountains and described them as *south-west coast avoiders* (Dahl, 1951, 1990).

A study of *Vaccinium myrtillus* in north-eastern Sweden (Ögren, 1996) noted that after a warmer

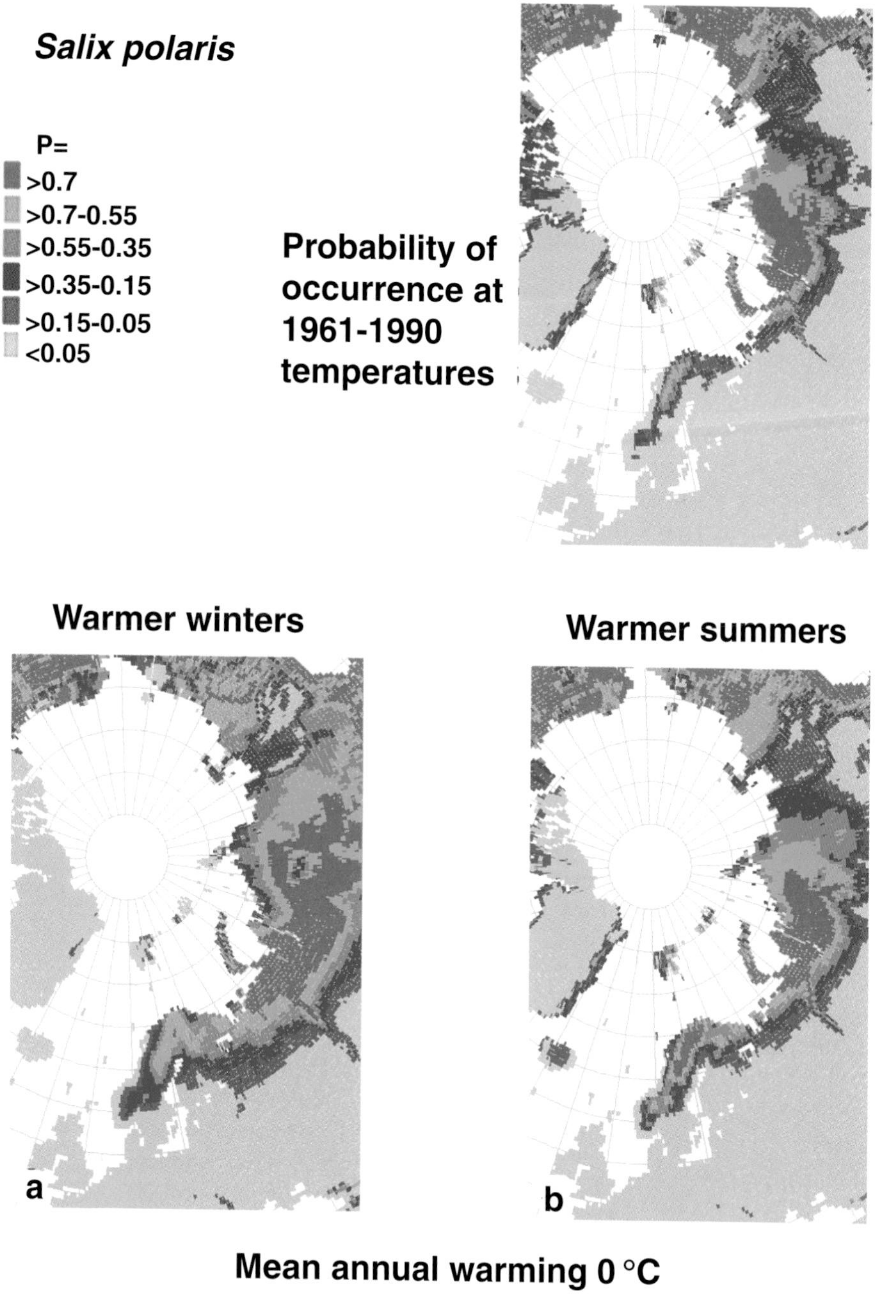

Salix polaris

P=
- >0.7
- >0.7-0.55
- >0.55-0.35
- >0.35-0.15
- >0.15-0.05
- <0.05

Probability of occurrence at 1961-1990 temperatures

Warmer winters

Warmer summers

a

b

Mean annual warming 0 °C

Fig. 9.24 *Salix polaris* and possible responses to seasonal differences in relation to climatic warming. For explanation see Fig. 9.21.

than usual winter this species suffered lethal injuries. During a mild winter it was found that rehydrated shoots were at their greatest degree of cold-hardiness when tested early in winter and that they gradually lost their frost tolerance as the mild winter weather progressed. This loss of frost tolerance was accompanied by a decrease in the solute content of the shoots, suggesting a progressive respiratory loss of cryo-protective sugars. Gas exchange measurements estimated that the initial carbohydrate reserves would have lasted for only four months if tissue water content remained high. When thin snow cover was coupled with clear skies then shoot dehydration could improve cold tolerance by 5–10 °C. However, in mild winters, long-term dehardening due to recurrent periods of mist and rain increased metabolic activity and resulted in a 100% increase in shoot damage.

9.8.3 Historical ecology of heathlands

The expansion of heathlands in north-western Europe began with the spread of Neolithic farming. The problems of maintaining soil fertility in an oceanic environment before the advent of deep ploughing and the use of inorganic fertilizers (see Chapters 1 and 10) inevitably led to soil impoverishment and the development of podzols, particularly in the Iron Age with the advent of more efficient ploughing. Consequently, more and more land was left to become either heath or bog depending on local conditions. This did not necessarily mean that no further agricultural use was made of the developing heaths. In many cases the moorlands became a significant agricultural asset for summer grazing as well as providing winter pasturage for hardy stock.

Given adequate manuring, excellent crops can be obtained from heathlands, usually through creating fertility by transporting turf and peat from other parts of the moor to the area for cultivation and incorporating as much manure as possible. This technique was practised throughout Europe and created *plaggen soils* (Dutch, layered). The creation of increased productivity in small areas, however, led to the further destruction of the agricultural potential of the region as a whole. Large barren areas were created where turf was taken from outfield areas such as moorlands and common-land forests, for bedding for livestock and then spreading the slurry-soaked bedding on the arable fields as fertilizer. Over time, this created very rich agricultural soils which in places could be over 1 m in depth, in contrast to many modern arable soils, which normally have a plough horizon depth of about 30 cm. However, the very extensive removal of turf over long periods of time led to the eventual destruction of the grazing value of the outfields (see Section 11.4.1).

In north-western Germany the plaggen method was used in the Lüneburger Heide (heath) from the Middle Ages with the result that over a very extensive area the soil was destroyed and the forest prevented from regenerating. Consequently severe erosion took place before strict laws were introduced to stop the practice. In Orkney and Shetland plaggen soils were created already in the twelfth to thirteenth centuries, and on some islands in Shetland these methods continued to be used until the 1960s.

9.9 NEW ZEALAND: A HYPEROCEANIC CASE STUDY

In New Zealand, there is a remarkable alpine and sub-alpine flora including families which in the northern hemisphere are mainly herbaceous, e.g. Rubiaceae, Asteraceae, Malvaceae and Scrophulariaceae, but in this southern location occur as woody-scrub species. A wealth of other genera create a woody flora which varies from creeping mat-forming species to almost tree-like branching woody shrubs with various species of *Dracophyllum*, *Halocarpus*, *Pittosporum* and other genera. Given the relatively recent uplift of the mountain chains in New Zealand the existence of this isolated yet species-rich alpine flora is a matter for botanical wonder.

Sixty million years ago, New Zealand was a low-lying, subtropical archipelago with a mild climate. The transformation to the mountainous topography that is now such a predominant feature of New Zealand's South Island is a result of a series of very rapid mountain-building events occurring mainly over the past three million years. In some places, the uplift rate is estimated to be as much as 100 m in 10 000 years. Over a three million year period this would have raised Mount Cook to five times its present elevation of 3764 m – three times that of Mount Everest were it not for the very high erosion rate which has removed two thirds of the uplift (Fleming, 1980).

As the land has risen, so has the vegetation changed to produce the modern species-rich alpine flora. Opinions vary as to whether this flora has evolved *in situ* from plants that inhabited the cool mist-covered rolling plains of an oceanic environment or whether they are recent immigrants from Asia via the mountains of New Guinea and Australia.

The isolation of the New Zealand Gondwana flora and the low number of genera, combined with a high number of endemic species, suggest that there has been rapid and recent evolution within certain groups. This is particularly the case with the alpine species where over 93% of the species are endemic to the New Zealand biological region, including the more southern and very isolated Sub-Antarctic Islands (Halloy & Mark, 2003). There are also extensive variations within a relatively small number of genera, with many closely related species which support further the *in situ*, recent-evolutionary hypothesis. One of the most striking examples is the dwarf woody mountain totara (*Podocarpus nivalis*) which with a stature of 1–3 m contrasts with its presumed ancestors in the giant *Podocarpus* forests of the lowlands (Fig. 9.25).

The alpine and subalpine scrublands of New Zealand are also the home of the world's smallest conifer, the pigmy pine (*Lepidothamnus laxifolius*), which grows as a trailing mat spreading over stones and gravel. An evaluation of these competing views (McGlone *et al.*, 2001) has suggested that the vascular plants reached New Zealand by long-distance transoceanic dispersal, probably during the Late Miocene to early Pleistocene period. Cooling climates and formation of a more mountainous and compact landscape after that time reduced the dispersal of woody plants and favoured herbaceous, wetland and highly dispersible species. Consequently, the woody alpine plants have evolved from a limited number of genera.

Fig. 9.25   The mountain totara (*Podocarpus nivalis*), a small prostrate New Zealand shrub, sometimes as large as a small tree that inhabits the upper forest margins and subalpine scrub from lat. 36° 50′ S.

Pleistocene speciation has probably also been aided by the development of a differentiated terrain and climates which have provided isolation and distinctive environments as well as creating greater opportunities for niche specialization.

Ecologically, the Southern Oceanic Islands off the coast of New Zealand, and in particular those that lie south of 50° parallel, namely the Auckland Islands, Campbell Island, and Macquarie Island (Fig. 9.26), with their relative lack of lack of human disturbance, provide a unique opportunity for studying the dynamics of forest and scrub survival in relation to climatic fluctuation. These highly oceanic habits have considerable peat deposits which have enabled the reconstruction of the Holocene history of these remote regions (McGlone *et al.*, 2000). Comparing the islands from north to south provides a comparative assessment of the response of woody vegetation to declining and variable climatic conditions.

The small, uninhabited subantarctic Auckland Island (*c.* 51° S) is the southernmost outpost of tall forest in the south-west Pacific while Macquarie Island at 54° S has no woody species (McGlone, 2002). The Auckland Islands are completely peat covered with extensive bogs. Low forest and scrub covers the lowland areas of the islands, with maritime tussock and herbfield associations on exposed coasts. Southern rata (*Metrosideros umbellata*) forest forms a coastal fringe in sheltered locations and is the southernmost limit of this typical New Zealand species. *Dracophyllum longifolium* forms a tall scrub at the forest margin that can be up to 5 m high, but this then grades into dense low shrubland less than 2 m in height. With increasing altitude low scrub and shrub–grassland predominate, while above 300 m, tussock grassland and fellfield are the norm (McGlone *et al.*, 2000).

Campbell Island at 600 km south of the New Zealand mainland (52° 33.7′ S) has an intermediate location between the Auckland Islands and the woodless scrub-deficient Macquarie Island. A series of photographic records starting from 1888, together with mid-nineteenth-century vegetation descriptions of the island, have made it possible to trace changes in the extent of scrub cover since the 1840s, and this together with pollen sampling of the peat has provided an in-depth analysis of the vegetation history of the island (Wilmshurst *et al.*, 2004). Campbell Island was extensively glaciated and *Dracophyllum* was uncommon until

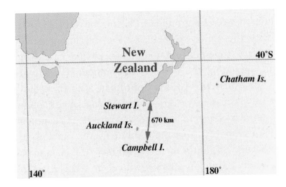

Fig. 9.26   (Above) Location of New Zealand's subantarctic islands. Auckland Island is the most southerly island in the south-western Pacific Ocean to have a forest cover. (Below) Enderby Island, a small island off the north coast of Auckland Island (50° 30′ S). View taken near sea level. Solitary tree of *Dracophyllum longifolium* emerging from *Myrsine divaricata* scrub. (Photo Dr M. S. McGlone.)

around 7000 BP and reached its greatest altitudinal and areal extent between then and 3000 BP. Warmer, drier summer conditions during the mid to late Holocene (+0.5 °C), favoured this scrub expansion (McGlone et al., 1997). However, cooling temperatures and increased south-westerly winds in the late Holocene were probably the cause of a reduction in both density and altitudinal range. The vegetation of Campbell Island now consists mainly of lowland *Dracophyllum* scrub and upland tussock grassland and tundra. Since the 1960s, the island has become warmer and drier. *Dracophyllum* scrub was restricted at the time of the first photographs (1888) and earlier written observations suggest that there had been little change between the 1840s and that time.

The recent spread of *Dracophyllum* scrub has been assisted by a pronounced shift to warmer, drier climates in the second half of the twentieth century. In terms of climatic limits to scrub survival, it is significant to note that the upper elevational limits of scrub have not increased, suggesting that factors other than summer temperature are controlling the altitudinal position of the scrubline in this hyperoceanic environment (Wilmshurst et al., 2004). It appears that in this intensely oceanic setting, warm, cloudy, low-radiation environments inhibited forest growth during the early Holocene, possibly by promoting saturated soils and reducing net photosynthesis. It was only in the later Holocene, when increased westerly windflow brought sunnier, but cooler and windier, climates, that forest expansion occurred on sheltered lowland sites. Pollen analysis has shown that the forest at the study site has collapsed to scrub at least twice within the last 2000 years, most likely because of extended periods of saturated soils (McGlone, 2002).

9.10 CONCLUSIONS

Throughout the world, dwarf woody plants living beyond the treeline are a significant component of the vegetation. They are found in a wide variety of habitats from coastal heaths to upland and alpine moorlands. With human disturbance of forest regeneration, beginning already in the Neolithic, their geographical range has been greatly extended. More recently, the last hundred years has seen the so-called reclamation of these peatlands with deep ploughing and reseeding. In many of the remaining moorlands overgrazing is now rapidly reducing the extent of scrub and heath communities. If left undisturbed it seems probable that woody scrub species will respond in a variety of ways should significant climatic warming continue in boreal and arctic regions. Climatic warming in northern latitudes will undoubtedly create new ecological opportunities for vegetation advance as ice sheets retreat and permanent snow cover is reduced. Changing thermal conditions will also be likely to alter the species composition of existing northern plant communities.

The nature of the migration of species into these vacated areas or changing communities is unlikely to be a mere latitudinal shift northwards of existing species assemblages. Depending on proximity to the ocean, and this includes also the Arctic Ocean, the degree of winter versus summer warming is likely to differ. In northern regions, the influence of oceanicity on plant distribution is very marked. The models described above suggest that among some of the commonest woody species winter warming may in some cases cause a significant retreat in areas where the climate is influenced by the ocean, while in other more continental areas the dwarf woody species may make significant advances. It appears therefore highly probable that species migration will be strongly influenced by *seasonality* of temperature change and any one species might migrate in opposite directions for the same patterns of change in different parts of its range. Attention will therefore have to be given to the roles of seasonality gradients as potential barriers to migration notwithstanding overall warming.

Fig. 10.1 Mountain isolation as seen at 1692 m on the summit of the Dedo de Deus (the Finger of God) in the Serra dos Órgãos, Teresópolis, Brazil. Plants that occur in this apparently challenging environment include bromeliads, orchids, Velloziaceae, and even Cactaceae which grow directly on the rocks or in association with moss colonies. In addition, several Ericaceae and Melastomataceae shrub species are present. In the Asteraceae, the genus *Baccharis* is widespread at high altitudes in south-east Brazil, and plants belonging to this genus are just visible in the photograph as a woody component of the scrub vegetation. (Photo Miguel D'Ávila de Moraes and botanical notes from Professor F. R. Scarano.)

10 · Plants at high altitudes

10.1 ALTITUDINAL LIMITS TO PLANT SURVIVAL

Plants that live at high altitudes demonstrate a remarkable ability to survive in some of the most challenging environments on Earth. The permanent snowline is generally taken as marking the principal upper altitudinal limit for flowering plant colonization (Fig. 10.2). There are however, peaks or horns of rock that rise above the snow, ice fields known internationally by the Inuit term *nunataks* where plants can manage to live above the snow. These emergent mountaintops have sides that are too steep to support snow and ice accumulation, and provide bare ledges and crevices which serve as microhabitats for a few hardy mountain plants. Below the permanent snowline is the subnival zone. In terms of a major altitudinal zonation where regular plant colonization can be encountered, the subnival zone is normally the uppermost altitudinal habitat for plants on snow-covered mountains. Even here, however, the terrain is fragmented by gullies that vary from year to year in their retention of permanent snow. Nevertheless, a snow-free growing season of only a few weeks in the year provides a marginal habitat where a few species can succeed in growing and sometimes reproducing.

Fig. 10.2   The nival zone as seen on Mt Sefton 3151 m, Southern Alps, New Zealand, photographed in late summer. Note the extensive areas of bare rock within the nival zone where local topography and wind reduce the snow cover creating potential plant habitats. Plants that live in this area are referred to as the nival flora.

As might be expected, the highest altitude records for flowering plants are found on tropical and subtropical mountains. In 1938 on Mount Everest (8848 m), Eric Shipton, the only veteran of all four 1930s British Everest expeditions, found *Saussurea gnaphalodes*, Asteraceae (Fig. 10.3), growing on scree at an altitude of 6400 m. Other species found on Mt Everest above 6100 m included *Ermania himalayensis*

(*Cheiranthus himalaicus*, Crucifereae) at 6300 m, and in the Caryophyllaceae, *Stellaria decumbens* and *Arenaria bryophylla* (Miehe, 1997).

These species survive at astonishing elevations on certain outstanding high mountains but this should not obscure the realization that every region has its own particular high-altitude flora, and plants that achieve altitude records in lower mountain ranges in temperate

Fig. 10.3 *Saussurea gnaphalodes*, the highest species on Mt Everest recorded in 1938 by Eric Shipton at 6400 m. Photograph taken in Sichuan Province by Dr D. Boufford on the north side of the pass at Zheduo Shan (30° 4′ 45″ N, 101° 48′ 27″ E). Elevation 4370–4485 m. (Photo taken during a botanical and mycological inventory of the Hengduan Mountains, China, a project funded by the U.S. National Science Foundation (grant no. DEB-0321846 to D. E. Boufford) and the National Natural Science Foundation of China (grants no. 40332021 and 30420120049 to H. Sun).)

and arctic regions are just as noteworthy, even if it is at a lower altitude. In the European Alps the high-altitude record for a flowering herb is *Saxifraga biflora*. This is a very local endemic species confined to high altitudes and often remains covered in snow until the second week in July; it thus escaped the notice of early alpine botanists, being described first by the Swiss biologist, physician and poet Albrecht von Haller in 1768. The highest altitude so far reported for *S. biflora* is at 4450 m on the Dom du Mischabel (Fig. 10.4). The species occurs also at 4200 m on the Matterhorn (Webb & Gornall, 1989) and in the French Alps (Chas *et al.*, 2006). This high-altitude record holder despite its high mountain locations and possible isolation shows considerable variation in flower colour and also produces the hybrid *S.×kochii* with *S. oppositifolia* (Fig. 10.4d).

Until the discovery of *Saxifraga biflora* on the Dom du Mischabel, the glacier buttercup *Ranunculus glacialis* was considered the highest growing European alpine species at 4270 m on the Finsterahorn. The glacier buttercup still retains the record for the highest growing flowering plant in Scandinavia, reaching 2370 m at Galdhøpiggen, the highest mountain in Norway and in northern Europe (2469 m a.s.l. – 62° N). Galdhøpiggen is also the site of the highest occurrence of *Saxifraga oppositifolia* in northern Europe (2350 m).

10.2 MOUNTAINTOP ISOLATION

High mountain peaks, wherever they lie, from the poles to the tropics, with or without snow, are places of isolation for plant life. Lacking the mobility of most birds and animals, the plants that grow in these upper altitude zones are confined there for life (Fig. 10.5). In temperate and tropical regions, the slopes and valleys below the mountain peaks or tablelands will usually be covered in a vegetation type that belongs to a totally alien environment as compared with the region above the treeline. Consequently, high-altitude plants often live in

Fig. 10.4 *Saxifraga biflora*, the species that holds the high-altitude record in Europe at 4450 m (see text). (a) *Saxifraga biflora* in the eastern Swiss Alps at Fuorcla da Faller. (b, c) *Saxifraga biflora* at 2500 m at Champsaur in the French Alps, showing diversity of flower colour. (d) *Saxifraga×kochii*, the hybrid between *S. biflora* and *S. oppositifolia* in the Swiss Alps. (Photos (a) and (d) ©Dr F. Gugerli; photos (b) and (c) © Franck Le Driant 2007, www.FloreAlpes.com.)

Fig. 10.5   Mount Roraima, a sandstone plateau rising above the surrounding savannas and forest and lying between Venezuela, Guyana and Brazil. This is the highest of the tepuis with an average height of 2500 m and a maximum elevation to 2810 m. These mountains are considered some of the oldest geological formations on Earth, dating back to the Precambrian Era. One third of the plant species are unique to the plateau. (Photo © Adrian Warren/www.lastrefuge.co.uk.)

what are essentially islands of seclusion from which migrations, either in or out, are not easily accomplished.

Isolation, both physical and biological, is a feature of many mountain-top plant populations. At the highest sites plant communities cannot benefit from an input of seeds and other propagules washing or blowing down the mountainside from vegetation that lives above them. Reproduction and persistence in such sites is therefore largely limited to the plants that are already there or else from those seeds that blow uphill or are transported there by animals. Immigration from other mountain sites will depend on the height and relative isolation of the mountain. It is therefore not surprising that endemic species and subspecies are a feature of many high mountain floras.

A classic study of endemism in Sri Lanka (Willis, 1922) noted that the commonest sites for endemic species were on mountain summits. Of the 809 species of flowering plants endemic to the island some 200 were confined to very small areas and half of these occurred on mountain tops or in small groups of mountains. This was particularly noticeable in the genus *Coleus* where many endemic species are restricted to localized mountain groups. In an earlier series of papers Willis had put forward his classic *Age and Area* theory (1915) which first advanced the hypothesis that the *larger the area covered by a species, the older is that species* (Willis, 1922). This work attracted much interest and comment when it was published, but is now largely disregarded as many factors other than age determine the area occupied by a species, including physical and climatic barriers, and the adaptability of species under different environments. In addition, the hypothesis failed to understand the complexity of endemism and that there exist two widely different types of endemic forms, 'relicts' and 'indigenes' (climatic relicts and recently evolved endemics – see Sections 1.6.1, 1.6.2). Plant species also differ widely in the rate at which they evolve, and hence in the rapidity at which endemic elements will arise. Even when the 'Age and Area' concept was first put forward it was considered untenable as 'no single hypothesis, however valuable it may be in explaining certain facts, can be used as a key to the whole problem' (Sinnott, 1917).

Despite the limitations of the 'Age and Area' theory, endemism in mountain species appears to be a phenomenon particularly associated with the warmer regions of the world where old established floras are preserved on ancient mountain tops with a climate that is radically different from the valleys below. Mountain endemics are few in the lands bordering the North Atlantic, possibly due to the Pleistocene impoverishment of the flora brought about by glaciation and the presence of geographical land bridges. In northern coastal regions, the early Holocene vegetation from shore to mountain consisted largely of arctic–alpine species as can be seen in Greenland today. Also, the mountains of Europe consist of ranges rather than isolated high peaks, the latter being also the case with East African volcanoes. Where endemics do occur in Europe they may not necessarily be due to *in situ* evolution. The increasing warmth of the last 12 000 years has clearly played a role in limiting the range of mountain species that were once more common. Consequently, many endemic species are likely to be late-glacial relicts rather than recently evolved endemics.

10.2.1 Inselbergs – isolated mountains

A particular form of isolated high-altitude terrain that occurs mainly in the tropics is the *inselberg* (Figs. 10.5–10.6). This usually consists of a monolithic block, mostly of granite or gneiss, but can be made of other hard rocks such as quartzite. Inselbergs are stable landscapes of considerable antiquity, mostly

Fig. 10.6   The plateau of Roraima (2500 m). Note bare crevassed quartzite rock and minimal plant cover. High rainfall has washed away soil and the resulting poor rooting conditions provide little plant nutrition. Many insectivorous species are, however, successful in this marginal environment. (Photo © Peter and Jackie Main.)

40–50 million years old, although some can be as young as 10 million years (Lüttge, 1997). Particularly striking are the inselbergs situated in the north-east of South America where Venezuela, Guyana and Brazil meet. The indigenous name *tepuis* is used to denote these enormous isolated South American inselbergs. Their height can vary, with the highest examples occurring in French Guiana.

Possibly the most famous inselberg is Roraima, 2810 m high, 16 km long, and up to 5 km wide, first climbed in 1884 by the British botanist Everard im Thurn, whose subsequent lecture in London on this feat inspired Arthur Conan Doyle (1912) to write his time-warp novel *The Lost World*, which imagined a re-entry by intrepid explorers to the era of the dinosaurs (Fig. 10.5).

The rock on the plateau of Roraima is a hard, ancient quartzite that has weathered into strange shapes and created a grotesque landscape of castellations and chasms (Fig. 10.6). Because of the frequent and heavy rainfall the quartzite, instead of having a white colour, is black due to a ubiquitous covering of cyanobacteria, which is a frequent occurrence on many inselbergs. The combination of hard rocks and constant leaching from heavy rainfall results in mineral-deficient sand and gravel soils which accumulate in the chasms and fissures in the rock surfaces. Species that flourish under such conditions are those with special modes of nutrition. Insectivorous species are common, such as the cosmopolitan sundews (*Drosera* spp.) and the pitcher plants (Sarraceniaceae). The pitcher plant genus *Heliamphora* (Fig. 10.7), consisting of five closely related species, is found in the tepuis of the Guyanan Highlands where Roraima is the only known locality of the primitive marsh pitcher plant (*Heliamphora nutans*) which has been suggested (see Mabberley, 1997) as the ancestral type for the North American pitcher plant species *Darlingtonia* and *Sarracenia*. Orchids, due to their mycorrhizal associations, also flourish in this nutrient-poor habitat.

The isolation of the tepuis both physically and climatologically has created a unique floral community rich in species that cannot survive in the forests and savannas that lie below. Equally, the species that live in the tropical lowlands below cannot survive on the tepuis. Such conditions favour the evolution of endemic species and approximately half the species that make up the tepuis flora are endemic, causing the tepuis to be

Fig. 10.7   The marsh pitcher plant of Roriama *Heliamphora nutans* (see text) is an example of insectivorous species that has evolved in relative isolation as the sandstone plateau eroded, leaving the tepuis. Among the eroded rocks of the tepuis are open marshy savannas kept cool and wet by rain, thunderstorms and the fog that habitually shrouds the summits. (Photo © Adrian Warren/www.lastrefuge.co.uk.)

described somewhat romantically as the 'Galapagos of the floral world' (Attenborough, 1995).

The peculiar biogeography of Roraima and other tepuis in the neotropical Guyana region of Venezuela has generated considerable debate regarding the factors that are thought to account for modern vegetation patterns in the South American inselbergs. The ethos of Conan Doyle's 'Lost World' may possibly have influenced the objectivity of some studies. Rull (2004a) has questioned the view that these seemingly isolated plant communities on the high-elevation summits of South America all have a long history of evolution in isolation. Early estimations of the number of endemic species claimed very high levels of endemism (90–95%) but this has now been revised to a smaller but still very high level of 33% (Rull, 2004a).

Examination of the pollen in Quaternary sediments of some tepuis gives an insight into past vertical migrations of vegetation in response to climate change, which supports the counterargument that the flora of at least some of the tepuis is not frozen in time as has been suggested by the more romantic 'lost world' view of their biogeographical history (Rull, 2004a,b). It would appear that a replacement of a high-altitude *Chimantaea* spp. paramo-like shrubland community by a lower elevation (<2300 m) *Stegolepis* meadow occurred about 2500 years BP (*Stegolepis* is a Rapateaceae genus confined to the sandstone tepuis of northern South America). Ecologically, these *Chimantaea* shrublands are a unique type of high-altitude vegetation, which has physiognomic and taxonomic affinities with the upper Andean *páramos* shrublands, due to the dominance of the caulirossulate (columnar rosette) growth forms belonging to the Asteraceae (Rull, 2004a).

A decrease in temperature and moisture led to the establishment of present conditions after *c.* 1450 BP. It may therefore be concluded that the highland vegetation of the tepuis has responded in the past to climate shifts with vertical displacements, supporting the hypothesis of vertical floristic mixing. However, a physiographical analysis also shows that at least during the major part of the Holocene approximately half of the tepuis would never have been connected with lowland plant communities. This conclusion is borne out in the positive correlation of the number of endemic species with the height of the tepuis (Fig. 10.8). Both hypotheses are needed to explain the origins of the summit flora in inselbergs. These two points of view of a 'Lost World' or a flora that is interchanging through vertical migration need not be mutually exclusive and it is probable that both should be invoked in explaining the botanical history of the tepuis (Rull, 2004b).

10.2.2 African inselbergs

African inselbergs provide a somewhat different view of the ecology and biogeographical history of tropical, high-mountain tablelands. The west African inselbergs south of the Equator in Guinea (20–30° S) in the Nama Karoo biome have been considered as ecologically important refugia for plant species with a high recolonization potential for the surrounding arid landscapes (Burke, 2003).

A survey in equatorial Guinea of characteristic plant communities and species occurring on granite inselbergs, sandstone outcrops, and ferricretes has

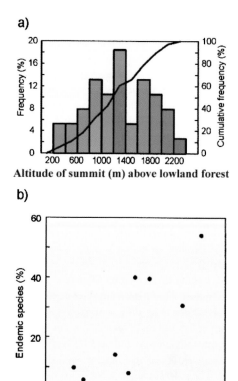

Fig. 10.8 (a) Distribution of the difference in height between tepuis summits and the surrounding lowland forests for the 48 most botanically important tepuis in Venezuela and Guyana. Bars indicate the relative frequency of each category and the solid line the cumulative frequency. (b) Relationship between difference in height between tepuis summit height and the surrounding lowland forests. Only the tepuis with a height difference above 1100 m and reliable data on endemic species have been depicted. (Reproduced with permission from Rull, 2004b.)

shown that local environmental conditions in West Africa can markedly affect the nature of the vegetation and the presence or absence of endemic species (Porembski, 2005). Examples of species restricted to the inselbergs are *Pitcairnia feliciana* (Bromeliaceae) and *Microdracoides squamosus* (Cyperaceae). *Pitcairnia feliciana* is of interest in that it is the only bromeliad species to have crossed the Atlantic.

Some sandstone outcrops are rich in species and endemics, which is probably due to the combined effects of vertical differentiation, large area, long-term climatic stability and isolation. By contrast, granite inselbergs and ferricretes were found to have no local endemics and a lower number of species. The species richness of these rock outcrops is determined largely by the existence of suitable resources, with the most diverse inselberg vegetation being found in Tanzania, Malawi, Mozambique, Zambia, Zimbabwe and Angola (Porembski & Barthlott, 2000).

10.3   ASPECTS OF HIGH-ALTITUDE HABITATS

High-altitude environments vary geographically, meteorologically, geologically and also historically. Despite this heterogeneity, they nevertheless have certain features in common. Mountain summits are generally the coldest habitats in their particular region, prone to erosion (Fig. 10.9), with poor soils and widely fluctuating day and night temperatures as well as being exposed to strong winds and high UV radiation, and in the temperate, boreal and arctic zones a short growing period. The problem of adjusting to the constant alternation of freezing nights followed with intense sunlight by day demands a greater physiological tolerance of temperature extremes than is normal in most flowering plants.

The atmosphere at high altitudes has a low water vapour content and this aridity factor has clearly imposed a selection for drought-resistant foliage in high mountain species. The low temperature and the stone covered soils of high mountains nevertheless protect the soil moisture from evaporation, which is a compensating factor for the dry atmospheric conditions (see also Chapter 3).

Whether or not high-altitude locations help plants avoid the dangers of herbivory is still an open question. The plants which have evolved under the limiting climatic and edaphic conditions on mountain tops share the characteristics that are associated with stress tolerance, namely slow growth, extended longevity, resource limitation and low palatability to herbivores. Under Grime's C-S-R theory (Grime, 2001) it would be expected that repeated biomass removals by herbivores would be a threat to plant persistence in alpine environments (see Section 3.4.2). Testing this hypothesis on populations of the alpine buttercup (*Ranunculus glacialis*) along an altitudinal

transect in the Central Alps of Austria found that between 15% and 26% of the *R. glacialis* plants in each population showed signs of grazing injury primarily by snow voles (*Microtus nivalis*) (Diemer, 1996). Only a small population, isolated by glaciers, at the highest site (3310 m) showed no traces of herbivory. Quantitative assessment of the extent of the herbivory at two high-altitude sites (2600 m and 3180 m) showed that there was considerable damage: on average nearly 25% of a plant's total leaf area was removed in one year as well as 65–85% of all flowering plants having their inflorescences removed. Despite the magnitude of these losses neither reproductive investment nor the number of leaves initiated per plant changed appreciably in the subsequent year.

It has been claimed that as *R. glacialis* populations and other similar grazed species (e.g. *Oxyria digyna*) are able to support populations of herbivores at the altitudinal limits for plant growth, without obvious reductions in vigour, then these plants cannot be considered as fitting the *stress-tolerator*, *competitor* and *ruderal* scheme proposed by Grime (see Section 3.4.2). The 'Grime theory' asserts that having to endure both extreme stress and disturbance is not compatible with survival (Diemer, 1996). However, this objection can be countered by the fact that glacier buttercups and similar high-altitude species are not stressed in their high-altitude sites. There is even the possibility that burrowing and grazing by small microtines can actually enhance plant diversity at high-altitude sites by increasing sites for regeneration.

In the high-altitude rangelands of the Trans-Himalaya, pastoralists consider that small mammals act as competitors for their livestock, causing rangeland degradation, and in many places actively eradicate a substantial portion of the vegetation. A study which investigated the effects of small herbivores like pikas (e.g. *Ochotona princeps*, the American pika) and voles (*Microtus* spp.) found that soil disturbance due to small mammals was associated with higher plant diversity without causing any dramatic decline in overall vegetation cover (Bagchi *et al.*, 2006).

Many high-altitude species are clonal and this can provide stability with access to reserves in high-altitude grasslands (Erschbaner *et al.*, 2003). It would appear therefore that high-altitude sites are not a refuge from herbivory and that despite the low growth of the high-altitude plants they nevertheless possess sufficient

Fig. 10.9   Mountain erosion as seen at the treeline in the Southern Alps, New Zealand. Remnants of the mountain beech forest (*Nothofagus solandri* var. *cliffortioides*) defying gravity in an eroding valley.

reserves to recover even from frequent grazing and can even benefit regeneration in these harsh exposed landscapes from the disturbance caused by burrowing herbivores.

Physical disturbance on mountains varies with the geology. The mass wasting of the habitat increases with the more friable sedimentary rocks and decreases where the rock is igneous. Disturbance from erosion, however, can have benefits as well as disadvantages for plant survival. Although erosion destroys vegetation it also provides fresh sites and nutrients, which promote colonization and growth. A sharp contrast in this effect can be seen when comparing mountain vegetation on the hard mountain rocks of the Caledonian system as seen in Norway, Scotland, and Newfoundland with the softer sedimentary rocks of the Southern Alps of New Zealand (Fig. 10.9). In the hard granites and metamorphic Caledonian rocks the soils of the upper mountain slopes are leached and nutrient poor, while the lower slopes are flushed and enriched with nutrients washed down from above. By contrast, in the hyperoceanic climate of New Zealand, the fast-weathering friable rocks provide available nutrients in the upper regions of the mountains. However, in contrast to the example of the hard Caledonian rocks the high rainfall and rapid erosion rates experienced in New Zealand rapidly wash out fine soils and nutrient from the valley bottoms leaving only gravel-filled river basins (Fig. 10.10; see also Fig. 11.30).

10.3.1 Geology and mountain floras

The very marked influence of geology on mountain vegetation has been noted ever since the beginnings of the systematic study of alpine Botany. In 1749 a thesis by H. F. Link was presented to the University of Göttingen entitled *Goettingensis specimen sistens vegetabilis saxo calcareo propria*, which described, probably for the first time, the difference between the flora of calcareous and siliceous rocks (Walter, 1960). Initially, this ecological distinction was interpreted as due to the physical rather than the chemical differences between the rocks. Calcareous rocks weather to produce soils that are warm and dry while soils based on siliceous rocks are colder and wetter; a distinction that is an important discriminatory feature in most mountain floras. More specifically plants differ in their reaction to soil chemistry as affected by pH. In acid soils aluminium (Al^{3+}) becomes increasingly soluble below pH 5. Species that live in acid soils are termed calcifuges and are able to sequester or exclude the potentially phytotoxic aluminium ions, together with iron and manganese in their soluble reduced state as ferrous and manganous ions.

Calcicole plants that live in soils where the pH values are high (> pH 7.0) escape the potential toxicity of Al^{3+}, Fe^{2+}, Mn^{2+} as these are only soluble in low pH soils. As calcicoles lack the chelating characteristics of the calcifuge species they are unable to survive in acid soils. The adaptations appear to be mutually exclusive. The highly efficient means for excluding iron and

Fig. 10.10   Upper waters of the Waimakariri River, South Island, New Zealand, illustrating extensive deposition of gravel eroding from fragile mountain ranges in the Southern Alps.

Table 10.1. *Examples of European alpine flowering plants which exist as contrasting vicarious species (calcicoles and calcifuges) occurring respectively on basic and acidic mountain soils*

Calcicole	Calcifuge
Achillea atrata	*A. moschata*
Carex curvula ssp. *rosae*	*C. curvula* ssp. *curvula*
Eritrichium nanum ssp. *jankkae*	*E. nanum*
Gentiana acaulis ssp. *clusii*	*G. acaulis* ssp. *kochiana*
Hutchinsia alpina	*H. brevicaulis*
Minuartia verna	*M. sedoides*
Primula auricula	*P. hirsuta*
Ranunculus alpestris	*Ranunculus glacialis*
Rhododendron hirsutum	*Rhododendron ferrugineum*
Salix retusa	*Salix herbacea*
Saxifraga aizoon	*Saxifraga cotyledon*
Saxifraga oppositifolia	*Saxifraga rupestris*
Sedum atratum	*Sedum montanum*
Sesleria caerulea	*Sesleria disticha*
Silene uniflora ssp. *petraea*	*Silene rupestris*
Soldanella alpina	*Soldanella pusilla*

Sources: various, including Walter (1960); Reisigl & Keller (1994); Jermyn (2005).

Fig. 10.11 *Rhododendron ferrugineum*, the calcifuge alpine rose. Also called the rusty alpine rose due to the iron-laden hairs in the momentum on the abaxial leaf surface (see inset). The removal of iron to the outside of the leaf presumably serves as an effective method for avoiding excessive accumulations of toxic ferrous iron within the leaf.

manganous ions in calcifuge species prevents them physiologically from accessing the necessary quantities of these ions. This together with the effects of high bicarbonate causes their foliage to become chlorotic – the condition generally referred to as lime-induced chlorosis (Marschner, 1995).

Many closely related species and subspecies differ genetically as to their preferences for calcareous or siliceous soils. This is particularly noticeable in mountains such as the Alps (Table 10.1) where the relationship between the underlying geology and the soils is not masked by deep soil development or overlying peat or glacial deposits.

One of the best known examples is the different forms of alpine rhododendron, the 'alpine roses', where *Rhododendron ferrugineum* occurs on acid soils while the hairy alpine rose (*R. hirsutum*) is confined to calcareous soils. Even specific sites such as snow patches differ in the species present depending on whether the underlying rocks are calcareous and siliceous (Figs. 10.11–10.12).

By contrast, in the very highest mountains, as in the Himalaya, geology appears to be less important than altitude, precipitation, and aspect. It is possible to pass from one rock formation to another without having any noticeable change in the vegetation. Thus the long-leaved pine (*Pinus roxburghii*) grows on acid or basic rocks without any apparent discrimination (Polunin & Stainton, 1984).

The basis for the differences between the floras of acidic and calcareous soils can be due in part to the very real physical divergence between calcareous and siliceous rocks in terms of their physical properties, as well as to the chemical properties of the soil.

Fig. 10.12 *Rhododendron hirsutum*, the calcicole alpine rose. (Photo Dr A. Gerlach.)

Calcareous soils are generally warm and dry and siliceous soils are cool and wet. This together with the different reactions to both soil chemistry and plant physiology in terms of pH, calcium and bicarbonate ions causes a pronounced distinction in plant distribution. This is particularly noticeable in mountain floras where altitudinal temperature limitations differentiate sharply between early and late sites for the resumption of growth in spring. Consequently, there is a marked phenological dimension to the advantages and disadvantages of warm dry calcareous soils as opposed to the generally cooler and wetter environments associated with siliceous soils.

10.3.2 Adiabatic lapse rate

The cooling effect of increasing altitude is influenced by the adiabatic lapse rate, which is the negative vertical gradient of temperature maintained by the vertical motion of air through surroundings in hydrostatic equilibrium (Calow, 1998). When the air is unsaturated it is known as the *dry adiabatic lapse rate* and has a value of $9.8\,^{\circ}\mathrm{C\,km^{-1}}$, provided the moving and ambient air are at nearly the same temperature (which is usually the case). If, however, saturation is maintained by the condensation of water vapour in rising air, then the release of latent heat reduces the adiabatic lapse rate (Fig. 10.13).

In the warm low troposphere (the lowermost portion of the Earth's atmosphere and the one in which most weather phenomena occur), the saturated adiabatic lapse rate may be as little as half the dry adiabatic lapse rate which prevails in the cold high troposphere (Calow, 1998). A global mean temperature lapse rate of $5.6\,^{\circ}\mathrm{C\,km^{-1}}$ has been reported for mountains (Körner, 2003). However, where there are strong temperature inversions, as along the edges of the major ice sheets in Greenland, values as high as $12\,^{\circ}\mathrm{C\,km^{-1}}$ have been recorded. Mixing of different air masses can modify the rate of temperature change with altitude. Oceanic regions of western Norway and Scotland frequently experience weather dominated by polar-maritime air. This air mass, which has been chilled by the Greenland ice cap, will be warmed in its lower levels as it passes over the North Atlantic Drift. Nevertheless, the upper air mass still remains cold. The mixture of low level warm air and high level cold air brings to these lands bordering the eastern shores of the North Atlantic a more rapid fall in temperature with altitude than would be normally predicted. In western Scotland, a decrease in temperature of 8–$9\,^{\circ}\mathrm{C\,km^{-1}}$ of altitude is not uncommon.

Ecologically, these different adiabatic rates have a profound effect on the zonation of mountain vegetation. In the hyperoceanic conditions of the Northern and Western Isles of Scotland the montane vegetation zonation is compressed due to the high adiabatic rate. It is possible in the Orkney Islands (59° N) to stand in montane tundra type vegetation on a summit at only 420 m, and view below crops of barley growing at sea level (Fig. 10.14).

10.3.3 Mountain topography and biodiversity

The potential floristic diversity of any particular mountain can often be predicted at a distance from its shape, geology and rock structure. Despite what might appear daunting prospects for plant colonization it is frequently surprising how many high mountains support a rich

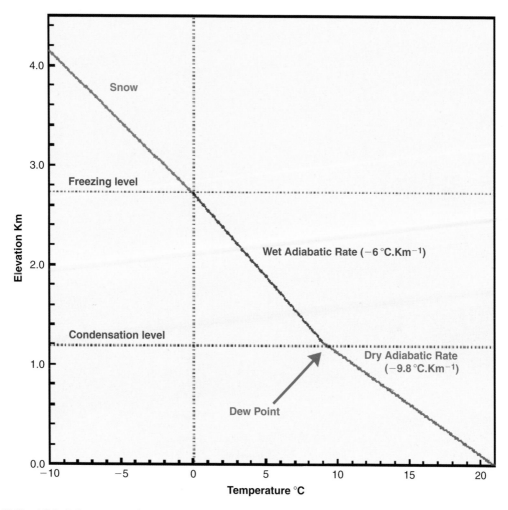

Fig. 10.13   Adiabatic lapse rate as a function of altitude. (Based on Strahler, 1963, pp. 246–252.)

alpine flora. Slopes, screes and terraces may look unstable but they can provide, depending on aspect and geological structure, varied and favourable habitats that are adequate for a range of suitably adapted species. The general roughness of mountain topography also makes available abundant microsites where localized environmental conditions can provide temperature regimes that can counteract the effects of altitude and latitude.

Mountain chains and their orientation are also very important biogeographically in providing scope for migration, particularly during periods of significant climate change. A north–south orientation is generally considered the most favourable for alterations in temperature. However east–west orientations as in the European Alps can accommodate changes from oceanic to continental climates. Not surprisingly it is often possible to predict the diversity of vegetation on many high mountains depending on whether the mountains can be described as concave or convex. Concave mountains provide reservoirs for soil, water, and nutrients, while convex mountains tend to be deficient in these resources. Consequently descriptors of relief curvature and roughness explain more of the variability in species distribution than 'classical' terrain attributes, such as elevation or exposure (Gottfried et al., 1998).

The negative features of being an isolated mountain are to some extent counteracted when the mountain is both large and high. Conversely, on small mountains,

Fig. 10.14   View from Ward Hill, Hoy, Orkney (59° N) illustrating the rapid change in climate and vegetation with increasing altitude in oceanic climates. Note the tundra vegetation in the foreground and active agriculture in practice only 305 m (1000 feet) below.

the bioclimatic vegetation zonation is compressed (low, middle and high alpine, see Section 10.5.1) as compared with larger mountains.

German-speaking ecologists refer to this phenomenon as *Massenerhebung* or the *Massenerhebung effekt* (mass elevation effect), where the size of the mountain can ameliorate the physical environment. The effect is most noticeable where large mountains are massed together. In the centre isotherm levels rise and create a more continental climate, usually with reduced cloud cover, which causes the nival zone (permanent snow)

level to retreat to a higher altitude. The uppermost alpine vegetation can therefore be found at higher altitudes on large mountains than on small mountains. The front ranges, however, do not differ greatly from smaller mountains.

As a consequence of this *Massenerhebung* effect there is also a greater production of seed at high altitudes, which provides a larger quantity of propagules to blow and wash down the mountainside and this in turn enhances the presence of alpine plants at lower elevations (Ellenberg, 1963).

10.4 PHYSIOLOGICAL IMPLICATIONS FOR PLANT SURVIVAL ON HIGH MOUNTAINS

10.4.1 Water availability at high altitudes

Water retention at high altitudes is highly dependent on topography and soil development, as well as on exposure and the seasonal distribution of precipitation. In common with other skeletal soils, such as sand dunes and deserts (see Section 7.1.2) the top surface is prone to desiccation and plants in these areas are dependent on being able to extract water from the deeper soil layers which are frequently absent in mountain habitats. The so-called dry calcareous scree-slopes on the southern flanks of the Alps have been shown to be well supplied with water as the coarse substrate prevents evaporation and traps moisture.

In tropical mountains soils above the cloud zone are progressively less well supplied with water with increasing altitude (see the current situation on Mt Kilimanjaro, Section 10.5.2). Fortunately, in cooler, high-altitude temperature regimes, despite the lower atmospheric pressure increasing evaporation, the ratio of evaporation/precipitation (E/P ratio) is reduced due to the lower temperatures with a consequent reduction of the risk of drought injury. Calculations based on predicted water needs in a temperate alpine zone, assuming a growing season length of 100 days and supposing a maximum soil depth of one metre, suggest that in this type of habitat there is no need for additional rainfall during the growing season, provided that the soil pore space is fully saturated in spring, which is usually the case immediately after snowmelt (Körner, 2003).

Not all mountains are endowed with a deep soil, or even with a starting point of total water saturation at the beginning of the growing season. Wind exposure, particularly on mountain ridges, not only increases water loss, it also denudes exposed ridges and slopes of soil particles and reduces soil fertility. Drought can also be a permanent feature of many high-altitude sites. When mountains are subject to warm dry catabatic winds (*föhn, chinook, mistral*, etc.) the lower slopes will suffer additional desiccation as the descending winds warm as they fall. Föhn, or valley wind, as it descends on the northern side of Mt Everest has a desiccating effect in the valleys of the north slope, creating an extreme asymmetry in the altitudinal vegetation belts between the south and north

slopes, with heavy rainfall on the former and extensive drought on the latter (Miehe, 1989). Also, in tropical and subtropical mountains, at altitudes that are above the main cloud zone, desert-style conditions can prevail for many months of the year.

Areas where snow accumulates and lies late into spring, although potentially limiting the length of the growing season, can nevertheless provide several advantages for plant survival and many mountain-plant species are known to be characteristic of snow patches. One of the principal compensations of snow patches for both plants and soils is the protection that is provided from winter winds, ablation and nutrient leaching. However, snow shelter on melting also ameliorates the water supply. The nature of the relief on the mountainside that determines where snow lies is usually coincidental with natural drainage patterns so that the same areas that gather snow in winter also benefit from summer precipitation. Much of the benefit that is derived by alpine plants from snow patches comes from the topography of the snow-patch area. Snow-accumulating hollows favour flushing rather than leaching, and therefore provide a natural reservoir for both water and nutrients. In addition, snow also traps dust and nutrients, which then settle in the patch as the snow melts. The same topography that gathers snow will also collect erosion sediments and as a result snow patches frequently have a greater soil depth with more favourable water and nutrient supplies than the more exposed parts of the mountain. In the modern world snow banks also accumulate air-borne pollutants and when they thaw release a surge of pollutants. Most noticeable is the lowering of pH in mountain streams as snow melts in polluted areas.

10.4.2 Adapting to fluctuating temperatures

The temperatures on high mountains and plateaux fluctuate widely. In the Bolivian Altiplano the diurnal range can frequently be $20\,^{\circ}C$ by day and $-5\,^{\circ}C$ by night. Throughout the world adaptation to these potentially stressful changes of temperature has resulted in remarkable examples of convergent evolution where plants of very different origin have evolved similar life-forms as they provide the optimal solution to this widespread phenomenon. The most striking example of this has been the evolution of the *pachycaul* plants (thick-stemmed species) in both South American and African alpine habitats. In the Afro-alpine region *Dendrosenecio* and *Lobelia* are typical examples as are *Puya* and *Espeletia* in South America. The pachycaul (thick stem) construction is also evident in desert environments where there are also large diurnal temperature fluctuations. Thick stems, such as are found in the large desert cacti, retard both heating and cooling by the mass of water they contain. However, the alpine pachycaul plants have an additional adaptation that is not found in the cacti through the nyctinastic movements of the leaves and bracts (closing at night) that envelope the stems.

Resisting heat by day is but one aspect of the adaptations needed for high mountain survival. Freezing tolerance that allows the plants to resume metabolic activity during the warm day is a highly specialized adaptation that has received considerable study. Not surprisingly, comparisons between different species in relation to relative use of different freezing-tolerance mechanisms can detect a diversity of responses which can be related to the size of the plants.

A study carried out on plants growing from two different elevations (3200 m and 3700 m) in a desert region of the high Andes (29° 45′ S, 69° 59′ W) found that all ground-level plants showed cellular freezing tolerance to be the main mechanism for resistance to freezing temperatures (Squeo *et al.*, 1996). Tall shrubs avoided freezing temperatures, mainly through supercooling, the phenomenon whereby a liquid can be cooled below its freezing point without freezing taking place. Alternatively, a saturated solution can be cooled without crystallization taking place, to form a supersaturated solution. In both cases supercooling is possible because of the lack of solid particles around which crystals can form. Crystallization rapidly follows the introduction of a small crystal (seed) or agitation of the supercooled solution. Supercooling was only present in plants occupying the lower elevation (3200 m).

In another study, arborescent forms (i.e. giant rosettes and small trees) showed avoidance mechanisms mainly through supercooling, while intermediate-height plants (shrubs and perennial herbs) exhibited both tolerance and avoidance mechanisms. Insulating tissues, which help to avoid temperature extremes, were present in both arborescent and cushion life-forms. It was therefore surmised that for high tropical mountain plants, a combination of freezing tolerance and avoidance by insulation is less expensive than supercooling alone in relation to resource utilization and also provides a more secure mechanism for avoiding cold injury (Squeo *et al.*, 1991).

High mountain plants are frequently highly pubescent. Dense well-developed leaf hairs can serve more than one purpose. They can shield vulnerable tissues against radiation and reduce the risk of heat injury. They can also dissipate heat into the atmosphere like flanges on a radiator. In addition, they have an important role in preventing stomata from becoming occluded by droplets of water that frequently condense on leaves in the cloud zone. Alpine plants that are adapted to living in the dry cold mountain air are very susceptible to moist conditions and the dense hairs give protection to the leaf and prevent water and potentially pathogenic fungal spores from settling on the epidermis (Figs. 10.15–10.16).

10.4.3 Protection against high levels of radiation at high altitudes and latitude

Cellular damage from high levels of radiation has been often suggested as a potential danger to both alpine and arctic plants. Early in the growing season the roots are still in a cold environment while the leaves may be exposed to high sunlight. Any limitation of growth at low temperatures while the leaves are exposed to full sunlight might lead to an accumulation of carbohydrates in leaves and could cause a feedback inhibition of carbon fixation. There could then follow a build-up in reducing power and an increase in light-generated reactive oxygen species with the potential for photo-damage to the chloroplasts.

At high altitudes increased levels of sunlight pose a risk of photo-oxidation damage. A study of a selection of alpine plants that occur at different altitudes in the subnival and nival zones of the Obergurgl (Ötztal, Austria) has shown marked increases in antioxidant content, principally ascorbic acid with increasing altitude (Wildi & Lütz, 1996). The contents of most compounds were found to follow a diurnal rhythm, with the maximum occurring at midday and the minimum during the night. This enhancement was mainly due to ascorbic acid contents. Each plant species displayed a specific reaction to the increase in stress

Fig. 10.15   An extreme example of hair development in a high-altitude specimen of *Tanacetum gossypinum* growing at 5000 m in the Himalayan Goyo valley. (Photo Professor R. M. Cormack.)

that accompanies an increase in altitude, resulting in a broad adaptation spectrum for these plants, which suggests that the combined effect of lower temperature and higher light intensity induces higher antioxidant contents (Figs. 10.17–10.18).

The overriding importance of plant form in relation to adaptation to high altitude conditions has been demonstrated in a series of studies carried out on the American marsh marigold (*Caltha leptosepala*) and

the yellow glacier lily (*Erythronium grandiflorum*; Fig. 10.19) which can be found in Utah at altitudes up to 3120 m (Germino & Smith, 2001). Both plants are perennials that commonly emerge from alpine snow banks where there is a combination of cool temperatures and strong reflected sunlight. *Caltha leptosepala* occurs in microsites where colder air accumulates, and has larger, less inclined and more densely clustered leaves compared with *E. grandiflorum* which has two steeply

Fig. 10.16   Edelweiss (*Leontopodium alpinum*), an example of a highly pubescent alpine plant which grows mainly on basic soils on alpine meadows between 1700 and 3400 m. (Photo Professor R. M. Cormack.)

inclined leaves. These differences in microsite and plant form make for differences in leaf temperature. *Caltha leptosepala*, which has the larger leaves, was observed to have leaf temperatures below 0 °C in 70% of nights during the summer growing season as compared with only 38% in *E. grandiflorum*, the species with smaller and more densely clustered leaves. In addition, the leaves of *C. leptosepala* warmed more slowly on mornings following frosts compared with *E. grandiflorum*, due to less aerodynamic coupling between leaf and air temperature, and also to a 45% smaller ratio of sunlit to total leaf area due to mutual shading among leaves. As a result, night frost did not affect subsequent CO_2 assimilation in *E. grandiflorum*, while in *C. leptosepala* frostless nights and warmer mornings led to a 35% greater CO_2 assimilation in the early morning. Greater daily carbon gain probably occurs for *E. grandiflorum* because of its plant form and microclimate, rather than through differences in photosynthetic efficiency (Germino & Smith, 2001).

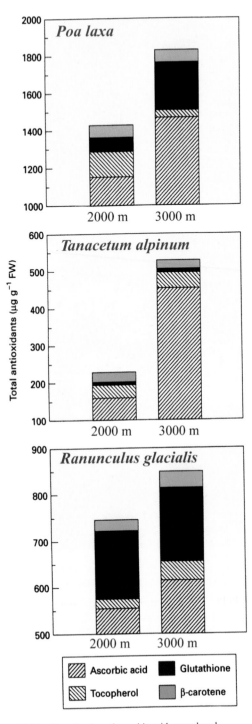

Fig. 10.17   Contribution of ascorbic acid, tocopherol, glutathione and beta-carotene in three species at two different altitudes on the Obergurgl, Austria. (Reproduced with permission from Wildi & Lütz, 1996.)

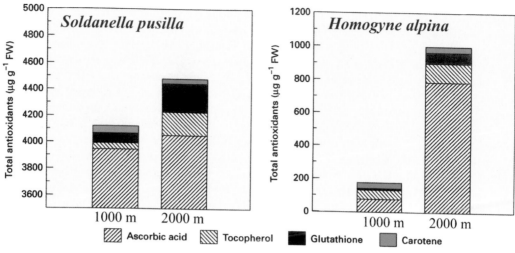

Fig. 10.18   Contribution of ascorbic acid, tocopherol, glutathione and beta-carotene to the total amount of antioxidants in *Soldanella pusilla* and *Homogyne alpina* at altitudes of 1000 and 2000 m on the Obergurgl, Austria. (Reproduced with permission from Wildi & Lütz, 1996.)

Fig. 10.19   The yellow glacier lily (*Erythronium grandiflorum*). Plants growing by side of glacier at 2000 m on Mt Rainier, Cascade National Park, Washington, USA. (Photo Professor R. M. Cormack.)

10.4.4   Effect of UV radiation on alpine vegetation

The depletion of the stratospheric ozone layer in recent years exposing the Earth's surface to increased levels of ultraviolet radiation has prompted considerable research into the potentially harmful effects of ultraviolet-B (UV-B, 280–320 nm) on plant tissues. High-altitude vegetation may be expected to be pre-adapted to this stress, particularly in mountains where the summits are frequently above the cloud zone. For plants, excessive UV-B radiation could, in theory, damage the photosynthetic apparatus and nucleic acids in the leaf mesophyll. Fibre-optic microprobes have been used to make direct measurements of the amount of UV-B reaching these potential targets in the mesophyll of intact foliage.

A comparison of foliage from a diverse group of Rocky Mountain plants showed that the foliage of some plant life-forms was more effective than others at screening UV-B radiation (Day *et al.*, 1992). The leaf epidermis of herbaceous dicots was found to be ineffective at attenuating UV-B, with epidermal transmittance ranging from 18% to 41% and UV-B reached 40–145 μm into the mesophyll or photosynthetic tissue. In contrast to the herbaceous dicots, the epidermis of one-year old conifer needles filtered out essentially all incident UV-B and virtually none of this radiation reached the mesophyll. Although on high elevation krummholz trees the epidermal layer was appreciably thinner in older needles (7 years), the epidermis still attenuated essentially all incident UV-B. The same epidermal screening effectiveness was observed after the removal of epicuticular waxes with chloroform. Leaves of woody dicots and grasses appeared to be intermediate between herbaceous dicots and conifers in their UV-B screening abilities with 3–12% of the incident UV-B reaching the mesophyll.

Evidence from other areas where UV-B may pose a risk to plants also suggests that little damage appears to take place. The southern part of Tierra del Fuego (Argentina, 55° S) is an area strongly affected by ozone depletion due to its proximity to Antarctica, and several investigations have been initiated to determine the biological impacts of the natural increase of solar UV-B on natural ecosystems in this region. Ambient UV-B has been found to have subtle but significant inhibitory effects on the growth of herbaceous and graminoid species, whereas no consistent inhibitory effects have been detected in woody perennials. The species investigated in greatest detail, the herbaceous *Gunnera magellanica*, showed increased levels of DNA damage in leaf tissue in the early spring which was correlated with the dose of weighted UV-B measured at ground level (Ballare *et al.*, 2001). However, an opposite effect has been noted in relation to herbivory of the Antarctic beech (*Nothofagus antarctica*) where it was found that insects consumed at least 30% less area from branches exposed to UV-B than from those with reduced UV-B exposure (Rousseaux *et al.*, 2004).

Further south in Antarctica, the performance of the only two Antarctic vascular plants, Antarctic pearl-wort (*Colobanthus quitensis*) and Antarctic hair grass (*Deschampsia antarctica*), has shown that UV-B leads to reductions in leaf longevity, branch production, cushion diameter growth, above-ground biomass, and thickness of the non-green cushion base and litter layer. However, exposure to UV-B accelerated the development of reproductive structures and increased the number of panicles in *D. antarctica* and of capsules in *C. quitensis*, when calculated in terms of per unit of ground surface area covered by the mother plants. However, this effect was offset by a tendency for these panicles and capsules to produce fewer spikelets and seeds. Ultimately, UV-B exposure did not affect the numbers of spikelets or seeds produced per unit of ground surface area. In relation to vegetative growth the relative reductions in leaf elongation rates increased over four seasons, suggesting that UV-B growth responses tended to be cumulative over successive years (Day *et al.*, 2001).

Not all these examples come from studies at high altitudes. Nevertheless, they illustrate the range of plant responses to UV-B radiation. Such is the range of the many individual effects that can be observed and investigated, it has to be concluded that the possibility of coming to any general conclusions as to their ultimate biological significance remains elusive.

The most common general effects noted so far of UV-B on plant growth is a reduction in plant height and possibly in severe cases a decrease in shoot mass and foliage area (Caldwell *et al.*, 2003). Viewed as a whole, therefore, it would appear that the survival capacity of terrestrial plants is usually unaffected by enhanced UV-B, even though reduced growth has been observed and may increase in magnitude over successive years. It is quite possible that survival of alpine species may be enhanced by reduction in growth and this should therefore not be regarded as a negative reaction.

10.4.5   Oceanic mountain environments

Paradoxically, the greatest danger to mountain floras from rising temperatures is probably not to be found in continental mountains, as in the central Swiss Alps, but in mountains with oceanic climates. It might have been expected that oceanic mountains would be buffered against climatic change due to a reduction in annual temperature range. The relatively species-poor mountain flora of the Scottish Highlands and south-west Norway is considered to be due, at least in part, to the mild periods of winter weather that encourage premature spring growth when there is still a risk of exposure to spring frosts. In Norway, the mountain species (Table 10.2) that are absent from more oceanic mountains have been described as *south-west coast avoiders* (Dahl, 1951, 1990).

The conflicting conditions of oceanicity versus continentality create differing upper and lower limits for the altitudinal range of species distributions. This interaction has been examined in a detailed study centred at 63° N on the Norwegian west coast and extending 135 km inland to cover the southern Fennoscandian mountain range – the southern Scandes (Holten, 2003). When species richness is plotted on a two-dimensional vertical projection of elevation and distance from the coast the maximum number of 70–80 arctic–alpine species is found in the Drividalen Knutshø Mountains (Fig. 10.20) and from there it decreased in all directions. The two-dimensional plot shows that the local maximum number of species

Table 10.2. *A selection of species described as south-west coast avoiders, grouped by their varying ability to extend their distribution south and west*

Boreal species that reach the Baltic but are restricted to high altitudes	Northern European and alpine species reaching the Alps and/or the British Isles but not extending to the Iberian Peninsula, Italy or the Balkans
Aconitium septentrionale	*Salix myrtilloides*
Rubus arcticus	*S. phylicifolia*
Epilobium hornemanni	*S. starkeana*
E. lactiflorum	*Stellaria calycantha*
Petasites frigidus	*Betula nana*
Glyceria lithuanica	*Rubus chamaemorus*
Carex disperma	*Polemonium caeruleum*
C. tenuiflora	*Tofeldia pusilla*
C. loliacea	

Data from Dahl (1998).

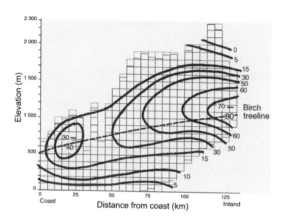

Fig. 10.20 Vertical projection of species richness of alpine vascular plants along a coast–inland transect. Note that although the number of species increases with distance from the sea the height of the mountains also increases. Thus, in addition to increasing continentality in the climate there will also be a *Massenerhebung effekt* (mass elevation effect; see Section 10.3.3). (Reproduced with permission from Holten, 2003.)

on steep coastal mountains is only half of that on inland mountains. It can also be seen that altitude difference between the upper and lower limits for the distribution of the arctic–alpine species decreases in moving towards the coast. An analysis of the species composition reveals two gradients: a humidity gradient that decreases on moving inland, and a temperature gradient that decreases with altitude. Furthermore, many of the species of the central Scandes show a negative correlation with winter temperatures as suggested by Dahl (1951) in the concept of *south-west coast avoiders*. Once again we have a clear indication that oceanic climates, especially when they impose mild winter temperatures, are generally unfavourable for alpine plants in boreal and temperate zones (see Chapter 1).

10.4.6   Phenological responses of mountain plants

A question that always arises in relation to climatic warming, particularly in areas where spring can be late and the growing season short, is whether or not photoperiodic control of phenology will limit the use that plants might be able to make of earlier warmer springs. An experiment to answer this question in Austria on a number of high-elevation species (2600–3200 m a.s.l.) found that the number of days between soil thawing and flowering was sensitive to photoperiod in just under half of the species. However, *Cerastium uniflorum*, *Ranunculus glacialis*, *Kobresia myosuroides*, *Saxifraga oppositifolia* and *S. seguieri* were found to be insensitive to both photoperiod and temperature and flowered as soon as released from the snow. These results suggest that about half of the tested alpine species are sensitive to photoperiod and may therefore not be able to utilize fully periods of earlier snowmelt (Keller & Körner, 2003).

Among the species that are insensitive to temperature or photoperiod are those that are found at the highest elevations (e.g. *Ranunculus glacialis*). In species adapted to the more extreme environments of high mountains, flowering proceeds as rapidly as possible after snowmelt and is not delayed for any further environmental signals.

Rapid flowering after snow and ice melt is a phenomenon that is common to both alpine and arctic floras (see Section 4.8). The purple saxifrage (*Saxifraga oppositifolia*) will flower 5–8 days after being released from snow (Larl & Wagner, 2006). This ability to produce flowers quickly is dependent on the possession of pre-formed flowering buds which are usually initiated as the day length shortens during the previous growing season. The speed with which these tissues can develop is remarkable.

A study of the dynamics of reproductive development from floral initiation to fruit maturation, as well as leaf turnover in vegetative short-stem shoots, of *Saxifraga oppositifolia* in the Austrian Alps found marked differences in the timing and progression of reproductive and vegetative development depending on altitudinal location. In an alpine population a four-month growing

season was required to complete reproductive development and initiate new flower buds, whereas individuals from later thawing subnival sites attained the same structural and functional state within only two and a half months (Larl & Wagner, 2006; Fig. 10.21). Reproductive and vegetative development were not strictly correlated because timing of flowering, seed development, and shoot growth depended mainly on the date of snowmelt, whereas the initiation of flower primordia was evidently controlled by photoperiod the previous season. Floral induction occurred during June and July, from which a critical day length for primary floral induction of about 15 hours could be inferred. Pre-formed flower buds overwinter in a pre-meiotic state and meiosis started immediately after snowmelt in spring. This study illustrates a striking differentiation between alpine and subnival populations in the timing and progression of reproduction.

As this study was carried out over a three-year period, during which there were differences in the degree of summer warmth, it was also possible to draw some conclusions as to how increasing climatic warming might affect the phenology of mountain plants. In 2003 when there was a higher number of day degrees accumulated at the subnival sites there was no further acceleration in the rate of development of these

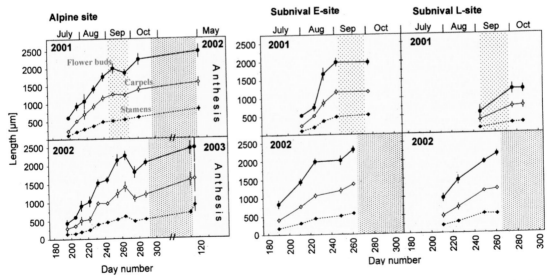

Fig. 10.21   Dynamics of reproductive growth of *Saxifraga oppositifolia* at alpine and subnival sites with early (E) and late thawing (L) over two years of observation. Values are the mean lengths of flower buds, carpels, and stamens. Light dotted shading, temporary snow cover; heavier shading, winter snow cover. (Reproduced with permission from Larl & Wagner, 2006.)

high-altitude populations. Such findings are in agreement with arctic studies on this same species where it was found that enclosing plants of *S. oppositifolia* in open-top chambers had little influence on phenology, flowering frequency, and reproductive success (Stenström *et al.*, 1997). It seems that for cold-tolerant plants that are indifferent to higher temperatures, e.g. species such as the purple saxifrage, which is capable of growing into late autumn even when covered with snow, climatic warming will be unlikely to evoke any phenological response.

For further discussion of reproduction in marginal areas see Chapter 4.

10.5 ALPINE VEGETATION ZONATION – CASE STUDIES

The definition of alpine vegetation as the plant communities that are found between the upper limit of tree growth and the snowline (nival zone) is used globally. As might be expected in any global categorization many differences can be found between the species that inhabit this broad zone in various parts of the world. Variation in the alpine environment is not just a question of temperature reduction with increasing altitude. Aspect, exposure, snow cover and water availability also play a role in controlling species distribution. In different parts of the world the nature and extent of these factors with increasing altitude impose their own particular set of controlling conditions on plant zonation within the alpine zone. Floras also differ in the number of species that have a wide altitudinal range and those in which the altitudinal range is more restricted.

Despite these regional differences there are usually a number of dominant species which can impose a structural change on the vegetation that differs in its impact in relation to altitude (Fig. 10.22).

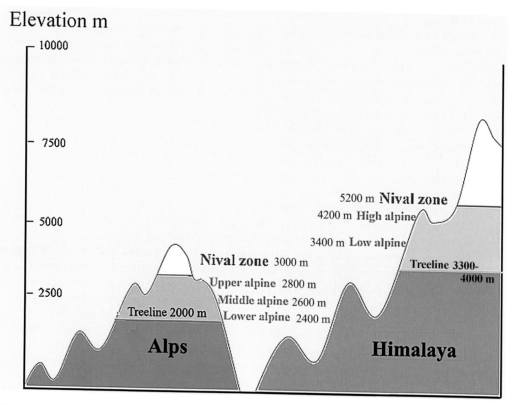

Fig. 10.22   Diagrammatic comparison of approximate altitudinal limits to vegetation zonation in the Alps and the Himalaya. (Data adapted from various sources.)

10.5.1   Temperate and boreal alpine zonation

In temperate and boreal regions the nature of the dominant woody and graminoid species makes for a distinct altitudinal pattern within the alpine zone. Frequently three sub-zones can be seen merely from visual inspection. In these regions the normal pattern is for rainfall and cloud cover either to increase or remain unchanged with increase in altitude, which causes the limits to the differentiation of plant communities in relation to altitude to be largely determined by temperature. The zones that are most frequently recognized are listed below.

(1) *A lower alpine zone with bush and tall herb communities*
    Compared with oceanic Scotland the vigorous and widespread mountain willow zone on exposed mountainsides and luxuriant tall herb communities in more sheltered areas in Scandinavia is very striking. This is especially so as attempts to restore the willow communities on Scotland's most species-rich mountain, Ben Lawers (1214 m, 3983 feet), have had very limited success.

(2) *A middle alpine zone*
    In this zone, graminoid (sedges and grasses) and heath species dominate, as can be seen in Scandinavia and Scotland. Heathlands are better developed in Scotland, due to the oceanic climate and the long-standing lack of forest. The large-scale removal of the upper forest and the montane willow zone has allowed the heath species of the middle alpine zone to extend downhill and create the *Calluna vulgaris* dominated heather moorland that typifies the Scottish Highland landscape. For some Scots the landscape is the glory of Scotland but for those with some perception of what has been lost biologically over the millennia it represents a wet desert, low in biodiversity and impoverished in its soils and productivity (see Darling & Boyd, 1964).

(3) *An upper alpine zone*
    At this upper level the vegetation is marked by loss of dominance of the graminoid species, which are replaced by dwarf herbaceous and prostrate woody plants together with lichens and mosses. The vegetation cover is reduced and more open ground is apparent. This is in effect a tundra-like vegetation that can be found well south of the Arctic Circle and descends to relatively low levels in hyperoceanic regions such as Scotland's Northern and Western Isles.

10.5.2   Tropical and subtropical mountains – East Africa

The zonation on tropical mountains is notably different and does not parallel that which is found in temperate regions. On tropical and subtropical mountains the entire zonation sequence is less abrupt. In East Africa the dry mountain forests typically flourish up to the top of the cloud zone, but above this region precipitation decreases at varying rates depending on location. In the Andes the eastern side of the mountain range descends into the Amazonian rainforest and precipitation is less limiting than on the Pacific side. In general, however, where there is a diminution in precipitation and water becomes a limiting factor, tree cover gradually diminishes.

Snow and ice-capped Kilimanjaro, the highest mountain in Africa (5895 m), still shows some volcanic activity with the presence of many fumaroles. In recent years, the permanent ice cover has been noted to be decreasing rapidly. Over the twentieth century, the areal extent of Kilimanjaro's ice fields decreased by 80% and if current climatological conditions persist, the remaining ice fields are likely to disappear between 2015 and 2020 (Thompson *et al.*, 2002). In March 2005, it was reported that the peak was almost bare of snow and ice, for the first time in 11 000 years (*The Guardian*, 14 March 2005).

Not surprisingly Kilimanjaro and its vanishing glaciers have become an 'icon' of global warming. However, the justification for this has been questioned. While many parts of the world have warmed in the last two decades, satellite data have shown that Kilimanjaro itself has cooled. The ice is still disappearing, and it is now thought that a local climate shift that occurred 120 years ago may be responsible. Field observations and climatic data suggest that factors other than air temperature directly control the ice recession. In particular it is suggested that a drastic drop in atmospheric moisture at the end of the nineteenth century and the ensuing drier climatic conditions are the likely factors rather than a rise in temperature in forcing glacier retreat on Kilimanjaro (Kaser *et al.*, 2004).

On Kilimanjaro, precipitation peaks with 2000 mm at an altitude of 2200 m and then drops sharply above the upper cloud zone to 220 mm at 4200 m producing an alpine desert. This pattern is also repeated on other high tropical mountains. The temperature conditions in this alpine zone are strikingly different from those on temperate and boreal mountains. In the latter, seasonal temperature differences are large and diurnal fluctuations small. On the Zugspitze (2964 m a.s.l.) in the Alps the mean diurnal temperature is about 5 °C with a seasonal monthly range of 13 °C. In boreal and temperate zones the large range of seasonal temperature differences causes a thermally induced seasonal phenology to differentiate the vegetation of the entire alpine zone into subregions that are adapted to different lengths of growing season. The result is the development of the lower, middle and upper alpine zones as described above. On tropical mountains, however, seasonal temperature differences are minimal and the seasons if any are distinguished on the basis of precipitation patterns and not temperature. Instead of a seasonally imposed stress there is a gradient of greater drought with increasing altitude coupled with freezing night temperatures at increasing frequency and strong insolation by day. These conditions impose gradual changes on the vegetation within the alpine zone. In these alpine deserts many species exhibit specialized morphological and physiological adaptations that aid drought and cold tolerance and were discussed above (Section 10.4).

On Kilimanjaro the upper level of the cloud forest lies at approximately 2800–3000 m. Above this lies a gradient of changing ecological zones (Hemp, 2006). Irrespective of the method of vegetation recording, four major altitudinal discontinuities can be recognized.

(1) An *Erica excelsa* dwarf forest (2800–3050 m) with many woody shrub species belonging to the Ericaceae and Proteaceae from 2800 m up to the lower edge of the alpine zone at 3300–4000 m. At the upper region of this zone there is a restricted elfinwood formation (a tropical mossy krummholz).

(2) A lower alpine zone (3300–4000 m), a region of lower rainfall with sparser vegetation. The lower altitudes are characterized by giant groundsels (*Senecio* spp.) and giant lobelias.

(3) An upper alpine desert (4000–5000 m), a harsh, dry, windy region, consisting mostly of bare rock and ice. Plant cover is minimal and consists mostly of lichens and small mosses. At certain times of the year snow covers the area.

(4) A summit zone from 4600 m consists of frost-shattered rock and mosses and lichens which are found right up to the summit (5895 m). Specimens of *Helichrysum newii* have been recorded close to a fumarole at an altitude as high as 5760 m.

Many species occur through more than one zone. Two distinct forms of giant groundsel occur on the upper mountain: *Senecio johnstonii* ssp. *cottonii*, endemic to the mountain and only occurring above 3600 m, and *S. johnstonii* ssp. *johnstonii* which occurs between 2450 m and 4000 m. At all altitudes *Senecio* favours the damper and more sheltered locations, and in the alpine bogs is associated with another conspicuous plant, growing up to 10 m tall, the endemic giant lobelia (*Lobelia deckenii*).

10.5.3   South America

The enormous north–south extension of the Andes from 9° N in northern Colombia to 56° S in the Tierra del Fuego provides a ready means of comparing tropical and cool temperate high-altitude areas in relation to the ecology of high-altitude vegetation. Despite many similarities between the mountain zonation in South America and East Africa there are also distinct differences reflecting the very different geological histories of the two continents. From Ethiopia to South Africa there is a disjunct series of individual peaks and ranges with different ages, origins, and geologies. African mountains tend to be more or less islands whereas in South America, the Andean Cordillera to the west forms a continuous range of mountains together with very extensive high altitude tablelands. To the east there is a notable collection of inselbergs with varying degrees of floristic isolation (Section 10.2.2).

Within the tropical zone a high-altitude grassland zone, dominated by bunch grasses (*Calamagrostis*, *Festuca* spp.) usually referred to as *páramo*, is widespread through Colombia, Venezuela, Ecuador, northern Peru and parts of northern Argentina where there is high rainfall and humidity throughout the year. This contrasts with the *puna* which has a prolonged dry season that increases in aridity from north to south and from east to west (Troll, 1959). The two types merge with each other and constitute the vast area of high-altitude Peru and Bolivia, popularly known as the *Altiplano*. In the Old World the equivalent of the Altiplano is found in the Ethiopian tableland and the high-altitude steppe of Tibet (Vuilluemier & Monasterio, 1988).

In the tropical to subtropical regions of the Andes, principally in Peru and Bolivia, the high-altitude zones have long been exploited for grazing. There are even historical records from the time of the Spanish Conquest (1528–31) that the flocks of llamas and alpacas were so large that grazing could not be found for them (Garcilaso de la Vega, 1608). Historically therefore, the upper limit of the treeline (4500 m) has been depressed by biotic factors and most of the land over 4000 m has been treeless for a very long time (Crawford *et al.*, 1970).

10.6   THE WORLD'S HIGHEST FORESTS

In the high-altitude plains of the Andes there is one outstanding exception to the general absence of trees and that is high-altitude forest patches dominated by *Polylepis* spp. The genus *Polylepis* comprises a group of tree species belonging to the rose family. Fifteen species of *Polylepis* grow in South America and occur from northern Venezuela to northern Chile and Argentina.

The highest number of species is found in Ecuador, Peru and Bolivia. The highest locations, however, occur on the long extinct Bolivian volcano Sajama (6542 m; Fig. 10.23), which can claim to have the world's highest forest with occurrences up to 5100–5200 m (Purcell *et al.*, 2004; Hoch & Körner, 2005; Figs. 10.24–10.26). This altitude record is well in excess of other possible arboreal alpinists. In southwest China, silver fir (*Abies squamata*) has been recorded at 4600 m with heights of 15.2 to 36.5 m, and in the Himalayas, the Indian paper birch (*Betula utilis*) has also been found at this altitude (Duncalf, 1976).

The growing season for the high-altitude *Polylepis* trees is extensive, extending over 265 days with minimal seasonal variation in soil temperature (Fig. 10.24) and no evidence of carbon depletion in their shoots despite the high altitude (Fig. 10.27).

The *Polylepis tarapacana* forests found in Bolivia are unique with respect to their altitudinal distribution. Given the extreme environmental conditions that characterize these altitudes, this species has to rely on distinct mechanisms to survive stressful temperatures. In this region there are two seasons, one dry and cold and the other warm and wet. The trees show a supercooling capacity (−3 to −6 °C for the cold dry and paradoxically −7 to −9 °C for the wet warm season) with tissue injury becoming apparent at temperatures of −18 to −23 °C in both seasons. The increase in supercooling capacity appears to be a consequence of

Fig. 10.23   The site of the highest elevated trees in the world. The Nevada Sajama (left) and the Cerro Jasasuni (right) in early morning light. Sajama is the highest mountain in Bolivia at 6542 m. *Polylepis tarapacana* grows up to about 5100–5300 m on Nevado Sajama. Cerro Jasasuni is about 4900 m and the summit is one of the sites established for the GLORIA study of long-term changes on alpine environments (www.gloria.ac.at). (Photo Dr L. Nagy.)

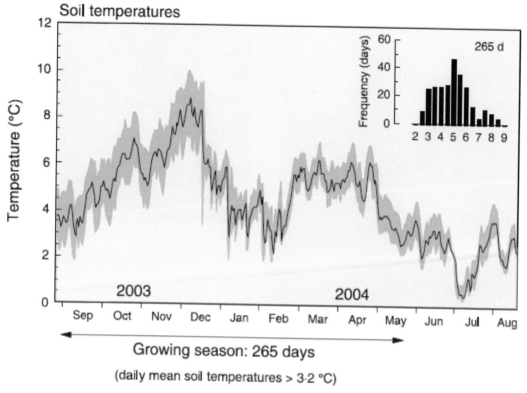

Fig. 10.24 Illustration of low seasonal range in root zone temperatures (−10 cm) under full canopy shade of *Polylepis* trees. Values are the mean of two independent measurements between 25 August 2003 and 1 September 2004 at 4780 and 4810 m a.s.l.) Black curve, daily mean temperatures; grey area, daily temperature amplitudes (diurnal minimum and maximum temperatures). Histogram shows frequency distribution of daily mean soil temperatures across the growing season. (Reproduced with permission from Hoch & Körner, 2005.)

Fig. 10.25 View of Cerro Jasasuni after a hailstorm that made the *Polylepis* trees readily visible. (Photo Dr L. Nagy.)

Fig. 10.26   Details of *Polylepis tarapacana*. (Above left) An isolated specimen surviving at *c*. 4700 m in tussock vegetation that has been subject to fire and vicuña grazing. (Above right) Detail of foliage. (Below) A tree island with prostrate, upright and layered forms growing at 4800 m near Sajama. (Photo Dr L. Nagy.)

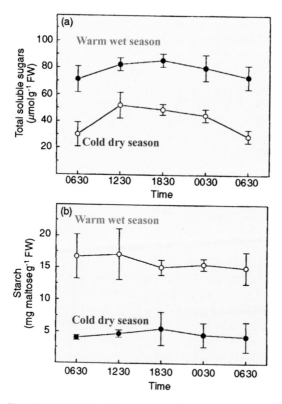

Fig. 10.27   (a) Total soluble sugar and (b) starch content in leaves of *Polylepis tarapacana* during a 24 h cycle in the wet warm and dry cold seasons. (Reproduced with permission from Rada *et al.*, 2001.)

the significant rise in total soluble sugar and proline contents during the wet warm season (Fig. 10.27). The trees in the altitude-record-holding location are therefore highly frost tolerant, and especially so during periods of metabolic activity in the warm wet season when night frosts are always a hazard at this altitude (Rada *et al.*, 2001).

10.6.1 The Peruvian Highlands

Taking the Peruvian Highlands as an example, most of the land above 4000 m is a high-altitude tableland generally described as the *puna* in which two major zones can be recognized (Smithsonian Museum, 2006).

(1) *Puna*, from 3500–4200 m: the predominant vegetation type consisting of grasslands with various species of bunchgrasses (*Calamagrostis*, *Festuca*

and *Stipa*) forming tussocks that can reach 1 m in height (if not burned or grazed). In the dry puna, *Festuca orthophylla* predominates. In the wet puna on the Eastern Cordillera, several species of *Cortaderia* can be important particularly along watercourses. Spaces around the tussocks are often filled by a number of herbs, including non-tussock-forming grasses and sedges, prostrate or low-growing forbs, lichens, mosses, and ferns and their allies such as *Jamesonia* and *Lycopodiella*. Up to 4400 m, especially in the moist puna of the Western Cordillera, some forest patches dominated by *Polylepis* spp. can also be found.

(2) *True puna*, above *c.* 4200 m: grasses begin to lose their dominance on well-drained substrates. The altitudinal zone between 4200 and 4800 m is sometimes described as the true puna. Here the predominant life-forms in grazed areas are prostrate, cushion and rosette herbs of genera such as *Azorella*, *Baccharis*, *Daucus*, *Draba*, *Echinopsis*, *Gentiana*, *Geranium*, *Lupinus*, *Nototriche*, *Plettkea*, *Valeriana* and *Werneria*. Elevations above 4800–5000 m have vegetation types with these same taxa dispersed as individual plants. Often bare ground predominates, especially on scree slopes made unstable by needle-ice (Smithsonian Museum, 2006).

As is evident from the above descriptions the species in these upland regions have wide altitudinal ranges. A detailed numerical analysis of seasonally dry mountain scrub carried out in the Vilcanota Valley in Peru identified the main constituents of high, middle, and low, altitude groups and showed that notwithstanding the existence of particular species combinations at different altitudes many species were widespread in their altitudinal distribution. Here again the dominant factors are moisture, grazing and erosion rather than temperature (Crawford *et al.*, 1970).

10.7 HIGH MOUNTAIN PLANTS AND CLIMATE CHANGE

Concern is often expressed for the future fate of mountain vegetation should there be sustained and marked climatic warming. The specific sources of danger fall into two categories:

(1) an upward migration of sedge heaths and forest that will eliminate the subnival and high-alpine

communities from mountain summits (Grabherr et al., 1994)

(2) the disappearance of snow patches and their associated species in the subnival zone (Guisan et al., 1995).

In the Central European Alps the vertical extent of the nival zone may be sufficient to accommodate much of the expected upward migration of the alpine flora and thus reduce any imminent danger of species extinctions in these regions. Also, within the alpine zone in higher mountains, there is probably sufficient variation in microclimate to accommodate the changes that may arise from global warming. It has therefore been argued that in these areas there are at present few examples of species that are likely to become extinct in the higher mountain ranges within the next century (Körner, 1995).

A situation similar to that observed in the Alps has also been observed in the Jotunheimen range of central Norway. A resurvey in 1998 of sites which had detailed site descriptions and species lists from 1930–31 showed that there had been an increase in species richness on 19 of the mountains, a tendency that was most pronounced at lower altitudes and in eastern areas (Klanderud & Birks, 2003). The greatest increases in abundance and altitudinal advances since 1930–31 were found in lowland species, and particularly in dwarf shrubs and species with wide altitudinal and ecological ranges. However, species with more restricted habitat demands, such as some hygrophilous snow-bed species, were found to have declined. High-altitude species have also disappeared from their lower-elevation sites and have increased their abundance at the highest altitudes.

The investigators suggested that recent climatic changes are likely to be the major driving factor for the changes observed (Klanderud & Birks, 2003). Nevertheless, in any assessment of change in mountain floras it is important to take into account that there have also been significant alterations in land use over the past century which are not without their effects on mountain vegetation. In particular, the reduction in grazing by goats in Norway has probably benefited biodiversity in many upland marginal habitats, especially around former summer grazing stations (saeter). In the Alps, this century has witnessed an increase in species richness on mountain summits. On Piz Linard (3411 m), a

pyramid-shaped mountain in the Swiss Alps, there was only one summit species of flowering plant in 1835, four in 1895 and 10 in 1992 (Pauli et al., 2003). A similar study on the summit flora of Mount Glungezer (2600 m) showed that in just 13 years, from 1986 to 2000, which included the 10 warmest years on record in central Europe, the total of 55 species recorded in 1986 had increased to 88 without any losses (Bahn & Körner, 2003).

Despite these observations of increasing biodiversity, which relate principally to the European Alps and Scandinavia, there are other regions of the world where climatic warming over the past 100 years has facilitated the invasion of mountain sites of modest elevation and lacking extensive nival zones by certain aggressive species, to the detriment of the diversity of the alpine flora. In particular the range extension uphill of grasses has resulted in increased competition at such sites, leading to a decreased abundance of the less competitive alpine species. Increased deposition of nitrogen during recent years and changes in grazing and tourism might also have influenced some of the species turnover. It is in these lower and less-studied mountains in particular that examples can be found of species that are on the verge of extinction. In the French Massif Central, the second highest mountain is the Mezenc (1754 m). Just under the summit of the mountain on its north-facing slope (Fig. 10.28, upper panel) grows the last French colony of the rare mountain Senecio leucophyllus. The few remaining plants do not exhibit any loss of vigour or reproductive capacity as a result of climatic warming, but nevertheless appear unlikely to survive due to the invasion of their preferred habitat by grasses (Fig. 10.28, lower panel). Fortunately, the species still exists in the eastern Pyrenees and its likely disappearance from its last French locality will not lead to its extinction.

Although the heterogeneity of the larger and higher mountain systems will provide adequate space for many mountain species even in a warmer alpine world there are nevertheless certain high-altitude areas where deleterious changes to high-altitude vegetation are already evident as a result of recent climatic alterations. Where the climate change is towards drier conditions, adverse effects are common irrespective of temperature change (see also Chapter 5). In the high alpine pastures of southern Tibet, between c. 5000 to 5300 m, up to 30 cm thick Kobresia pygmaea mats

Fig. 10.28   Species extinction on the isolated summit of the Mezenc (1754 m) in the French Massif Central. (Above) Just under the north side of the summit the arrow marks the last location in France of the rare *Senecio leucophyllus*. Increasing grass cover and advancing trees threaten to cover the open scree habitat on which this species grows. (Below) *Senecio leucophyllus* on the verge of extinction due to environmental change favouring the expansion of grasses onto montane scree.

dominate the south-eastern humid quarter of the Tibetan Highlands where there is a water surplus (Miehe, 1989). This formation developed during younger more humid Holocene phases (the 'Kobresia pygmaea age') but has now been widely destroyed by the Himalayan föhn (warm, dry catabatic wind on the lee side of a mountain range) which creates conditions only suitable for semi-arid alpine steppe on stone pavements. There is a recently observed decreased vigour and lack of regeneration of this alpine turf which is a relict of a past climatic optimum. In the Karakorum Mountains the decrease is particularly evident in plant vigour at climatically sensitive vegetation borders. It is at the upper treeline that drought limits forest survival, as well as in the transition zone between humid alpine mats and alpine steppe. Diminishing winter precipitation during the last decades of the twentieth century has adversely affected these high-altitude plant communities (Miehe, 1996).

10.7.1 Indirect effects of increased temperature on alpine vegetation – reduction in winter snow cover

The second major danger to alpine plants from climatic warming is indirect, and is related to the reduction in snow cover that might take place should global temperatures rise. Plants that are adapted to survive under snow are generally considered to be less frost resistant than those that live in exposed habitats.

There are few detailed studies to corroborate this general assumption. However, a New Zealand study of the annual course of frost resistance of species of native alpine plants from southern New Zealand that are normally buried in snow banks over winter (e.g. *Celmisia haastii*, *C. prorepens*, *Hebe mora*) with other species, typical of more exposed areas, that are relatively snow-free (e.g. *Celmisia viscosa*, *Poa colensoi*, *Dracophyllum muscoides*) has confirmed the assumption that species from snow banks or sheltered areas have the least frost resistance (Bannister *et al.*, 2005).

Away from the snow banks it is the loss of resistance in late winter to early summer (August–December in New Zealand) that is most likely to expose the plants to injury (e.g. *Poa colensoi* and *Dracophyllum muscoides*). However, in the principal species examined, seasonal frost resistance was more strongly related to day length than to temperature. This phenological control appears

to be sufficient to ensure that frost resistance is likely to be unaltered by climatic warming as the relationship of frost resistance to day length prevents frost damage at any time of year (Bannister *et al.*, 2005).

Notwithstanding the above conclusion, absence or early loss of winter snow is likely to affect water availability to mountain plants and increase the risks of desiccation injury as the season advances and especially if summer temperatures also rise (see below). In Sweden, where the snow-bank-dependent *Vaccinium myrtillus* suffered lethal injuries during a 5 °C warmer-than-average winter, the damage was associated with a reduction over winter in shoot solute content brought about by the progressive respiratory loss of cryo-protective sugars (Ögren, 1996). There would appear therefore to be two classes of mountain plants in relation to susceptibility to warmer winters at high altitudes: one from exposed ridges where frost tolerance is photoperiodically induced and will therefore persist even in a warmer climate, and a second where cover under snow is essential and where exposure as a result of warmer winters will lead to a loss of frost tolerance due to desiccation.

10.7.2 Effects of increased atmospheric CO_2 on high mountain vegetation

The current change in atmospheric greenhouse gases inevitably raises the question as to whether increasing global levels of carbon dioxide will be beneficial to high-altitude vegetation. The low growth rates and small stature of plants from high altitudes do not suggest that growth is limited by carbon dioxide availability. In addition, comparisons of related species from low and high altitude sites in the European Alps have shown that the possession by high-altitude species of thicker leaves, with well-developed palisade layers more than one cell deep (see Section 3.2.1), together with a higher protein content (largely ribulose carboxylase, RuBisCO) suggests that the alpine species are well adapted to utilize current levels of atmospheric carbon dioxide even when growing at high altitudes where the partial pressure of carbon dioxide is reduced.

Morphological adaptations are noticeable in relation to gas exchange in high-altitude plants. Alpine species show higher levels of both ab- and ad-axial stomatal density. These morphological and physiological adaptations enable plants to compensate metabolically for low temperatures, and also the brevity of the growing season. A similar phenomenon with increased enzymatic capacity is seen in the low temperature muscular physiology of fish from cold waters and is described as *capacity enhancement* irrespective of whether it is due to morphological or metabolic adaptations or both (Hochachka & Somero, 1973). Due to capacity enhancement, many species from high altitudes have higher rates of CO_2 uptake as determined on a leaf area basis than their related lowland species (Körner *et al.*, 1989). The greater efficiency of thicker leaves in alpine plants in taking up carbon dioxide at high altitudes can be likened to the greater vital capacity of the lungs of Quechua Indians who have been born and raised at high altitudes, in that they have a greater lung vital capacity (the volume change of the lung between a full inspiration and a maximal expiration) which aids their uptake of oxygen (Heath & Williams, 1977).

Notwithstanding the efficiency of high-altitude plants in taking up carbon dioxide there still remains the question of whether or not increased levels of carbon dioxide will have any effect on carbohydrate supply and growth on high-altitude vegetation. Comparisons between related high and low altitude species, e.g. *Ranunculus glacialis* and *R. acris* and the similar altitudinally distinct *Geum reptans* and *G. rivale* grown in CO_2-enriched atmospheres (Fig. 10.29), have demonstrated that alpine plants may be able to obtain, at least initially, greater carbon gains from increased carbon dioxide availability than comparable lowland plants (Körner & Diemer, 1994). Whether or not increases in RuBisCO activity will lead automatically to increased productivity is open to question. Decreased rates of starch accumulation in leaves may actually allow a more efficient use of fixed carbon as lower amounts of RuBisCO can make available nitrogen that would otherwise be limiting growth due to sequestration in Rubisco protein. In perennial plants from arctic and montane habitats, growth is not directly related to photosynthesis and many plants retain a small stature despite their photosynthetic activity. Futile cycles that oxidize sugars without generating ATP can operate and reduce metabolic efficiency, a process that has been studied in the boreal cold desert shrub *Erotia lanata* (Thygerson *et al.*, 2002). Thus, even though alpine plants may increase photosynthetic activity with elevated CO_2 levels, this carbon gain may be dissipated by futile cycles and not necessarily translated into higher biomass production. It follows also that if lower levels of RuBisCO can increase growth then higher levels of RuBisCO may cause a reduction. In many habitats a reduction in growth and an increase in stored carbohydrate from increased photosynthetic activity could have a significant effect on survival.

In uncertain environments, such as the neighbourhood of snow banks that do not always melt promptly in spring, starch accumulation without growth is important in aiding plants to overcome long non-productive periods. High levels of soluble carbohydrates are also necessary to provide cryo-protectants for plants in cold regions. Some plants that live in snow patches can survive 2–3 years without emerging from the snow bank (Pielou, 1994) and such a feat is dependent on adequate carbohydrate reserves. If, in addition to being covered in snow the plants have to endure the hazard of also being encased in ice, this can induce a prolonged period of anoxia which places additional demands on carbohydrate reserves (see Section 3.6.3).

It would appear that the enhanced capacity adaptation (whether of enzymatic or morphological origin) that ensures that mountain species have adequate supplies of carbon dioxide even at high altitudes would make it unlikely that any increased in levels of atmospheric carbon dioxide will have any great effect either on growth or survival given their intrinsic low growth rates. In a number of ways the plants that live at high altitudes present an example of the long known phenomenon of the Montgomery effect, which asserts that 'ecological advantage is conferred by low growth rates in areas of low environmental potential' (Montgomery, 1912)'.

In this respect it is relevant to examine the results that have been obtained experimentally by providing increased levels of atmospheric carbon dioxide to trees growing at the treeline. Historically, it has been argued that carbon, through a shortage of photo-assimilates, limits the growth of trees at the upper altitudinal treeline. Re-examination of this possibility in a wide range of alpine habitats has failed to reveal any clear metabolic evidence of carbohydrate limitation to plant growth either in alpine vegetation or in trees at the upper limits of distribution at the treeline (Körner, 2003).

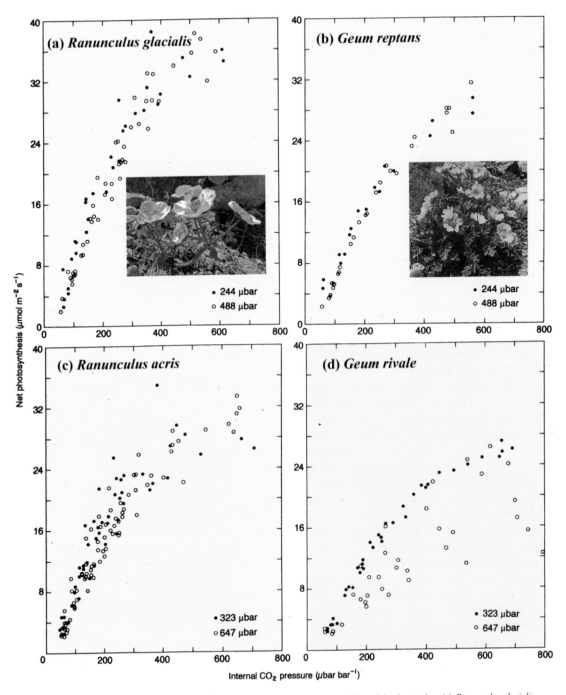

Fig. 10.29   Effects of increased atmospheric CO_2 on pairs of comparable high and low altitude species. (a) *Ranunculus glacialis* and (b) *Geum reptans* from high altitudes; (c) *R. acris* and (d) *G. rivale* from low altitudes. (Reproduced with permission from Körner & Diemer, 1994.)

Despite this apparent satiation of the current needs for photo-assimilates in high-altitude plants there nevertheless remains the possibility that future increases in atmospheric concentrations of carbon dioxide may stimulate plant growth at high altitudes where the reduction in partial pressure of all constituents of the atmosphere might make additional CO_2 beneficial. In a three-year free-air CO_2 enrichment (FACE) experiment, two species of 30-year-old alpine conifers (*Larix decidua* and *Pinus uncinata*) were studied *in situ* in the Swiss Central Alps (2180 m above sea level). Elevated CO_2 enhanced photosynthesis and increased non-structural carbohydrate (NSC) concentrations in the needles of both species (Fig. 10.30). While the deciduous larch trees showed longer needles and a stimulation of shoot growth over all three seasons when grown *in situ* under elevated CO_2, pine trees showed no such responses. The study also involved the removal of needles to determine if defoliation stimulated photosynthesis either in the

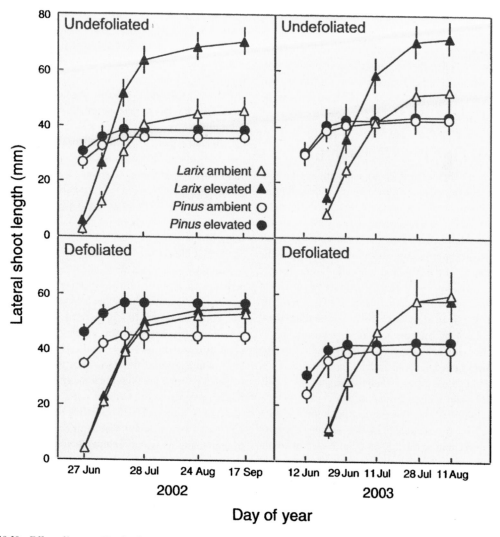

Fig. 10.30   Effect of increased levels of atmospheric carbon dioxide on lateral shoot extension ($n = 5$) over 2 years in undefoliated and defoliated trees of *Larix decidua* and *Pinus uncinata* growing at the timberline in the Swiss Alps. (Reproduced with permission from Handa *et al.*, 2005.)

current-year needles or those produced in the following year. Defoliated larch trees had fewer and shorter needles with reduced NSC concentrations in the year following defoliation and showed no stimulation in shoot elongation when exposed to elevated CO_2. By contrast, defoliation of the evergreen pine trees had no effect on needle NSC concentrations, but stimulated shoot elongation when defoliated trees were exposed to elevated CO_2.

The conclusion after three years of this study suggests that deciduous larch is carbon limited at treeline, while evergreen pine is not (Handa *et al.*, 2005). Whether this extends to other treelines where both deciduous and evergreen trees can be found is an intriguing question. If the response to additional CO_2 depends on whether the trees are evergreen or deciduous it is probable that other factors such as growing season length, temperature, moisture and nutrients will also be involved.

10.8 ALPINE FLORAL BIOLOGY

No account of mountain vegetation is complete without some discussion of alpine floral biology. Alpine flowers attract attention even from the most casual mountain visitor. A similar situation exists in the Arctic where the flora is less diverse and individual plants are somewhat smaller than those in alpine regions. Nevertheless, the visual impact of the Alps and tundra in flower in early summer is an impressive tribute to the sexual activity of plants in cold climates. Why plants of these thermally limited marginal areas have such outstanding floral displays has long been a source of wonder and speculation. In many habitats, such as cliffs and screes, the plant cover in general is sparse and therefore the floral parts of individual plants and clonal patches are vividly displayed against a background of bare rock or gravel.

Visibility of the flowers to the human eye is enhanced by the reduction in size of the vegetative organs in relation to the flowers (Fig. 10.31). It would appear that reduction in leaf size reduces the hazards of exposure but a similar reduction in flower size would result only in a flower that would not fulfil its primary function in attracting insect pollinators. To the human observer and also to insects the visibility of alpine and arctic flowers is enhanced as they are frequently borne in clustered displays on creeping and cushion plants.

The ability of flowering parts to fulfil their normal function of facilitating outcrossing and seed production could be problematical in many marginal alpine and arctic habitats. As already discussed, there are limits to the ability of plants to reproduce sexually, which may arise from numerous causes, grouped under two headings, pre-zygotic and post-zygotic limitations (see Chapter 4).

In pre-zygotic limitations there is the possible lack of suitable vectors for pollination and even if pollination does take place, fertilization of the ovule may not be achieved, either because the pollen grain does not germinate or else the pollen tube fails to reach the ovary. Under post-zygotic causes of reproductive failure there may be low seed viability, failure to germinate and destruction by predators. In marginal areas seeds frequently do not mature in short growing seasons, and even if they do eventually germinate, the young seedlings may fail to become established in the plant community.

In cold regions it might therefore be expected that there may not be sufficient insect pollen vectors and that wind might provide a more reliable form of pollen dispersal. Surprisingly, wind pollination is not as prevalent in alpine and arctic floras as might be imagined. However, insects are ubiquitous. They succeed in reaching some of the coldest regions of the Earth, where they adapt their lifestyles to mitigate the effects of the diurnal temperature oscillations of mountain habitats or seasonal activities to profit from the brief periods of summer warmth in the arctic and alpine habitats.

Pollen transport that is targeted in its delivery is much more efficient than that which is thrown into the wind. It is sometimes suggested that the open nature of high mountains and tundra is ideal for the wind dispersal of pollen. However, the ferocity of the winds and the open expanses of rock and water that make up so much of these terrains will inevitably make wind pollen dispersal a very wasteful process. The thermal climate of alpine and arctic environments when viewed from the standpoint of insects and vegetation is not so adverse during the flowering season as might be expected. The diminutive vegetation lives in a microclimate close to the ground that is noticeably warmer than the air temperature. This is the case especially for bulky organs such as stems, cones and buds, which heat up significantly in bright sunshine, as has been shown

Fig. 10.31   The alpine forget-me-not or king of the Alps (*Eritrichium nanum*). A species of high mountains occurring between 2500 and 3600 m in the Alps and a noted example of maintaining the alpine habit of having a highly visible flower display which when in full flower almost totally obscures the diminutive foliage. (Photo by courtesy of Professor R. M. Cormack.)

in pine trees growing near the treeline in Scotland by inserting extremely fine thermocouples into the tree buds in their natural environment. For mature pines the bud temperatures were 4 °C warmer than the air by day, and even warmer in the dwarf *krummholz* pines near the treeline (Grace, 2006). The insects of these regions also inhabit this same thermal space, where they crawl over the plants rather than fly. This restricted movement not only aids their metabolic activity, but also reduces the risk of being blown away by wind. Sun-tracking flowers such as the glacier buttercup (*Ranunculus glacialis*) maximize their sun trapping capacity with the parabolic reflectance of the flowers concentrating radiation on the reproductive organs and thus enticing insect pollination (Fig. 10.32). The flowers are white while they are required as reflectors but after fertilization the petals close to protect the developing seed and turn red, which will increase radiation absorbance (Fig. 10.33). Some controversy has taken place as to whether there are both red and white forms of *R. glacialis* or whether the red colour is a post-fertilization phenomenon when the

Fig. 10.32   Sun-tracking (heliotropic) flowers of *Ranunculus glacialis*. The flowers act as parabolic reflectors concentrating radiation in the centre of the flower and thus attracting pollinators. (Photo Professor R. M. Cormack.)

petals close around the flower and absorption rather than reflectance has a greater efficacy in absorbing heat. In Fig. 10.33 the plant on the left is still showing some white open flowers while the closing flowers are turning red. It appears, however, that both cases exist with some plants having red flowers before they are fertilized. (See also Chapter 4.)

Although insect pollinators are present at astonishingly high altitudes and latitudes, and even the purple saxifrage (*Saxifraga oppositifolia*), the earliest flowering plant in many high mountain sites, is an insect-pollinated species, their abundance is much diminished compared with temperate habitats. It is

therefore not surprising that the flowering plants of these regions have a number of adaptations that can compensate for the potential dangers of this deficiency. Among the most common is self-compatibility. However, persistent self-fertilization will reduce genetic variation which could be unfavourable for population persistence in highly stochastic environments. An evaluation of these contradictory scenarios for the advantages of self-pollination as opposed to out crossing was carried out with *in situ* pollination experiments on the predominantly out crossing species *Saxifraga oppositifolia* in the Swiss Alps at sites where pollinator limitation had been detected (Gugerli, 1998).

Fig. 10.33 Diversity of colour in *Ranunculus glacialis*. Some controversy has taken place as to whether there are both red and white forms of *R. glacialis* or whether the red colour is a post-fertilization phenomenon when the petals close around the flower and absorption rather than reflectance has a greater efficacy in warming the developing seeds. The plant on the left is still showing some white open flowers while the closing flowers are turning red. It appears, however, that both cases exist with some plants having red flowers before they are fertilized. (See also Section 3.1 and Fig. 3.3.) (Photo by courtesy of Professor R. M. Cormack.)

Hand-crossings and hand-selfings yielded seed set at both elevations that were studied. It was concluded that this constant pattern of the breeding system in *S. oppositifolia* indicated that selective factors must exist that lead to the maintenance of a high level of out crossing even in high-elevation populations. Such findings can be claimed as contradicting the hypothesis that inimical environmental factors in alpine or arctic habitats necessarily select for increased selfing rates in a preferentially out crossing species. Further discussion of the capacity for sexual reproduction in marginal areas including alpine and arctic habitats can be found in Chapter 4.

10.9 CONCLUSIONS

It has been said that evolution defies the laws of physics and that enzymatic processes liberate living organisms from the dictates of the Arrhenius equation (Barcroft, 1934). If such a statement is justified it might be claimed to be an achievement of high mountain vegetation. It is one of the wonders of nature that plants can survive on the highest mountains in the world while depending on a metabolism based on the making and breaking of the covalent bonds of carbon, which take place spontaneously only at temperatures of 60–70 °C. This achievement has to be judged not in terms of growth rates or biomass accumulation – these are only attributes of interest to potential herbivores (the human race included). From the point of view of the plant itself, growth rates and biomass are characteristics that can be modified to suit their environment. They are also characteristics that can be acted on selectively with relative ease. Twenty million years ago New Zealand was a collection of low-lying tropical islands with a forest dominated by the giant *Podocarpus* trees. In a space of only ten million years, as the mountains of New Zealand's southern island rose ever higher, so did the flora evolve to occupy these new mountain climates. The same genus that contains the giants of the lowland forest also has evolved its high-altitude counterparts in the diminutive snow totara (*Podocarpus nivalis*; see Fig. 9.25). Given liquid water and light for a few weeks in the year there is scarcely any habitat to which the flowering plants can not adapt, such is their ability to overcome the apparent obstacles of the physical world.

Fig. 11.1   The Icelandic upland farm of Hraun in Öxnadalur (northern Iceland) in June. Birthplace of the Icelandic poet and botanist Jónas Hallgrímsson (1807–1845). A modern example of tenacity of occupation in a marginal environment where late springs and unpredictable growing seasons were a particular feature during the Little Ice Age.

11 · Man at the margins

11.1 HUMAN SETTLEMENT IN PERIPHERAL AREAS

Adaptability is a feature of mankind, with the result that even the most marginal habitats are seldom devoid of human settlements. Furthermore, once having populated a peripheral area the community often appears reluctant to abandon what they have come to consider as their home territory, even when deteriorating environmental conditions have brought prolonged periods of hardship. Ecologically, the range of habitats in which prehistoric human settlements developed knew few boundaries. There is increasing evidence that the first human beings to cross into North America used coastal migration routes even before the late Pleistocene ice had retreated from the Beringian shores (Shreeve, 2006). It is therefore not surprising that many early human communities can be found on offshore islands, marshes, and flood-prone deltas. Archaeological records from the poles to the tropics have revealed the rapidity with which our early ancestors spread through an astonishing range of marginal environments.

Peripheral areas can present both opportunities and risks for their human inhabitants. On one hand, historical and archaeological records provide numerous examples where fluctuating climatic conditions opened up areas of opportunity, which attracted migration and settlement. On the other hand, fluctuating conditions could lead to crop failure or the collapse of other resources such as hunting or fishing and force the abandonment of areas that were formerly attractive. Such events are taking place today at an ever-increasing rate, particularly in regions where large human populations are having their agricultural and pastoral activities destroyed by prolonged droughts, possibly connected with global warming. For others, rising sea levels threaten to erode or inundate their lands.

Current climatic warming is, however, not universally damaging for agriculture. In many northern regions there is a feeling that at last they are free from the adverse effects of the Little Ice Age (Fig. 11.1). The growing season for grass is extending and wheat and other crops are being cultivated further north than ever before. Boreal forest timber production is also benefiting in many locations due to increased tree growth.

In marginal settlements, irrespective of whether they are inland or coastal, there have always been special characteristics which enable the inhabitants to survive in potentially insecure and variable habitats. The essential aspect common to them all is the ability to plan for the future and foresee the possibility that times of plenty can be followed by times of hardship. Man's unique ability to anticipate the future is particularly relevant in the Arctic. The hunter-gatherer communities that live in the far north are obliged to store food for the winter and so have had to develop a large number of social strategies, including the control of property, together with scheduling, and co-ordinating labour to a degree that is not required in tropical or temperate climates (Ingold, 1982).

The development of a viable lifestyle for these early pre-agricultural communities, wherever they were, was influenced by the nature of the vegetation and the animals that it supported. This is well illustrated in the Late Glacial and early Holocene period when the inhabitants of polar regions, with the exception of the coastal regions where marine resources were available, relied on the larger herbivores (mammoth, reindeer, musk oxen) for their nutritional demands, and these animals in turn required substantial plant forage.

The archaeological history of *Homo sapiens* in the Late Pleistocene reveals that hunters spread across northern Eurasia and Alaska to the edges of the ice sheets in pursuit of the 'megafauna'. These large herbivorous mammals were dependent on a single hyper-zone of vegetation described botanically as *tundra–steppe* that existed where today there is modern steppe, taiga, and tundra (see below).

With the development of agriculture, many examples can be found where prehistoric peoples showed considerable resourcefulness in adapting to habitats in areas that today might be considered marginal for agriculture. In the semi-arid montane regions of South America, ancient, specialized agricultural traditions maintained extensive human populations in a region where readily cultivable land and rainfall is severely restricted. In the Andes, a tradition of irrigation and terrace-farming long predates the Inca civilization, with evidence of terrace construction beginning probably as early as 2400 BC (see Denevan, 2001).

Many other marginal habitats have long histories of human habitation, where either early agriculture was practised using specialized techniques, or where use was made of the natural plant communities to compensate for a limited ability to grow crops. Marshes, mangrove swamps, storm-swept coasts, and drylands in many parts of the world have supported individualistic forms of human settlement. Striking examples of human ingenuity of extracting crop productivity out of marginal habitats are found in the raised fields of the wetlands around Lake Titicaca in the Bolivian Highlands (Denevan, 2001), the lazy beds of the Celtic western fringe of the British Isles and Ireland, and the ancient settlements (possibly dating back to the time of Sumerians in the fourth millennium BC) of the Marsh Arabs on the wetlands of the Tigris-Euphrates alluvial salt marshes.

Africa provides numerous and varied cases of peoples who live in semi-arid lands with the constant risk of rain failure and ensuing starvation. The transition zones between forest and savanna, and savanna and semi-desert, are physically very fragile and delicately balanced ecosystems. In these marginal areas, human activity can damage the ecosystems beyond their limits for recovery. Excessive soil trampling of the ground by livestock compacts the soil, increases the proportion of fine material, and reduces the percolation of rainfall. The increasing run-off when the rains return brings about erosion, first by water and then by wind. Excessive grazing and the collection of firewood in periods of drought reduce or even eliminate the plants that help to bind the soil. Consequently, in many regions it is the increased human population and livestock pressure that has caused desertification and not climate change.

In times of drought nomads tend to move to less arid areas and thus disrupt the local ecosystem and increase the rate of erosion of the land. In trying to escape the desert the nomads bring the desert with them. Sometimes the habitat degradation that is caused by the drought is described as desertification. However, this is a misconception and can be misleading, as the natural vegetation of these semi-deserts is well adapted to withstand periodic rain failure and prolonged drought. Well-managed lands can recover from drought when the rains return. It is land abuse during drought that causes the degradation and desertification.

Despite these evident hazards, marginal areas have nevertheless always attracted human occupation. In most cases the inhabitants are well aware of the dangers and have developed lifestyles that minimize the impact of the risks to which they are periodically exposed. The motivation for accepting these conditions can vary. Escape from oppression or persecution, avoidance of competition with other peoples and tribes, and a sense of independence and pride in their own fortitude in overcoming obstacles that are a deterrent to less adventurous neighbours, are all factors that contribute to the continued occupancy of marginal regions.

The following pages examine some individual case histories in an attempt to assess their relative vulnerability and the role of plants in sustaining human occupation of these peripheral areas. How long such human communities will survive in a modern world with a global economy and changing climatic conditions will depend in great measure on the strength of cultural attachment of the inhabitants to these peripheral lands and the value that is placed on the maintenance of their individualistic lifestyles, cultures, and perceived freedom.

11.2   PAST AND PRESENT CONCEPTS OF MARGINALITY

The concept of marginality in relation to land use varies historically. In the modern world, where local economies are affected by the existence of competitors in distant parts of the globe, the successful pursuit of agriculture depends on more than just the fluctuations of local environmental conditions. The concept of marginality in relation to modern land use therefore requires some elaboration. An area that may be marginal for wheat production may not be marginal for other cereals such as oats or barley. In relation to farming *marginal* is a comparative term and any particular situation can be considered as marginal only where there is a related core area where yields are greater and the risks of crop failure smaller than at the putative margin. This aspect of marginality is particularly relevant when comparing the historical and prehistoric use of land with modern farming practices. The remains of Neolithic settlements in parts of northern Europe, as in northern Norway, the Northern Isles of Scotland (Fig. 11.2), demonstrate that despite an agriculturally challenging environment, these early settlers maintained a viable existence. In Orkney, there were sufficient resources to support a population that could create substantial buildings and long-lasting monuments (Fig. 11.3). Estimates of the size of the Neolithic settlement of Orkney suggest a population of

Fig. 11.2 The Knap of Howard, Papa Westray, Orkney (see inset). These two oblong stone-built houses, preserved by wind-blown sand are the earliest North European dwellings known, dating back to 3800 BC. They were occupied by Neolithic farmers for 500 years, furnished with hearths, pits, stores, stone and possibly wooden benches. From midden remains the mode of subsistence was primarily pastoral, rearing cattle, sheep and pigs. There is some evidence of cereal cultivation and harvesting of fish and shellfish (see www.papawestray.co.uk). The easily cultivable soils and the additional ready access to marine resources would have made the first farmers self-sufficient in the Orkneys and their agricultural viability should not be considered as in any way marginal.

Fig. 11.3   The Ring of Brodgar, Orkney. The monument was erected probably between 4500 and 4000 BP and covers an area of 8435 square metres (90 790 square feet). Estimates for the labour required to excavate the massive ditch surrounding the Ring which is 3 m deep have been placed at 75 000 man-hours (Renfrew, 1979). These structures and the extent of other Neolithic remains suggest that the population of Orkney at this time may have been around 20 000 which is similar to that of today (Wickham-Jones, 2006).

possibly 20 000 (Wickham-Jones, 2006), which is not unlike that of today (*c.* 19 200).

A modern economic burden that did not exist in the past stems from the fact that farmers in peripheral regions (by comparison with other more densely settled regions) can become economically marginalized as they have to compete with areas which are more productive, as well as with regions where lower labour costs reduce world prices. Increased expectations in the quality of life, and a dependence on transport for goods and people, are additional modern problems. In these cases, communities that have become agriculturally and economically marginal due to external factors will fail unless financial support is obtained either in the form of subsidies, or by marketing of non-agricultural activities, such as shooting, fishing and other recreational pursuits that can be exploited commercially (Figs. 11.4–11.5).

11.2.1   Agricultural sustainability in marginal areas

Marginal areas frequently show signs that their limited resources are being over exploited, or else managed in such a way as to cause a gradual and sustained deterioration of the environment. This is not a recent phenomenon. Evidence of soil exhaustion around early

agricultural settlements is often noted in archaeological excavations. The proximate cause of abandonment around 1500 BC of an early settlement at the Scord of Brouster in Shetland (see Section 1.8) has been attributed to a combination of soil erosion, podzol formation and increased stoniness of the remaining eroded soil (Whittle *et al.*, 1986).

During the second millennium BC, in the Bronze Age, a considerable expansion of agriculture took place. In Scotland this was particularly notable in upland areas which are inherently vulnerable to even minor fluctuations in climate and human mismanagement through soil impoverishment from clearing, grazing and over exploitation (Cowie & Shepherd, 1997). Thus, a climatic downturn in the latter part of the second millennium, when the climate of Scotland became much colder and wetter, resulted in the abandonment of many of the earlier Bronze Age settlements. Studies of tree-ring chronologies have suggested that the catalyst in bringing about this retreat from upland areas may have been a series of cataclysmic volcanic eruptions. Volcanic eruptions, wherever they occur, spread their dust worldwide in the upper atmosphere from where it disappears only slowly. Despite the global spread the dust produces its most severe effects on climate at higher latitudes, where due to the low angle of incidence of solar radiation towards the poles any

Fig. 11.4   Fair Isle, an isolated island lying midway between Orkney and Shetland with a population of approximately 70
inhabitants. Marginal farming as practised here and elsewhere in the Highlands and Islands of Scotland is termed *crofting* and
provides legal security of tenure for small tenant farmers. The economy of this isolated island is aided by the fame of its knitwear
and the opportunities it provides for bird-watching.

additional attenuation can effect a marked climatic
alteration.

   In more recent times Scotland, together with
Finland, Scandinavia, Estonia and Iceland suffered the

'seven ill years' due to volcanic eruptions. This began in
AD 1693 with an enormous eruption of Hekla, which
had a wide effect across the North Atlantic and into
Scandinavia. This was followed in the same year with

Fig. 11.5   Crofting on Fair Isle. Hay drying on the trestles illustrates the difficulties of securing good-quality hay in a humid oceanic environment.

Serua in Indonesia and Komaga-Take in Japan, and in 1694 by Aboina in Indonesia (Luterbacher *et al.*, 2001). These eruptions took place in the Little Ice Age during a time referred to as the *Maunder Minimum c.* 1645–1715 when sun spot activity was particularly low and solar radiation was already reduced (Langematz *et al.*, 2005). The result at the end of the seventeenth century was a succession of harvest failures that caused widespread famine and large losses of population in all these northern lands. The great famine of Estonia in 1694–97

is particularly remembered for the extensive loss of human life from winter starvation. Such was Scotland's plight that it finally brought about political union with England and the departure of its Parliament to London.

Through historical palaeontological research it is possible to quantify the deterioration of agricultural areas either from records of reduction in productivity or else loss of biodiversity. In oceanic areas with high rainfall, the expansion of poor acid grasslands as a result of a long history of mismanagement and overstocking and erosion has long been a problem. In more arid parts of the world population pressure on the land, both for cultivation and for wood for fuel, has also led to soil erosion and desertification. In many areas limited water resources aggravate this degradation. The alternation of wet and dry seasons can clear the land of plant cover after harvest which then leads to rapid erosion when the rains return.

In conclusion to this study of plants in marginal areas, it is pertinent to consider the varying consequences of human settlement in areas where climatic or soil resources are potentially limiting and where the dual impact of climatic warming and increasing human exploitation may combine to irreversibly damage the plant communities on which the landscape depends for its stability. The inclusion of an historical element in this appraisal is an essential background to understanding the present-day needs for conserving species, habitats and productivity. Marginal areas also attract considerable conservation interest due to their perceived wilderness status and presumed fragility in relation to climatic change. Unfortunately, few marginal areas are true wildernesses, and when they show signs of ecological fragility it is frequently due to denial of sufficient space and time for the proper function of natural regeneration cycles.

11.3   MAN IN THE TERRESTRIAL ARCTIC

Human beings, as noted above, are no newcomers to the tundra. Late Palaeolithic hunters roamed over a landscape that enjoyed ecological conditions very different from those that exist in the tundra today. The warmth that caused the ice retreat after the Last Glacial Maximum produced higher temperatures at the ice–vegetation interface than those that now prevail. Consequently, there developed a more luxuriant vegetation, the *tundra–steppe*, which comprised a combination of plant communities containing both steppe and tundra species (including prostrate shrubs) that had a productivity great enough to support the nutritional needs of the Pleistocene megafauna (Pielou, 1991). Although the climate was predominantly cold and dry there was a much greater diversity of herbaceous vegetation (grasses, sedges and other herbaceous species) than is found in the modern tundra. Dry watersheds and slopes had cold-tolerant herbaceous and prostrate

shrub communities. Meadows in valleys and on slope-pediments were the most productive as pastures for ungulates due to the redistribution of moisture and nutrients within the landscape (Yurtsev, 2001). Vegetation zonation was also different. The latitudinal differentiation into High and Low Arctic had not developed and there was a mosaic of plant communities which added both to habitat productivity and species diversity.

The last remains of this vegetation can still be found on Wrangel Island off the north-east coast of Siberia (Figs. 11.6–11.8) where a continental climate and a lack of moisture kept the land free of extensive glacial ice during the Last Glacial Maximum (Gualtieri et al., 2005). The persistence on Wrangel Island of productive arctic vegetation may explain why a dwarf form of the mammoth was able to survive here at least up to 3700 BP (Kuz'min et al., 2000).

The unrelenting and successful pursuit of mammoth, and woolly rhinoceros, in both the Old and the New World was first described as the *Pleistocene overkill* in the 1960s and suggested as the prime cause for their extinction (Mithen, 2003). However, the dramatic contraction in mammoth range *c.* 12 kyr ago, after which known populations were confined to northern Siberia (mainly the Taymyr Peninsula and Wrangel Island; Figs. 11.6–11.8), has been shown to correlate well with the extensive spread of trees after the Allerød phase (14–13 000 BP) of the Late Glacial Interstadial. The return of open tundra–steppe in the Younger Dryas cold phase, *c.* 11–10 000 BP, saw a limited mammoth re-expansion into north-east Europe, followed by retraction and apparent extinction of mainland populations, which matches the marked loss of open habitats in the early Holocene (Stuart, 2005). When evidence from palaeontology, climatology, archaeology, and ecology are combined it would appear that human hunting was not solely responsible for the pattern of extinction everywhere. Instead, it is more probable that it was the interaction of human disturbance along with the impact of pronounced climatic change on the vegetation that brought about their gradual demise (Barnosky et al., 2004). Irrespective of whether or not early man was responsible for the disappearance of the megafauna it was nevertheless extensively exploited by the Palaeolithic cultures as they migrated across the Siberian tundra–steppe 40–30 000 years ago. Ultimately, this migration reached

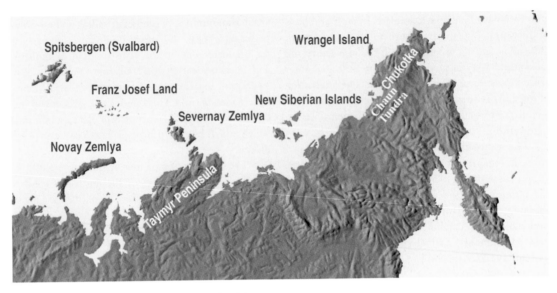

Fig. 11.6   Location of Russian arctic regions mentioned in text.

Fig. 11.7   Wrangel Island at 72° N lies between the Chukchi and East Siberian Seas and supports a rich and varied vegetation even though the Island is surrounded by ice for most of the year. During the last glaciation Wrangel Island also remained largely ice-free. It has been shown that mammoths survived on Wrangel Island until 3700 BP which is the most recent survival of all known mammoth populations (see text). However, due to limited food supply, they were much smaller than the typical mammoth. The flora still resembles the tundra–steppe that supported the mega fauna of the Late Glacial period (see text). The 417 species of vascular plants present is double that of any other arctic tundra territory of comparable size, and more than any other arctic island. For these reasons the island was proclaimed the northernmost World Heritage Site in 2004. (Satellite image by courtesy of NASA.)

northern Scandinavia, while other populations pressed eastwards across the Bering Strait to make an early and rapid settlement of North America, possibly between 20 000 and 15 000 BP (Shreeve, 2006).

After the disappearance of the megafauna the early Holocene warm period continued and the vegetation still had sufficient productivity to support considerable populations of musk oxen. In the most northerly land in the world (Peary Land – far north of Greenland) Palaeo-Eskimo campsites have been dated from charcoal remains to between 3000 and 4000 BP. Judging from the remains found in these camp sites these early

Fig. 11.8   Wrangel Island in summer (Photo Professor R. L. Jefferies.)

arctic dwellers were able to survive mainly by hunting the non-migratory musk oxen with little use of marine resources (McGhee, 1996). Such an achievement at these high latitudes points to a vegetation that was more productive in this *Post-Glacial Warm Period* than that which exists today.

Various cultures have subsequently migrated through these arctic regions. The climatic cooling that took place after the Hypsithermal made the migratory reindeer (with their ability to extract nutrition from lichens and mosses) the predominant large arctic herbivore. The reindeer hunting and herding peoples who still survive today (Sami, Nenets, Chukchi, Inuit, etc.; Fig. 11.9) are the modern equivalent of the ancient Palaeolithic hunter-gatherers that once peopled northern Europe, Arctic Siberia, and Beringia.

The success of these early polar human inhabitants and their never-ending search for food gives an historical perspective to the popular conception of modern man pressing ever further towards the margins of the habitable world and beyond in an unrelenting quest for natural resources. Resource exploration and exploitation is clearly a human characteristic with a long history. Despite the climatic vicissitudes of the Arctic the pursuit of resources has continued unabated, but instead of the ancient quest for fish, furs, walrus ivory (mors), and whale oil, the list is now minerals, oil and gas together with a much more thorough exploitation of the biological wealth of the waters of both the Arctic and Antarctic. This coupled with the marked climatic warming in these regions is creating an ever-growing risk of irreversible ecological damage to the polar biota from pollution and disturbance as well as from overexploitation.

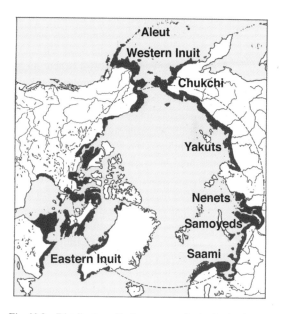

Fig. 11.9   Distribution of indigenous peoples in the Arctic (various sources).

11.3.1   Acquisition of natural resources at high latitudes

Arctic peoples, who live off the land as opposed to the maritime food chain, have always had to contend with uncertainty in relation to their future food supplies. The unpredictable environmental fluctuations in the Arctic have profound effects on plant and animal populations. Polar animal populations constantly run the gamut from super-abundance to near extinction and back. The human race's intelligent provisioning for future needs is a particular advantage in combating the inherent variation of arctic ecosystems. Man's unique ability to anticipate the future is particularly relevant in the Arctic. The hunter-gatherer cultures of the far north are obliged to store food for the winter and so have had to develop a large number of social attributes, such as a control of property, scheduling, and co-ordinated labour, to a degree that is not required in tropical or temperate climates (Ingold, 1982).

The acquisition of natural resources in the Arctic has two fundamental aspects that distinguish the exploitation of this habitat from boreal and temperate regions. First is the need for rapid harvesting. For the early hunting peoples, short runs in the annual cycle of available game meant that sometimes a few weeks or even days had to provide hunters, their families and the entire community with their food supplies for possibly a whole year. Secondly, the uncertainty of the arctic environment makes any prediction of when future resources may become available unreliable. Land hunters do not have the certainty that exists with marine sources (such as annual salmon runs) that next year's hunting will be successful. Many traditional arctic hunting techniques were therefore designed to seize an entire herd or stock of animals when it was encountered so as to obtain a maximum catch. Consequently, some means of food storage is essential in these situations (Ingold, 1982; Krupnik, 1993).

For arctic plants there are similar risks that the availability of resources may fail in certain years and a strategy that conserves carbohydrate reserves is essential. Snow cannot be guaranteed to melt every year and there may be many patches where the underlying vegetation will have to endure one or more years of uninterrupted snow cover. In some years, even when the snow does melt, the summer season may lack the warmth that is necessary to provide a positive carbon balance. A notable conservation strategy that is found in arctic flowering plants is a high photosynthetic rate coupled with an ability to store enough carbohydrate in their perennating organs. In established plants this can ensure survival even when there is a total failure to make any carbon gain due to one or more failed growing seasons (see Chapter 3).

Since the demise of the Pleistocene megafauna reindeer have always been the most important animal for the aboriginal peoples of the Arctic, both in the past for hunting communities, and for the present day herders of the far north of Eurasia. The remarkable ability of reindeer to survive on lichens and mosses gives them unsurpassed advantage in the conversion of vegetation into flesh. Just when reindeer hunting gave way to reindeer herding is difficult to date with any precision due to a lack of reliable historical resources. A distinction also has to be made between domestication of the reindeer and the later development of large-scale intensive herding, events which appear to have been separated by several hundred years (Krupnik, 1993).

One clear date is, however, recorded. The Norwegian chieftain Ottar from Halogoland in northern Norway is recorded as having visited the Anglo-Saxon royal court of King Alfred in England in c. AD 890 (Meriot, 1984). During his visit he gave King Alfred an account of life in the north of Norway which provides a remarkable insight into arctic life in the ninth century. Ottar said that he owned a herd of 600 domesticated reindeer. Visible evidence of the wealth of this marginal region of Europe can be seen in the reconstruction of the ancient Longhouse at Borg in the Lofoten islands (part of Halogoland), a chieftain's dwelling dating from c. AD 700 (Fig. 11.10).

It is not until the time of Russian contacts with the northern aboriginal peoples in the 1600s that there is any further written evidence of small-scale reindeer herding being part of a complex system dominated by hunting and fishing, as prevalent among the inhabitants of the Eurasian tundra. Such a way of life was practised by the Sami of the Kola Peninsula in the west, the Nenets of the Taymyr Peninsula, and extended to the nomadic Chukchis at the eastern limits of the Asiatic Arctic. The Chukchi are the largest group of the north-eastern Palaeoasiatics and are most probably the descendants of the earliest reindeer breeding nomads to move northward toward the North Pacific coast. Another century was to elapse, however, before there is evidence of extensive reindeer herding. The 1700s saw the beginning of a significant growth in reindeer stocks, with the onset of this change taking place virtually simultaneously in the western and eastern fringes of the Eurasian Arctic (Krupnik, 1993). As these domestic stocks grew the hunting of wild reindeer virtually disappeared.

The eighteenth century was one of the coldest periods of the Little Ice Age. This would have provided a climate in which reindeer might be expected to flourish. Reindeer are poorly adapted to high summer temperatures. If the temperature rises above 10 °C

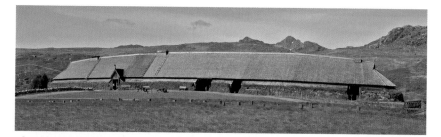

Fig. 11.10   Reconstruction of the Longhouse at Borg in the Lofoten islands (part of Halogoland), a chieftain's dwelling dating from
c. AD 700, showing the exterior and interior.

reindeer are noticeably uncomfortable and when temperatures reach 15 °C the animals suffer various physiological disorders. Under such conditions reindeer fail to thrive in the short summer season and do not make the gain in body weight that is essential for successful overwintering.

Warm dry summers are also disadvantageous for the lichens and mosses that make up a large part of the reindeer diet. The lichens become brittle and in consequence are more easily damaged by reindeer trampling. The danger of fire also increases and many decades are required for recovery. At the height of the Little Ice Age northern summers were marked in many regions by cool moist conditions which would have been ideal for both the reindeer and the growth of their fodder. It is therefore perhaps not surprising that this period saw a sustained increase in the size of domesticated herds (Krupnik, 1993).

The apparent end of the Little Ice Age signals an uncertain future for reindeer herding. In Russia the rain belts are moving north and there is a noticeable tendency for the climate in the western regions of northern Siberia to become more oceanic (see Section 5.2, Fig. 5.6). This will favour the growth of lichens and mosses. Unfortunately, there is also a downside to oceanic conditions in relation to winter conditions. Milder winters can result in periods when warmer weather intrudes for short periods into the Arctic. This can deposit rain on frozen ground which then becomes encased in ice, with the result that the vegetation becomes inaccessible to the reindeer and remains this way until the following spring. Such events are catastrophic for both wild and herded reindeer and bring about dramatic population reductions (Figs. 11.11–11.12).

Herds that are close to the tundra–taiga interface can possibly avoid the worst effects of this catastrophe by migration into the forests and feeding on the lichens hanging from tree branches. In the High Arctic, however, native reindeer do not have this alternative source of arboreal lichens. During the winter of 1993–4 the reindeer population in the Brøgger Peninsula (Spitsbergen 78° 55′ N) fell from 360 individuals to less than 80. Despite this 77% population reduction the reindeer have shown themselves able to recover gradually from this diminution in their numbers and by 1998 the Brøgger Peninsula population had regained approximately half their pre-1994 population (Aanes et al., 2000).

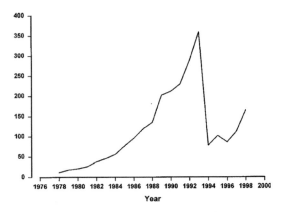

Fig. 11.11   The population development of Svalbard reindeer on Brøggerhalvøya since their introduction to the peninsula in 1978 to 1998. Numbers are from annual counting of reindeer during April. (Reproduced with permission from Aanes et al., 2000.)

Many high arctic plant populations are highly resistant to ice encasement and the consequent imposition of long periods of oxygen deprivation (anoxia). The ability of even the shoots and leaves to withstand total anoxia is a distinct feature of high arctic populations of some native plant species, and is in contrast to populations of the same species from further south which are not anoxia tolerant (Crawford et al., 1994).

Human occupation of the inland tundra by its aboriginal peoples is therefore inescapably linked with the plant productivity of this region. As already noted (Section 5.1) the tundra–steppe of the late Pleistocene was a more productive environment than the tundra of the late Holocene. It would not have been possible for mammoth to have survived on the lichen–moss communities that support reindeer populations today.

In the Eurasian Arctic the north–south extent of the tundra as measured from the shores of the Arctic Ocean to the northern fringes of the boreal forest varies from west to east. In the western Timan district it has a latitudinal extension of 150–200 km while in the eastern regions from the Bolshaya Zemlya tundra to the northern Trans Urals it can be as much as 450–600 km from north to south. Still further east in the Chukotka Peninsula the tundra zone is only 50 to 100 km deep (see Fig. 5.3). Where the tundra zone is narrow, lengthy east–west migrations are necessary and therefore the

Fig. 11.12   The Spitsbergen reindeer (*Rangifer tarandus platyrhynchus*) belong to a distinct subspecies that is endemic to Spitsbergen where they are the most northerly cervid population in the world. The animals are small (approximately 150 cm high) and have short legs and a short head. They are more closely related to the Canadian caribou (reindeer) than to the Scandinavian or Siberian reindeer (see Fig. 2.6). When the tundra becomes encased in ice after the re-freezing of melted snow, grazing becomes impossible and many reindeer die (see Fig. 11.11).

Chukchi reindeer hunters have a longer annual migration route as they move further into continental Siberia. The length of the Chukchi reindeer herders' annual migration increases from 50 to 100 km in the Chuckchi Peninsula to 200 to 400 km in the narrow tundra belt in the western Chaun tundra (see map Fig. 11.6). As the availability of pasture is reduced so is the population density of the individual herding peoples. Among the European Nenets, population densities, as estimated in the former USSR, varied between 0.025 and 0.038 people per square kilometre, while the eastern Chukchi had a lower density of 0.008 to 0.013 (Krupnik, 1993).

11.3.2   Future prospects for the tundra and its native peoples

The future looks uncertain for large-scale reindeer herding as developed over the last 200–300 years, and particularly so for the very large herds that were developed as a consequence of the collectivization that was enforced in Eurasia by Soviet Russia. During the 1960s the collectivization of herding brought the number of reindeer on the Chukotka Peninsula to over 100 000, exceeding the capacity of the winter range (Krupnik, 1993). Consequently, the essential lichens were seriously overgrazed and have not fully recovered.

Another present threat to the tundra, particularly to lichens on which reindeer depend, comes from atmospheric pollution generated locally and in distant regions. Future threats may arise due to climate change. The forests that have retreated southwards from the arctic shores over the last 6000 years (see Fig. 5.4) may return with climatic warming to narrow the coastal belt where reindeer grazing developed during the Little Ice Age. Where the forests do not return it is probable that the terrain will be occupied by extensive bogs (see Chapter 5) due to the process of paludification favoured by the onset of mild, wet winters. The recent increases in oceanicity and nitrogen input from air-borne pollutants will also promote the growth of bogs.

Human presence in the Arctic will undoubtedly continue, but it will be the mineral, gas, and oil resources that will dominate the economy. These activities may put the tundra at risk from the disturbance that is caused by the movement of equipment and installation of roads and pipelines, as well as the hazards of pollution (Crawford, 1997b).

11.4 MAN ON COASTAL MARGINS

In a world without roads coastlines provide convenient migration routes as well as access to the resources of the land and the sea. The early Mesolithic and later Neolithic coastal settlements that are found from the extreme north of Norway to the Atlantic islands of the Hebrides are a clear testimony to an early exploitation of the advantages of northern oceanic environments. The maritime conditions caused by the ameliorating influence of the North Atlantic Drift on Europe's north-western littoral allow pastoral and arable farming to be pursued at higher latitudes than would otherwise be the case. In particular, the cultivation of winter crops and the outdoor overwintering of animals have supported relatively large human populations as already referred to in relation to Neolithic Orkney (Section 11.2). The long history of human settlement in these regions has in turn produced marked changes in the landscape through human impact on vegetation.

Island woody vegetation appears to be particularly vulnerable to human disturbance both in the Atlantic and Pacific Oceans. Easter Island with its monumental statuary has now been found to have been settled later (c. AD 1200) and deforested more rapidly than

previously thought (Hunt & Lipo, 2006). The islands of Macaronesia (the Canary Islands, Madeira and the Azores) all had their tree cover rapidly altered by human settlement. In contrast to the Canary Islands, Madeira and the Azores had no autochthonous human populations and were first settled with the arrival of Europeans in the early fifteenth century. The immediate effect of the human settlement was the large-scale removal of forest. In Madeira there is evidence that much of the forest was removed by devastating fires (Sziemer, 2000). Iceland also suffered serious loss of tree cover in the centuries after its settlement c. AD 870 (see below).

In oceanic territories such as Orkney and Shetland (Bunting, 1996) western Norway (Kaland, 1986) and the Hebrides (Tipping, 1994), the advent of Neolithic farming was marked by the rapid and extensive replacement of trees by heathlands to an extent that did not take place in more continental areas. The relative importance of climate and human disturbance in the formation of modern heaths has been the subject of extensive discussion for many years. The existence of moorlands in western Europe can be traced to pre-Pleistocene times (Stevenson & Birks, 1995) and there is also evidence for the initial replacement of trees by moorlands and mires in a number of oceanic habitats even before the arrival of the first farmers. In Unst (Shetland), some opening of the tree canopy is detectable before the arrival of the first Neolithic settlers (Bennett et al., 1992). The new Holocene heathlands in north-west Europe differ from those of earlier times in having a pronounced dominance of *Calluna vulgaris*. However, it was with the arrival of Neolithic farmers that heathland development became much more extensive along the European Atlantic seaboard. In Orkney most of the tree cover appears to have been removed within 500 years of the arrival of the first Neolithic farmers (Bunting, 1996).

The length of time that the more oceanic areas of Scotland have been without any substantial tree cover has sometimes been assumed to mean that these peripheral and exposed islands of the Outer Hebrides, Orkney, and Shetland were always treeless. Early accounts (McVean & Ratcliffe, 1962) depicted many of the western islands of Scotland as naturally treeless. However, subsequent research has shown that these hyperoceanic regions once supported extensive areas of birch woodland with *Corylus avellana* (Fig. 11.13),

Fig. 11.13 Monoecious flowers on hazel (*Corylus avellana*). The male flowers are in the pendent catkins while the female flowers with conspicuous red styles are borne separately in clusters. Hazelnuts have been found in large numbers in Mesolithic sites and would have made a valuable contribution to the protein content of the winter diet of early settlers throughout Europe.

Salix spp., *Populus* spp., and *Sorbus aucuparia* even near their coastal fringes, while more central and eastern parts of the islands may have had stands with more warmth-demanding species (Tipping, 1994). The warmer conditions that existed in Mesolithic times are shown by the substantial remains of hazelnuts found on archaeological excavations in the Hebrides. Remains of over 100 000 nuts were found on a Mesolithic site on Colonsay (Inner Hebrides) which were dated to around 6700 BC. Pollen sampling from a nearby loch also showed an almost complete collapse of the woodland just after the intense hazelnut harvests had taken place (Mithen, 2003). Hazel is particularly sensitive to climatic factors for flowering and fruiting (Tallantire, 2002) and crops best when the hazel bushes form the forest canopy. At present hazel does not produce commercially harvestable quantities of nuts in Scotland. This may be due to the fact that in present-day forests hazel is mainly found as an under-storey component, which can reduce flowering and

fruiting. The copious remains of hazelnuts found in Mesolithic sites may therefore reflect the more dominant status of hazel in woodlands before the Atlantic oak woods became dominant, as well as the warmer climate of this period. The abundance of hazelnuts would have been especially advantageous nutritionally as they are rich in protein and easily conserved for winter use.

In the Outer Hebrides blanket peat began to appear between 9000 and 8000 BP (Fossitt, 1996). A marked climatic deterioration (Klitgaard-Kristensen *et al.*, 1998), commonly termed 'the 8200 BP event' (probably due to freshwater fluxes in the final degla-ciation of the Laurentide ice sheet), appears to have been accompanied by some reduction in tree cover throughout western Europe. In western Lewis (Outer Hebrides), there was a progressive replacement of trees by blanket peat which began about 7900 BP and continued with a further reduction of tree cover between 5200 and 4000 BP (Fig. 11.14). This later forest decline was associated with early Neolithic settlements (Fossitt, 1996).

In evaluating the relative roles of climate and human interference in removing tree cover, the late Norse settlement of Iceland (*c.* AD 870) provides a useful comparison with the longer settled regions of the North Atlantic. Studies of pollen and plant macro-fossils show that dense birch forest was present in Iceland from 6900 BP onwards (Rundgren, 1998).

Accurate association of vegetation changes imme-diately before and just after the time of the *landnám* (agricultural settlement) has now been possible with dating of the landnám tephra to AD 875 from the GRIP ice core (Pilcher *et al.*, 2005). Differences in the pollen record below and above the tephra layer show that the decline of birch was greatly accelerated after the settlement and was accompanied by an expansion of waterlogged soils and mires (Hallsdóttir, 1987). The pollen record therefore shows that Iceland, despite exposure to the hyperoceanic conditions of the North Atlantic, retained much of its tree cover throughout most of the Holocene, and it was only after the Norse settlement that forest disappeared with great rapidity.

Ari 'the Learned', writing in AD 1120–30, states in the *Islendingabók* that, 'at that time [the time of the settlement] Iceland was covered by woodland from the mountains to the coast'. It is unlikely that the treeline in Iceland ever exceeded 300–400 m (Hallsdóttir, 1987),

Fig. 11.14   Paludification accelerated by human settlement in Harris (Outer Hebrides, UK) as seen at the head of Loch Maaruig in 1966. Harris and Lewis were once tree-covered and are now largely bog and moorland. The continued retreat of agriculture and the advance of heathland can be seen on the far hillside where abandoned lazy beds are now covered in heather.

yet Ari's description nevertheless indicates that there had been a marked change by the twelfth century from that at the settlement in the nineth century. The *Landnámabók* also records that the early settlers had to clear trees to make it suitable for farming (Palsson & Edwards, 1972). Icelandic land-claim laws contained in the Grágás Lawbook (as recorded in codexes from about 1270) pertaining certainly to the twelfth century and probably also the eleventh (P. Foote, pers. com.) define very precise rules governing forest utilization for fuel and building. Careful distinctions are made as to whether wood is to be taken by cutting or pulling and whether or not meadow is allowed to replace woodland. Conservation was also an issue as there are regulations concerning joint owners of woodland who may have had disputes with regard to woodlands being overused. There are even penalties for hacking notches in a tree or causing scrapes that result in damage. Similarly, the penalties for browsing another man's trees are greater than those exacted for encroaching on his grassland. There are also precise laws governing the exploitation of new growth from old trees (Dennis *et al.*, 1980). The detail of these laws tells us that by the twelfth century woodland was regarded as a very important asset and not something which farmers had an automatic right to remove.

In England similar strictures conserving trees also appear in King Alfred's law code, clause 12: 'If one man burns or fells the tree of another, without permission, he shall pay five shillings for each big tree, and five pence for each of the rest, however many there may be; and 30 shillings as a fine' (King Alfred, AD 849–899).

11.4.1   Human acceleration of soil impoverishment in oceanic regions

The absence of prolonged periods of frost in coastal regions results in the leaching of soils of their nutrients by high rainfall. On arable lands this is particularly severe in winter when the lack of plant cover deprives the soil of the principal means of nutrient retention. In areas near the sea drenching with salt spray can accelerate ion exchange and further deplete soils of nutrients. Nevertheless, for early farmers oceanic pastures made it possible to overwinter greater numbers of mature cattle. It is possible that it was this facility to keep some cattle outdoors in winter that promoted the development of dairy farming with cattle in Ireland in the fifth to sixth centuries AD as opposed to milking only sheep and goats (McCormick, 1995). Cattle kept mainly for dairying, rather than just for meat, give a greater yield of energy and protein per hectare and this may have further contributed to the increase in the human population in oceanic areas.

The dependence of early agriculture on animal manure for nitrogen and phosphorus frequently caused what has been described as a nutrient *flow trap* (Dodgshon, 1994a). Some archaeological studies suggest that as a result of favourable conditions for grass production and outdoor overwintering of livestock, the growth of the human population frequently exceeded the carrying capacity of the land (Dodgshon, 1994b). When such populations expanded, the increased pressure for more land for arable crops reduced the area of land available for pasture. In continental climates the loss of lowland pasture can be made good by clearing more forest further uphill. However, the high lapse rate in oceanic regions (0.8 to 1 °C 100 m^{-1} of altitude) usually limits winter pasturage to land below 150 m (Fig. 11.15). The reduction in pasture would then have caused a diminished input of nutrients from animal manure to cultivated land and a consequent lowering of crop yields.

In addition to leaching and high rainfall, waterlogging impedes mineralization and nitrogen fixation, and frequently aggravates soil impoverishment in oceanic environments. The development of ploughing technology in the Iron Age would also have increased soil leaching in oceanic areas leading eventually to the podzolization (Fig. 11.16) of many soils – a process which had already begun in many western European locations with extensive production of heathlands in the early Neolithic (Behre, 1988). A particular case of oceanic heathland development is found in the 25 km wide belt of heathlands that extends along the coast of western Norway to the Arctic Circle. Some of the oldest heaths are found in the most westerly peninsulas and islands north-west of Bergen, where they have been dated to the Neolithic (4300 BP) with later extensions around AD 0, and again during the Viking age (Kaland, 1986).

The inevitable formation of iron pans, leading to drainage impairment and the development of gley soils and then the subsequent bog growth, further

Fig. 11.15   Langdale in the Lake District showing the strict limitation of improved pasture to the valley floor (105 m above sea level) and the neglect of potential adjacent hillside grazing in a region with an oceanic environment.

Fig. 11.16   A typical podzol soil exposed by a road cutting through a Scottish moorland. Podzol comes from the Russian roots *pod* underneath and *zola* ashes, and expresses an ancient and erroneous belief that the bleached layer was due to ashes from former forest fires. The light-coloured soil is, however, produced by the leaching of minerals such as iron and alumina which then leaves a bleached zone, which is often also depleted of clay. Podzols are found predominantly under coniferous forests and on moorlands in cool regions where rainfall exceeds evaporation. The leached minerals can accumulate lower down the soil profile to form a hard, impermeable layer (an iron pan) which eventually restricts drainage.

decreases the inputs of nitrogen and phosphorus and thus contributes to the general nutrient impoverishment of coastal lands that results from agricultural over-exploitation. In many maritime regions this entire scenario can be described as *paludification* (bog growth), as former mineral soils with unimpeded drainage were gradually converted to bogs causing many Neolithic and

Iron Age farms to be abandoned. Frequently, all that remains of this early farming activity on mineral soils are ancient walls, as the farmers retreated as a result of soil exhaustion and the subsequent advance of the bog.

The island of Papa Stour on the west coast of Shetland (Fig. 11.17) practised transporting turf and peat from large parts of the island to a relatively limited

Fig. 11.17   The hyperoceanic island of Papa Stour (Shetland, UK) showing a hill dyke dividing cultivated land to the right from the outfield to the left which has been completely stripped of all turf. The island has a 5000-year history of human occupation (see also Fig. 11.18) with a population of over 300 in the mid nineteenth century (now approximately 16).

sheltered area for cultivation creating the *plaggen soils* (Dutch, layered; Section 9.8.3). However, the very extensive removal of turf over a long period has led to the eventual destruction of the grazing value of the outfields (Fig. 11.17). This stripping of the peat and turf has, however, revealed the remains of many ancient walls that provide evidence of a well-developed pastoral economy over the greater part of the island in prehistoric times (Fig. 11.18).

In Ireland there are over 50 known locations of such prehistoric farms, which became engulfed by peat through paludification. The most famous is the early Neolithic site at Céide Fields (County Mayo) where an extensive Neolithic walled field system, now dated to before 5000 BP, was eventually buried by 4 m of peat (Mitchell & Ryan, 1998). The stumps of ancient pines (*Pinus sylvestris*) are found *in situ* at many locations within the blanket bog that covered the site. In most cases the pine roots are either on the surface of the mineral soil under the peat, or at an intermediate level in the peat itself. The age of the trees in the bog overlying Céide Fields has therefore been of great significance for the dating of the fields, as the trees must be younger than the bog in which they are growing, which in turn must be younger than the field system beneath (Caulfield *et al.*, 1998). The results of the

dendro-chronological studies suggest that the dates for the construction and period of use of Céide Fields and other Neolithic pre-bog field systems in North Mayo are older than was originally thought and that the initiation of blanket bog in many parts of North Mayo began more than 5000 years ago – which is also older than previously estimated. Further, the range of dates of the pine stumps indicates a synchronic event contemporary with a similar phenomenon observed in Scotland (Caulfield *et al.*, 1998). A range of detailed palaeo-environmental analyses carried out on a series of three peat profiles from Achill Island, Co. Mayo, western Ireland, where after a period of relatively dry climate Neolithic communities expanded in the region, found evidence for an extreme climatic event, probably a storm or series of storms, around 5200–5100 BP. This event is possibly linked to human abandonment of the area, comparable to that observed at nearby Céide Fields (Caseldine *et al.*, 2005). Irrespective of the exact chronology at different sites, the general trend therefore in this hyperoceanic region would appear to be for paludification starting to displace agriculture about 5000 years ago.

During periods of generally lower temperatures, as in the Little Ice Age, marginal areas that relied on agriculture were particularly susceptible to sudden

Fig. 11.18   Prehistoric walls on the island of Papa Stour (Shetland, UK) which have been exposed by removal of peat that grew since the walls were built. Evidence from sites such as this and archaeological investigations of Neolithic/Bronze Age settlements in Scotland and Ireland (see text) strongly suggest that peat formation became extensive in these regions as a result of human activity and tree removal, illustrating the sensitivity of oceanic habitats to environmental alteration.

climatic oscillations. An interdisciplinary study, combining the climate record from the Greenland Ice Sheet Project 2 (GISP2) with historical documents, has related the ending of the Norse Western Settlement in Greenland (mid fourteenth century) to evidence for lowered temperatures and severe weather in the North Atlantic. It appears that periods of unfavourable climatic fluctuations exposed the cultural vulnerability of the Norse farming settlers to environmental change to a greater extent than their maritime-based Inuit neighbours (Barlow et al., 1997).

11.4.2 Sustainable agriculture in oceanic climates: Orkney – an oceanic exception

Living in proximity to the ocean clearly has both advantages and disadvantages for farming. For early settlers in north-western Europe the advantages of mild winters and warmer conditions for plant growth in summer in early Neolithic times were clearly advantageous both for pastoral farming and cropping. However, the development of heathlands and the growth of peat deposits subsequent to human settlement show that oceanic conditions can create problems for sustaining soil fertility, especially when the human population grows and makes further demands for arable land that can only be made available by reducing the proportion of land kept for grazing. Although this has been a frequent occurrence in many oceanic areas with hard acid rocks there still remain regions where a favourable geology coupled with some shelter from some of the excesses of oceanic rainfall can permit a sustainable and productive agriculture. Such a situation is seen in the Orkney Islands off the north of Scotland.

The Orkneys never fail to astonish summer visitors for the verdure of the landscape and the density of cattle grazing on highly productive pastures (Fig. 11.19). This scene of vigorous pastoral activity is particularly striking for those who cross the Pentland Firth having travelled across the bleak moorlands of Sutherland or having visited the desolate expanse of the Flow Country of north-east Scotland. Here in Orkney is a country capable of supporting an active agricultural economy and a high level of human settlement. The Orkney Islands probably owe much of their good fortune to a combination of historical and physical factors. The archipelago lies in a rain-shadow area to the north of Scotland and this location, coupled with the excellent drainage provided by Old Red Sandstone geology, protects the islands from the worst dangers of excessive precipitation. The Devonian sandstones and shales weather to provide fertile soils often with access to ground-water springs with high pH values. Acidification and paludification are therefore less common than in areas with acidic rocks, although they always remain a risk in abandoned land in an oceanic environment. Being mostly low-lying arable land Orkney did not have significant extents of overpopulated uplands in the hands of large estate owners, and thus avoided the 'Clearances' of the Scottish Highlands where small tenant farmers maintaining a subsistence existence were removed and replaced by large sheep-runs.

Orkney continues to prosper agriculturally. The extension of the growing season for grass due to warmer autumn weather and milder winters is an enormous benefit. In summer it has now become easier to achieve two cuts of silage together with easier haymaking. The shortening of the time that cattle have to be tended

Fig. 11.19   Black cattle grazing by the 5000-year-old burial mound of Maes Howe in Orkney: the finest Neolithic chambered cairn and passage grave in north-west Europe. Despite Orkney's northern location (59° N) and a hyperoceanic environment there is a thriving cattle economy. The archipelago lies in a rain-shadow area to the north of Scotland and this location coupled with the excellent drainage provided by Old Red Sandstone geology, protects the islands from the worst dangers of excessive precipitation.

under cover is also a welcome development following the improved autumn temperatures in recent years.

11.5   EXPLOITING THE WETLANDS

Wetlands in cool temperate climates are a hindrance to agriculture, particularly in those regions of the world where there is a long winter. In tropical climates wetlands can be used for growing rice and in many areas swamp forests and bottomland forests provide excellent growing conditions for tree growth. In lands where there is a prolonged dormant season waterlogged soils are generally harmful to overwintering crops as well as being an obstacle to tilling and harvesting. Drainage has therefore been the general agricultural solution to the farming of wetlands. The low-lying marshes that surround the mouths and deltas of major rivers are not readily drained due to periodic flooding and tidal activity.

Wetlands can, however, be highly productive with the natural vegetation of reeds, rushes, sedges and willows that colonize the nutrient-rich sediments deposited where major rivers meet the ocean. The oxidation of organic waste by oxygen diffusing out from plant root systems together with the activity of the soil microflora purifies the water. From these wetlands clean water can then eventually recharge underground aquifers causing such regions to be referred to as 'the kidneys of the landscape'. In addition to their nutrient-absorbing capacity they also form land which protects coastlines, and buffers the potentially damaging effects of both flooding and drought. Given these properties it is therefore not surprising that there has been a long history of human usage and occupation of tidal marshes and that they have provided a home since prehistoric times for peoples who could adopt a lifestyle which was not entirely dependent on agriculture and who were prepared to accept the physical risks of their marginal location.

11.5.1   Coastal wetlands

Europe's largest coastal wetlands lie along the southern coast of the North Sea from Holland and North Germany through to the west coast of Denmark. The Wadden Sea, and its coastal marshes, are enriched by the effluent from the great rivers that cross the North German Plain, the Ems, the Weser, and the Elb. The highly productive marshes have a 7500-year record of human settlement which over the years has adapted to the considerable risks involved in harvesting the outstanding productivity of the salt marshes and swamps. The earliest Neolithic communities in the Wadden Sea region were confined to summer camps. Gradually, Neolithic and Bronze Age settlers learned to make increased use of the fertile salt marshes for pasturage through a form of transhumance with grazing on the marshes in summer and retreating with their cattle to higher ground in winter (Knottnerus, 2005).

While the Neolithic settlers used the wetlands as they found them, the Bronze Age farmers began to modify their immediate surroundings. Gradually, instead of limiting habitation to areas just above high-tide level, the inhabitants began to raise mounds (*Wurten*; Fig. 11.20) from sods and dung on which they

Fig. 11.20   Farm buildings on raised mounds (*Wurten*) which are typical of an early pattern of agricultural settlement in the low marsh region of East Friesland. (Photo Dr Felix Bittmann, Niedersaechsisches Institut für historische Kuestenforschung.)

Fig. 11.21   Early cereals. (left) Bere barley (*Hordeum vulgare* var. *tetrastichon* and (right) black oats (*Avena strigosa*).

situated their farms and infields. In summer they would take hay and graze their animals on the marshes where they also cultivated salt-resistant summer crops, such as field beans (*Vicia faba* var. *minor*), the four-rowed bere barley (*Hordeum vulgare* var. *tetrastichon*; Fig. 11.21, left panel) and oats (*Avena sativa*). With the arrival of the winter rains they would then retreat to their mounds along with their cattle and sheep, followed by rodents, hedgehogs, weasels and stoats. By Roman times the riverbanks were densely populated and salt-making was actively pursued. From time to time there were periods of maritime regression and sites were abandoned as a result of flooding and storm surges. Particularly during the period AD 450–600 tribal wars and malaria drastically reduced the population and new settlers then migrated into the area from the east.

From the ninth and tenth centuries onwards drainage and dyking began to transform the landscape with the former salt marshes being protected by sea walls and eventually the incorporation of sluices for retaining fresh water and repelling floods. Crops that would thrive on the peat soils that had now developed on the marshes included rye (*Secale cereale*), black oats (*Avena strigosa*; Fig. 11.21, right panel) and buckwheat (*Fagopyrum esculentum*). This was partly interrupted in the fourteenth and fifteenth centuries by large-scale inundations of marshland. However, the progressive construction of embankments and sea walls (AD 1500–1800) did

succeed in transforming much of the amphibic transition zone into arable land and freshwater lakes. In unreclaimed areas the nutrient-rich coastal ecosystem with its highly productive salt marshes continue to provide summer pasturage and hay meadows that support extensive grazing for both cattle and sheep. The drainage and dyking has also caused new plant communities to develop, and wet pastures with *Molinia caerulea* and waterlogged *Glyceria* meadows now extend the range of grazing lands.

Exploiting these areas has always required considerable effort as the nature of the terrain prevented the use of horses and carts. The rewards for this labour were considerable, and much use was made of rafts and barges to transport cattle to the pastures and harvest the herb-rich hay.

As with many estuaries and deltas throughout the world in recent times, dredging to provide shipping canals has caused the natural vegetation at the land–sea interface to lose its ability to retain and filter river sediments. Riverine loads are now flushed directly into the North Sea. This is also the case with the Mississippi delta where the loss of sediment that would have been spread over adjoining marshland and by transfer to deep water in the Gulf of Mexico has led to coastal habitat degradation and a decline in natural habitat diversity, falling shorelines, and increased risks of flooding (Reise, 2005). Rising sea levels are therefore likely to bring about substantial habitat reconfiguration on these sedimentary coasts where the marshlands have been starved of the sediment that might have enabled them to rise along with the sea level.

11.5.2   Human settlement in reed beds

In southern Iraq the Shatt al-Arab Marshes and their reed beds that straddle the confluence and lower reaches of the Tigris and Euphrates rivers have also provided a habitat for marsh peoples that have preserved a unique way of life for the past 7000 years. The area is now shrinking as upstream dams and irrigation projects siphon off vast volumes of water and lower the water levels of the rivers. The marshes are highly heterogeneous. Some parts are permanent marsh while others are only flooded in spring and summer, which enables water buffalo to be driven through the reed beds when the water is low. However, feeding the buffalo is very labour intensive as the young green shoots of the reeds, rushes and sedges have to be cut by hand and fed to the animals on firm land. In the past the Marsh Arabs have grown rice, wheat and barley in the seasonal marshland, but this is no longer common (Hemminger, 2005).

The highest altitude reed-bed settlements in the world are found around Lake Titicaca lying at an altitude of 3810 m (12 500 feet) above sea level in the Andes. The remnants of an ancient people, the Uros, still live on floating mats made of dried totora reeds. These reeds are a widespread American species of bulrush or club-rush (*Schoenoplectus* (*Scirpus*) *californicus*) that grows in dense stands around the edges of the lake. On these floating reed rafts huts are also made from the same totora reeds and are still home to several hundred people who are partly descended from the Uro Indians who retreated onto the lake with the spread of the Inca Empire. The Uros and other lake dwellers also make from the totora reeds their distinctive balsa-boats with fashioned bundles of dried bulrushes lashed together (Figs. 11.22–11.23). As well as providing a home, shelter and a means of transport the young stems of the bulrushes are also eaten. Although the lake-dwelling culture of Lake Titicaca has outlasted the Inca civilization it is now under threat as modern life encroaches.

The bulrushes or club-rushes (*Schoenoplectus lacustris*, *S. tabernaemontani*) together with other closely related species of *Scirpus* are a particularly notable group of plants as their nutritive value has been recognized not only by man, but also by geese and rodents. *Schoenoplectus* spp. are a preferred forage source for geese in the marshes of the Doñana in southern Spain and for muskrats in the swamps of the Mississippi delta. The genus *Schoenoplectus* is also distinctive for the tolerance of its rhizomes to total anoxia, which appears to be linked to their high carbohydrate content (Crawford, 2003). It is very striking therefore that plants with high levels of anoxia tolerance are sought out in marsh-dwelling animals, presumably because anoxia tolerance is linked to a supply of readily metabolized carbohydrates.

11.5.3   Agricultural uses of wetlands

Mires, marshes and bogs have some of the highest levels of productivity to be found in any natural ecosystem. Although not generally part of mainstream agriculture they have long been skilfully diverted to human needs, particularly in areas of marginal farming. Marshes that dry out sufficiently in summer can provide grazing or in some cases a crop of hay. Well-managed water meadows

Fig. 11.22   An Indian island dwelling on Lake Titicaca (3810 m). The present-day descendants of the Uro Indians who retreated onto islands on the lake due to the expansion of the Inca Empire still make use of these islands where they construct houses and build boats made of dried bundles of the South American bulrush, the totora (*Schoenoplectus californicus*).

Fig. 11.23   Shallow water navigation of Lake Titicaca. Manoeuvring a boat made from totora reeds (*Schoenoplectus californicus*) by means of punting with a pole.

can add greatly to grass production for hay in areas where gross production is limited by drought in summer. However, the management of these water meadows requires careful attention and they are now much less used than in the past.

An outstanding example of extensive water meadows that are still in use is found along the banks of the Shannon in Ireland. The River Shannon traces a long S-shaped course through the centre of Ireland where it receives the waters of 12 counties. Along its 205 km journey to the sea it falls only 12 m, which gives it the shallowest gradient of any large river in Europe. Consequently, given this extensive catchment area and Ireland's plentiful rainfall, the Shannon pours nutrient-rich floodwaters across 35 square kilometres of fields every winter. The flooded fields are called the *Callows* (Irish *Caladh* means either *water meadow* or *meeting place* depending on context) and offer a refuge for large numbers of wildfowl and wading birds in winter while in summer they provide hay and pasturage for cattle (Heery, 1993).

Peat bogs, although in large measure created by human disturbance through deforestation, have eventually earned a wilderness conservation status as well as being a useful source of fuel for inhabitants of marginal areas. Peat has provided fuel for fires and bedding for livestock for centuries. Cutting, drying and carrying peat is very labour intensive, and peat cutting for domestic use is much reduced from what it was in earlier centuries. In the British Isles extensive clearing of peatlands took place with land improvement schemes in the seventeenth and eighteenth centuries of

which Flanders Moss at the head of the River Forth is a well-documented Scottish example. The eradication of smallpox caused the human population of neighbouring areas to the north in the Scottish Highlands to increase to such an extent that there was an acute shortage of land. Highlanders were offered land on the bog rent-free for life if they cleared the peat. The peat provided material for building their homes, fuel for their fires and bedding for their animals. What was not needed was cast into the upper waters of the Forth. This eventually proved detrimental to downstream salmon fishings, and further clearance was halted leaving the considerable expanse of undisturbed raised bog that is now preserved by conservation regulations. Modern peat harvesting for industrial use unfortunately leaves a wetland that is of very little use either for agriculture or wildlife (Fig. 11.24).

When properly managed the interface between wet and dry land can provide a double resource that has been exploited since the beginnings of human settlement. When fenland or raised bogs are drained and cleared of peat a nutrient-rich organic soil of great agricultural potential is left. The neighbouring wetland provides grazing in dry summers when the surface of the bog is dry enough to bear livestock. Marshes and peatlands also provide ample opportunity for dietary supplements from wildfowling and fishing. Such a multi-use habitat is typified by the Fenlands of southeast England which were formed by the silting-up of a bay of the North Sea to form a flat lowland extending west and south of the Wash and becoming the largest swampland in England. The lower peat levels were

Fig. 11.24   Effect of the industrial removal of peat from a bog in Caithness (Scotland) – a landscape in which the only remaining commercial resource is wind.

deposited from 7500 BC onwards but widespread coverage did not occur until *c*. 3500 BC. Due to the wetland being at a maximum during the Iron Age human settlements were limited to the edge of the Fenlands although lowering of the water table by the Romans who attempted drainage and built a few roads across the Fens brought about a denser occupation of the edges and islands – only to be followed by drowning of many settlements after their departure (Hall & Coles, 1994). Cornelius Vermuyden, a Dutch engineer, developed the first effective drainage systems in the seventeenth century. Drainage and construction of

dikes and channels in the various sections or 'levels' continued through the nineteenth century despite problems of land sinkage, water accumulation, and periodic flooding. Agriculture is now plentiful on the fertile alluvial soils, with vegetables, fruit and wheat being the principal crops.

A somewhat different interaction of people with the edges of wetlands is found in south-west England on the area of marshes and moors known as the Somerset Levels. With the rise in sea level at the end of the last glaciation this area was flooded and remained under water until *c*. 4500 BC when peat deposits began to form in the salt marsh, leading gradually to the development of fens and raised bogs. To cross these wetlands and reach the islands of rock and sand in the valleys, prehistoric people built wooden trackways, remains of which survive to the present day due to the waterlogging of the peat. The oldest is the Sweet Track which was a raised walkway built in 3806 BC. In the Roman period the first sea and river defences were built, but as with the Norfolk Fenlands, the end of the Roman period saw large-scale flooding. Despite the constant risk of flooding, islands in the moors were chosen as sites for monastic centres and for pasturing sheep on the flood-enriched summer grazing. It was this principal use of the wetlands in summer that was probably the origin of the regional name 'Somerset' from the Anglo-Saxon *'Sumersatan'* – land of the Summer People (Dunning, 2004).

11.5.4   Recent developments in bog cultivation

An increase in the demand for fruit juices in a modern health-conscious world has caused a marked increase in the use of bogs for the cultivation of cranberries. The American cranberry (*Vaccinium macrocarpon*) is indigenous to northern America where it grows naturally in acid peat bogs and has a distribution that includes areas that are prone to frost at any time of the year. Commercial cranberry production started in the mid nineteenth century and is now a major industry occupying about 14 000 acres in Wisconsin alone. Cranberry production as a cultivated crop has until recently been mainly concentrated in northern America but is now being developed also in Finland and Poland.

Because cranberries are sensitive to frost damage and natural winter protection by snow cover is often unpredictable and insufficient, the cranberry vines are grown in specially constructed cranberry bogs which are flooded in winter (Fig. 11.25). The water freezes to form an ice blanket, which prevents drying of the evergreen vines by cold winds and minimizes fluctuations in temperature. While frost damage is avoided by this treatment, the ice impairs oxygen diffusion. Consequently, the submerged plants are prone to oxygen deficiency stress, particularly when the ice is covered by snow which reduces the light reaching the submerged plants and prevents the evergreen leaves from alleviating their interrupted oxygen supply through photosynthetic activity.

Ice encasement of vegetation occurs regularly under subcontinental and northern maritime climates and has been a serious hazard in these regions for a number of overwintering crops. Healthy cranberry plants can survive for up to six months under winter ice in flooded bogs provided the ice is not covered in snow. A deep snow cover excludes light from penetrating to the submerged plants, and by depriving them of photosynthetic activity stops the generation of oxygen, and consequently anoxic conditions frequently develop to a degree that damages the plants under the ice. Especially dangerous are warmer periods during flooding, when respiration rises, and carbohydrate reserves in the plant are rapidly depleted making the plants more sensitive to subsequent ice encasement. The carbohydrate status of the overwintering vines has also a direct influence on fruit set. In spring, a rapid resumption of growth requires adequate carbohydrate supplies, and insufficient carbohydrate levels may be responsible for the low fruit set which is sometimes observed. Experimental studies on the effects of oxygen deprivation have shown that during anoxia (the total withdrawal of available oxygen) the mature leaves exhibit a marked down-regulation of metabolism. Carbohydrate consumption and energy metabolism stabilize at low levels soon after the switch from aerobic to anaerobic pathways. Pathways such as TCA cycle or photosynthesis, which are non-operating during the anoxia treatment, are severely affected but still measurable after 28 days of anoxia. Most remarkable, however, is the rapid recovery (Fig. 11.26) in aerobic respiration and photosynthesis which takes place when the plants are returned to air (Schlüter & Crawford, 2003).

The many medical benefits of cranberries have created a strong demand for this fruit and its juices. Cranberries contain proanthocyanidins (PACs) which can prevent the adhesion of certain bacteria associated with urinary tract infections, including *Escherichia coli*. The anti-adhesion properties of cranberry may also inhibit the bacteria associated with gum disease and stomach ulcers. Such is the success of the American cranberry (*Vaccinium macrocarpon*) it has stimulated the market for similar fruits from Europe with an increasing demand for the European cranberry (*Vaccinium oxycoccos*) and the small cranberry (*V. microcarpon*), lingon berries or cowberry (*V. vitis-idaea*), and blueberries or bilberries (*V. myrtillus*).

Fig. 11.25   Cranberry bog in Massachusetts in late autumn just after the vines have been flooded for winter frost protection. (Inset) Cranberry plant (*Vaccinium macrocarpon*) which has been kept under total anoxia for 21 days followed by 3 days post-anoxic recovery in air. The only damage visible is to the younger leaves, which turned pale yellow on return to air.

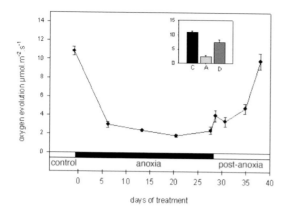

Fig. 11.26   Effect of imposition of prolonged anoxia in the dark on cranberry leaves and their capacity to resume photosynthesis when restored to air and light after 30 days anoxia. The small histograms show the comparison between control (C), 28 days anoxia in darkness treatment (A) and 28 days darkness in air (D). (Reproduced from Schlüter & Crawford, 2003.)

11.5.5   Future uses for wetlands

Although the use of wetlands as a zone for marginal pastoral activity has declined there is increasing awareness of their value as 'the kidneys of the landscape' for their capacity to absorb wastewaters, disperse floods and detoxify organic wastes. The growth of wetlands in coastal habitats, which receive sediment loads from periodic flooding, can help to raise shorelines, especially where they are adjacent to salt marshes. If undisturbed, they remain one of the few true wilderness areas, where plant and animal wildlife, especially waterfowl and wading birds, live in communities that reflect a landscape which is timeless and provides a glimpse of scenes with which our ancestors would have been familiar in past centuries, including the continuation of wildfowling. The cultural and ecological value of wetlands are now increasingly being recognized and they are much better appreciated as wildlife reserves as well as for natural overflow areas in diverting floodwater. This latter aspect is assuming increasing importance with the danger of rising sea levels and the greater intensity of precipitation that is now becoming more frequent.

11.6   MAN IN THE MOUNTAINS

Mountain peoples are outstanding throughout the world for their strong affiliation to their homeland, combined with a tolerance of an environment that is usually noticeably poorer in resources than in the adjacent lowland plains. Historically, mountains have provided a sense of independence to their inhabitants. The resulting long-term survival of groups of peoples in montane retreats has preserved the customs and languages of many ancient cultures from assimilation by larger and more powerful neighbours. These minority groups and races can be considered as peripheral human populations, living in marginal habitats. In the British Isles, the Welsh and the Scots both owe something to the mountainous nature of their respective territories for their cultural survival. Elsewhere, Tibetans, Kurds, Georgians, Armenians, Chechens, Basques, and even the Swiss, have had their identity preserved from that of their surrounding neighbours due to the elevation of their residential circumstances.

11.6.1   Transhumance

As a natural resource, mountains wherever they are have provided over the millennia one common feature that has aided the agricultural survival of mountain people, and that is the phenomenon of *transhumance*. The practice of seasonal migration with sheep and cattle to summer pastures in alpine habitats can be found across Europe and Asia. These summer grazings have their own specific names in many languages. In the Celtic world of Scotland and Ireland it is *sheiling*, in Norway *saeter* or *støl*, in Switzerland, and in neighbouring German-speaking regions, *Alp* or *Alm*, and in Turkey *Yayla*. One of the great advantages of the summer grazings, apart from releasing lower altitude lands for crops and hay meadows, is their freedom from drought. Mountains attract clouds and precipitation and when pastures at lower altitudes have their summer productivity reduced by water shortages the summer grazings can prove valuable sources of green lush pasture. In the past mountain pastures were particularly useful for dairy farming as butter and cheese could be made in the mountain and later carried back down to the valley. However, there was always one inherent practical disadvantage in the summer pasture transhumance system, namely the loss of manure which might have been more usefully applied to the lower altitude arable fields.

The ancient human use of mountain slopes for hay meadows has contributed greatly to the colourful appearance of the European Alps in early summer.

The suppression of forest in favour of meadows and pasture opened up, particularly in calcareous regions, a terrain that could support a herb-rich flora. Transhumance has been practised for over 5000 years in the Alps and coupled with the practice of cutting hay from alpine meadows has caused the evolution of a mountain grassland flora that is still to be seen on the alpine slopes above the treeline in early summer. Five millennia of haymaking have selected for a flora that has had to flower and set seed before the mowing season began. To this we owe the spectacular early summer floral display of the Alps.

Sadly, current trends in Europe reveal a large-scale abandonment of alpine pastures with trees readvancing over pastures that have provided summer grazings for centuries and probably millennia. Even in Norway, with its limited area of lowland pastures and long tradition of summer pasturage in the mountain saeter (støl), trees are now advancing up mountainsides that nineteenth-century photographs showed as treeless.

Fig. 11.27   Cutting hay in the traditional manner on an alpine meadow near Innsbruck (Austria). Hay cutting after the bulk of the alpine flora has flowered is essential for the preservation of the floristic diversity of these alpine meadows and the farmers receive a subsidy for continuing this practice.

In Europe at least, the time has passed when shepherds, cowherds and milkmaids took to the mountains for the summer, returning in the autumn with their flocks, herds and dairy produce. Subsidies are, however, given to farmers in several countries to preserve some remnants of this upland way of life. In Switzerland payments are made to farmers for mowing the alpine meadows and thus preserving the grassland flora of these alpine meadows. Much of this cutting is still done by hand with a scythe, by a rural population that is sympathetic to this cause. How long such a relationship will endure for a laborious task not founded in necessity

is questionable (Fig. 11.27). However, should climatic conditions change, and increased temperatures prevail, and the migration of rain belts impose summer droughts in the valleys, then alpine pastures and transhumance might once again become a valuable pastoral resource.

11.6.2   Terrace farming

Although use of upland areas in Europe is declining there are other regions of the world where mountainsides are used intensively for agriculture, usually due to

Fig. 11.28   Terraces on the mountain below the Lost Inca City of Machu Picchu near Cuzco (Peru). The development of terracing in Andean agriculture reached its peak in the Inca Empire and produced a form of mountain horticulture that has never been surpassed in terms of soil and water conservation. (Photo Barbara Crawford.)

population pressure and a shortage of level land. In highly settled mountainous areas one of the most sophisticated agricultural techniques is the use of terracing. It is often assumed that the terracing has been put in place with the sole purpose of reducing erosion. Terracing as developed in the Andes not only reduces soil erosion but serves also the equally important purpose of maximizing water use by the planted crops (Fig. 11.28). The use of terraces and irrigation for growing crops at high altitudes in the Andean cloud zone should perhaps be described as horticultural

rather than agricultural. The terraces were not just revetments of the hillside but were specially constructed stone-lined troughs into which fertile soil was transported from the valley bottoms (Fig. 11.28).

The construction of terraces in the Andes along with a tradition of irrigation and terrace farming long predates the Inca civilization, with evidence of terrace farming beginning probably as early as 2400 BC. In the Colca valley in southern Peru evidence has emerged of terrace farming which began probably as early as 2400 BC (Denevan, 2001). Over the subsequent millennia

Fig. 11.29   Produce from a mountain farm in the Andes. A farmer near Sicuani (Peru) with a collection of maize and various tubers harvested from his farm. Note the variation in the maize and tubers that come from his one farm. The unconscious selection of maize, a C4 plant with an enhanced carbon dioxide harvesting system and high water use efficiency, was a fortuitous event for Andean terrace farming (see text).

terrace construction took various forms depending on the terrain. Small narrow terraces predominate on steep mountainsides but on the lower slopes where the gradient is reduced broad gently sloping terraces like small fields were still constructed with irrigation canals to maximize production. In the Andes precipitation is a limiting factor and the careful construction of irrigation canals over the centuries made it possible to grow crops further up the mountainsides than would have otherwise been possible.

Terrace cultivation reached its maximum with the Inca Empire (AD 1200–1535) extending along the Andes from the Equator to the Pacific coast of Chile with an extensive range of crops including maize (*Zea mays*), quinoa (*Chenopodium quinoa*), various species of potato (*Solanum tuberosum*, *S. acaule*, etc.), as well as other Andean tubers such as caldas (*Ullucus tuberosus*), anu (*Tropaeolum tuberosum*), oca (*Oxalis tuberosa*) and ullucu (*Ullucus tuberosus* – Basellaceae), an Andean plant with potato-like tubers. Images of *Oxalis tuberosa* and *Ullucus tuberosus* have been documented on ceramics from the early pre-conquest era in southern Peru (Vargas, 1981). One highland archaeological site, Tres Ventanas at 3925 m, has yielded material of *Ullucus tuberosus* reputed to be 10 000 years old (Sperling &

King, 1990). The highest cultivated plot in the world is recorded from near Lake Titicaca – a field of barley growing at a height of 4700 m (15 420 feet), too high for the grain to ripen but the stalks furnished forage for llamas and alpacas (www.crystalinks.com).

The domestication of maize (Fig. 11.29) and therefore the unconscious selection of a C4 plant as a cereal crop was particularly appropriate for the Andean situation. Not only are C4 plants economical in their use of water, their carbon dioxide harvesting system is ideally suited for the fixation of carbon in the high-altitude rarefied atmospheric environments. How and why the South American civilizations were able to have made the fortuitous selection of a C4 plant as a cereal provider is an intriguing question. A possible answer may be found in the palaeo-botanical history of the region. It has been argued that the altitudinal vegetation distribution in the northern Andes during glacial times differed from the present-day conditions as a result of temperature and precipitation changes. It is possible that reduced atmospheric partial CO_2 pressure would alter the competitive balance between C3 and C4 plants in favour of the latter (Boom *et al.*, 2001). In this connection it is also of interest that in the high plateaux of Tibet there is a marked predominance of C4 plants

Fig. 11.30   Northern Iceland showing eroding hillsides filling a level plain with soil, creating a potential for agriculture that did not exist at the time of the original ninth century settlement (landnám). (Photograph Dr Bjarni Gudliefsson.)

Fig. 11.31   A marginal farm in early summer in the Geiranger Fiord, Norway.

with 95% of the C4 species being found at altitudes above 3000 m (Wang, 2003).

11.7   CONCLUSIONS

When we look at regions that are now considered marginal and wonder why they frequently have had a long history of human settlement it is necessary to try and imagine the environmental conditions that existed for early settlers. At the time of the early development of agriculture valley bottoms which are now favoured for agriculture would have been filled with lakes, gravel beds, rivers and marshes as they had not yet received the sediment loads from mountain erosion that have now created extensive flat and fertile alluvial plains. The amelioration of valley bottoms in some cases is relatively recent and still taking place. In Iceland the flat fields that lie below the mountains only appeared in the one thousand years since the first settlement of Iceland and after the upland soils had eroded as a result of excessive human disturbance (Fig. 11.30). Similarly, islands and other maritime areas that now seem remote and exposed had coastal fringes where soils were more easily drained and woodlands more readily cleared than in the interior regions with dense vegetation. Many

early migrations both in North America and Europe were made by sea and therefore the concept of geographical inaccessibility that we now have in relation to offshore islands and isolated archipelagos is not relevant to these earlier landscapes.

With the passage of time other sites would have become available that appear potentially more productive and rewarding. However, history demonstrates that competition for fertile land is intense, and subsequent migrants, possibly better equipped than the autochthonous populations, may have been more successful in capturing the better land. In South America the high-altitude mountain valleys remain the preserve of the Quechua-speaking population of the Andes. A similar situation exists for the Sherpas of Nepal. The more ancient indigenous populations having adapted to their original chosen habitat are able to survive there in what other people might consider suboptimal or marginal conditions. We must admire the territorial tenacity of those who continue to people peripheral regions in that they enrich the world culturally and preserve the diversity of the human race (Fig. 11.31). In just the same manner, the plant populations that survive in marginal habitats throughout the world enrich the flora and preserve these unique habitats.

Fig. 12.1   The Athabasca Glacier ice front in late July 2006. The glacier flows from the Columbian ice field (altitude 3491 m) of the Canadian Rockies and has retreated 1500 m in the last century. Since 1980 there has been a noted acceleration of this retreat. (Photo Professor R. J. Abbott.)

12 · Summary and conclusions

12.1 SIGNS OF CHANGE

We live in times of change and the impact of climatic warming can already be seen in many marginal areas. Effects that appear to be directly attributable to climate change are most noticeable in polar and alpine regions with the retreat of glaciers and snow and ice cover regions (Fig. 12.1). Coastal erosion as a result of rising sea levels is also having a noticeable impact. Archaeological rescue excavations of ancient coastal settlement sites exposed by erosion are now numerous. Several Scottish coastal golf courses have had to be redesigned as sections have been totally removed by the advancing sea. A coastal Scottish National Nature Reserve (Tentsmuir) that had been growing steadily seawards for centuries is now in places retreating rapidly by as much as 200 m depth of dunes over the last 20 years, returning sections of the coastline to where they were in the 1920s (Fig. 12.2).

More positive changes include an increase in the number of plant species to be found on the summits of European mountains (see Section 10.7), and an earlier flowering of the vernal flora with an extension of the growing season into the autumn. Northern agriculture is also profiting from the passing of the Little Ice Age in both crop production and length of the grazing season. However, in many parts of the world deleterious changes are taking place which are being aggravated by increasing human disturbance. Desertification in arid grasslands is certainly influenced by increasing

Fig. 12.2   Coastal erosion at the Scottish National Nature Reserve, Tentsmuir, Fife. The line of concrete blocks were laid down as coastal defences on the high tide line in 1940. They became completely buried in this region of the reserve by accreting sand dunes that advanced 50–75 m seawards from the line of blocks. Large 8–12 m high dunes have been removed in the last 20 years and the coastline in this region is now where it is estimated to have been in the early 1920s (Crawford & Wishart, 1966).

drought, but what is perhaps more serious is that the innate ability of the vegetation to recover is hindered by the excessive trampling of livestock (see Section 11.1).

Drought may also be an increasing hazard for arctic plants. There are already reports of arctic berry-bearing plants such as cloudberry (*Rubus chamaemorus*; Fig. 9.18) and bog whortleberry (*Vaccinium uliginosum*) suffering from drought. Drought is also affecting trees in the non-coastal parts of Alaska (see Section 5.3.1). This contrasts with the more oceanic regions of the Arctic, such as the West Siberian Lowlands where the treeline is already depressed south of its temperature limitations. The rain belts are moving north in Russia and therefore paludification in arctic and subarctic regions with oceanic climates may be accelerated. The rate of bog growth due to increased precipitation will undoubtedly be augmented by the increased nitrogen content that these rains now carry northwards from the northern hemisphere's industrial regions (Section 5.3.1). These diverging scenarios of drought versus paludification make it impossible to generalize on the migration of high-latitude vegetation zones. Thus, model predictions of treeline advance based on temperature increase in the Arctic suggesting that by AD 2100 there will be a more than 500 km northward migration of the treeline, unless prevented by anthropogenic disturbance, is unlikely to be a general occurrence (Arctic Climate Impact Assessment report, Callaghan *et al.*, 2005).

12.2 VEGETATION RESPONSES TO CLIMATE CHANGE

An underlying question throughout this book has been whether or not marginal areas will be sensitive to climate change. Although treelines may be slow to advance, just as they are slow to retreat, there are many other species which have a capacity to invade and which may therefore readily migrate with a changing climate. Changes in both location and species composition of some marginal communities are one of the possible consequences of climatic change. A threat that cannot be ignored is that rapid climatic change may outpace the ability of plants either to migrate or adapt (Jump & Penuelas, 2005).

Even if the threat of extinction is absent, frequent species advances and retreats might impoverish biodiversity. This has been discussed mainly in relation to

animal species (Hewitt, 1996). During periods of rapid climatic change population extinctions might be presumed to outnumber immigrations. In addition, leading edge colonization during a rapid expansion could be expected to be leptokurtic (similar to a normal distribution, but with a higher frequency of values near the mean) and thus lead to increased homozygosity (Hewitt, 1996).

Both past history and current climatic trends mark out the Arctic as an area that has endured an unending series of climatic fluctuations over the past 2–3 million years. It might therefore be expected that repeated mass extinctions and immigrations have reduced polar plant population heterozygosity. Although there are examples of some species in arctic and subarctic regions with low levels of genetic diversity it is not a universal phenomenon and many examples exist of plant species at high latitudes which are just as diverse as those at lower latitudes (Crawford, 2005). This may be due to the ability of many plant species to maintain viable populations in favoured refugia during times of climatic change. Plants have a number of long-term survival strategies that are denied to most animals (see Section 4.1). Being able to dispense with sexual reproduction in favour of asexual reproduction is a major benefit in times of climatic adversity. There are viable aspen populations in Scotland that are now beginning to produce seed, possibly for the first time since the onset of the Little Ice Age (see Section 4.13, Fig. 4.40). Many plant species have reserves of biodiversity in their seed banks and this together with clonal longevity, which can be measured in thousands of years (Steinger *et al.*, 1996), can do much to ensure survival through periods of climatic adversity.

The application of molecular techniques to the biogeography of high-latitude populations is providing more and more evidence to refute the oversimplistic concept of a *tabula rasa* at high latitudes as a result of the Pleistocene ice ages and in particular the Last Glacial Maximum. Studies of high-latitude populations of *Saxifraga oppositifolia*, particularly in Greenland and Beringia (Abbott & Comes, 2004), and white spruce (*Picea glauca*) in Alaska (Anderson *et al.*, 2006) have demonstrated the Pleistocene survival of arctic populations north of the major ice sheets. Such a feat is a striking demonstration of the ability of marginal plant populations to withstand extreme and repeated climatic fluctuations. Having survived substantial climatic oscillations in the past, which sometimes took place with great rapidity, it is not unreasonable to assume that they may be able to adapt to future changes.

12.3   PRE-ADAPTATION OF PLANTS IN MARGINAL AREAS TO CLIMATIC CHANGE

There is an alternative hypothesis to consider in relation to marginal plant populations and their responses to climatic change. As climatic fluctuations are frequent in marginal areas regions the vegetation may be pre-adapted to climatic change. If this is the case the plant communities have to be robust in maintaining their identity and biodiversity and if they do migrate there may be a considerable lag phase before this becomes noticeable.

The habitat tenacity of the arctic flora, and the ability to withstand genetic depauperization, raises the question of how plant populations in marginal areas are able to withstand climatic change. Such a phenomenon has been discussed in relation to both plants and birds in comparisons of populations at the core and the periphery of species distributions (Safriel et al., 1994). The argument asserts that as environmental conditions outside the periphery of a species' distribution prevent population persistence, the peripheral populations must live under conditions different from those of core populations. Peripheral areas are therefore characterized by variable and unstable conditions, relative to core areas. Populations in marginal areas can therefore be expected to be genetically more diverse, since the variable conditions induce fluctuating selection, which maintains high genetic diversity. It is also likely that peripheral populations evolve resistance to extreme conditions. It has even been suggested that peripheral populations rather than core ones may be resistant to environmental extremes and changes, such as global climate change induced by the anthropogenically emitted 'greenhouse gases', and that they should be treated as a biogenetic resource used for rehabilitation and restoration of damaged ecosystems. In addition, it has been claimed that as climatic transition zones are characterized by a high incidence of species represented by peripheral populations, they should be conserved as repositories of these resources, to be used in the future for mitigating undesirable effects of global climate change (Safriel et al., 1994).

Evidence for the support of this argument can also be found at the upper limits for tree growth. A distinct feature in many treelines is the degree of variation that exists in populations as the upper limit for tree survival is approached. In New Zealand mountain beech (Nothofagus solandri) forms hybrids with other Nothofagus species more commonly at the treeline than in the forest core. In the European Alps, the variation found in the treeline krummholz pine (Pinus mugo) is particularly noticeable (Section 5.2.2, Fig. 5.12). Phenological plasticity is also a feature of plants at the limit of their distribution. Cliff-top populations of plantains (e.g. Plantago maritima and P. lanceolata) show great variation in size as well a high degree of pubescence that is not found away from the edge of the cliff (Crawford, 1989). In the Himalaya the genus Saussurea (the alpine saw worts – see Fig. 10.3) are so variable that species identification can be difficult (Mani, 1978). It might also be concluded that populations that have adapted to survive periods of climatic adversity in marginal areas provide a degree of collective homeostasis to the vegetation of peripheral areas. Plants in these regions where extinction and recolonization are frequent occurrences can be conceived as existing in stable metapopulations.

As already argued in Section 2.2.2, divergent adaptations do not normally exist in the same individual, e.g. it is rare for an individual plant to be both flood and drought tolerant. Divergent adaptations can however exist in a metapopulation with the proportion showing any one particular adaptation fluctuating from year to year in response to climatic fluctuations. In wild sunflower (Helianthus annuus) populations the iso-enzyme complement for alcohol dehydrogenase is able to vary depending on whether at the time of sampling there has been a series of wet or dry years (Torres & Diedenhofen, 1979). Such genetic variation confers a wider degree of environmental tolerance on the metapopulation than that found in any one individual.

This same argument is made for the concept of *suspended speciation* as it applied to arctic species (Murray, 1995). In this concept wide-ranging polymorphic species with only low levels of polyploidy and readily capable of hybridizing confer a physiological plasticity in metapopulations that enables the species as a whole to withstand the fluctuations of polar climates. This phenomenon is well illustrated in arctic populations of the purple saxifrage (see Section 2.5, Fig. 2.24).

12.4 PHYSICAL FRAGILITY VERSUS BIOLOGICAL STABILITY AND DIVERSITY

The sight of major physical disturbance inevitably creates alarm and anxiety for the well-being of the biota. When widespread areas are devastated by fire or erosion these anxieties may be well founded. However, an area may be physically fragile and subject to periodic disturbance such as flooding, erosion, drought or insect attack yet nevertheless be biologically diverse as these episodic events often serve to reduce the presence of dominant species. Dune and slack systems are more diverse when prone to physical disturbance. A prolonged period of coastal accretion may appear on the surface to be beneficial in extending the terrain for coastal species, but as distance from the sea increases so can diversity be lost.

The Scottish National Nature Reserve at Tentsmuir already referred to above was in serious decline in terms of its biodiversity during a period of rapid coastal accretion. Birch invaded the dune slacks and increased the process of eutrophication which led to the dominance of a few aggressive grass species (Crawford, 1996b). Fortunately, rising sea levels eroded the major dune front, and inundation with seawater killed the invading birch and restored the diversity of the halophytic vegetation.

Dune systems can look physically fragile and give an appearance of imminent coastal retreat. Such has been the case on the island of Vatersay in the Scottish Hebrides where the fragile nature of dunes on either side of a strip of much grazed machair has long given an impression of imminent disappearance. An early nineteenth century geological visitor described the island of Vatersay (Fig. 12.3) 'as two green hills united by a sandy bar where the opposite seas nearly meet. Indeed if the water did not perpetually supply fresh sand to replace what the wind carries off, it would very soon form two islands; nor would the tenant have much cause for surprise, if on getting up some morning he should find that he required a boat to milk his cows.' However, comparison of the front line of these apparently badly eroded dune fronts with photographs taken 80 years apart (Fig. 12.4) shows little sign of any imminent retreat (Crawford, 2001). The dissected dunes are constantly undergoing a cycle of erosion down to a level where the water table is high enough to enable regeneration and stabilize the dune system first with sand couch grass (*Elytrigia repens*) and then marram (*Ammophila arenaria*).

Salt marshes which consist of a number of small fragmented islands frequently give an impression of physical instability, suggesting that they are in imminent danger of erosion. Here again, as with the tombola beach on Vatersay, photographic records can demonstrate a

Fig. 12.3   The machair and the tombola beaches on the Island of Vatersay (Outer Hebrides). Despite fears that were first recorded nearly 200 years ago for the stability of the tombola, the natural cycle of machair regeneration (see text) has proved effective in prolonging its existence.

Fig. 12.4 Sandy beach on the east side of Vatersay, Outer Hebrides, photographed from the north (upper) on 20 July 1922 by Robert M. Adam, (lower) on 7 May 2000 by the author. Despite evidence of continuing erosion activity there has been no significant retreat of the position of the dunes. Cycles of erosion and regeneration appear to be operating. (Robert Adam photograph reproduced with permission from the Archives and Muniments of the University of St Andrews.)

remarkable stability even of small fragments of marsh (Fig. 12.5).

Similarly, plant communities that live near areas with fluctuating water tables are typically diverse with zonations of various species of reeds, sedges, rushes, and flood-tolerant willow and alder scrub. Each of these groups differs in their ability to tolerate flooding in relation to depth, duration, and seasonality (Chapter 8). When the flooding stress is reduced by drainage then the floristic diversity of the regions is reduced. Reed beds (*Phragmites australis*) are normally found as monospecific stands at the edges of lakes and rivers and suggest a homogeneous type of vegetation. However, within large reed beds on larger lakes molecular studies have shown that there are both morphological and physiological variation between the plants in the centre of a reed bed as compared with those at the edge of the stand, with the former being more tolerant of anoxia than the plants at the periphery (Keller, 2000). Thus a marginal single species plant population can structure itself genetically so as to increase the physiological fitness of the stand as a whole. These selected examples and others mentioned in the foregoing chapters demonstrate an innate biological robustness in many marginal plant communities.

12.5 MARGINAL AREAS AND CONSERVATION

When nature reserves are examined in detail it is often apparent that the greatest areas of species richness are localized. In many cases the regions with greater biodiversity are at margins or ecotones where one community merges with another. In the case illustrated above (Fig. 12.2) at the Tentsmuir Nature Reserve, the immediately visible loss from erosion was the disappearance of 10 m high dunes and replacing them with a deeply cut bay reaching back to where the shoreline had been almost a century ago. The physical loss of the dune did not deprive the reserve of any particularly rare species. However, when the erosion reached the ecotone zone between the flood-line alder community (Fig. 12.6) and the adjacent dune slack it removed a large part of the former territory of the coralroot orchid (*Corallorrhiza trifida*; Fig. 12.7). This was a regrettable loss as Tentsmuir once hosted one of the largest populations of this rare orchid in the British Isles.

The above example illustrates that often it is not any one community in a reserve that is important for maintaining biodiversity but the boundaries and margins between them are.

Ecotones are not only rich in species but also in hybrids between species which gives these areas a special significance in facilitating both plant adaptation and migration (see Chapter 4). It has frequently been pointed out that it is not just individual plants that migrate as seeds but genes also do through pollen dispersal. A time of climatic change is also a time for migration. Ectones, with the niches that they provide for hybrids, will also facilitate gene migration and therefore have a vital conservation role for the movement of plant populations. On a mountainside, high-altitude populations of dwarf birch will hybridize with the common birch that grows at a lower elevation. During a period of climatic warming the F_1 hybrid generation will survive and may backcross with the lower altitude common birch, resulting eventually in the lowland birch replacing the mountain birch (see Chapter 4). Surrogate motherhood by the mountain birch has facilitated by gene migration the movement of common birch to higher altitudes (Fig. 12.8). Marginal situations where the parent species are less viable than the hybrid are ideal areas for preserving the mobility of plant communities, especially in a modern world where intensive land use confines many plant species to isolated wilderness habitats in a matrix of managed land.

12.5.1 Regeneration and the role of margins

Many forest trees have difficulty in regenerating under the shade of the parent trees. Fruiting of hazel (*Corylus avellana*) as discussed in Section 11.4 is much reduced when hazel becomes an understorey tree. Young saplings therefore tend to arise either in clearings or at the margins of the forest. A romantic and controversial hypothesis has even suggested (Vera, 2000) that light-demanding trees and thorny shrubs in temperate plant communities may reflect adaptations to now-extinct large grazers, such as aurochs and tarpans (see Section 2.4.1). The hypothesis suggests that grazing induces a shifting mosaic of grassland, shrub thickets and woodland, and that thorny scrub margins of woodlands might be favoured places for tree regeneration. The hypothesis goes further and claims that the biodiversity of forest trees declines in the absence of large grazers.

Fig. 12.5 Salt marsh at the head of Long Tongue, Sutherland (Northern Scotland). (Upper) Photograph taken in 1949. (Lower) Photograph taken in 2002. Note that even the small islands at the head of the major spits have not visibly altered. (Upper photograph is reproduced from the Valentine collection with permission from the Archives and Muniments of the University of St Andrews.)

Fig. 12.6   Eroding margins on the National Nature Reserve at Tentsmuir due to record high tide levels. Photograph taken in October 2006 showing where erosion has removed an extensive area of dune and slack to breech the flood-line alder association. The vanished margin of the alder stand from the foreground to the middle distance was a former site of the coralroot orchid (*Corallorrhiza trifida*).

Fig. 12.7   The coralroot orchid (*Corallorrhiza trifida*) – a saprophytic herb that inhabits shaded damp *Salix* and alder carr as well as freshwater dune slacks with *Salix repens*. (Photo T. Cunningham.)

The validity of this assertion has been vigorously challenged, as in many parts of the world, from New Zealand to Scotland, grazing by animals such as deer has greatly reduced forest cover and diversity. However, for Scotland at least it has to be remembered that the present red deer population is probably greater than at any time in the past. Deer grazing would have been less severe in times when native wolves controlled the population. There is therefore, despite the obvious criticisms of the Vera hypothesis, some truth in the assertion that thorny thickets around woodlands could protect young saplings from large herbivores and that forest clearings would allow vigorous regeneration. Managers of the *Nothofagus* forests of Patagonia favour wild pigs for regeneration as their rooting of the soil encourages germination and seedling establishment. Winter grazing by sheep increases heathland biodiversity through the opening up of the turf by the impact of their trotters on wet soils. It has even been shown to be particularly helpful in maintaining populations of the rare endemic Scottish primrose (*Primula scotica*; Section 2.2.3). Grazing is essential in herbaceous communities to preserve diversity so there is no a-priori reason that it cannot also aid the regeneration of forest species.

Marginal areas, whether they are forest or any other type of vegetation due to their fluctuating

Fig. 12.8   Birchwood near the treeline in Sutherland (northern Scotland). The dominant form of birch is *Betula intermedia* (*B. pubescens* × *B. nana*), a variable form where the rounded shape and sweet smell of the small leaves (see inset) indicates hybridization with dwarf birch.

Fig. 12.9   Oak (*Quercus petraea*) established in pine (*Pinus sylvestris*) forest in Glen Affric, Scotland. The mild winters of recent years are already favouring the spread of oak in many Scottish glens.

environment, provide opportunities for regeneration and also modification of the species composition. In clearings in pine forests in Scotland, oak establishment (*Quercus petraea*) is now taking place as a result of warmer winters (Fig. 12.9). These marginal areas are even raising the possibility that the native pine forests of the Scottish Highlands are now experiencing very marginal conditions for their survival and may be gradually replaced by oak if the present climatic warming trend continues. The importance of margins in facilitating vegetation succession in response to climatic change should not be underestimated.

12.6   FUTURE PROSPECTS FOR MARGINAL AREAS

In a time of climate change the problems of conservation are increased. Nature reserves as they are commonly constituted are usually centred on areas that are biologically outstanding, either as representing examples of valued plant communities or as sites that have a special interest for the species they harbour. Whatever the reason for the decision that brought about their designation they are normally classified as representing a recognized community. In the British cases this usually carries a reference to an acknowledged national reference system (e.g. Rodwell, 1992). It is also customary to attach a management plan designed to keep this reserve or site in the manner in which it was found. This is not to imply that those involved in nature conservation are unaware of climate change or indeed the innate dynamic processes of vegetation succession that operates even without any climatic alteration. Nevertheless reserves are classified in relation to that part of the natural scene that it is hoped they will preserve. The majority of significant reserves are run either by government-funded agencies or other bodies that are dedicated to preserving our natural heritage and it is therefore understandable that their mission statements can be justified by having a list of sites duly labelled as dune systems, oakwoods, marshes, or bogs, even if it subordinates the inherent dynamic nature of vegetation to a bureaucratic inventory. It would be difficult in any national system to propose the inclusion of areas of ecological interest in the likelihood that they will change. This would imply the probability of their territorial migration away from the designated area and in a culture based on property rights would nullify any hope of legal protection.

Management plans therefore have great difficulty in accommodating dynamic ecosystems. Again this reserve at Tentsmuir in south-east Scotland illustrates this problem. The entire reserve was in danger of being colonized by naturally seeding birch and pine from nearby plantations. This could have been regarded as a natural succession and left to find its own ecological equilibrium. Would a tree-covered dune system have resisted erosion better than naked dunes is a pertinent question. The trees would have reduced ground-level wind speed and if the dunes did erode their timber remains would probably have stabilized the foredunes and the coralroot orchid site might not have vanished from this particular site.

Looking further afield to the Arctic where the greatest degree of climatic warming is expected, many marginal regions for species occurrence can be found and it is already apparent that some wide-ranging scheme of protection may be needed to encompass the range of different habitats that will be dependent on each other in a time of migrating biota (Usher, 2005). The organization for the Conservation of Arctic Flora and Fauna (CAFF) is well aware of the need to enhance ecosystem integrity in the Arctic and to avoid habitat fragmentation and degradation (CAFF, 2002). It must be hoped that both here and elsewhere margins are no longer regarded as mere edge effects. Instead they should be valued as an inherent and vital part of the interacting network of plant communities with special properties that have an essential role in the adjustment of the world's vegetation to climatic change.

References

Aanes, R., Saether, B.-E. & Øritsland, N. A. (2000) Fluctuations of an introduced population of Svalbard reindeer: the effects of density dependence and climatic variation. *Ecography*, **23**, 437–443.

Aanes, R., Saether, B. E., Solberg, E. J., Aanes, S., Strand, O. & Oritsland, N. A. (2003) Synchrony in Svalbard reindeer population dynamics. *Canadian Journal of Zoology-Revue Canadienne de Zoologie*, **81**, 103–110.

Aas, B. & Faarlund, T. (2001) The holocene history of the nordic mountain birch belt. In *Nordic Mountain Birch Ecosystems* (ed. F. E. Wielgolaski), pp. 5–22. New York: Parthenon Publishing.

Abbott, R. J. & Brochmann, C. (2003) History and evolution of the arctic flora: in the footsteps of Eric Hultén. *Molecular Ecology*, **12**, 299–313.

Abbott, R. J. & Comes, H. P. (2004) Evolution in the Arctic: a phylogeographic analysis of the circumarctic plant, *Saxifraga oppositifolia* (purple saxifrage). *New Phytologist*, **161**, 211–224.

Abbott, R. J. & Forbes, D. G. (2002) Extinction of the Edinburgh lineage of the allopolyploid neospecies, *Senecio cambrensis* Rosser (Asteraceae). *Heredity*, **88**, 267–269.

Abbott, R. J., Smith, L. C., Milne, R. I., *et al.* (2000) Molecular analysis of plant migration and refugia in the Arctic. *Science*, **289**, 1343–1346.

Abbott, R. J., Ireland, H. E., Joseph, L., Davies, M. S. & Rogers, H. J. (2005) Recent plant speciation in Britain and Ireland: origins, establishment and evolution of four new hybrid species. *Proceedings of the Royal Irish Academy*, **105B**, 173–181.

Aboal, J. R., Jimenez, M. S., Morales, D. & Gil, P. (2000) Effects of thinning on throughfall in Canary Islands pine forest – the role of fog. *Journal of Hydrology*, **238**, 218–230.

ACIA (2005) *Arctic Climate Impact Assessment*. Cambridge: Cambridge University Press.

Adams, R. P., Pandey, R. N., Leverenz, J. W., Dignard, N., Hoegh, K. & Thorfinnsson, T. (2003) Pan-Arctic variation in *Juniperus communis*: historical biogeography based on DNA fingerprinting. *Biochemical Systematics and Ecology*, **31**, 181–192.

Aerts, R., Cornelissen, J. H. C. Dorrepaal, E., van Logtestijn, R. S. P. & Callaghan, T. V. (2004) Effects of experimentally imposed climate scenarios on flowering phenology and flower production of subarctic bog species. *Global Change Biology*, **10**, 1599–1609.

Aiken, S. G., Dallwitz, M. J., Consaul, L. L., *et al.* (1999 onwards) *Flora of the Canadian Arctic Archipelago: Descriptions, Illustrations, Identification, and Information Retrieval*. Version: 29 April 2003, online: www.mun.ca/biology/delta/arcticf/

Alatalo, J. M. & Totland, O. (1997) Response to simulated climatic change in an alpine and subarctic pollen-risk strategist, *Silene acaulis*. *Global Change Biology*, **3**, 74–79.

Alberdi, M., Bravo, L. A., Gutierrez, A., Gidekel, M. & Corcuera, L. J. (2002) Ecophysiology of Antarctic vascular plants. *Physiologia Plantarum*, **115**, 479–486.

Albrethsen, S. E. & Keller, C. (1986) The use of the saeter in medieval Norse farming in Greenland. *Arctic Anthropology*, **23**, 91–107.

Alekseev, A. S. & Soroka, A. R. (2002) Scots pine growth trends in Northwestern Kola Peninsula as an indicator of positive changes in the carbon cycle. *Climatic Change*, **55**, 183–196.

Allaby, M. (1998) *Oxford Dictionary of Plant Sciences*. Oxford: Oxford University Press.

Allen, N., Nordlander, M., McGonigle, T., Basinger, J. & Kaminskyj, S. (2006) Arbuscular mycorrhizae on Axel Heiberg Island (80 degrees N) and at Saskatoon (52 degrees N) Canada. *Canadian Journal of Botany-Revue Canadienne de Botanique*, **84**, 1094–1100.

Allen, R. B. & Platt, K. H. (1990) Annual seedfall variation in *Nothofagaus solandri* var. *cliffortiodes*. *Oikos*, **57**, 119–206.

Allison, M. A., Khan, S. R., Goodbred, S. L. & Kuehl, S. A. (2003) Stratigraphic evolution of the late Holocene Ganges-Brahmaputra lower delta plain. *Sedimentary Geology*, **155**, 317–342.

Anamthawat-Jonsson, K. & Thorsson, A. T. (2003) Natural hybridisation in birch: triploid hybrids between *Betula nana* and *B. pubescens*. *Plant Cell Tissue and Organ Culture*, **75**, 99–107.

Anderson, A. O. & Anderson, M. O. (1991) *Adomnán's Life of Columba*. Oxford: Clarendon Press.

Anderson, L. L., Hu, F. S., Nelson, D. M., Petit, R. J. & Paige, K. N. (2006) Ice-age endurance: DNA evidence of a white spruce refugium in Alaska. *Proceedings of the National Academy of Sciences of the United States of America*, **103**, 12447–12450.

Andreev, A. A., Tarasov, P. E., Siegert, C., *et al.* (2003) Late Pleistocene and Holocene vegetation and climate on the northern Taymyr Peninsula, Arctic Russia. *Boreas*, **32**, 484–505.

Antonovics, J., Newman, T. J. & Best, B. J. (2001) Spatially explicit studies on the ecology and genetics of population margins. In *Integrating Ecology and Evolution in a Spatial Context* (ed. J. Silvertown & J. Antonovics), pp. 97–116. Oxford: Blackwell Science.

Anttila, C. K., King, R. A., Ferris, C., Ayres, D. R. & Strong, D. R. (2000) Reciprocal hybrid formation of *Spartina* in San Francisco Bay. *Molecular Ecology*, **9**, 765–770.

Arft, A. M., Walker, M. D., Gurevitch, J., *et al.* (1999) Responses of tundra plants to experimental warming: meta-analysis of the international tundra experiment. *Ecological Monographs*, **69**, 491–511.

Armesto, J. J., Casassa, I. & Dollenz, O. (1992) Age structure and dynamics of Patagonian beech forests in Torres del Paine National Park, Chile. *Vegetatio*, **98**, 13–22.

Armstrong, J., Armstrong, W. & Beckett, P. M. (1992) *Phragmites australis*: Venturi-induced and humidity-induced pressure flows enhance rhizome aeration and rhizosphere oxidation. *New Phytologist*, **120**, 197–207.

Armstrong, J., Jones, R. E. & Armstrong, W. (2006) Rhizome phyllosphere oxygenation in *Phragmites* and other species in relation to redox potential, convective gas flow, submergence and aeration pathways. *New Phytologist*, **172**, 719–731.

Armstrong, W. & Armstrong, J. (2005) Stem photosynthesis not pressurized ventilation is responsible for light-enhanced oxygen supply to submerged roots of alder (*Alnus glutinosa*). *Annals of Botany*, **96**, 591–612.

Armstrong, W., Braendle, R. & Jackson, M. B. (1994) Mechanisms of flood tolerance in plants. *Acta Botanica Neerlandica*, **43**, 307–358.

Armstrong, W., Armstrong, J. & Beckett, P. M. (1996) Pressurized ventilation in emergent macrophytes: the mechanism and mathematical modelling of humidity-induced convection. *Aquatic Botany*, **54**, 121–135.

Arseneault, D. & Payette, S. (1992) A postfire shift from lichen spruce to lichen tundra vegetation at tree line. *Ecology*, **73**, 1067–1081.

Arseneault, D. & Payette, S. (1997) Landscape change following deforestation at the Arctic tree line in Quebec, Canada. *Ecology*, **78**, 693–706.

Asselin, H. & Payette, S. (2005) Late Holocene opening of the forest tundra landscape in northern Quebec, Canada. *Global Ecology and Biogeography*, **14**, 307–313.

Atkin, O. K. (1996) Reassessing the nitrogen relations of arctic plants: a mini-review. *Plant, Cell and Environment*, **19**, 695–704.

Atkin, O. K., Botman, B. & Lambers, H. (1996) The causes of inherently slow growth in alpine plants: an analysis based on the underlying carbon economies of alpine and lowland *Poa* species. *Functional Ecology*, **10**, 698–707.

Attenborough, D. (1995) *The Private Life of Plants*. London: BBC Books.

Ayyad, M. A., Fakhry, A. M. & Moustafa, A. R. A. (2000) Plant biodiversity in the Saint Catherine area of the Sinai Peninsula, Egypt. *Biodiversity and Conservation*, **9**, 265–281.

Baddeley, J. A., Woodin, S. J. & Alexander, I. J. (1994) Effects of increased nitrogen and phosphorus availability on the photosynthesis and nutrient relations of 3 arctic dwarf shrubs from Svalbard. *Functional Ecology*, **8**, 676–685.

Bagchi, S., Namgail, T. & Ritchie, M. E. (2006) Small mammalian herbivores as mediators of plant community dynamics in the high-altitude arid rangelands of Trans-Himalaya. *Biological Conservation*, **127**, 438–442.

Bahn, M. & Körner, C. (2003). Recent increases in summit flora caused by warming in the Alps. In *Alpine Biodiversity in Europe* (ed. L. Nagy, G. Grabherr,

C. Körner & D. B. A. Thompson), pp. 437–441. Berlin: Springer.

Ballantyne, C. K., McCarroll, D., Nesje, A., *et al.* (1998) The last ice sheet in north-west Scotland: reconstruction and implications. *Quaternary Science Reviews*, **17**, 1149–1184.

Ballare, C. L., Rousseaux, M. C., Searles, P. S., *et al.* (2001) Impacts of solar ultraviolet-B radiation on terrestrial ecosystems of Tierra del Fuego (southern Argentina) – an overview of recent progress. *Journal of Photochemistry and Photobiology*, **62**, 67–77.

Bannister, P. (1981) Carbohydrate concentration of heath plants of different origins. *Journal of Ecology*, **69**, 769–780.

Bannister, P., Maegli, T., Dickinson, K. J. M., *et al.* (2005) Will loss of snow cover during climatic warming expose New Zealand alpine plants to increased frost damage? *Oecologia*, **144**, 245–256.

Barber, K. E. (1981) *Peat Stratigraphy and Climate Change*. Rotterdam: A. A. Balkema.

Barclay, A. M. & Crawford, R. M. M. (1983) The effect of anaerobiosis on carbohydrate levels in storage tissues of wetland plants. *Annals of Botany*, **51**, 255–259.

Barclay, A. M. & Crawford, R. M. M. (1984) Seedling emergence in the rowan (*Sorbus aucuparia*) from an altitudinal gradient. *Journal of Ecology*, **72**, 627–636.

Barcroft, J. (1934) *Features in the Architecture of Physiological Function*. Cambridge: Cambridge University Press.

Barlow, L. K., Sadler, J. P., Ogilvie, A. E. J., *et al.* (1997) Interdisciplinary investigations of the end of the Norse western settlement in Greenland. *Holocene*, **7**, 489–499.

Barnosky, A. D., Koch, P. L., Feranec, R. S., Wing, S. L. & Shabel, A. B. (2004) Assessing the causes of Late Pleistocene extinctions on the continents. *Science*, **306**, 70–75.

Barton, N. H. & Hewitt, G. M. (1985) Analysis of hybrid zones. *Annual Review of Ecology and Systematics*, **16**, 113–148.

Baskin, C. C. & Baskin, J. M. (1998) *Seeds, Ecology, Biogeography, and Volution of Dormancy and Germination*. San Diego: Academic Press.

Bauert, M. R. (1996) Genetic diversity and ecotypic differentiation in arctic and alpine populations of *Polygonum viviparum*. *Arctic and Alpine Research*, **28**, 190–195.

Baxter-Burrell, A., Yang, Z. B., Springer, P. S. & Bailey-Serres, J. (2002) RopGAP4-dependent Rop GTPase rheostat control of *Arabidopsis* oxygen deprivation tolerance. *Science*, **296**, 2026–2028.

Becker, P., Asmat, A., Mohamad, J., Moksin, M. & Tyree, M. T. (1997) Sap flow rates of mangrove trees are not unusually low. *Trees–Structure and Function*, **11**, 432–435.

Beerling, D. J. (1998) *Salix herbacea* L. *Journal of Ecology*, **86**, 872–895.

Behre, K. E. (1988) The role of man in European vegetation history. In *Vegetation History* (ed. B. Huntley & T. Webb III), pp. 633–672. Dordrecht: Kluwer.

Bell, M. & Walker, M. J. C. (1992) *Late Quaternary Environmental Change*. New York: Longman Scientific and Technical.

Bennett, K. D., Boreham, S., Sharp, M. J. & Switsur, V. R. (1992) Holocene history of environment, vegetation and human settlement on Catta Ness, Lunnasting, Shetland. *Journal of Ecology*, **80**, 241–273.

Bennike, O. & Böcher, J. (1990) Forest-Tundra neighboring the North-Pole – plant and insect remains from the Pliopleistocene Kap Kobenhavn Formation, North Greenland. *Arctic*, **43**, 331–338.

Bennington, C. C., McGraw, J. B. & Vavrek, M. C. (1991) Ecological genetic variation in seed banks. II. Phenotypic and genetic differences between young and old populations of *Luzula parviflora*. *Journal of Ecology*, **79**, 627–643.

Bertrand, A., Castonguay, Y., Nadeau, P., *et al.* (2001) Molecular and biochemical responses of perennial forage crops to oxygen deprivation at low temperature. *Plant Cell and Environment*, **24**, 1085–1093.

Bertrand, A., Castonguay, Y., Nadeau, P., *et al.* (2003) Oxygen deficiency affects carbohydrate reserves in overwintering forage crops. *Journal of Experimental Botany*, **54**, 1721–1730.

Bierzychudek, P. (1985) Patterns in plant parthogenesis. *Experimentia*, **41**, 1255–1263.

Billings, W. D. & Mooney, H. A. (1968) The ecology of arctic and alpine plants. *Biological Reviews*, **43**, 481–529.

Biswas, J. K., Ando, H. & Kakuda, K. (2002) Seedling establishment of anoxia-tolerant rice (*Oryza sativa* L.) as affected by anaerobic seeding in two different soils. *Soil Science and Plant Nutrition*, **48**, 95–99.

Bitonti, M. B., Cozza, R., Wang, G., *et al.* (1996) Nuclear and genomic changes in floating and submerged buds and leaves of heterophyllous waterchestnut (*Trapa natans*). *Physiologia Plantarum*, **97**, 21–27.

Blair, B. (2001) Effect of soil nutrient heterogeneity on the symmetry of belowground competition. *Plant Ecology*, **156**, 199–203.

Blasco, F. (1977) Outlines of ecology, botany and forestry of the mangals of the Indian subcontinent. In *Wet Coastal Ecosystems*, Vol. 1 (ed. V. J. Chapman), pp. 241–260. Amsterdam: Elsevier.

Bliss, L. C. (1977) Vascular plants of Truelove Lowland and adjacent areas including their relative importance. In *Truelove Lowland, Devon Island, Canada: A High Arctic Ecosystem* (ed. L. C. Bliss), pp. 697–698. Edmonton: University of Alberta Press.

Bliss, L. C. & Matveyeva, N. V. (1992) Circumpolar arctic vegetation. In *Arctic Ecosystems in a Changing Climate* (ed. F. S. Chapin III, R. L. Jefferies, J. F. Reynolds, G. R. Shaver, J. Svoboda & E. W. Chu), pp. 59–89. San Diego: Academic Press.

Blom, C., Voesenek, L., Banga, M., *et al.* (1994) Physiological ecology of riverside species – adaptive responses of plants to submergence. *Annals of Botany*, **74**, 253–263.

Boamfa, E. I., Ram, P. C., Jackson, M. B., Reuss, J. & Harren, F. J. M. (2003) Dynamic aspects of alcoholic fermentation of rice seedlings in response to anaerobiosis and to complete submergence: relationship to submergence tolerance. *Annals of Botany*, **91**, 279–290.

Bond, W. J. & van Wilgen, B. W. (1996) *Fire and Plants*. London: Chapman and Hall.

Bond, W. J., Woodward, F. I. & Midgley, G. F. (2005) The global distributions of ecosystems in a world without fire. *New Phytologist*, **165**, 525–538.

Boom, A., Mora, G., Cleef, A. M. & Hooghiemstra, H. (2001) High altitude C-4 grasslands in the northern Andes: relicts from glacial conditions? *Review of Palaeobotany and Palynology*, **115**, 147–160.

Bradley, P. M. & Morris, J. T. (1990) Influence of oxygen and sulfide concentration on nitrogen uptake kinetics in *Spartina alterniflora*. *Ecology*, **7**, 282–287.

Braendle, R. (1991) Flooding resistance of rhizomatous amphibious plants. In *Plant Life Under Oxygen Stress: Ecology, Physiology and Biochemistry* (ed. M. B. Jackson, D. D. Davies & H. Lambers), pp. 35–46.

The Hague, The Netherlands: SPB Academic Publishing.

Braendle, R. & Crawford, R. M. M. (1999) Plants as amphibians. *Perspectives in Plant Evolution and Systematics*, **2**, 56–78.

Braithwaite, M. E., Ellis, R. W. & Preston, C. D. (2006) *Change in the British Flora*. London: Botanical Society of the British Isles.

Brando, P. M. & Durigan, G. (2004) Changes in cerrado vegetation after disturbance by frost (Sao Paulo State, Brazil). *Plant Ecology*, **175**, 205–215.

Bravo, L. A., Ulloa, N., Zuniga, G. E., Casanova, A., Corcuera, L. J. & Alberdi, M. (2001) Cold resistance in Antarctic angiosperms. *Physiologia Plantarum*, **111**, 55–65.

Brewer, S., Hely-Alleaume, C., Cheddadi, R., *et al.* (2005) Post-glacial history of Atlantic oakwoods: context, dynamics and controlling factors. *Botanical Journal of Scotland*, **57**, 41–57.

Briffa, K. R., Jones, P. D., Schweingruber, F. H., Shiyatov, S. G. & Cook, E. R. (1995) Unusual 20th-century summer warmth in a 1,000-year temperature record from Siberia. *Nature*, **376**, 156–159.

Brochmann, C., Soltis, D. E. & Soltis, P. S. (1992) Electrophoretic relationships and phylogeny of Nordic polyploids in *Draba* (Brassicaceae). *Plant Systematics and Evolution*, **182**, 35–70.

Brochmann, C., Gabrielsen, T. M., Nordal, I., Landvik, J. Y. & Elven, R. (2003) Glacial survival or tabula rasa? The history of North Atlantic biota revisited. *Taxon*, **52**, 417–450.

Brochmann, C., Brysting, A. K., Alsos, I. G., *et al.* (2004) Polyploidy in arctic plants. *Biological Journal of the Linnean Society*, **82**, 521–536.

Brown, J. (1997) Disturbance and recovery of permafrost terrain. In *Disturbance and Recovery in Arctic Lands: An Ecological Perspective* (ed. R. M. M. Crawford), pp. 167–168. Dordrecht: Kluwer Academic Publishers.

Brown, J., Ferrians, J. O. J., Heginbottom, J. A. & Melnikov, E. S. (1995) *Circum-arctic Map of Permafrost and Ground-ice Conditions*. Reston. VA: U.S. Geological Survey.

Brubaker, L. B., Anderson, P. M., Edwards, M. E. & Lozhkin, A. V. (2005) Beringia as a glacial refugium for boreal trees and shrubs: new perspectives from mapped pollen data. *Journal of Biogeography*, **32**, 833–848.

Brunstein, F. C. & Yamaguchi, D. K. (1992) The oldest known Rocky-Mountain bristlecone-pines

(*Pinus–Aristata* Engelm). *Arctic and Alpine Research*, **24**, 253–256.

Bryson, R. A. (1966) Air masses, streamlines, and the boreal forest. *Geographical Bulletin*, 8, 228–269.

Brysting, A. K., Gabrielsen, T. M., Sørlibråten, O., Ytrehorn, O. & Brochmann, C. (1996) The purple saxifrage, *Saxifraga oppositifolia*, in Svalbard: two taxa or one? *Polar Research*, 15, 93–105.

Bucher, M., Braendle, R. & Kuhlemeier, C. (1996) Glycolytic gene expression in amphibious *Acorus calamus* L. under natural conditions. *Plant and Soil*, **178**, 75–82.

Buchmann, N., Kao, W. Y. & Ehleringer, J. R. (1996) Carbon dioxide concentrations within forest canopies: variation with time, stand structure, and vegetation type. *Global Change Biology*, 2, 421–432.

Buckland, P. & Dugmore, A. J. (1991) "If this is a refugium, why are my feet so bloody cold?" The origins of the Icelandic biota in the light of recent research. In *Environmental Change in Iceland: Past and Present* (ed. J. K. Maizels & C. Caseldine), pp. 107–125. Dordrecht: Kluwer Academic Publishers.

Bunting, M. J. (1996) Holocene vegetation and environment of Orkney. In *The Quaternary of Orkney: Field Guide* (ed. A. M. Hall), pp. 20–29. Cambridge: Quaternary Research Association.

Burdick, D. & Konisky, R. A. (2003) Determinants of expansion for *Phragmites australis*, common reed, in natural and impacted coastal marshes. *Estuaries*, 26, 407–416.

Burke, A. (2003) Inselbergs in a changing world – global trends. *Diversity and Distributions*, 9, 375–383.

CAFF (2002) Arctic flora and fauna: recommendations for conservation. In *Conservation of Arctic Flora and Fauna*, Akureyri: International Secretariat.

Cairns, D. M. (2001) Patterns of winter desiccation in krummholz forms of *Abies lasiocarpa* at treeline sites in Glacier National Park, Montana, USA. *Geografiska Annaler Series a–Physical Geography*, 83A, 157–168.

Cairns, D. M. (2005) Simulating carbon balance at treeline for krummholz and dwarf tree growth forms. *Ecological Modelling*, 187, 314–328.

Caldwell, M. M., Ballare, C. L., Bornman, J. F., *et al.* (2003) Terrestrial ecosystems increased solar ultraviolet radiation and interactions with other climatic change factors. *Photochemical and Photobiological Sciences*, 2, 29–38.

Callaghan, T. V., Carlsson, B. A., Jonsdottir, I. S., Svensson, B. M. & Jonasson, S. (1992) Clonal plants and

environmental-change: introduction to the proceedings and summary. *Oikos*, 63, 341–347.

Callaghan, T. V., Crawford, R. M. M., Eronen, M., *et al.* (2002) The dynamics of the tundra-taiga boundary: an overview and suggested coordinated and integrated approach to research. *Ambio*, Special issue 12, 3–5.

Callaghan, T. V. *et al.* (2005) Arctic tundra and polar desert ecosystems. In *Arctic Climate Impact Assessment*, pp. 244–352. Cambridge: Cambridge University Press.

Calow, P., ed. (1998) *The Encyclopedia of Ecology and Environmental Management*. Oxford: Blackwell Science.

Campbell, D. R. & Waser, N. M. (2001) Genotype-by-environment interaction and the fitness of plant hybrids in the wild. *Evolution*, 55, 669–676.

Campbell, I. D. & McAndrews, J. H. (1993) Forest disequilibrium caused by rapid Little Ice Age cooling. *Nature*, 366, 336–338.

Caseldine, C., Thompson, G., Langdon, C., & Hendon, D. (2005) Evidence for an extreme climatic event on Achill Island, Co. Mayo, Ireland around 5200–5100 cal. yr BP. *Journal of Quaternary Science*, 20, 169–178.

Caulfield, S., O'Donnell, R. G. & Mitchell, P. I. (1998) C-14 dating of a Neolithic field system at Ceide Fields, County Mayo, Ireland. *Radiocarbon*, 40, 629–640.

Chambers, F. M., Barber, K. E., Maddy, D. & Brew, J. (1997) A 5500-year proxy-climate and vegetation record from blanket mire at Talla Moss, Borders, Scotland. *Holocene*, 7, 391–399.

Chambers, R. M., Osgood, D. T., Bart, D. J & Montalto, F. (2003) *Phragmites australis* invasion and expansion in tidal wetlands: interactions among salinity, sulfide, and hydrology. *Estuaries*, 26, 398–406.

Chapin, F. S. & Shaver, G. R. (1996) Physiological and growth-responses of arctic plants to a field experiment simulating climatic-change. *Ecology*, 77, 822–840.

Chapin, F. S., III, Johnson, D. A. & McKendrick, J. D. (1980) Seasonal movement of nutrients in plants of differing growth form in an Alaskan tundra ecosystem: implications for herbivory. *Journal of Ecology*, 68, 189–209.

Chapin, F. S., III, Moilanen, L. & Keilland, K. (1993) Preferred use of organic nitrogen for growth by non-mycorrhizal arctic sedge. *Nature*, 361, 150–153.

Chapin, F. S., Shaver, G. R., Giblin, A. E., Nadelhoffer, K. J. & Laundre, J. A. (1995) Responses of arctic tundra to experimental and observed changes in climate. *Ecology*, 76, 694–711.

Chapin, F. S., Sturm, M., Serreze, M. C., *et al.* (2005) Role of land-surface changes in Arctic summer warming. *Science*, **310**, 657–660.

Chapman, H. & Bicknell, R. (2000) Recovery of a sexual and an apomictic hybrid from crosses between the facultative apomicts *Hieracium caespitosum* and *H. praealtum*. *New Zealand Journal of Ecology*, **24**, 81–85.

Chas, E., Le Driant, F., Dentant, C., *et al.* (2006) *Atlas des Plantes Rares ou Protéges des Hautes Alpes*. Turriers: Naturalia Publications.

Chernov, Y. I. & Matveyeva, N. V. (1997) Arctic ecosystems in Russia. In *Polar and Alpine Tundra* (ed. F. E. Wielgolaski), pp. 361–507. Amsterdam: Elsevier.

Childe, V. G. (1928) *The Most Ancient East: The Oriental Prelude to European Prehistory*. London: Routledge and Kegan Paul.

Christensen, K. K., Jensen, H. S., Andersen, F. O., Wigand, C. & Holmer, M. (1998) Interferences between root plaque formation and phosphorus availability for isoetids in sediments of oligotrophic lakes. *Biogeochemistry*, **43**, 107–128.

Clemmensen, K. E. & Michelsen, A. (2006) Integrated long-term responses of an arctic-alpine willow and associated ectomycorrhizal fungi to an altered environment. *Canadian Journal of Botany–Revue Canadienne de Botanique*, **84**, 831–843.

Clevering, O. A., Blom, C. & VanVierssen, W. (1996) Growth and morphology of *Scirpus lacustris* and *S. maritimus* seedlings as affected by water level and light availability. *Functional Ecology*, **10**, 289–296.

Colwell, R. K. & Coddington, J. A. (1995) Estimating terrestrial biodiversity through extrapolation. In *Biodiversity: Measurement and Estimation* (ed. D. L. Hawksworth). London: Chapman and Hall.

Comtois, P., Simon, J. P. & Payette, S. (1986) Clonal constitution and sex ratio in northern populations of balsam poplar *Populus balsamifera*. *Holarctic Ecology*, **9**, 251–260.

Conner, W. H., Gosselink, J. G. & Parrondo, R. T. (1981) Comparison of the vegetation of three Louisiana swamp sites with different flooding regimes. *American Journal of Botany*, **68**, 320–331.

Conrad, V. (1946) Usual formulas of continentality and their limits of validity. *Transactions American Geophysical Union*, **27**, 663–664.

Constable, J. V. H. & Longstreth, D. J. (1994) Aerenchyma carbon-dioxide can be assimilated in *Typha latifolia* L. leaves. *Plant Physiology*, **106**, 1065–1072.

Constable, J. V. H., Grace, J. B. & Longstreth, D. J. (1992) High carbon dioxide concentrations in aerenchyma of *Typha latifolia*. *American Journal of Botany*, **79**, 415–418.

Convey, P. & Smith, R. I. L. (2006) Responses of terrestrial Antarctic ecosystems to climate change. *Plant Ecology*, **182**, 1–10.

Cooper, A. (1997) Plant species coexistence in cliff habitats. *Journal of Biogeography*, **24**, 483–494.

Cooper, E. J. & Wookey, P. A. (2003) Floral herbivory of *Dryas octopetala* by Svalbard reindeer. *Arctic Antarctic and Alpine Research*, **35**, 369–376.

Cooper, E. J., Alsos, I. G., Hagen, D., Smith, F. M., Coulson, S. J. & Hodkinson, I. D. (2004) Plant recruitment in the High Arctic: seed bank and seedling emergence on Svalbard. *Journal of Vegetation Science*, **15**, 115–124.

Cornelissen, J. H. C., Aerts, R., Cerabolini, B., Werger, M. J. A. & van der Heijden, M. G. A. (2001) Carbon cycling traits of plant species are linked with mycorrhizal strategy. *Oecologia*, **129**, 611–619.

Costich, D. E. & Meagher, T. R. (1992) Genetic variation in *Ecballium elateriumi* (Cucurbitaceae): breeding system and geographic distribution. *Journal of Evolutionary Biology*, **5**, 589–601.

Coutts, M. P. & Philipson, J. J. (1987) Structure and physiology of Sitka spruce roots. *Proceedings of the Royal Society of Edinburgh*, **93B**, 131–144.

Cowie, T. G. & Shepherd, I. G. (1997) The Bronze Age. In *Scotland: Environment and Archaeology, 8000 BC–AD 1000* (ed. K. J. Edwards & I. B. M. Ralston), pp. 151–193. Chichester: John Wiley.

Cowling, R. M. & Richardson, D. (1995) *Fynbos: South Africa's Unique Floral Kingdom*. Vlaeberg: Fernwood Press.

Cowling, R. M., Holmes, P. M. & Rebelo, A. G. (1992) Plant diversity and endemism. In *The Ecology of Fynbos* (ed. R. M. Cowling), pp. 62–110. Cape Town: Oxford University Press.

Cowling, R. M., Witkowski, E. T. F., Milewski, A. V. & Newbey, K. R. (1994) Taxonomic, edaphic and biological aspects of narrow plant endemism on matched sites in Mediterranean South Africa and Australia. *Journal of Biogeography*, **21**, 651–664.

Crawford, R. M. M. (1989) *Studies in Plant Survival*. Oxford: Blackwell Scientific Publications.

Crawford, R. M. M. (1992) Oxygen availability as an ecological limit to plant distribution. *Advances in Ecological Research*, **23**, 93–185.

Crawford, R. M. M. (1996a) Whole plant adaptations to fluctuating water tables. *Folia Geobotanica et Phytotaxonomica*, **31**, 7–24.

Crawford, R. M. M. (1996b) Tentsmuir Point: a national nature reserve in decline? In *Fragile Environments: The Use and Management of Tentsmuir* (ed. G. W. Whittington), pp. 65–88. Aberdeen: Scottish Cultural Press.

Crawford, R. M. M. (1997a) Habitat fragility as an aid to long term survival in arctic vegetation. In *Ecology of Arctic Environments* (ed. S. J. Woodin & M. Marquiss), Special Publication 13 of the British Ecological Society, pp. 113–136. Oxford: Blackwell Scientific.

Crawford, R. M. M., ed. (1997b) *Disturbance and Recovery in Arctic Lands: An Ecological Perspective.* Dordrecht: Kluwer Academic Publishers.

Crawford, R. M. M. (1999) The Arctic as a peripheral area. In *The Species Concept in the High North – A Panarctic Flora Initiative* (ed. I. Nordal & V. Y. Razzhivin), pp. 131–153. Oslo: The Norwegian Academy of Science and Letters.

Crawford, R. M. M. (2000) Ecological hazards of oceanic environments. *New Phytologist*, **147**, 257–281.

Crawford, R. M. M. (2001) Plant community responses to Scotland's changing environment. *Botanical Journal of Scotland*, **53**, 77–105.

Crawford, R. M. M. (2003) Seasonal differences in plant responses to flooding and anoxia. *Canadian Journal of Botany–Revue Canadienne de Botanique*, **81**, 1224–1246.

Crawford, R. M. M. (2004) Long-term plant survival at high latitudes. *Botanical Journal of Scotland*, **56**, 1–23.

Crawford, R. M. M. (2005) Peripheral plant population survival in polar regions. In *Mountain Ecosystems: Studies in Treeline Ecology* (ed. G. Broll & B. Keplin), pp. 43–76. Berlin: Springer.

Crawford, R. M. M. & Balfour, J. (1983) Female predominant sex ratios and physiological differentiation in arctic willows. *Journal of Ecology*, **71**, 149–160.

Crawford, R. M. M. & Balfour, J. (1990) Female-biased sex ratios and differential growth in arctic willows. *Flora*, **183**, 291–302.

Crawford, R. M. M. & Braendle, R. (1996) Oxygen deprivation stress in a changing climate. *Journal of Experimental Botany*, **47**, 145–159.

Crawford, R. M. M. & Jeffree, C. E. (2007) Northern climates and woody plant distribution. In *Arctic Alpine Ecosystems and People in a Changing Environment* (ed. J. B. Ørbk, R. Kallenborn, I. Tombre, E. N. Hegseth, S. Falk-Petersen & A. H. Hoel), pp. 85–104. Berlin: Springer-Verlag.

Crawford, R. M. M. & Palin, M. A. (1981) Root respiration and temperature limits to the north–south distribution of four perennial maritime species. *Flora*, **171**, 338–354.

Crawford, R. M. M. & Smith, L. C. (1997) Responses of some high Arctic shore plants to variable lengths of growing season. *Opera Botanica*, **132**, 201–214.

Crawford, R. M. M. & Wishart, D. (1966) A multivariate analysis of the development of dune slack vegetation in relation to coastal accretion at Tentsmuir, Fife. *Journal of Ecology*, **54**, 729–743.

Crawford, R. M. M., Wishart, D. & Campbell, R. M. (1970) A numerical analysis of high altitude scrub vegetation in relation to soil erosion in the eastern Cordillera of Peru. *Journal of Ecology*, **58**, 173–191.

Crawford, R. M. M., Monk, L. S. & Zochowski, Z. M. (1987) Enhancement of anoxia tolerance by removal of volatile products of anaerobiosis. In *Plant Life in Aquatic and Amphibious Habitats* (ed. R. M. M. Crawford), pp. 375–384. Oxford: Blackwell Scientific Publications.

Crawford, R. M. M., Studer, C. & Studer, K. (1989) Deprivation indifference as a survival strategy in competition: advantages and disadvantages of anoxia tolerance in wetland vegetation. *Flora*, **182**, 189–201.

Crawford, R. M. M., Chapman, H. M., Abbott, R. J. & Balfour, J. (1993) Potential impact of climatic warming on arctic vegetation. *Flora*, **43**, 367–381.

Crawford, R. M. M., Chapman, H. M. & Hodge, H. (1994) Anoxia tolerance in high Arctic vegetation. *Arctic and Alpine Research*, **26**, 308–312.

Crawford, R. M. M., Chapman, H. M. & Smith, L. C. (1995) Adaptation to variation in growing season length in arctic populations of *Saxifraga oppositifolia* L. *Botanical Journal of Scotland*, **41**, 177–192.

Crawford, R. M. M., Jeffree, C. E. & Rees, W. G. (2003) Paludification and forest retreat in northern oceanic environments. *Annals of Botany*, **91**, 213–226.

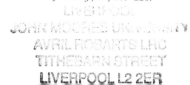

Crawley, M. J., ed. (1997) *Plant Ecology*. Oxford: Blackwell Science.

Cruzan, M. B. (2005) Patterns of introgression across an expanding hybrid zone: analysing historical patterns of gene flow using nonequilibrium approaches. *New Phytologist*, **167**, 267–278.

Currey, D. R. (1965) An ancient bristlecone pine stand in eastern Nevada. *Ecology*, **46**, 564–566.

Dahl, E. (1951) On the relation between summer temperatures and the distribution of alpine plants in the lowlands of Fennoscandinavia. *Oikos*, **3**, 22–52.

Dahl, E. (1990) Probable effects of climatic change due to the greenhouse effect on plant productivity and survival in North Europe. In *Effects of Climate Change on Terrestrial Ecosystems* (ed. J. I. Holtne), pp. 7–18. Trondheim: Norwegian Institute for Nature Research.

Dahl, E. (1998) *The Phytogeography of Northern Europe*. Cambridge: Cambridge University Press.

Dalen, L. & Hofgaard, A. (2005) Differential regional treeline dynamics in the Scandes Mountains. *Arctic, Antarctic and Alpine Research*, **37**, 284–296.

Dalpe, Y. & Aiken, S. G. (1998) Arbuscular mycorrhizal fungi associated with *Festuca* species in the Canadian high arctic. *Canadian Journal of Botany–Revue Canadienne de Botanique*, **76**, 1930–1938.

Dalton, D. A., Kramer, S., Azios, N., Fusaro, S., Cahill, E. & Kennedy, C. (2004) Endophytic nitrogen fixation in dune grasses (*Ammophila arenaria* and *Elymus mollis*) from Oregon. *Fems Microbiology Ecology*, **49**, 469–479.

Darling, F. F. & Boyd, J. M. (1964) *The Highlands and Islands*. London: Collins.

Davis, M. A., Grime, J. P. & Thompson, K. (2000) Fluctuating resources in plant communities: a general theory of invasability. *Journal of Ecology*, **88**, 528–534.

Dawson, T. E. (1998) Fog in the California redwood forest: ecosystem inputs and use by plants. *Oecologia*, **117**, 476–485.

Dawson, T. E. & Bliss, L. C. (1989a) Intraspecific variation in the water relations of *Salix arctica*, an arctic–alpine dwarf willow. *Oecologia*, **79**, 322–331.

Dawson, T. E. & Bliss, L. C. (1989b) Patterns of water use and the tissue water relations in the dioecious shrub, *Salix arctica*: the physiological; basis for habitat partitioning between the sexes. *Oecologia*, **79**, 332–343.

Dawson, T. E. & Bliss, L. C. (1993) Plants as mosaics – leaf-level, ramet-level, and gender-level variation in the physiology of the dwarf willow, *Salix arctica*. *Functional Ecology*, **7**, 293–304.

Dawson, T. E. & Pate, J. S. (1996) Seasonal water uptake and movement in root systems of Australian phraeatophytic plants of dimorphic root morphology: a stable isotope investigation. *Oecologia*, **107**, 13–20.

Dawson, T. E., Ward, J. K. & Ehleringer, J. R. (2004) Temporal scaling of physiological responses from gas exchange to tree rings: a gender-specific study of *Acer negundo* (Boxelder) growing under different conditions. *Functional Ecology*, **18**, 212–222.

Day, T. A., Vogelmann, T. C. & Delucia, E. H. (1992) Are some plant life forms more effective than others in screening out ultraviolet-B radiation? *Oecologia*, **92**, 513–519.

Day, T. A., Ruhland, C. T. & Xiong, F. S. (2001) Influence of solar ultraviolet-B radiation on Antarctic terrestrial plants: results from a 4-year field study. *Journal of Photochemistry and Photobiology B–Biology*, **62**, 78–87.

DeGroot, W. J. & Wein, R. W. (1999) *Betula glandulosa* Michx. response to burning and postfire growth temperature and implications of climate change. *International Journal of Wildland Fire*, **9**, 51–64.

DeGroot, W. J., Thomas, P. A. & Wein, R. W. (1997) *Betula nana* L. and *Betula glandulosa* Michx. *Journal of Ecology*, **85**, 241–264.

Denevan, W. M. (2001) *Cultivated Landscapes of Native Amazonia and the Andes*. Oxford: Oxford University Press.

Dennis, A., Foote, P. & Perkins, R., eds. (1980) *Laws of Early Iceland*. Winnipeg: University of Manitoba Press.

Desmet, P. G. & Cowling, R. M. (1999) Biodiversity, habitat and range-size aspects of a flora from a winter-rainfall desert in north-western Namaqualand, South Africa. *Plant Ecology*, **142**, 23–33.

Di Tomaso, J. M. (1998) Impact, biology, and ecology of saltcedar (*Tamarix* spp.) in the southwestern United States. *Weed Technology*, **12**, 326–336.

Dickinson, G. & Randall, R. E. (1979) An interpretation of machair vegetation. *Proceedings of the Royal Society of Edinburgh*, **77B**, 267–278.

Diemer, M. (1996) The incidence of herbivory in high-elevation populations of *Ranunculus glacialis*: a re-evaluation of stress-tolerance in alpine environments. *Oikos*, **75**, 486–492.

Diggle, P. K. (1997) Extreme preformation in alpine *Polygonum viviparum*: an architectural and developmental analysis. *American Journal of Botany*, **84**, 154–169.

Dodgshon, R. A. (1994a) Budgeting for survival: nutrient flow and traditional Highland farming. In *The History of Soils and Field Systems* (ed. S. Foster & T. C. Smout), pp. 83–93. Aberdeen: Scottish Cultural Press.

Dodgshon, R. A. (1994b) Rethinking highland field systems. In *The History of Soils and Field Systems* (ed. S. Foster & T. C. Smout), pp. 53–65. Aberdeen: Scottish Cultural Press.

Dormann, C. F. & Skarpe, C. (2002) Flowering, growth and defence in the two sexes: consequences of herbivore exclusion for *Salix polaris*. *Functional Ecology*, **16**, 649–656.

Doyle, A. C., Sir (1912) *The Lost World: Being an Account of the Recent Amazing Adventures of Professor E. Challenger, Lord John Roxton, Professor Summerlee and Mr Ed Malone of the 'Daily Gazette'*. London: Hodder and Stoughton.

Duncalf, W. G. (1976) *The Guinness Book of Plant Facts and Feats*. London: Guinness Superlatives.

Dunning, R. W. (2004) *A History of the County of Somerset*. Vol. 8, *The Polden and the Levels (The Victoria History of the Counties of England)*. London: Boydell & Brewer.

Edwards, M. E. & Barker, E. D. (1994) Climate and vegetation in Northeastern Alaska 18,000 yr Bp – Present. *Palaeogeography, Palaeoclimatology, Palaeoecology*, **109**, 127–135.

Ehlers, J. & Gibbard, P. (2004) *Quaternary Glaciations: Extent and Chronology*. Amsterdam: Elsevier.

Eiten, G. (1972) The cerrado vegetation of Brazil. *The Botanical Review*, **38**, 201–341.

Ellenberg, H. (1963) *Vegetation Mitteleuropas mit den Alpen*. Stuttgart: Eugen Ulmer.

Elvebakk, A. (1999) Bioclimatic delimitation and subdivision of the Arctic. In *The Species Concept in the High North – A Panarctic Flora Initiative* (ed. I. Nordal & V. Y. Razzhivin), pp. 81–112. Oslo: The Norwegian Academy of Science and Letters.

Emmerson, M. (2000) Remedial habitat creation: does *Nereis diversicolor* play a confounding role in the colonisation and establishment of the pioneering saltmarsh plant, *Spartina anglica*? *Helgoland Marine Research*, **54**, 110–116.

Emms, S. K. & Arnold, M. L. (1997) The effect of habitat on parental and hybrid fitness: transplant experiments with Louisiana irises. *Evolution*, **51**, 1112–1119.

Ennos, R. A., Cowie, N. R., Legg, C. J. & Sydes C. (1997) Which measures of genetic variation are relevant in plant conservation? In *The Role of Genetics in Conserving Small Populations* (ed. T. E. Tew, J. W. Crawford, J. W. Spencer, *et al.*), pp. 73–79. Peterborough: Joint Nature Conservation Committee.

Eriksen, B. & Topel, M. H. (2006) Molecular phylogeography and hybridization in members of the circumpolar *Potentilla* sect. Niveae (Rosaceae). *American Journal of Botany*, **93**, 460–469.

Eriksson, O. & Ehrlen, J. (2001) Landscape fragmentation and the viability of populations. In *Integrating Ecology and Evolution in a Spatial Context* (ed. J. Silvertown & J. Antonovics), pp. 157–175. Oxford: Blackwell Science.

Erkamo, V. (1956) Untersuchungen über die Pflanzenbiologischen und einige andere Folgeerscheinungen der neuzeitlichen Klimaschwankung in Finnland. *Annales Botanici Societas-Zoologicae Botanicae Fennicae 'Vanamo'*, **28**, 1–283.

Erschbaner, B., Virtanen, R. & Nagy, L. (2003) The impacts of vertebrate grazers on vegetation in European high mountains. In *Alpine Biodiversity in Europe* (ed. L. Nagy, G. Grabherr, C. Körner & D. B. A. Thompson), pp. 377–396. Berlin: Springer-Verlag.

Ewel, K. C. & Odum, H. T., eds. (1984) *Cypress Swamps*. Gainesville: University of Florida Press.

Fjeldsa, J., Ehrlich, D., Lambin, E. & Prins, E. (1997) Are biodiversity 'hotspots' correlated with current ecoclimatic stability? A pilot study using the NOAA-AVHRR remote sensing data. *Biodiversity and Conservation*, **6**, 401–422.

Fleisher, A. & Fleisher, Z. (2004) Study of *Dictamnus gymnostylis* volatiles and plausible explanation of the "burning bush" phenomenon. *Journal of Essential Oil Research*, **16**, 1–3.

Fleming, C. A. (1980) *The Geological History of New Zealand and its Life*. Auckland: Auckland University Press.

Forbes, J. C. (1976) Influence of management and environmental factors on distribution of marsh ragwort *Senecio aquaticus* hybrids in agricultural grassland in Orkney. *Journal of Applied Ecology*, **13**, 985–990.

Forbes, J. C. & Kenworthy, J. B. (1973) Distribution of two species of birch forming stands in Deeside,

Aberdeenshire. *Transactions of the Botanical Society of Edinburgh*, **42**, 101–110.

Forchhammer, M. C., Post, E., Berg, T. B. G., Hoye, T. T. & Schmidt, N. M. (2005) Local-scale and short-term herbivore-plant spatial dynamics reflect influences of large-scale climate. *Ecology*, **86**, 2644–2651.

Fossitt, J. A. (1996) Late quaternary vegetation history of the Western Isles of Scotland. *New Phytologist*, **132**, 171–196.

Fox, G., McCallan, N. R. & Ratcliffe, R. G. (1995) Manipulating cytoplasmic pH under anoxia: a critical test of the role of pH in the switch from aerobic to anaerobic metabolism. *Planta*, **195**, 324–330.

Frankel, O. H., Brown, A. H. D. & Burdon, J. J. (1995) *The Conservation of Plant Biodiversity*. Cambridge: Cambridge University Press.

Fredskild, B. (1991) The genus *Betula* in Greenland: Holocene history, present distribution and synecology. *Nordic Journal of Botany*, **11**, 393–412.

Freeman, D. C., Klikoff, L. G. & Harper, K. T. (1976) Differential resource utilisation by the sexes of dioecious plants. *Science*, **193**, 597–599.

French, C. E., French, J. R., Clifford, N. J. & Watson, C. J. (2000) Sedimentation-erosion dynamics of abandoned reclamations: the role of waves and tides. *Continental Shelf Research*, **20**, 1711–1733.

Fukao, T., Kennedy, R. A., Yamasue, Y. & Rumpho, M. E. (2003) Genetic and biochemical analysis of anaerobically-induced enzymes during seed germination of *Echinochloa crus-galli* varieties tolerant and intolerant of anoxia. *Journal of Experimental Botany*, **54**, 1421–1429.

Fürtig, K., Ruegsegger, A., Brunold, C. & Braendle, R. (1996) Sulphide utilisation and injuries in hypoxic roots and rhizomes of reed plants (*Phragmites australis*). *Folia Geobotanica et Phytotaxonomica*, **31**, 143–151.

Gamache, I. & Payette, S. (2005) Latitudinal response of subarctic tree lines to recent climate change in eastern Canada. *Journal of Biogeography*, **32**, 849–862.

Gansert, D. (2003) Xylem sap flow as a major pathway for oxygen supply to the sapwood of birch (*Betula pubescens* Ehr.). *Plant Cell and Environment*, **26**, 1803–1814.

Garcia, D., Zamora, R., Gomez, J. M., Jordano, P. & Hodar, J. A. (2000) Geographical variation in seed production, predation and abortion in *Juniperus communis* throughout its range in Europe. *Journal of Ecology*, **88**, 436–446.

Garcilaso de la Vega (1608) *The Royal Commentaries of the Incas*. English edition (1869) by the Hakluyt Society, London.

Gee, D. & Anderson, L. W. J. (1998) Influence of leaf age on responsiveness of *Potamogeton nodosus* to ABA-induced heterophylly. *Plant Growth Regulation*, **24**, 119–125.

Geoghegan, I. E. & Sprent, J. I. (1996) Aluminium and nutrient contents in species native to central Brazil. *Communications in Soil Science and Plant Analysis*, **27**, 2925–2934.

Gerloff, L. M., Hills, L. V. & Osborn, G. D. (1995) Postglacial vegetation history of the Mission Mountains, Montana. *Journal of Paleolimnology*, **14**, 269–279.

Germino, M. J. & Smith, W. K. (2001) Relative importance of microhabitat, plant form and photosynthetic physiology to carbon gain in two alpine herbs. *Functional Ecology*, **15**, 243–251.

Gervais, B. R., MacDonald, G. M., Snyder, J. A. & Kremenetski, C. V. (2002) *Pinus sylvestris* treeline development and movement on the Kola Peninsula of Russia: pollen and stomate evidence. *Journal of Ecology*, **90**, 627–638.

Gibbens, R. P. & Lenz, J. M. (2001) Root systems of some Chihuahuan Desert plants. *Journal of Arid Environments*, **49**, 221–263.

Gillibrand, P. A. & Balls, P. W. (1998) Modeling salt intrusion and nitrate concentrations in the Ythan estuary. *Estuarine Coastal and Shelf Science*, **47**, 695–706.

Gimingham, C. H. (1964) The maritime zone. In *The Vegetation of Scotland* (ed. J. H. Burnett), pp. 67–142. Edinburgh: Oliver and Boyd.

Girardin, M. P., Tardif, J. & Bergeron, Y. (2001) Radial growth analysis of *Larix laricina* from the Lake Duparquet area, Quebec, in relation to climate and larch sawfly outbreaks. *Ecoscience*, **8**, 127–138.

Glover, B. J. & Abbott, R. J. (1995) Low genetic diversity in the Scottish endemic *Primula scotica* Hook. *New Phytologist*, **129**, 147–153.

Glueck, N. (1959) *Rivers in the Desert*. London: Weidenfeld & Nicolson.

Godwin, H. (1975) *History of the British Flora: A Factual Basis for Phytogeography*, 2nd edn. Cambridge: Cambridge University Press.

Goetz, S. J., Bunn, A. G., Fiske, G. J. & Houghton, R. A. (2005) Satellite-observed photosynthetic trends across boreal North America associated with climate and fire disturbance. *Proceedings of the National Academy of Sciences of the United States of America*, **102**, 13521–13525.

Goldberg, D. & Novoplansky, A. (1997) On the relative importance of competition in unproductive environments. *Journal of Ecology*, **85**, 409–418.

Goldblatt, P. & Manning, J. (2000) *Cape Plants: A Conspectus of the Cape Flora of South Africa*. Cape Town: National Botanical Institute of South Africa.

Gottfried, M., Pauli, H. & Grabherr, G. (1998) Prediction of vegetation patterns at the limits of plant life: a new view of the alpine-nival ecotone. *Arctic and Alpine Research*, **30**, 207–221.

Gould, W. (2000) Remote sensing of vegetation, plant species richness, and regional biodiversity hotspots. *Ecological Applications*, **10**, 1861–1870.

Gould, W. A. & Walker, M. D. (1997) Landscape-scale patterns in plant species richness along an arctic river. *Canadian Journal of Botany–Revue Canadienne de Botanique*, **75**, 1748–1765.

Gouyon, P. H., Fort, P. & Caraux, G. (1983) Selection of seedlings of *Thymus vulgaris* by grazing slugs. *Journal of Ecology*, **71**, 299–306.

Grabherr, G., Gottfried, M. & Pauli, H. (1994) Climate effects on mountain plants. *Nature*, **369**, 448.

Grace, J. (2006) The temperature of buds may be higher than you thought. *New Phytologist*, **170**, 1–3.

Grace, J., Berninger, F. & Nagy, L. (2002) Impacts of climate change on the tree line. *Annals of Botany*, **90**, 537–544.

Grace, J. B. (1990) On the relationship between plant traits and competitive ability. In *Perspectives on Plant Competition* (ed. J. B. Grace & D. Tilman), pp. 51–65. New York: Academic Press.

Grace, J. B. (1993) The effects of habitat productivity on competition density. *Trends in Ecology and Evolution*, **8**, 229–230.

Graham, E. A. & Andrade, J. L. (2004) Drought tolerance associated with vertical stratification of two co-occurring epiphytic bromeliads in a tropical dry forest. *American Journal of Botany*, **91**, 699–706.

Grant, M. C. & Mitton, J. B. (1979) Elevational gradients in adult sex ratios and sexual differentiation in vegetative growth rates of *Populus tremuloides* Michx. *Evolution*, **33**, 914–918.

Graumlich, L. J. (1987) Precipitation variation in the Pacific Northwest (1675–1975) as reconstructed from tree rings. *Annals of the Association of American Geographers*, **77**, 19–29.

Gray, A. J., Marshall, D. F. & Raybould, A. F. (1991) A century of evolution in *Spartina anglica*. *Advances in Ecological Research*, **21**, 1–62.

Greig-Smith, P. (1983) *Quantitative Plant Ecology*, 2nd edn. Oxford: Blackwell Science.

Grime, J. P. (2001) *Plant Strategies, Vegetation Processes and Ecosystem Properties*, 2nd edn. Chichester: John Wiley.

Grime, J. P., Hodgson, J. G. & Hunt, R. (1988) *Comparative Plant Ecology*. London: Unwin Hyman.

Grosse, W., Büchel, H. H. & Tiebel, H. (1991) Pressurized ventilation in wetland plants. *Aquatic Botany*, **39**, 89–98.

Grosswald, M. G. (1998) Late-Weichselian ice sheets in Arctic and Pacific Siberia. *Quaternary International*, **45**, 3–18.

Grubb, P. J. (1977) The maintenance of species richness in plant communities: the importance of the regeneration niche. *Biological Reviews*, **52**, 107–145.

Grundt, H. H., Kjølner, S., Borgen, L., Rieseberg, L. H. & Brochmann, C. (2006) High biological species diversity in the arctic flora. *Proceedings of the National Academy of Sciences*, **103**, 972–975.

Gualtieri, L., Vartanyan, S. L., Brigham-Grette, J. & Anderson, P. M. (2005) Evidence for an ice-free Wrangel Island, northeast Siberia during the Last Glacial Maximum. *Boreas*, **34**, 264–273.

Gudleifsson, B. E. (1997) Survival and metabolite accumulation by seedlings and mature plants of timothy grass during ice encasement. *Annals of Botany*, **79**, 93–96.

Gugerli, F. (1998) Effect of elevation on sexual reproduction in alpine populations of *Saxifraga oppositifolia* (Saxifragaceae). *Oecologia*, **114**, 60–66.

Guisan, A., Holten, J. L., Spichiger, R. & Tessier, L., eds. (1995) *Potential Ecological Impacts of Climate Change in the Alps and Fennoscandian Mountains*. Geneva: Ville de Génève.

Gutterman, Y. (2000) Seed dormancy as one of the survival strategies in annual plant species occurring in deserts. In *Dormancy in Plants* (ed. J.-D. Viémont & J. Crabbé), pp. 139–159. Wallingford: CABI Publishing.

Gutterman, Y. (2002) Minireview: survival adaptations and strategies of annuals occurring in the Judean and Negev Deserts of Israel. *Israel Journal of Plant Sciences*, **50**, 165–175.

Hadley, J. L. & Smith, W. K. (1989) Wind erosion of leaf surface wax in timberline conifers. *Arctic, Antarctic and Alpine Research*, **21**, 392–398.

Hagström, J., James, W. M. & Skene, K. R. (2001) A comparison of structure, development and function in cluster roots of *Lupinis albus* L. under phosphate and iron stress. *Plant and Soil*, **232**, 81–90.

Haldemann, C. & Braendle, R. (1986) Jahrzeitliche Unterschiede im Reservstoffgehalt und von Gärungsprozessen in Rhizomen von Sumpf- und Röhrlichenpflanzen aus dem Freiland. *Flora*, **178**, 307–313.

Haldemann, C. & Braendle, R. (1988) Amino acid composition in rhizomes of wetland species in their natural habitat and under anoxia. *Flora*, **180**, 407–411.

Hall, D. & Coles, J. (1994) *Fenland Survey*, London: English Heritage.

Haller, Albrecht von (1768) *Historia Stirpium Indigenarum Helvetiae inchoata*. Bern.

Halliday, G. (2002) The British flora in the Arctic. *Watsonia*, **24**, 133–144.

Halloy, S. R. P. & Mark, A. F. (2003) Climate-change effects on alpine plant biodiversity: a New Zealand perspective on quantifying the threat. *Arctic, Antarctic and Alpine Research*, **35**, 248–254.

Hallsdóttir, M. (1987) Pollen analytical studies of human influence on vegetation in relation to the Landám tephra layer in southwest Iceland. Ph.D. thesis, University of Lund, Lund.

Handa, I. T., Korner, C. & Hattenschwiler, S. (2005) A test of the tree-line carbon limitation hypothesis by in situ CO_2 enrichment and defoliation. *Ecology*, **86**, 1288–1300.

Hanhijärvi, A. M. & Fagerstedt, K. V. (1995) Comparison of carbohydrate utilization and energy-charge in the yellow flag iris (*Iris pseudacorus*) and garden iris (*Iris germanica*) under anoxia. *Physiologia Plantarum*, **93**, 493–497.

Hanski, I. (1999) *Metapopulation Ecology*. Oxford: Oxford University Press.

Haraguchi, H., Ishikawa, H. & Kubo, I. (1997) Antioxidative action of diterpenoids from *Podocarpus nagi*. *Planta Medica*, **63**, 213–215.

Harper, J. L. (1977) *Population Biology of Plants*. London: Academic Press.

Harris, S. A. (2002) Introduction of Oxford ragwort, *Senecio squalidus* L. (Asteraceae), to the United Kingdom. *Watsonia*, **24**, 31–43.

Hättenschwiler, S. (2001) Tree seedling growth in natural deep shade: functional traits related to interspecific variation in response to elevated CO_2. *Oecologia*, **129**, 31–42.

Havström, M., Callaghan, T. V. & Jonasson, S. (1993) Differential growth responses of *Cassiope tetragona*, an arctic dwarf-shrub, to environmental perturbations among three contrasting high- and subarctic sites. *Oikos*, **66**, 389–402.

Hawksworth, D. L., ed. (1995) *Biodiversity Measurement and Estimation*. London: Chapman and Hall.

Headley, A. D., Callaghan, T. V. & Lee, J. A. (1985) The phosphorus economy of the evergreen tundra plant *Lycopodium annotinum*. *Oikos*, **45**, 235–245.

Heath, D. & Williams, D. R. (1977) *Man at High Altitude*. Edinburgh: Churchill Livingstone.

Hedberg, O. (1997) The genus *Koenigia* L. emend. Hedberg (Polygonaceae). *Botanical Journal of the Linnean Society*, **124**, 295–330.

Heery, S. (1993) *The Shannon Floodlands*. Kinvara, County Galway: Tír Eolas.

Heide, O. M. (1992) Flowering strategies of the high-Arctic and high-alpine snow bed grass species *Phippsia algida*. *Physiologia Plantarum*, **85**, 606–610.

Heide, O. M. (1997) Environmental control of flowering in some northern *Carex* species. *Annals of Botany*, **79**, 319–327.

Heide, O. M. & Gauslaa, Y. (1999) Developmental strategies of *Koenigia islandica*, a high-arctic annual plant. *Ecography*, **22**, 637–642.

Heinz, C., Figueiral, I., Terral, J. F. & Claustre, F. (2004) Holocene vegetation changes in the northwestern Mediterranean: new palaeoecological data from charcoal analysis and quantitative eco-anatomy. *Holocene*, **14**, 621–627.

Hemminger, P. (2005) The Iraqi drain. *Journal of Soil and Water Conservation*, **60**, 86A–89A.

Hemp, A. (2006) Continuum or zonation? Altitudinal gradients in the forest vegetation of Mt. Kilimanjaro. *Plant Ecology*, **184**, 27–42.

Henkel, T. W., Mayor, J. R. & Woolley, L. P. (2005) Mast fruiting and seedling survival of the ectomycorrhizal, monodominant *Dicymbe corymbosa* (Caesalpiniaceae) in Guyana. *New Phytologist*, 167, 543–556.

Henry, H. A. L. & Jefferies, R. L. (2002) Free amino acid, ammonium and nitrate concentrations in soil solutions of a grazed coastal marsh in relation to plant growth. *Plant Cell and Environment*, 25, 665–675.

Henzi, T. & Brändle, R. (1993) Long term survival of rhizomatous species under oxygen deprivation. In *Interacting Stresses on Plants in a Changing Climate* (ed. M. B. Jackson & C. R. Black), pp. 305–314. Berlin: Springer-Verlag.

Hewitt, G. M. (1996) Some genetic consequences of ice ages, and their role in divergence and speciation. *Biological Journal of the Linnean Society*, 58, 247–276.

Hewitt, G. M. (1999) Post-glacial re-colonization of European biota. *Biological Journal of the Linnean Society*, 68, 87–112.

Hitier, H. (1925) Condensateurs des vapeurs atmosphériques dans l'antiquité. *Comptes Rendus de l'Académie d'Agriculture de France*, 11.

Hobbs, R. J., Mallik, A. U. & Gimingham, C. H. (1984) Studies on fire in Scottish heathland communities. III. Vital attributes of the species. *Journal of Ecology*, 72, 963–976.

Hoch, G. & Körner, C. (2005) Growth, demography and carbon relations of *Polylepis* trees at the world's highest treeline. *Functional Ecology*, 19, 941–951.

Hoch, G., Richter, A. & Korner, C. (2003) Non-structural carbon compounds in temperate forest trees. *Plant Cell and Environment*, 26, 1067–1081.

Hochachka, P. W. & Somero, G. N. (1973) *Strategies of Biochemical Adaptation*. Philadelphia: W. B. Saunders Company.

Hofgaard, A. (1997) Inter-relationships between treeline position, species diversity, land use and climate change in the central Scandes Mountains of Norway. *Global Ecology and Biogeography Letters*, 6, 419–429.

Hogg, E. H., Brandt, J. P. & Kochtubajda, B. (2005) Factors affecting interannual variation in growth of western Canadian aspen forests during 1951–2000. *Canadian Journal of Forest Research–Revue Canadienne de Recherche Forestière*, 35, 610–622.

Holderegger, R., Stehlik, I., Smith, R. I. L. & Abbott, R. J. (2003) Populations of Antarctic hairgrass (*Deschampsia antarctica*) show low genetic diversity. *Arctic Antarctic and Alpine Research*, 35, 214–217.

Holmgren, B. & Tjus, M. (1996) Summer air temperatures and tree line dynamics at Abisko. In *Plant Ecology in the Subarctic Swedish Lapland* (ed. P. S. Karlsson & T. V. Callaghan), Ecological Bulletins No. 45, pp. 159–169. Copenhagen: Munksgaard.

Holten, J. I. (2003) Altitude ranges and spatial patterns of alpine plants in Northern Europe. In *Alpine Biodiversity in Europe* (ed. L. Nagy, G. Grabherr & C. Körner), pp. 173–184. Berlin: Springer.

Holtmeier, F.-K. (1995) Waldgrenzen und Klimaschwankun-gen. Ökologische Aspekte eines vieldiskutierten Phänomens. *Geoökodynamik*, 16, 1–24.

Holtmeier, F.-K. (2002) *Tiere in der Landschaft*. Stuttgart: Eugen Ulmer.

Holtmeier, F.-K. (2003) *Mountain Timberlines*. Dordrecht: Kluwer.

Holtmeier, F.-K. & Broll, G. (2005) Sensitivity and response of northern hemisphere altitudinal and polar treelines to environmental change at landscape and local scales. *Global Ecology and Biogeography*, 14, 395–410.

Hoosbeek, M. R., van Breemen, N., Berendse, F., Grosvernier, P., Vasander, H. & Wallen, B. (2001) Limited effect of increased atmospheric CO_2 concentration on ombrotrophic bog vegetation. *New Phytologist*, 150, 459–463.

Houghton, J. T., Ying, Y., Griggs, D. J., et al. (2001) *Climate Change 2001: The Scientific Basis*. Cambridge: Cambridge University Press.

Hueck, K. (1966) *Die Wälder Südamerikas*. Stuttgart: Gustav Fischer.

Hughes, R. G. (1999) Saltmarsh erosion and management of saltmarsh restoration; the effects of infaunal invertebrates. *Aquatic Conservation–Marine and Freshwater Ecosystems*, 9, 83–95.

Hughes, R. G., Lloyd, D., Ball, L. & Emson, D. (2000) The effects of the polychaete *Nereis diversicolor* on the distribution and transplanting success of *Zostera noltii*. *Helgoland Marine Research*, 54, 129–136.

Hulme, M., Conway, D., Jones, P. D., Jiang, T., Barrow, E. M. & Turney, C. (1995) Construction of a 1961–1990 European climatology for climate change modelling and impact applications. *International Journal of Climatology*, 15, 1333–1363.

Hultén, E. (1971) *Atlas över växternas utbredning i norden*. Stockholm: Almqvist and Wiksell.

Hultén, E. & Fries, M. (1986) *Atlas of North European Vascular Plants North of the Tropic of Cancer*. Königstein: Koeltz Scientific Books.

Hunt, T. L. & Lipo, C. P. (2006) Late colonization of Easter Island. *Science*, 311, 1603–1606.

Hunter, M. I., Hetherington, A. M. & Crawford, R. M. M. (1983) Lipid peroxidation: a factor in anoxia intolerance in *Iris* species. *Phytochemistry*, **22**, 1145–1147.

Huntley, B. (1990) European post-glacial forests: compositional changes in response to climatic change. *Journal of Vegetation Science*, **1**, 507–518.

Huntley, B. & Birks, H. J. B. (1983) An atlas of past and present pollen maps for Europe: 0–13000 years ago. Cambridge: Cambridge University Press.

Huntley, B. & Webb, T., III (1988) *Vegetation History*. Dordrecht: Kluwer Academic Publishing.

Hurrell, J. W. & VanLoon, H. (1997) Decadal variations in climate associated with the north Atlantic oscillation. *Climatic Change*, **36**, 301–326.

Imbert, E. & Houle, G. (2000) Ecophysiological differences among *Leymus mollis* populations across a subarctic dune system caused by environmental, not genetic, factors. *New Phytologist*, **147**, 601–608.

Ims, R. A., Yoccoz, N. G. & Hagen, S. B. (2004) Do sub-Arctic winter moth populations in coastal birch forest exhibit spatially synchronous dynamics? *Journal of Animal Ecology*, **73**, 1129–1136.

Ingold, T. (1982) The significance of storage in hunting societies. *Man (N. S.)*, **18**, 553–571.

Ingólfson, O. & Forman, S. L. (1997) Late Quaternary ice sheets in the Barents and Kara Seas. *Polarforsknings-sekretariatets årsbok*, 53–55.

Ingram, H. A. P. (1978) Soil layers in mires: function and terminology. *Journal of Soil Science*, **29**, 224–227.

Jacobs, J. D., Headley, A. N., Maus, L. A., Mode, W. N. & Simms, E. L. (1997) Climate and vegetation of the interior lowlands of southern Baffin Island: long-term stability at the low arctic limit. *Arctic*, **50**, 167–177.

Jacoby, G. C., Darrigo, R. D. & Davaajamts, T. (1996) Mongolian tree rings and 20th-century warming. *Science*, **273**, 771–773.

James, J. K. & Abbott, R. J. (2005) Recent, allopatric, homoploid hybrid speciation: the origin of *Senecio squalidus* (Asteraceae) in the British Isles from a hybrid zone on Mount Etna, Sicily. *Evolution*, **59**, 2533–2547.

Jano, A. P., Jefferies, R. L. & Rockwell, R. F. (1998) The detection of vegetational change by multitemporal analysis of LANDSAT data: the effects of goose foraging. *Journal of Ecology*, **86**, 93–99.

Jansen, C., Van de Steeg, H. M. & De Kroon, H. (2005) Investigating a trade-off in root morphological responses to a heterogeneous nutrient supply and to flooding. *Functional Ecology*, **19**, 952–960.

Jarvis, P. G. & Leverenz, J. W. (1983) Productivity of temperate, deciduous and evergreen trees. In. *Physiological Plant Ecology*, Vol. IV (ed. O. L. Lange, P. S. Nober, C. B. Osmond & H. Ziegler), pp. 233–280. Berlin: Springer-Verlag.

Jefferies, R. L. (1997) Long-term damage to sub-arctic coastal ecosystems by geese: ecological indicators and measures of ecosystem dysfunction. In *Disturbance and Recovery in Arctic Lands* (ed. R. M. M. Crawford), pp. 151–165. Dordrecht: Kluwer.

Jefferies, R. L. & Gottlieb, L. D. (1983) Genetic variation within and between populations of the asexual plant *Puccinellia × phryganodes*. *Canadian Journal of Botany*, **61**, 774–779.

Jefferies, R. L. & Rockwell, R. F. (2002) Foraging geese, vegetation loss and soil degradation in an Arctic salt marsh. *Applied Vegetation Science*, **5**, 7–16.

Jefferies, R. L., Rockwell, R. F. & Abraham, K. E. (2004) Agricultural food subsidies, migratory connectivity and large-scale disturbance in arctic coastal systems: a case study. *Integrative and Comparative Biology*, **44**, 130–139.

Jeffree, C. E. & Jeffree, E. P. (1996) Redistribution of the potential geographical ranges of mistletoe and Colorado beetle in Europe in response to the temperature component of climate change. *Functional Ecology*, **10**, 562–577.

Jermyn, J. (2005) *Alpine Plants of Europe: a Gardener's Guide*. Portland, OR: Timber Press.

Johansen, S. & Hytteborn, H. (2001) A contribution to the discussion of biota dispersal with drift ice and driftwood in the North Atlantic. *Journal of Biogeography*, **28**, 105–115.

Johnson, E. A. (1975) Buried seed populations in the subarctic forest east of Great Slave lake, Northwest Territories. *Canadian Journal of Botany*, **10**, 2933–2941.

Joly, C. A. (1994) Flooding tolerance: a reinterpretation of Crawford's metabolic theory. *Proceedings of the Royal Society of Edinburgh Section B – Biological Sciences*, **102**, 343–354.

Jonasson, S. (1992) Plant responses to fertilization and species removal in tundra relate to community structure and clonality. *Oikos*, **63**, 420–429.

Jónsdóttir, I. S. & Callaghan, T. V. (1990) Interclonal translocation of ammonium and nitrate nitrogen in *Carex bigelowii* using 15N and nitrate reductase assays. *New Phytologist*, **114**, 419–428.

Jónsdóttir, I. S., Augner, M., Fagerstrom, T., Persson, H. & Stenström, A. (2000) Genet age in marginal populations of two clonal *Carex* species in the Siberian Arctic. *Ecography*, **23**, 402–412.

Juday, G. P. (2005) Forests, land management, and agriculture. In *Arctic Climate Impact Assessment*, pp. 781–862. Cambridge: Cambridge University Press.

Jump, A. S. & Penuelas, J. (2005) Running to stand still: adaptation and the response of plants to rapid climate change. *Ecology Letters*, **8**, 1010–1020.

Jutila, H. M. (2001) Effect of flooding and draw-down disturbance on germination from a seashore meadow seed bank. *Journal of Vegetation Science*, **12**, 729–738.

Kahkonen, M. P., Hopia, A. I., Vuorela, H. J., *et al.* (1999) Antioxidant activity of plant extracts containing phenolic compounds. *Journal of Agricultural and Food Chemistry*, **47**, 3954–3962.

Kaland, P. E. (1986) The origin and management of Norwegian coastal heathlands as reflected by pollen analysis. In *Anthropogenic Indicators in Pollen Diagrams* (ed. K.-E. Behre), pp. 19–36. Rotterdam: Balkema.

Kallio, P. & Lehtonen, J. (1973) Birch forest damage caused by *Oporinia autumnata* (Bkh.) in 1965–1966 in Utsjoki, N. Finland. *Reports from the Kevo Subarctic Research Station*, **10**, 55–69.

Kallio, P., Laine, U. & Mäkinen, Y. (1971) Vascular flora of Inari Lapland. 2. Pinaceae and Cupressaceae. *Reports from the Kevo Subarctic Research Station*, **8**, 73–100.

Kaser, G., Hardy, D. R., Molg, T., Bradley, R. S. & Hyera, T. M. (2004) Modern glacier retreat on Kilimanjaro as evidence of climate change: observations and facts. *International Journal of Climatology*, **24**, 329–339.

Kato-Noguchi, H. (2002) Ethanol sensitivity of rice and oat coleoptiles. *Physiologia Plantarum*, **115**, 119–124.

Keeley, J. E. (1998) CAM photosynthesis in submerged aquatic plants. *Botanical Review*, **64**, 121–175.

Keller, B. E. M. (2000) Genetic variation among and within populations of *Phragmites australis* in the Charles River watershed. *Aquatic Botany*, **66**, 195–208.

Keller, F. & Körner, C. (2003) The role of photoperiodism in alpine plant development. *Arctic, Antarctic and Alpine Research*, **35**, 361–368.

Kelly, D. & Sork, V. L. (2002) Mast seeding in perennial plants: why, how, where? *Annual Review of Ecology and Systematics*, **33**, 427–447.

Kelly, D. L. & Iremonger, S. F. (1997) Irish wetland woods: the plant communities and their ecology. *Biology and Environment–Proceedings of the Royal Irish Academy*, **97B**, 1–32.

Kent, M., Owen, N. W. & Dale, M. P. (2005) Photosynthetic responses of plant communities to sand burial on the machair dune systems of the Outer Hebrides, Scotland. *Annals of Botany*, **95**.

Kielland, K. & Chapin, F. S., III (1992) Nutrient absorption and acumulation in arctic plants. In *Arctic Ecosystems in a Changing Climate* (ed. F. S. Chapin III, R. L. Jefferies, J. F. Reynolds, G. R. Shaver, J. Svoboda & E. W. Chu), pp. 321–335. San Diego: Academic Press.

Kienast, F., Schirrmeister, L., Siegert, C. & Tarasov, P. (2005) Palaeobotanical evidence for warm summers in the East Siberian Arctic during the last cold stage. *Quaternary Research*, **63**, 283–300.

Kirk, H., Vrieling, K. & Klinkhamer, P. G. L. (2005) Maternal effects and heterosis influence the fitness of plant hybrids. *New Phytologist*, **166**, 685–694.

Klak, C., Reeves, G. & Hedderson, T. (2004) Unmatched tempo of evolution in Southern African semi-desert ice plants. *Nature*, **427**, 63–65.

Klanderud, K. & Birks, H. J. B. (2003) Recent increases in species richness and shifts in altitudinal distributions of Norwegian mountain plants. *Holocene*, **13**, 1–6.

Klasner, F. L. & Fagre, D. B. (2002) A half century of change in alpine treeline patterns at Glacier National Park, Montana, USA. *Arctic, Antarctic and Alpine Research*, **34**, 49–56.

Klein, D. R. (1999) The role of climate and inularity in establishment and persistence of *Rangifer tarandus* populations in the high Arctic. *Ecological Bulletins*, **47**, 96–104.

Klinger, L. F. (1996) Coupling of soils and vegetation in peatland succession. *Arctic and Alpine Research*, **28**, 380–387.

Klitgaard-Kristensen, D., Sejrup, H. P., Haflidason, H., Johnsen, S. & Spurk, M. (1998) A regional 8200 cal. yr BP cooling event in northwest Europe, induced by final stages of the Laurentide ice-sheet deglaciation? *Journal of Quaternary Science*, **13**, 165–169.

Knottnerus, O. S. (2005) History of human settlement, cultural change and interference with the marine environment. *Helgoland Marine Research*, **59**, 2–8.

Kobayashi, T., Nakagawa, Y., Tamaki, M., Hiraki, T. & Aikawa, M. (2001) Cloud water deposition to forest canopies of *Cryptomeria japonica* at Mt. Rokko, Kobe, Japan. *Water, Air and Soil Pollution*, **130**, 601–606.

Koenig, W. D. & Knops, J. M. H. (2005) The mystery of masting in trees. *American Scientist*, **93**, 340–347.

Kolb, R. M., Rawyler, A. & Braendle, R. (2002) Parameters affecting the early seedling development of four neotropical trees under oxygen deprivation stress. *Annals of Botany*, **89**, 551–558.

Koller, D. (1969) The physiology of dormancy and survival of plants in desert communities. *Symposium of the Society for Experimental Biology*, **23**, 449–469.

Koncalova, H. (1990) Anatomical adaptations to waterlogging in roots of wetland graminoids: limitations and drawbacks. *Aquatic Botany*, **38**, 127–134.

Köppen, W. (1931) *Grundriss der Klimakunde*, 2te Verbesserte Auflagender *Klimate der Der Erde*. Berlin: Walter Gruyter.

Köppitz, H. (1999) Analysis of genetic diversity among selected populations of *Phragmites australis* world-wide. *Aquatic Botany*, **64**, 209–221.

Körner, C. (1995) Alpine plant diversity: a global survey and functional interpretations. In *Arctic and Alpine Biodiversity: Patterns, Causes, and Ecosystem Consequences* (ed. F. S. Chapin III & C. Körner). Berlin: Springer-Verlag.

Körner, C. (1999) *Alpine Plant Life*. Berlin: Springer-Verlag.

Körner, C. (2003) Carbon limitation in trees. *Journal of Ecology*, **91**, 4–17.

Körner, C. & Diemer, M. (1994) Evidence that plants from high altitudes retain their greater photosynthetic efficiency under elevated CO_2. *Functional Ecology*, **8**, 58–68.

Körner, C. & Larcher, W. (1988) Plant life in cold climates. In *Plants and Temperature* (ed. S. P. Long & F. I. Woodward), pp. 25–57. *Symposia of the Society for Experimental Biology*, **42**, 25–57.

Körner, C. & Paulsen, J. (2004) A world-wide study of high altitude treeline temperatures. *Journal of Biogeography*, **31**, 713–732.

Körner, C., Neumayer, M., Pelaez Menendez-Riedl, S. & Smeets-Scheel, A. (1989) Functional morphology of mountain plants. *Flora*, **182**, 353–383.

Korpelainen, H. (1998) Labile sex expression in plants. *Biological Reviews*, **73**, 157–180.

Kozlowski, T. T. & Pallardy, S. G. (1997) *Physiology of Woody Plants*. San Diego: Academic Press.

Kremenetski, C. V., Sulerzhitsky, L. D. & Hantemirov, R. (1998) Holocene history of the northern range limits of some trees and shrubs in Russia. *Arctic and Alpine Research*, **30**, 317–333.

Kriuchkov, V. V. (1971) Woodland communities in tundra, possibilities for its establishment and dynamics. *Ekologiya (Soviet Journal of Ecology)*, **6**, 9–19 (in Russian).

Krupnik, I. (1993) *Arctic Adaptations*. Hanover: University Press of New England.

Kullman, L. (1999) Early holocene tree growth at a high elevation site in the northernmost Scandes of Sweden (Lapland): a palaeobiogeographical case study based on megafossil evidence. *Geografiska Annaler Series A–Physical Geography*, **81A**, 63–74.

Kullman, L. (2002a) Boreal tree taxa in the central Scandes during the Late-Glacial: implications for Late-Quaternary forest history. *Journal of Biogeography*, **29**, 1117–1124.

Kullman, L. (2002b) Rapid recent range-margin rise of tree and shrub species in the Swedish Scandes. *Journal of Ecology*, **90**, 68–77.

Kullman, L. (2003) Recent reversal of Neoglacial climate cooling trend in the Swedish Scandes as evidenced by mountain birch tree-limit rise. *Global and Planetary Change*, **36**, 77–88.

Kuwabara, A., Tsukaya, H. & Nagata, T. (2001) Identification of factors that cause heterophylly in *Ludwigia arcuata* Walt. (Onagraceae). *Plant Biology*, **3**, 98–105.

Kuz'min, Y. V., Orlova, L. A., Zol'nikov, I. D. & Igol'nikov, A. E. (2000) The history of mammoth (*Mammuthus primigenius* Blum.) population in Siberia and adjacent areas (based on radiocarbon data). *Geologiya I Geofizika*, **41**, 746–754.

Kwant, C. (2006) *The Ginkgo Pages*. online: www.xs4all.nl/~Kwanten/.

Kytoviita, M. M. (2005) Asymmetric symbiont adaptation to Arctic conditions could explain why high Arctic plants are non-mycorrhizal. *Fems Microbiology Ecology*, **53**, 27–32.

La Peyre, M. K. G., Grace, J. B., Hahn, E. & Mendelssohn, I. A. (2001) The importance of competition in regulating plant species abundance along a salinity gradient. *Ecology*, **82**, 62–69.

Laberge, M. J., Payette, S. & Bousquet, J. (2000) Life span and biomass allocation of stunted black spruce clones in the subarctic environment. *Journal of Ecology*, **88**, 584–593.

Laderman, A. D., ed. (1998a) *Coastally Restricted Forests*. New York: Oxford University Press.

Laderman, A. D. (1998b) Freshwater forests of continental margins: overview and synthesis. In *Coastally Restricted Forests* (ed. A. D. Laderman), pp. 3–35. New York: Oxford University Press.

Landvik, J. Y., Brook, E. J., Gualtieri, L., Raisbeck, G., Salvigsen, O. & Yiou, F. (2003) Northwest Svalbard during the last glaciation: ice-free areas existed. *Geology*, **31**, 905–908.

Langematz, U., Claussnitzer, A., Matthes, K. & Kunze, M. (2005) The climate during the Maunder Minimum: a simulation with the Freie Universitat Berlin Climate Middle Atmosphere Model (FUB-CMAM). *Journal of Atmospheric and Solar-Terrestrial Physics*, **67**, 55–69.

Lanner, R. M. & Connor, K. F. (2001) Does bristlecone pine senesce? *Experimental Gerontology*, **36**, 675–685.

Laporte, M. M. & Soule, J. D. (1996) Sex-specific physiology and source-sink relations in the dioecious plant *Silene latifolia*. *Oecologia*, **106**, 63–72.

Larl, I. & Wagner, J. (2006) Timing of reproductive and vegetative development in *Saxifraga oppositifolia* in an alpine and a subnival climate. *Plant Biology*, **8**, 155–166.

Larson, R. A. (1988) The antioxidants of higher plants. *Phytochemistry*, **27**, 969–978.

Lavoie, C. & Payette, S. (1994) Recent fluctuations of the lichen-spruce forest limit in subarctic Québec. *Journal of Ecology*, **82**, 725–734.

Lavoie, C. & Payette, S. (1996) The long-term stability of the boreal forest limit in sub-arctic Québec. *Ecology*, **77**, 1226–1233.

Lee, H. S. J., Overdieck, D. & Jarvis, P. G. (1998). Biomass, growth and carbon allocation. In *European Forests and Global Change* (ed. P. G. Jarvis), pp. 126–191. Cambridge: Cambridge University Press.

Legesse, N. & Powell, A. A. (1992) Comparisons of water-uptake and imbibition damage in 11 cowpea cultivars. *Seed Science and Technology*, **20**, 173–180.

Legesse, N. & Powell, A. A. (1996) Relationship between the development of seed coat pigmentation, seed coat adherence to the cotyledons and the rate of imbibition during the maturation of grain legumes. *Seed Science and Technology*, **24**, 23–32.

Lehouerou, H. N. (1995) The Sahara from the bioclimatic viewpoint: definition and limits. *Annals of Arid Zone*, **34**, 1–16.

Lévesque, E. & Svoboda, J. (1999) Vegetation re-establishment in polar 'lichen-kill' landscapes: a case study of the Little Ice Age impact. *Polar Research*, **18**, 221–228.

Levitt, J. (1980) *Responses of Plants to Environmental Stress*. New York: Academic Press.

Lewington, A. & Parker, E. (1999) *Ancient Trees*. London: Collins & Brown.

Lewis, P. O. & Crawford, D. J. (1995) Pleistocene refugium endemics exhibit greater allozymic diversity than widespread congeners in the genus *Polygonella* (Polygonaceae). *American Journal of Botany*, **82**, 141–149.

Lid, J. & Lid, D. T. (1994) *Norsk Flora*, 6th edn (ed. R. Elven). Oslo: Det Norske Samlaget.

Lightfoot, D. R. (1994) Morphology and ecology of lithic-mulch agriculture. *Geographical Review*, **84**, 172–185.

Lincoln, R. J., Boxshall, G. A. & Clark, P. F. (1998) *A Dictionary of Ecology, Evolution and Systematics*, 2nd edn. Cambridge: Cambridge University Press.

Linder, H. P., Meadows, M. E. & Cowling, R. M. (1992) History of the Cape flora. In *The Ecology of Fynbos* (ed. R. M. Cowling), pp. 113–134. Cape Town: Oxford University Press.

Linder, H. P. & Hardy, C. R. (2004) Evolution of the species-rich Cape flora. *Philosophical Transactions of the Royal Society B*, **359**, 1623–1632.

Linhart, Y. B. & Baker, G. (1973) Intra-population differentiation of physiological responses to flooding in a population of *Veronica peregrina* L. *Nature*, **242**, 275.

Linhart, Y. B., Keefover-Ring, K., Mooney, K. A., Breland, B. & Thompson, J. D. (2005) A chemical polymorphism in a multitrophic setting: thyme monoterpene composition and food web structure. *American Naturalist*, **166**, 517–529.

Little, L. R. & Maun, M. A. (1996) The '*Ammophila* problem' revisited: a role for mycorrhizal fungi. *Journal of Ecology*, **84**, 1–7.

Llewellyn, P. J. & Shackley, S. E. (1996) The effects of mechanical beach-cleaning on invertebrate populations. *British Wildlife*, **7**, 147–155.

Lloyd, A. H. (2005) Ecological histories from Alaskan tree lines provide insight into future change. *Ecology*, **86**, 1687–1695.

Lloyd, A. H. & Fastie, C. L. (2002) Spatial and temporal variability in the growth and climate response of treeline trees in Alaska. *Climatic Change*, **52**, 481–509.

Lloyd, C. R. (2001) On the physical controls of the carbon dioxide balance at a high Arctic site in Svalbard. *Theoretical and Applied Climatology*, **70**, 167–182.

Lloyd, D. G. (1974) Female predominant sex ratios in angiosperms. *Heredity*, **32**, 35–44.

Lowe, A., Unsworth, C., Gerber, S., *et al.* (2005) Route, speed and mode of oak post-glacial colonisation across the British Isles: integrating molecular ecology, palaeoecology and modeling approaches. *Botanical Journal of Scotland*, **57**, 59–81.

Luterbacher, J., Rickli, R., Xoplaki, E., *et al.* (2001) The Late Maunder Minimum (1675–1715): a key period for studying decadal scale climatic change in Europe. *Climatic Change*, **49**, 441–462.

Lüttge, U. (1997) *Physiological Ecology of Tropical Plants*. Berlin: Springer Verlag.

Lynn, D. E. & Waldren, S. (2002) Physiological variation in populations of *Ranunculus repens* L. (creeping buttercup) from the temporary limestone lakes (turloughs) in the west of Ireland. *Annals of Botany*, **89**, 707–714.

Lynn, D. E. & Waldren, S. (2003) Survival of *Ranunculus repens* L. (creeping buttercup) in an amphibious habitat. *Annals of Botany*, **91**, 75–84.

Maarel, D. E., von (1979) Environmental management of coastal dunes in the Netherlands. In *Ecological Processes in Coastal Environments* (ed. R. L. Jefferies & A. J. Davy), pp. 543–570. Oxford: Blackwell Scientific.

Mabberley, D. J. (1997) *The Plant Book*. Cambridge: Cambridge University Press.

MacArthur, R. H. & Wilson, E. O. (1967) *The Theory of Island Biogeography*. Princeton, NJ: Princeton University Press.

MacDonald, G. M., Velichko, A. A., Kremenetski, C. V., *et al.* (2000) Holocene treeline history and climate change across northern Eurasia. *Quaternary Research*, **53**, 302–311.

Mallik, A. U., Gimingham, C. H. & Rahman, A. A. (1984) Ecological effects of heather burning. 1. Water infiltration, moisture retention and porosity of surface soil. *Journal of Ecology*, **72**, 767–776.

Mangerud, J., Dokken, T., Hebbeln, D., *et al.* (1998) Fluctuations of the Svalbard Barents Sea ice sheet during the last 150 000 years. *Quaternary Science Reviews*, **17**, 11–42.

Mani, M. S. (1978) *Ecology and Phytogeography of High-altitude Plants of the Northwest Himalaya*. London: Chapman and Hall.

Mannheimer, S., Bevilacqua, G., Caramaschi, E. P. & Scarano, F. R. (2003) Evidence for seed dispersal by the catfish *Auchenipterichthys longimanus* in an Amazonian lake. *Journal of Tropical Ecology*, **19**, 215–218.

Mantovani, A. & Vieira, R. C. (2000) Leaf micromorphology of antarctic pearlwort *Colobanthus quitensis* (Kunth) Bartl. *Polar Biology*, **23**, 531–538.

Mark, A. F. (1970) Floral initiation and development in New Zealand alpine plants. *New Zealand Journal of Botany*, **8**, 67–75.

Marone, L. & Horno, M. (1997) Seed reserves in the central Monte Desert, Argentina: implications for granivory. *Journal of Arid Environments*, **36**, 661–670.

Marrs, R. H. & Watt, A. S. (2006) Biological flora of the British Isles: *Pteridium aquilinum* (L.) Kuhn. *Journal of Ecology*, **94**, 1272–1321.

Marschner, M. (1995) *Mineral Nutrition of Higher Plants*. London: Academic Press.

Marshall, J., Dawson, T. & Ehleringer, J. (1993) Gender-related differences in gas exchange are not related to host quality in the xylem-tapping *Phoradendron juniperinum* (Viscaceae). *American Journal of Botany*, **80**, 641–645.

Mauquoy, D. & Barber, K. (1999) A replicated 3000 year proxy-climate record from Coom Rigg Moss and Felecia Moss, the Border Mires, northern England. *Journal of Quaternary Science*, **14**, 263–275.

Mayr, S., Schwienbacher, F. & Bauer, H. (2003) Winter at the alpine timberline. Why does embolism occur in Norway spruce but not in stone pine? *Plant Physiology*, **131**, 780–792.

McCormick, F. (1995) Cows, ringforts and the origins of early Christian Ireland. *Emania*, **13**, 33–37.

McDonald, M. P., Galwey, N. W. & Colmer, T. D. (2002) Similarity and diversity in adventitious root anatomy as related to root aeration among a range of wetland and dryland grass species. *Plant Cell and Environment*, **25**, 441–451.

McFadden, M. A., Patterson, W. P., Mullins, H. T. & Anderson, W. T. (2005) Multi-proxy approach to long- and short-term Holocene climate-change: evidence from eastern Lake Ontario. *Journal of Paleolimnology*, **33**, 371–391.

McGhee, R. (1996) *Ancient People of the Arctic*. Vancouver: University of British Columbia.

McGlone, M. S. (2002) The late quaternary peat, vegetation and climate history of the southern oceanic islands of New Zealand. *Quaternary Science Reviews*, **21**, 683–707.

McGlone, M. S., Moar, N. T., Wardle, P. & Meurk, C. D. (1997) Late-glacial and Holocene vegetation and environment of Campbell Island, far southern New Zealand. *Holocene*, **7**, 1–12.

McGlone, M. S., Wilmshurst, J. M. & Wiser, S. K. (2000) Lateglacial and Holocene vegetation and climatic change on Auckland Island, Subantarctic New Zealand. *Holocene*, **10**, 719–728.

McGlone, M. S., Duncan, R. P. & Heenan, P. B. (2001) Endemism, species selection and the origin and distribution of the vascular plant flora of New Zealand. *Journal of Biogeography*, **28**, 199–216.

McGraw, J. B. (1987) Experimental ecology of *Dryas octopetala* ecotypes. V. Field photosynthesis of reciprocal transplants. *Holarctic Ecology*, **10**, 303–311.

McGraw, J. B. & Antonovics, J. (1983) Experimental ecology of *Dryas octopetala* ecotypes. I. Ecotypic differentiation and life cycle stages of selection. *Journal of Ecology*, **71**, 879–897.

McGraw, J. B. & Day, T. A. (1997) Size and characteristics of a natural seed bank in Antarctica. *Arctic and Alpine Research*, **29**, 213–216.

McNally, A. & Doyle, G. J. (1984) A study of subfossil pine layers in a raised bog complex in the Irish Midlands. *Proceedings of the Royal Irish Academy*, **84B**, 57–70.

McVean, D. N. & Ratcliffe, D. A. (1962) *Plant Communities of the Scottish Highlands*. London: HMSO.

Menard, C., Duncan, P., Fleurance, G., *et al.* (2002) Comparative foraging and nutrition of horses and cattle in European wetlands. *Journal of Applied Ecology*, **39**, 120–133.

Meriot, C. (1984) The Saami peoples from the time of the voyage of Ottar to Thomas Von Westen. *Arctic*, **37**, 373–384.

Meusel, H. & Jäger, E. J. (1992) *Vergleichende Chorologie der Zentraleuropäischen Flora*. Jena: Gustav Fischer Verlag.

Michaelis, P. (1934) Ökologische Studien an der Baumgrenze V. Osmotischer Wert und Wassergehalt während des Winters in den verschiedenen Höhenlagen. *Jahrbuch für wissenschaftliche Botanik*, **80**, 169–243.

Miehe, G. (1989) Vegetation patterns on Mount Everest as influenced by monsoon and föhn. *Vegetatio*, **79**, 21–32.

Miehe, G. (1996) On the connexion of vegetation dynamics with climatic changes in High Asia. *Palaeogeography Palaeoclimatology, Palaeoecology*, **120**, 5–24.

Miehe, G. (1997) Alpine vegetation types of the central Himalaya. In *Polar and Alpine Tundra* (ed. F. E. Wielgolaski), pp. 161–184. Amsterdam: Elsevier.

Milne, R. I. (2004) Phylogeny and biogeography of *Rhododendron* subsection *Pontica*, a group with a tertiary relict distribution. *Molecular Phylogenetics and Evolution*, **33**, 389–401.

Milne, R. I. & Abbott, R. J. (2000) Origin and evolution of invasive naturalized material of *Rhododendron ponticum* L. in the British Isles. *Molecular Ecology*, **9**, 541–556.

Milne, R. I. & Abbott, R. J. (2002) The origin and evolution of tertiary relict floras. In *Advances in Botanical Research*, **38**, 281–314.

Milne, R. I., Terzioglu, S. & Abbott, R. J. (2003) A hybrid zone dominated by fertile F(1)s: maintenance of species barriers in *Rhododendron*. *Molecular Ecology*, **12**, 2719–2729.

Minchinton, T. E. (2002) Precipitation during El Niño correlates with increasing spread of *Phragmites australis* in New England, USA, coastal marshes. *Marine Ecology–Progress Series*, **242**, 305–309.

Mitchell, F. & Ryan, M. (1998) *Reading the Irish Landscape*. Dublin: Town House.

Mitchell, F. J. G. (1993) The biogeographical implications of the distribution and history of the strawberry tree, *Arbutus unedo*, in Ireland. In *Occasional Publications of the Irish Geographical Society* (ed. M. J. Costello & K. S. Kelly), Vol. 2, pp. 35–44, Dublin.

Mitchell, F. J. G. (2005) How open were European primeval forests? Hypothesis testing using palaeoecological data. *Journal of Ecology*, **93**, 168–177.

Mithen, S. (2003) *After the Ice*. London: Weidenfeld & Nicolson.

Molau, U. (1992) On the occurrence of sexual reproduction in *Saxifraga cernua* and *S. foliolosa* (Saxifragaceae). *Nordic Journal of Botany*, **12**, 197–203.

Molau, U. (1993a) Relationships between flowering phenology and life-history strategies in tundra plants. *Arctic and Alpine Research*, **25**, 391–402.

Molau, U. (1993b) Reproductive ecology of the three Nordic *Pinguicula* species (Lentibularieaceae). *Nordic Journal of Botany*, **13**, 149–157.

Molau, U., Nordenhall, U. & Eriksen, B. (2005) Onset of flowering and climate variability in an alpine landscape: a 10-year study from Swedish Lapland. *American Journal of Botany*, **92**, 422–431.

Möller, P., Bolshiyanov, D. & Bersten, H. (1999) Weichselian geology and palaeoenvironmental history of the central Taymyr Peninsula, Siberia, indicating no glaciation during the last global glacial maximum. *Boreas*, **28**, 92–114.

Mommer, L. (2005) Gas exchange under water. Ph.D. thesis, Radboud University, Nijmegen.

Mommer, L., Pedersen O. & Visser, E. J. W. (2004) Acclimation of a terrestrial plant to submergence facilitates gas exchange under water. *Plant Cell and Environment*, **27**, 1281–1287.

Mommer, L., de Kroon, H., Pierik, R., Bogemann, G. M. & Visser, E. J. W. (2005) A functional comparison of acclimation to shade and submergence in two terrestrial plant species. *New Phytologist*, **167**, 197–206.

Monk, L. S., Fagerstedt, K. V. & Crawford, R. M. M. (1987) Superoxide dismutase as an anaerobic polypeptide. A key factor in recovery from oxygen deprivation in *Iris pseudacorus*. *Plant Physiology*, **85**, 1016–1020.

Monteith, J. L. (1963) Dew: facts and fallacies. In *Water Relations of Plants* (ed. A. J. Rutter & F. H. Whitehead), pp. 37–56. Oxford: Blackwell Scientific Publications.

Montgomery, E. G. (1912) Competition in cereals. *Bulletin of the Nebraska Argricultural Station*, **26**, 1–12.

Mooney, H. A. & Billings, W. D. (1965) Effect of altitude on the carbohydrate content of mountain plants. *Ecology*, **46**, 750–751.

Moore, P. D. (1996) Mystery of moribund marram. *Nature*, **380**, 285–286.

Moreira, A. G. (2000) Effects of fire protection on savanna structure in Central Brazil. *Journal of Biogeography*, **27**, 1021–1029.

Muraoka, H., Uchida, M., Mishio, M., Nakatsubo, T., Kanda, H. & Koizumi, H. (2002) Leaf photosynthetic characteristics and net primary production of the polar willow (*Salix polaris*) in a high arctic polar semi-desert, Ny-Alesund, Svalbard. *Canadian Journal of Botany–Revue Canadienne de Botanique*, **80**, 1193–1202.

Murray, D. F. (1995) Causes of arctic plant diversity: origin and evolution. In *Arctic and Alpine Biodiversity: Patterns, Causes and Ecosystem Consequences* (ed. F. S. Chapin & C. Körner), pp. 21–32. Heidelberg: Springer.

Murray, D. F. (1997) Regional and local vascular plant diversity in the Arctic. *Opera Botanica*, **132**, 9–18.

Myers, N. (2003) Biodiversity hotspots revisited. *BioScience*, **53**, 916–917.

Myneni, R. B., Dong, J., Tucker, C. J., *et al.* (2001) A large carbon sink in the woody biomass of Northern forests. *Proceedings of the National Academy of Sciences of the United States of America*, **98**, 14784–14789.

Nabben, R. H. M. (2001) Metabolic adaptations to flooding-induced oxygen deficiency and post-anoxia stress in *Rumex* species. Ph.D. thesis, Catholic University of Nijmegen, Nijmegen.

Nadelhoffer, K. J., Giblin, A. E., Shaver, G. R. & Laundre, J. L. (1992) Effects of temperature and substrate quality on element mineralization in six arctic soils. *Ecology*, **72**, 242–253.

Nakatsubo, T., Bekku, Y., Kume, A. & Koizumi, H. (1998) Respiration of the belowground parts of vascular plants: its contribution to total soil respiration on a successional glacier foreland in Ny-Alesund, Svalbard. *Polar Research*, **17**, 53–59.

Nelson, C. D., Weng, C., Kubisiak, T. L., Stine, M. & Brown, C. L. (2003) On the number of genes controlling the grass stage in longleaf pine. *Journal of Heredity*, **94**, 392–398.

Nesje, A. & Kvamme, M. (1991) Holocene glacier and climate variations in Western Norway: evidence for early Holocene glacier demise and multiple neoglacial events. *Geology*, **19**, 610–612.

Neuvonen, S., Ruohomäki, K., Bylund, H. & Kaitaniemi, P. (2001) Insect herbivores and herbivory effects on mountain birch dynamics. In *Nordic Mountain Birch Ecosystems* (ed. F. E. Wiellgolaski), pp. 207–222. New York: Parthenon Publishing.

New, M., Hulme, M. & Jones, P. (1999) Representing twentieth-century space-time climate variability. Part I. Development of a 1961–1990 mean monthly terrestrial climatology. *Journal of Climate*, **12**, 829–856.

New, M., Hulme, M. & Jones, P. (2000) Representing twentieth-century space-time climate variability. Part II. Development of 1901–1996 monthly grids of terrestrial climate. *Journal of Climate*, **13**, 2217–2238.

Nordal, I. (1987) Tabula rasa after all? Botanical evidence for ice-free refugia in Scandinavia reviewed. *Journal of Biogeography*, **14**, 377–388.

Odasz, A. M. (1994) Nitrate reductase activity in vegetation below an arctic bird cliff, Svalbard, Norway. *Journal of Vegetation Science*, **5**, 913–920.

Ogden, J., Stewart, G. H. & Allen, R. B. (1996) Ecology of New Zealand *Nothofagus* forests. In *The Ecology and Biogeography of Nothofagus Forests* (ed. T. T. Veblen, R. S. Hill & J. Read), pp. 25–82. New Haven, CT: Yale University Press.

Ögren, E. (1996) Premature dehardening in *Vaccinium myrtillus* during a mild winter: a cause for winter dieback? *Functional Ecology*, **10**, 724–732.

Ojeda, F., Arroyo, J. & Maranon, T. (1995) Biodiversity components and conservation of Mediterranean heathlands in southern Spain. *Biological Conservation*, **72**, 61–72.

Oksanen, L. (1990) Predation, herbivory and plant strategies along gradients of primary productivity. In *Perspectives on Plant Competition* (ed. J. B. Grace & D. Tilman), pp. 445–474. San Diego: Academic Press.

Oksanen, L. (1993) Plant strategies and environmental stress: a dialectic approach. In *Plant Adaptation to Environmental Stress* (ed. L. Fowden, T. A. Mansfield & J. L. Stoddart), pp. 313–333. London: Chapman and Hall.

Oksanen, L. & Ranta, E. (1992) Plant strategies along vegetational gradients on the mountains of Iddonjarga: a test of two theories. *Journal of Vegetation Science*, **3**, 175–186.

Oksanen, L. & Virtanen, R. (1997) Adaptation to disturbance as a part of the strategy of Arctic and Alpine plants. In *Disturbance and Recovery in Arctic Lands* (ed. R. M. M. Crawford), pp. 91–113. Dordrecht: Kluwer.

Ostendorp, W. (1993) Schilf als Lebensraum. *Beihefte zu den Veröffentlichungen für Natur- und Landschaftspflege in Baden-Würtemberg*, **68**, 173–280.

Ovstedal, D. O. & Mjaavatten, O. (1992) A multivariate comparison between 3 new European populations of *Artemisia norvegica* (Asteraceae) by means of chemometric and morphometric data. *Plant Systematics and Evolution*, **181**, 21–32.

Palsson, H. & Edwards, P., eds. (1972) *The Book of Settlements*. Winnipeg: University of Manitoba.

Pate, J. S., Verboom, W. H. & Galloway, P. D. (2001) Co-occurrence of Proteaceae, laterite and related oligotrophic soils: coincidental associations or causative inter-relationships? *Australian Journal of Botany*, **49**, 529–560.

Patten, K. (1997). *Proceedings of the Second International Spartina Conference, Olympia, Washington*. Long Beach, WA: Washington State University.

Pauli, H., Gottfried, M., Dirnböck, T., Dullinger, S. & Grabherr, G. (2003) Assessing the long-term dynamics of endemic plants at summit habitats. In *Alpine Biodiversity in Europe* (ed. L. Nagy, G. Grabherr, C. Körner & D. B. A. Thompson), pp. 195–207. Berlin: Springer.

Pavlidis, Y. A., Dunayev, N. N. & Shcherbakov, F. A. (1997) The Late Pleistocene palaeogeography of Arctic Eurasian shelves. *Quaternary International*, **41**, 3–9.

Payette, S. (1984) Peat inception and climatic change in northern Québec. In *Climatic Change on a Yearly to Millennial Basis* (ed. N. A. Mörner & W. Karlén), pp. 173–179. Dordrecht: Reidel.

Payette, S. & Delwaide, A. (2004) Dynamics of subarctic wetland forests over the past 1500 years. *Ecological Monographs*, **74**, 373–391.

Payette, S. & Filion, L. (1985) White spruce expansion at the tree line and recent climatic change. *Canadian Journal of Forest Research–Revue Canadienne de Recherche Forestière*, **15**, 241–251.

Payette, S., Fortin, M. J. & Gamache, I. (2001) The subarctic forest-tundra: the structure of a biome in a changing climate. *BioScience*, **51**, 709–718.

Peck, J. R., Yearsley, J. M. & Waxman, D. (1998) Explaining the geographic distribution of sexual and asexual populations. *Nature*, **391**, 889–892.

Person, B. T., Herzog, M. P., Ruess, R. W., *et al.* (2003) Feedback dynamics of grazing lawns: coupling vegetation change with animal growth. *Oecologia*, **135**, 583–592.

Peteet, D., Andreev, A., Bardeen, W. & Mistretta, F. (1998) Long-term Arctic peatland dynamics, vegetation and climate history of the Pur-Taz region, Western Siberia. *Boreas*, **27**, 115–126.

Petit, R. J., Csaikl, U. M., Bordacs, S., *et al.* (2002) Chloroplast DNA variation in European white oaks: phylogeography and patterns of diversity based on data from over 2600 populations. *Forest Ecology and Management*, **156**, 5–26.

Pfanz, H. (1999) Photosynthetic performance of twigs and stems of trees with and without stress. *Phyton–Annales Rei Botanicae*, **39**, 29–33.

Philipp, M., Böcher, J., Mattsson, O. & Woodell, S. R. J. (1990) A quantitative approach to the sexual reproductive biology and population structure in some arctic flowering plants: *Dryas integrifolia, Silene acaulis* and *Ranunculus nivalis. Meddelelser om Gronland: Bioscience*, **34**, 3–60.

Pielou, E. C. (1991) *After the Ice Age*. Chicago: University of Chicago Press.

Pielou, E. C. (1994) *A Naturalist's Guide to the Arctic*. Chicago: University of Chicago Press.

Pigott, C. D. (1981) Nature of seed sterility and natural regeneration of *Tilia cordata* near its northern limit in Finland. *Annales Botanici Fennici*, **18**, 255–263.

Pigott, C. D. (1991) Biological flora of the British Isles No. 174. *Tilia cordata* Miller. *Journal of Ecology*, **79**, 1147–1207.

Pigott, C. D. & Taylor, K. (1964) The distribution of some woodland herbs in relation to the supply of phosphorus and nitrogen in the soil. *Journal of Ecology*, **52**, 175–185.

Pilcher, J., Bradley, R. S., Francus, P. & Anderson, L. (2005) A Holocene tephra record from the Lofoten Islands, Arctic Norway. *Boreas*, **34**, 136–156.

Plantlife International (2004) *Bluebells for Britain: A Report on the 2003 Bluebells for Britain survey*. Salisbury: Plantlife International.

Pollmann, W. & Veblen, T. T. (2004) *Nothofagus* regeneration dynamics in south-central Chile: a test of a general model. *Ecological Monographs*, **74**, 615–634.

Polunin, O. & Stainton, A. (1984) *Flowers of the Himalaya*. Delhi: Oxford University Press.

Porembski, S. (2005) Floristic diversity of African and South American inselbergs: a comparative analysis. *Acta Botanica Gallica*, **152**, 573–580.

Porembski, S. & Barthlott, W. (2000) Granitic and gneissic outcrops (inselbergs) as centers of diversity for desiccation-tolerant vascular plants. *Plant Ecology*, **151**, 19–28.

Porsild, A. E. & Cody, W. J. (1980). *Vascular Plants of the Continental Northwest Territories, Canada*. Ottawa: National Museum of Canada.

Porter, S. C., Sauchyn, D. J. & Delorme, L. D. (1999) The ostracode record from Harris Lake, southwestern Saskatchewan: 9200 years of local environmental change. *Journal of Paleolimnology*, **21**, 35–44.

Post, E. (2003) Large-scale climate synchronizes the timing of flowering by multiple species. *Ecology*, **84**, 277–281.

Prance, G. T. & Lovejoy, T. E., eds. (1985) *Amazonia*. Oxford: Pergamon.

Preston, C. D. (1997) *Aquatic Plants in Britain and Ireland*. Colchester: Harley Books.

Preston, C. D., Pearman, A. & Dines, T. D., eds. (2002) *New Atlas of the British and Irish Flora*. Oxford: Oxford University Press.

Price, M. V. & Joyner, J. W. (1997) What resources are available to desert granivores: seed rain or soil seed bank? *Ecology*, **78**, 764–773.

Purcell, J., Brelsford, A. & Kessler, M. (2004) The world's highest forest. *American Scientist*, **92**, 454–461.

Rada, F., Garcia-Nunez, C., Boero, C., *et al.* (2001) Low-temperature resistance in *Polylepis tarapacana*, a tree growing at the highest altitudes in the world. *Plant Cell and Environment*, **24**, 377–381.

Raffaelli, D. (2000) Interactions between macro-algal mats and invertebrates in the Ythan estuary, Aberdeenshire, Scotland. *Helgoland Marine Research*, **54**, 71–79.

Ratcliffe, R. G. (1997) In vivo NMR studies of the metabolic response of plant tissue to anoxia. *Annals of Botany*, **79**, 39–48.

Ratter, J. A., Ribeiro, J. F. & Bridgewater, S. (1997) The Brazilian cerrado vegetation and threats to its biodiversity. *Annals of Botany*, **80**, 223–230.

Reggiani, R., Giussani, P. & Bertani, A. (1990) Relationship between the accumulation of putrescine and the tolerance to oxygen-deficit stress in Gramineae seedlings. *Plant Cell Physiology*, **31**, 489–494.

Reise, K. (2005) Coast of change: habitat loss and transformations in the Wadden Sea. *Helgoland Marine Research*, **59**, 9–21.

Reisigl, H. & Keller, R. (1994) *Alpenpflanzen im Lebensraum*. Stuttgart: Gustav Fischer.

Renfrew, A. C. (1979) Investigations in Orkney. *Reports of the Research Committee of the Society of Antiquaries of London*, **38**, 234.

Richter-Menge, J., Overland, J., Proshutinsky, A., *et al.* (2006) *State of the Arctic*. Seattle: National Oceanic & Atmospheric Administration (NOAA), Pacific Marine Environmental Laboratory.

Rieley, G., Welker, J. M., Callaghan, T. V. & Eglinton, G. (1995) Epicuticular waxes of 2 arctic species: compositional differences in relation to winter snow cover. *Phytochemistry*, **38**, 45–52.

Rieseberg, L. H., Wood, T. E. & Baack, E. J. (2006) The nature of plant species. *Nature*, **440**, 524–527.

Ritchie, J. C. (1987) *Postglacial Vegetation of Canada*. Cambridge: Cambridge University Press.

Robe, W. E. & Griffiths, H. (1998) Adaptations for an amphibious life: changes in leaf morphology, growth rate, carbon and nitrogen investment, and reproduction during adjustment to emersion by the freshwater macrophyte *Littorella uniflora*. *New Phytologist*, **140**, 9–23.

Robe, W. E. & Griffiths, H. (2000) Physiological and photosynthetic plasticity in the amphibious, freshwater plant, *Littorella uniflora*, during the transition from aquatic to dry terrestrial environments. *Plant Cell and Environment*, **23**, 1041–1054.

Rodwell, J., ed. (1992) *British Plant Communities*. Vol. 3: *Grasslands and Montane Communities*. Cambridge: Cambridge University Press.

Rodwell, J. S., ed. (1991) *British Plant Communities*. Vol. 2: *Mires and Heaths*. Cambridge: Cambridge University Press.

Rønning, O. I. (1996) *The Flora of Svaldbard*. Oslo: Norwegian Polar Institute.

Rousseaux, M. C., Julkunen-Tiitto, R., Searles, P. S., Scopel, A. L., Aphalo, P. J. & Ballare, C. L. (2004) Solar UV-B radiation affects leaf quality and insect herbivory in the southern beech tree *Nothofagus antarctica*. *Oecologia*, **138**, 505–512.

Roy, V., Bernier, P. Y., Plamondon, A. P. & Ruel, J. C. (1999) Effect of drainage and microtopography in forested wetlands on the microenvironment and growth of planted black spruce seedlings. *Canadian Journal of Forest Research–Revue Canadienne de Recherche Forestière*, **29**, 563–574.

Rull, V. (2004a) Is the 'Lost World' really lost? Palaeoecological insights into the origin of the peculiar flora of the Guayana Highlands. *Naturwissenschaften*, **91**, 139–142.

Rull, V. (2004b) An evaluation of the Lost World and Vertical Displacement hypotheses in the Chimanta Massif, Venezuelan Guayana. *Global Ecology and Biogeography*, **13**, 141–148.

Rundel, P. W. & Dillon, M. O. (1998) Ecological patterns in the Bromeliaceae of the lomas formations of coastal Chile and Peru. *Plant Systematics and Evolution*, **212**, 261–278.

Rundgren, M. (1998) Early-Holocene vegetation of northern Iceland: pollen and plant macrofossil evidence from the Skagi peninsula. *Holocene*, **8**, 553–564.

Safriel, U. N., Volis, S. & Kark, S. (1994) Core and peripheral populations and global climate-change. *Israel Journal of Plant Sciences*, **42**, 331–345.

Sage, R. F. (1995) Was low atmospheric CO_2 during the Pleistocene a limiting factor for the origin of agriculture? *Global Change Biology*, **1**, 93–106.

Salmon, A., Ainouche, M. L. & Wendel, J. F. (2005) Genetic and epigenetic consequences of recent hybridization and polyploidy in Spartina (Poaceae). *Molecular Ecology*, **14**, 1163–1175.

Sanz-Elorza, M., Dana, E. D., Gonzalez, A. & Sobrino, E. (2003) Changes in the high-mountain vegetation of the central Iberian peninsula as a probable sign of global warming. *Annals of Botany*, **92**, 273–280.

Scarano, F. R., Pereira, T. S. & Rocas, G. (2003) Seed germination during floatation and seedling growth of *Carapa guianensis*, a tree from flood-prone forests of the Amazon. *Plant Ecology*, **168**, 291–296.

Schlesinger, W. H. & Andrews, J. A. (2000) Soil respiration and the global carbon cycle. *Biogeochemistry*, **48**, 7–20.

Schlüter, U. & Crawford, R. M. M. (2001) Long-term anoxia tolerance in leaves of *Acorus calamus* L. and *Iris pseudacorus* L. *Journal of Experimental Botany*, **52**, 2213–2225.

Schlüter, U. & Crawford, R. M. M. (2003) Metabolic adaptation to prolonged anoxia in leaves of American cranberry (*Vaccinium macrocarpon*). *Physiologia Plantarum*, **117**, 492–499.

Schröter, C. (1908) *Lebensgeschichte der Blütenpflanzen: Mitteleuropas*. Stuttgart: Ulmer.

Scuderi, L. A. (1994) Solar influence in Holocene treeline altitude variability in the Sierra-Nevada. *Physical Geography*, **15**, 146–165.

Seel, W. E., Hendry, G. A. F. & Lee, J. A. (1992) Effects of desiccation on some activated oxygen processing enzymes and anti-oxidants in mosses. *Journal of Experimental Botany*, **43**, 1031–1037.

Senn, J. (1999) Tree mortality caused by *Gremmeniella abietina* in a subalpine afforestation in the central Alps and its relationship with duration of snow cover. *European Journal of Forest Pathology*, **29**, 65–74.

Serebryanny, L. & Malyasova, E. (1998) The Quaternary vegetation and landscape evolution of Novaya Zemlya in the light of palynological records. *Quaternary International*, **45**, 59–70.

Serratovalenti, G., Devries, M. & Cornara, L. (1995) The hilar region in *Leucaena leucocephala* Lam (De Wit) seed: structure, histochemistry and the role of the lens in germination. *Annals of Botany*, **75**, 569–574.

Shaver, G. R. & Chapin, F. S. (1995) Long-term responses to factorial, NPK fertilizer treatment by Alaskan wet and moist tundra sedge species. *Ecography*, **18**, 259–275.

Shen-Miller, J. (2002) Sacred lotus, the long-living fruits of China Antique. *Seed Science Research*, **12**, 131–143.

Shen-Miller, J., Schopf, J. W., Harbottle, G., *et al.* (2002) Long-living lotus: germination and soil gamma-irradiation of centuries-old fruits, and cultivation, growth, and phenotypic abnormalities of offspring. *American Journal of Botany*, **89**, 236–247.

Sher, A. V. (1996) Late-Quaternary extinction of large mammals in northern Eurasia: a new look at the

Siberian contribution. In *Past and Future Rapid Environmental Changes* (ed. B. Huntley, W. Cramer, A. V. Morgan, H. C. Prentice & J. R. M. Allen), pp. 319–339. Berlin: Springer.

Shreeve, J. (2006) The greatest journey. *National Geographic*, **209**, 60–69.

Sibly, R. & Antonovics, J. (1992) Life history evolution. In *Genes in Ecology* (ed. R. J. Berry, T. J. Crawford & G. M. Hewitt), pp. 87–122. Oxford: Blackwell Scientific Publications.

Sibthorp, J. (1794) *Flora Oxoniensis, exhibens plantas in agro Oxoniensis sponte crescentes, secundum Systema Sexuale Distributas. Oxoni Typis Academicus*, Oxford.

Sieber, M. & Braendle, R. (1991) Energy metabolism in rhizomes of *Acorus calamus* (L.) and in tubers of *Solanum tuberosum* (L.) with regard to their anoxia tolerance. *Botanica Acta*, **104**, 279–282.

Siebel, H. N., VanWijk, M. & Blom, C. (1998) Can tree seedlings survive increased flood levels of rivers? *Acta Botanica Neerlandica*, **47**, 219–230.

Simard, S. W., Perry, D. A., Jones, M. D., Myrold, D. D., Durall, D. M. & Molina, R. (1997) Net transfer of carbon between ectomycorrhizal tree species in the field. *Nature*, **388**, 579–582.

Simon, M. F. & Proenca, C. (2000) Phytogeographic patterns of *Mimosa* (Mimosoideae, Leguminosae) in the cerrado biome of Brazil: an indicator genus of high-altitude centers of endemism? *Biological Conservation*, **96**, 279–296.

Simpson, I. A., Dugmore, A. J., Thomson, A. & Vesteinsson, O. (2001) Crossing the thresholds: human ecology and historical patterns of landscape degradation. *Catena*, **42**, 175–192.

Sinnott, E. W. (1917) The 'age and area' hypothesis and the problem of endemism. *Annals of Botany*, **31**, 209–216.

Skarpe, C. & van der Wal, R. (2002) Effects of simulated browsing and length of growing season on leaf characteristics and flowering in a deciduous Arctic shrub, *Salix polaris*. *Arctic, Antarctic and Alpine Research*, **34**, 282–286.

Skene, K. R. (1998) Cluster roots: some ecological considerations. *Journal of Ecology*, **86**, 1060–1064.

Skene, K. R. (2003) The evolution of physiology and development in the cluster root: teaching an old dog new tricks? *Plant and Soil*, **248**, 21–30.

Skre, O., Baxter, R., Crawford, R. M. M., Callaghan, T. V. & Fedorkov, A. (2002) How will the tundra-taiga

interface respond to climate change? *Ambio*, Special Issue **12**, 37–46.

Smith, D. M., Jackson, N. A., Roberts, J. M. & Ong, C. K. (1999) Reverse flow of sap in tree roots and downward siphoning of water by *Grevillea robusta*. *Functional Ecology*, **13**, 256–264.

Smith, T. M., Shugart, H. H. & Woodward, F. I., eds. (1997) *Plant Functional Types*. Cambridge: Cambridge University Press.

Smithsonian Museum (2006) *South America*. www.nmnh.si. edu/botany/.

Smolders, A. J. P., Tomassen, H. B. M., Pijnappel, H. W., Lamers, L. P. M. & Roelofs, J. G. M. (2001) Substrate-derived CO_2 is important in the development of *Sphagnum* spp. *New Phytologist*, **152**, 325–332.

Snyder, K. A. & Williams, D. G. (2000) Water sources used by riparian trees varies among stream types on the San Pedro River, Arizona. *Agricultural and Forest Meteorology*, **105**, 227–240.

Southwood, T. R. E. (1988) Tactics, strategies and templates. *Oikos*, **52**, 3–18.

Sperling, C. R. & King, S. R. (1990). Andean tuber crops: worldwide potential. In *Advances in New Crops* (ed. J. Janick & J. E. Simon), pp. 428–435. Portland, OR: Timber Press.

Spiecker, H., Mielikäinen, K., Köhl, M. & Skovsgaard, J., eds. (1996) *Growth Trends in European Forests*. Berlin: Springer.

Squeo, F. A., Rada, F., Azocar, A. & Goldstein, G. (1991) Freezing tolerance and avoidance in high tropical Andean plants: is it equally represented in species with different plant height? *Oecologia*, **86**, 378–382.

Squeo, F. A., Rada, F., Garcia, C., Ponce, M., Rojas, A. & Azocar, A. (1996) Cold resistance mechanisms in high desert Andean plants. *Oecologia*, **105**, 552–555.

Stace, C. A. (1997) *New Flora of the British Isles*, 2nd edn. Cambridge: Cambridge University Press.

Stanley, D. J. & Hait, A. K. (2000) Holocene depositional patterns, neotectonics and Sundarban mangroves in the western Ganges-Brahmaputra delta. *Journal of Coastal Research*, **16**, 26–39.

Stebbins, G. L. (1971) *Chromosomal Evolution in Higher Plants*. London: Edward Arnold.

Stecher, G., Schwienbacher, F., Mayr, S. & Bauer, H. (1999) Effects of winter-stress on photosynthesis and antioxidants of exposed and shaded needles of *Picea abies* (L.) Karst. and *Pinus cembra* L. *Phyton–Annales Rei Botanicae*, **39**, 205–211.

Steinger, T., Körner, C. & Schmid, B. (1996) Long-term persistence in a changing climate: DNA analysis suggests very old ages of clones of alpine *Carex curvula*. *Oecologia*, **105**, 94–99.

Stenström, A., Jonsson, B. O., Jonsdottir, I. S., Fagerstrom, T. & Augner, M. (2001) Genetic variation and clonal diversity in four clonal sedges (*Carex*) along the Arctic coast of Eurasia. *Molecular Ecology*, **10**, 497–513.

Stenström, M. & Molau, U. (1992) Reproductive ecology of *Saxifraga oppositifolia*: phenology, mating system, and reproductive success. *Arctic and Alpine Research*, **24**, 337–343.

Stenström, M., Gugerli, F. & Henry, G. H. R. (1997) Response of *Saxifraga oppositifolia* L. to simulated climate change at three contrasting latitudes. *Global Change Biology*, **3**, 44–54.

Stevenson, A. C. & Birks, H. J. B. (1995) Heaths and moorlands: long-term ecological changes and interactions with climate and people. In *Heaths and Moorlands: Cultural Landscapes* (ed. D. B. A. Thompson, A. J. Hester & M. B. Usher), pp. 224–239. Edinburgh: Scottish Natural Heritage.

Stewart, W. S. & Bannister, P. (1973) Seasonal changes in carbohydrate content of three *Vaccinium* species with particular reference to *V. uliginosum* and its distribution in the British Isles. *Flora*, **162**, 134–155.

Strahler, A. N. (1963) *The Earth Sciences*. New York: Harper & Row.

Stuart, A. J. (2005) The extinction of woolly mammoth (*Mammuthus primigenius*) and straight-tusked elephant (*Palaeoloxodon antiquus*) in Europe. *Quaternary International*, **126**, 171–177.

Stuart, S. A., Choat, B., Martin, K. C., Holbrook, N. M. & Ball, M. C. (2007) The role of freezing in setting the latitudinal limits of mangrove forests. *New Phytologist*, **173**, 576–583.

Studer, C. & Braendle, R. (1984) Sauerstoffkonsum und Versorgung der Rhizome von *Acorus calamus* L., *Glyceria maxima* (Hartman) Holber, *Menyanthes trifoliata* L., *Phalaris arundinacea* L., Trin. und *Typha latifolia* L. *Botanica Helvetica*, **94**, 23–31.

Studer-Ehrensberger, K., Studer, C. & Crawford, R. M. M. (1993) Competition at community boundaries: mechanisms of vegetation structure in a dune-slack complex. *Functional Ecology*, **7**, 156–168.

Sturm, M., Schimel, J., Michaelson, G., *et al.* (2005) Winter biological processes could help convert arctic tundra to shrubland. *BioScience*, **55**, 17–26.

Sveinbjörnsson, B. (2000) North American and European treelines: external forces and internal processes controlling position. *Ambio*, **29**, 388–395.

Sveinbjornsson, J., Hofgaard, A. & Lloyd, A. (2002) Natural causes of the tundra-taiga boundary. *Ambio*, Special Issue **12** , 23–29.

Svoboda, J. & Freedman, B., eds. (1992) *Ecology of a Polar Oasis*. Toronto: Captus University Publication.

Svoboda, J. & Henry, G. H. R. (1987) Succession in marginal Arctic environments. *Arctic and Alpine Research*, **19**, 373–384.

Szeicz, J. M. & MacDonald, G. M. (1995) Recent white spruce dynamics at the Sub-Arctic alpine treeline of North-Western Canada. *Journal of Ecology*, **83**, 873–885.

Sziemer, P. (2000) *Madeira's Natural History in a Nutshell*. Funchal: Ribiero.

Tallantire, P. A. (2002) The early-Holocene spread of hazel (*Corylus avellana* L.) in Europe north and west of the Alps: an ecological hypothesis. *Holocene*, **12**, 81–96.

Tamura, F., Tanabe, K., Katayama, M. & Itai, A. (1996) Effects of flooding on ethanol and ethylene production by pear rootstocks. *Journal of the Japanese Society for Horticultural Science*, **65**, 261–266.

Teeri, J. A. (1972) Microenvironmental adaptations of local populations of *Saxifraga oppositifolia* in the high Arctic. Ph.D. thesis, Duke University, Durham, NC.

Teeri, J. A. (1973) Polar desert adaptations of a high Arctic plant species. *Science*, **179**, 496–497.

Telewski, F. W. & Zeevaart, J. A. D. (2002) The 120-yr period for Dr. Beal's seed viability experiment. *American Journal of Botany*, **89**, 1285–1288.

Tenow, O. & Bylund, H. (2000) Recovery of a *Betula pubescens* forest in northern Sweden after severe defoliation by *Epirrita autumnata*. *Journal of Vegetation Science*, **11**, 855–862.

Tenow, O., Bylund, H. & Holmgren, B. (2001) Impact on mountain birch forests in the past and future of outbreaks of two geometrid insects. In *Nordic Mountain Birch Ecosystems* (ed. F. E. Wielgolaski), pp. 223–239. New York: Parthenon Publishing.

Thompson, K., Bakker, J. P. & Bekker, T. M. (1997) *The Soil Seed Banks of North West Europe*. Cambridge: Cambridge University Press.

Thompson, L. G., Mosley-Thompson, E., Davis, M. E., *et al.* (2002) Kilimanjaro ice core records: evidence of

Holocene climate change in tropical Africa. *Science*, **298**, 589–593.

Thygerson, T., Harris, J. M., Smith, B. N., Hansen, L. D., Pendleton, R. L. & Booth, D. T. (2002) Metabolic response to temperature for six populations of winterfat (*Eurotia lanata*). *Thermochimica Acta*, **394**, 211–217.

Tilman, D. (1987) On the meaning of competition and the mechanisms of competitive superiority. *Functional Ecology*, **1**, 304–315.

Tilman, D. (1988) *Plant Strategies and Dynamics and Structure of Plant Communities.* Princetown, NJ: Princetown University Press.

Tipping, R. (1994) The form and fate of Scotland's woodlands. *Proceedings of the Society of Antiquaries of Scotland*, **124**, 1–54.

Tolmatchev, A. I. (1966) Die Evolution der Pflanzen in arktischen-Eurasien während und nach der quaternären Vereisung. *Botanisk Tidsskrift*, **62**, 27–36.

Tomback, D. F. (2001) Clark's nutcracker: agent of regeneration. In *Whitebark Pine Communities: Ecology and Restoration* (ed. D. F. Tomback, S. F. Arno & R. E. Keane), pp. 89–104. Washington, DC: Island Press.

Tomback, D. F. (2005) The impact of seed dispersal by Clark's nutcracker on whitebark pine: multi-scale perspective on a high mountain mutualism. In *Mountain Ecosystems: Studies in Treeline Ecology* (ed. G. Broll & B. Keplin), pp. 181–201. Berlin: Springer.

Tomback, D. F., Anderies, A. J., Carsey, K. S., Powell, M. L. & Mellmann-Brown, S. (2001) Delayed seed germination in whitebark pine and regeneration patterns following the yellowstone fires. *Ecology*, **82**, 2587–2600.

Torres, A. M. & Diedenhofen, U. (1979) Baker sunflower populations revisited. *Journal of Heredity*, **70**, 275–276.

Tranquillini, W. (1979) *Physiological Ecology of the Alpine Timberline.* Berlin: Springer Verlag.

Troll, C. (1959) Die tropischen Gebirge. *Bonner geographische Abhandlungen*, **25**, 1–93.

Tucker, C. J. (1979) Red and photographic infrared linear combinations for monitoring vegetation. *Remote Sensing of Environment*, **8**, 127–150.

Tveranger, J., Astakhov, V., Mangerud, J. & Svendsen, J. I. (1999) Surface form of the south-western sector of the last Kara Sea ice sheet. *Boreas*, **28**, 81–91.

UNEP World Conservation Monitoring Centre (2003) website: www.unep-wcmc.org/

Usher, M. B. (2005). Principles of conserving the Arctic's biodiversity. In *Arctic Climate Impact Assessment* (ed. ACIA), pp. 539–596. Cambridge: Cambridge University Press.

Vajda, A. & Venalainen, A. (2005) Feedback processes between climate, surface and vegetation at the northern climatological tree-line (Finnish Lapland). *Boreal Environment Research*, **10**, 299–314.

Valentini, R., Dore, S., Marchi, G., *et al.* (2000) Carbon and water exchanges of two contrasting central Siberia landcape types: regenerating forest and bog. *Functional Ecology*, **14**, 87–96.

van der Merwe, M., Winfield, M. O., Arnold, G. M. & Parker, J. S. (2000) Spatial and temporal aspects of the genetic structure of *Juniperus communis* populations. *Molecular Ecology*, **9**, 379–386.

van der Wal, R., Egas, M., van der Veen, A. & Bakker, J. (2000) Effects of resource competition and herbivory on plant performance along a natural productivity gradient. *Journal of Ecology*, **88**, 317.

Vanhinsbergh, D. P. & Chamberlain, D. E. (2001) Habitat associations of breeding Meadow Pipits *Anthus pratensis* in the British uplands. *Bird Study*, **48**, 159–172.

Väre, H. (2001) Mountain birch taxonomy and floristics of mountain birch woodlands. In *Nordic Mountain Birch Ecosystems* (ed. F. E. Wielgolaski) pp. 35–46. New York: Parthenon Publishing.

Vargas, C. (1981) Plant motifs on Inca ceremonial vases from Peru. *Botanical Journal of the Linnean Society*, **82**, 313–325.

Vavrek, M. C., McGraw, J. B. & Bennington, C. C. (1991) Ecological genetic variation in seed banks. III: Phenotypic and genetic differences between young and old seed populations of *Carex bigelowii*. *Journal of Ecology*, **79**, 645–662.

Velichko, A. A., Kononov, Y. M. & Faustova, M. A. (1997) The last glaciation of Earth: size and volume of ice-sheets. *Quaternary International*, **41**, 43–51.

Velichko, A. A., Kremenetski, C. V., Borisova, O. K., Zelikson, E. M., Nechaev, V. P. & Faure, H. (1998) Estimates of methane emission during the last 125,000 years in Northern Eurasia. *Global and Planetary Change*, **17**, 159–180.

Vera, F. W. M. (2000) *Grazing Ecology and Forest History.* Wallingford: CABI Publishing.

Vlassova, T. K. (2002) Human impacts on the tundra-taiga zone dynamics: the case of the Russian lesotundra. *Ambio*, Special Issue **12**, 30–36.

Voesenek, L., Rijnders, J., Peeters, A. J. M., Van de Steeg, H. M. V. & De Kroon, H. (2004) Plant hormones regulate fast shoot elongation under water: from genes to communities. *Ecology*, **85**, 16–27.

Vorren, T. O., Vorren, K. D., Alm, T., *et al.* (1988) The last deglaciation (20,000 to 11,000 BP) on Andøya, Northern Norway. *Boreas*, **17**, 41–77.

Vuilluemier, F. & Monasterio, M. (1988) *High Altitude Tropical Biogeography*. Oxford: Oxford University Press.

Wagner, J. & Reichegger, B. (1997) Phenology and seed development of the alpine sedges *Carex curvula* and *Carex firma* in response to contrasting temperatures. *Arctic and Alpine Research*, **29**, 291–299.

Walker, D. A., Auerbach, N. A., Bockheim, J. G., *et al.* (1998) Energy and trace-gas fluxes across a soil pH boundary in the arctic. *Nature*, **394**, 469–472.

Walker, D. A., Gould, W. A., Maier, H. A. & Raynolds, M. K. (2002) The Circumpolar Arctic Vegetation Map: AVHRR-derived base maps, environmental controls, and integrated mapping procedures. *International Journal of Remote Sensing*, **23**, 4551–4570.

Walker, D. A., Raynolds, M. K., Daniels, F. J. A., *et al.* (2005) The Circumpolar Arctic Vegetation Map. *Journal of Vegetation Science*, **16**, 267–282.

Walter, H. (1960) *Grundlagen der Pflanzenverbreitung I Teil Standortslehre (analytisch-ökologische Geobotanik)*, 2nd edn. Stuttgart: Eugen Ulmer Verlag.

Walters, M. B. & Reich, P. B. (1999) Low-light carbon balance and shade tolerance in the seedlings of woody plants: do winter deciduous and broad-leaved evergreen species differ? *New Phytologist*, **143**, 143–154.

Walther, G. R. (1999) Distribution and limits of evergreen broad-leaved (laurophyllous) species in Switzerland. *Botanica Helvetica*, **109**, 153–167.

Walther, G. R. (2002) Weakening of climatic constraints with global warming and its consequences for evergreen broad-leaved species. *Folia Geobotanica*, **37**, 129–139.

Wang, R. Z. (2003) C-4 plants in the vegetation of Tibet, China: their natural occurrence and altitude distribution pattern. *Photosynthetica*, **41**, 21–26.

Wardle, P. (1991) *Vegetation of New Zealand*. Cambridge: Cambridge University Press.

Wardle, P. & Coleman, M. C. (1992) Evidence for rising upper limits of 4 native New-Zealand forest trees. *New Zealand Journal of Botany*, **30**, 303–314.

Weathers, K. C., Lovett, G. M., Likens, G. E. & Caraco, N. F. M. (2000) Cloudwater inputs of nitrogen to forest ecosystems in southern Chile: forms, fluxes, and sources. *Ecosystems*, **3**, 590–595.

Webb, D. A. & Gornall, R. J. (1989) *Saxifrages of Europe*. London: Christopher Helm.

Weber, M. & Braendle, R. (1994) Dynamics of nitrogen-rich compunds in roots, rhizomes ad leaves of the Sweet Flag (*Acorus calamus*). *Flora*, **189**, 63–68.

Weber, M. & Braendle, R. (1996) Some aspects of the extreme anoxia tolerance of the Sweet Flag (*Acorus calamus*) L. *Folia Geobotanica et Phytotaxonomica*, **31**, 37–46.

Weis, I. M. & Hermanutz, L. A. (1993) Pollination dynamics of arctic dwarf birch (*Betula glandulosa* Betulaceae) and its role in the loss of seed production. *American Journal of Botany*, **80**, 1021–1027.

Werner, P. A. (1979) Competition and coexistence of similar species. In *Topics in Population Biology* (ed. O. T. Solbrig, S. Jain, G. B. Johnson & P. H. Raven), pp. 287–310. New York: Columbia University Press.

White, K. D. (1970) *Roman Farming*. London: Thames and Hudson.

Whitlock, C. (1993) Postglacial vegetation and climate of Grandteton and Southern Yellowstone National Parks. *Ecological Monographs*, **63**, 173–198.

Whittle, A., Keith-Lucas, M., Miles, A., Noddle, B., Rees, S. & Romans, J. (1986) *Scord of Brouster: An Early Agricultural Settlement on Shetland*. Oxford: Oxford Committee for Archaeology.

Wickham-Jones, C. (2006) *Between the Wind and the Water: World Heritage Orkney*. Macclesfield: Windgather Press.

Wiegolaski, F. E. & Nilsen, J. (2001) Coppicing and growth of various provenances of mountain birch in relation to nutrients and water. In *Nordic Mountain Birch Ecosystems* (ed. F. E. Wiegolaski), pp. 77–92. New York: Parthenon Publishing.

Wielgolaski, F. E. & Sonesson, M. (2001). Nordic mountain birch ecosystems: a conceptual overview. In *Nordic Mountain Birch Ecosystems* (ed. F. E. Wielgolaski), pp. 377–384. New York: Parthenon Publishing.

Wieser, G. (1997) Carbon dioxide gas exchange of cembran pine (*Pinus cembra*) at the alpine timberline during winter. *Tree Physiology*, **17**, 473–477.

Wieser, G., Gigele, T. & Pausch, H. (2005) The carbon budget of an adult *Pinus cembra* tree at the alpine timberline in the Central Austrian Alps. *European Journal of Forest Research*, **124**, 1–8.

Wiessner, A., Kuschk, P. & Stottmeister, U. (2002) Oxygen release by roots of *Typha latifolia* and *Juncus effusus* in laboratory hydroponic systems. *Acta Biotechnologica*, **22**, 209–216.

Wildi, B. & Lütz, C. (1996) Antioxidant composition of selected high alpine plant species from different altitudes. *Plant Cell and Environment*, **19**, 138–146.

Williams, K., Caldwell, M. M. & Richards, J. H. N.A. (1993) The influence of shade and clouds on soil water potential: the buffered behaviour of hydraulic lift. *Plant and Soil*, **157**, 83–95.

Williamson, M. (2003) Species-area relationships at small scales in continuum vegetation. *Journal of Ecology*, **91**, 904–907.

Willis, A. J. (1985) Plant diversity and change in a species-rich dune system. *Transactions of the Botanical Society of Edinburgh*, **44**, 291–308.

Willis, J. C. (1922) *Age and Area*. Cambridge: Cambridge University Press.

Wilmking, M. & Juday, G. P. (2005) Longitudinal variation of radial growth at Alaska's northern treeline: recent changes and possible scenarios for the 21st century. *Global and Planetary Change*, **47**, 282–300.

Wilmking, M., Juday, G. P., Barber, V. A. & Zald, H. S. J. (2004) Recent climate warming forces contrasting growth responses of white spruce at treeline in Alaska through temperature thresholds. *Global Change Biology*, **10**, 1724–1736.

Wilmshurst, J. M. & Higham, T. F. G. (2004) Using rat-gnawed seeds to independently date the arrival of Pacific rats and humans in New Zealand. *Holocene*, **14**, 801–806.

Wilmshurst, J. M., Bestic, K. L., Meurk, C. D. & McGlone, M. S. (2004) Recent spread of *Dracophyllum* scrub on subantarctic Campbell Island, New Zealand: climatic or anthropogenic origins? *Journal of Biogeography*, **31**, 401–413.

Wilson, E. O. (1959) Adaptive shift and dispersion in a tropical ant fauna. *Evolution*, **13**, 122–144.

Wisheu, I. C., Rosenzweig, M. L., Olsvig-Whittaker, L. & Shmida, A. (2000) What makes nutrient-poor Mediterranean heathlands so rich in plant diversity? *Evolutionary Ecology Research*, **2**, 935–955.

Wolff, K., El-Akkad, S. & Abbott, R. J. (1997) Population substructure in *Alkanna orientalis* (Boraginaceae) in the Sinai desert, in relation to its pollinator behaviour. *Molecular Ecology*, **6**, 365–372.

Woodward, F. I., Lake, J. A. & Quick, W. P. (2002) Stomatal development and CO_2: ecological consequences. *New Phytologist*, **153**, 477–484.

Wookey, P. A. & Robinson, C. H. (1997) Interpreting environmental manipulation experiments in arctic ecosystems: are 'disturbance' perspectives properly accounted for? In *Disturbance and Recovery in Arctic Lands: An Ecological Perspective* (ed. R. M. M. Crawford), pp. 115–134. Dordrecht: Kluwer.

Wookey, P. A., Robinson, C. H., Parsons, A. N., *et al.* (1995) Environmental constraints on the growth, photosynthesis and reproductive development of *Dryas octopetala* at a high arctic polar semidesert, Svalbard. *Oecologia*, **102**, 478–489.

Wuebker, E. F., Mullen, R. E. & Koehler, K. (2001) Flooding and temperature effects on soybean germination. *Crop Science*, **41**, 1857–1861.

Xiong, F. S., Ruhland, C. T. & Day, T. A. (1999) Photosynthetic temperature response of the Antarctic vascular plants *Colobanthus quitensis* and *Deschampsia antarctica*. *Physiologia Plantarum*, **106**, 276–286.

Yallop, A. R., Thacker, J. I., Thomas, G., *et al.* (2006) The extent and intensity of management burning in the English uplands. *Journal of Applied Ecology*, **43**, 1138–1148.

Yurtsev, B. A. (2001) The Pleistocene 'Tundra-Steppe' and the productivity paradox: the landscape approach. *Quaternary Science Reviews*, **20**, 165–174.

Author index

Aanes 145, 394
Aas 322, 323
Abbott 125, 126, 211, 212, 224, 244, 422
Aboal 94
ACIA 178
Adams 328
Aerts 326
Aiken 140, 217
Alatolo 136
Alberdi 246, 248
Albrethsen 323
Alekseev 188
Allaby 14, 15
Allen 102, 145
Allison 263
Anamthawat-Jonsson 319, 323
Anderson 266, 422
Andreev 194
Antonovics 9, 17
Anttila 127
Arft 221
Armesto 156
Armstrong 98, 281, 283, 284, 286, 297, 300, 302, 303
Arseneault 178, 324
Asselin 163
Atkin 98, 99
Attenborough 346
Ayyad 56

Baddeley 314
Bagchi 348
Bahn 370
Ballantyne 206
Ballare 360
Bannister 73, 372
Barber 229
Barclay 146, 299
Barcroft 379
Barlow 403
Barnosky 389
Barton 119
Baskin 133, 134, 279
Bauert 154, 241
Baxter-Burrell 279

Becker 260
Beerling 314
Behre 399
Bell 25
Bennett 175, 396
Bennike 163
Bennington 147
Bertrand 279, 280
Bierzychudek 154
Billings 153
Biswas 279
Bitonti 291
Blair 71
Blasco 263
Bliss 199, 219
Blom 288
Boam fa 279
Bond 42, 79
Boom 415
Bradley 300
Braendle 86, 279, 287, 289, 293, 304
Braithwaite 131
Bravo 248
Brewer 17
Briffa 187
Brochmann 57, 120, 216
Brown 172, 199
Brubaker 163
Bryson 167
Brysting 212, 214
Bucher 297
Buchmann 97
Buckland 206
Bunting 396
Burdick 252
Burke 347

CAFF 431
Cairns 190, 191
Caldwell 360
Callaghan 156, 183, 202
Calow 14, 31, 81, 199, 352
Campbell 119, 175
Caseldine 402
Caulfield 402

Chamberlain 43
Chambers, F. 229
Chambers, R. 252
Chapin 84, 99, 106, 107, 175, 223
Chapman 158
Chas 343
Chernov 169
Childe 25
Christensen 282
Clemmensen 102
Clevering 291
Colwell 32
Comtois 151
Conner 301
Conrad 229
Constable 98
Convey 248
Cooper 39, 106, 148, 253
Cornelissen 101
Costich 148
Coutts 304
Cowie 386
Cowling 32, 45, 46, 47
Crawford 7, 13, 14, 17, 36, 59, 73, 74, 75, 76, 80, 85, 86, 88, 89, 97, 114, 137, 148, 149, 151, 153, 156, 181, 185,193, 203, 210, 214, 279, 280, 288, 290, 293, 303, 304, 305, 317, 318, 324, 329, 366, 369, 394, 396, 406, 421, 422, 423, 424
Crawley 37
Cruzan 120
Currey 155

Dahl 332, 360
Dalen 175, 188
Dalpe 102
Dalton 264
Darling 364
Davis 131
Dawson 80, 92, 94, 148, 151
Day 359, 360
DeGroot 323, 324
Denevan 383, 384, 414
Dennis 398
Desmet 56

Di Tomaso 92
Dickinson 267
Diemer 348
Diggle 137
Dodgshon 399
Dormann 317
Duncalf 366
Dunning 409

Edwards 6
Ehlers 208
Eiten 51
Ellenberg 354
Elvebakk 205, 206
Emmerson 251
Emms 120
Ennos 35
Eriksen 57
Eriksson 18
Erkamo 6, 7, 146
Erschbaner 348
Ewel 301

Fjeldsa 219
Fleisher 50
Fleming 334
Forbes 118, 319
Forchhammer 181
Fossitt 397
Fox 286
Frankel 33
Fredskild 324
Freedman 220
Freeman 151
French 251
Fries 5, 173, 218, 237, 314, 326
Fukao 279

Gaelic 225
Gamache 170, 171
Gansert 303
Garcia 328
Garcilaso de la Vega 366
Gee 291
Geoghegan 51
Gerloff 166
Germino 357, 358
Gervais 175
Gibbens 92
Gillibrand 250
Gimingham 270
Girardin 185
Glover 35
Glueck 94
Godwin 325
Goetz 202
Goldberg 85

Goldblatt 45
Gottfried 353
Gould 220
Gould 219
Gouyon 50
Grabherr 370
Grace 77, 84, 172, 191, 377
Graham 87
Grant 151
Graumlich 187
Gray 123
Greig-Smith 32
Grime 15, 19, 77, 81, 83, 85, 103, 106,
    153, 348
Grosse 284
Grosswald 209
Grubb 14, 15, 89
Grundt 36
Gualtieri 389
Gudleifsson 185
Gugerli 378
Guisan 370
Gutterman 135

Hadley 191
Hagström 103
Haldemann 297
Hall 408
Halloy 335
Hallsdóttir 397
Handa 190, 376
Hanhijärvi 279
Hanski 14, 19
Haraguchi 50
Harper 14, 144, 151
Harris 123, 125
Hättenschwiler 84
Haustrôm 106
Hawksworth 33
Headley 106
Heath 373
Hedberg 140
Heery 407
Heide 66, 136, 141
Heinz 50
Hemminger 406
Hemp 365
Henkel 144, 145
Henry 100
Henzi 297
Hewitt 17, 36, 422
Hitier 94
Hobbs 43
Hoch 74, 366, 367
Hochachka 70, 373
Hofgaard 188
Hogg 177

Holderegger 248
Holmgren 175
Holten 360, 361
Holtmeier 11, 39, 167, 171, 175, 177, 187,
    188, 191, 194, 328
Hoosbeek 96
Houghton 227
Hueck 51, 56, 96, 155
Hughes 250, 251
Hulme 113
Hultén 5, 218, 237, 314, 326
Hunt 396
Hunter 279
Huntley 6, 174
Hurrell 182

Imbert 240
Ims 145
Ingold 383, 392
Ingólfson 209
Ingram 277

Jacobs 324
Jacoby 187
James 123, 125, 126
Jansen 289
Jarvis 13
Jefferies 39, 41, 217
Jeffree 113, 192, 193
Johansen 147, 228
Joly 288
Jonasson 105
Jónsdóttir 68, 105
Juday 178, 188
Jump 422

Kahkonen 325
Kaland 396, 399
Kallio 319, 328
Kaser 364
Kato-Noguchi 279, 280
Keeley 98
Keller 362, 426
Kelly 145, 277
Kent 267
Kielland 102, 106
Kienast 209
Kirk 120
Klak 47, 48
Klanderud 370
Klasner 177
Klein 39
Klinger 184
Knottnerus 404
Kobayashi 94
Koenig 145
Koller 134

Koncalova 289
Korpelainen 153
Köppen 14, 167
Köppitz 156
Körner 11, 13, 107, 167, 190, 352, 355, 370, 373, 374
Kremenetski 6, 166, 181, 185
Kriuchkov 168
Krupnik 392, 394, 395
Kullman 175, 322, 323
Kuwabara 291
Kuz'min 39
Kwant 20, 21
Kytoviita 102

Laberge 178
Laderman 21, 22, 227, 229
Landvik 206, 208
Langematz 388
Lanner 155
Larl 362
Larson 297
Lavoie 178, 187
Lee 10
Legesse 280
Lehouerou 56
Lévesque 204
Levins 14
Levitt 89
Lewington 155
Lewis 58
Lid 17, 77, 154, 314, 326
Lightfoot 94
Lincoln 14, 15, 32, 65
Linder 45, 46
Linhart 50, 292
Little 264
Llewellyn 237
Lloyd 97, 148, 177, 178, 187
Longstreth 98
Lovejoy 51
Lowe 17
Luterbacher 388
Lüttge, 346
Lütz 359
Lynn 292, 294

Maarel 93
Mabberley 346
MacArthur 17, 36
MacDonald 174, 185
Mallik 37, 178
Mangerud 206
Mani 423
Mannheimer 118
Mantovani 248
Mark 143

Marone 148
Marrs 131
Marschner 351
Mauquoy 229, 232
Mayr 191
McCormick 398
McDonald 282
McFadden 181
McGhee 391
McGlone 335, 337
McGraw 17, 59, 148, 248
McNally 281
McVean 324, 328, 396
Menard 41
Meriot 392
Meusel 5, 231
Michaelis 191
Miehe 342, 355, 372
Milne 20, 22, 120, 124, 129
Minchinton 252
Mitchell 22, 38, 402
Mithen 389, 397
Molau 137, 140, 154
Mommer 289
Monk 287
Monteith 94
Montgomery 14, 16, 373
Mooney 11
Moore 264
Moreira 42
Muraoka 312
Murray 216, 423
Myers 44, 47, 57
Myneni 97

Nadelhoffer 98
Nakatsubo 314
Nelson 86
Nerem 227
Nesje 166
Neuvonen 320
New 330
Nordal 206

Odasz 99
Ogden 15, 146, 332, 372
Ögren, 15, 332, 372
Ojeda 48
Oksanen 82, 83
Ostendorp 301
Ovstedal 111

Palsson 398
Pate 103
Patten 122
Pauli 370
Pavlidis 209

Payette 169, 175, 181
Peck 154
Person 41
Peteet 182
Pfanz 302
Philipp 135
Pielou 88, 373, 389
Pigott 101, 111
Pilcher 397
Plantlife 127
Platt 146
Pollmann 114
Polunin 351
Porembski 347, 348
Porter 166
Portuguese 51
Post 145
Preston 129, 131, 290
Price 148
Purcell 366

Rada 369
Raffaelli 250
Ranta 83
Ratcliffe 286
Ratter 51
Reggiani 286
Reise 405
Renfrew 386
Richardson 46
Rieley 214
Rieseberg 19
Ritchie 174
Robe 98, 291
Rodwell 324, 328, 431
Rønning 331
Rousseaux 360
Roy 185
Rull 346, 347
Rundel 56
Rundgren 397

Safriel 31, 58, 423
Sage 25
Salmon 122
Scarano 118
Schlüter 88, 279, 280, 288, 409, 411
Schlesinger 97
Schröter 295
Scuderi 175
Seel 263
Senn 194
Serebryanny 209
Serratovalenti 134
Shakespeare 252
Shaver 106
Shen-Miller 147

Sher 184
Shreeve 383, 390
Sibly 19
Sibthorp 125
Siebel 297, 301
Sieber 297
Simard 87
Simon 51
Simpson 25
Sinnott 345
Skarpe 316
Skene 103
Skre 169
Smith, D. M. 90, 93
Smith, T. M. 34, 80
Smithsonian Museum 369
Smolders 96
Snyder 93
Southwood 83
Sperling 415
Spiecker 187
Squeo 356
Stace 118, 326
Stanley 263
Stebbins 217
Stecher 193
Steinger 153, 422
Stenström 137
Stevenson 396
Stewart 11
Stuart 260, 389
Studer 288, 297
Studer-Ehrensburger 295
Sturm 202
Sveinbjörnsson 167, 187
Svoboda 204, 219

Szeicz 175
Sziemer 396

Tallantire 397
Tamura 279
Teeri 59, 139, 214
Telewski 147
Tenow 319, 322
Thompson 147, 364
Thygerson 373
Tilman 77, 83
Tipping 181, 396, 397
Tolmatchev 212, 213
Tomback 114, 115, 116
Torres 423
Tranquillini 145, 146, 191
Troll 366
Tucker 167
Tveranger 209

Usher 431

Vahinsbergh 43
Vajda 179
Valentini 178
van der Merwe 328
VanLoon 182
Vargas 415
Väre 171, 319
Velichko 183, 209
Vera 38, 426
Virtanen 84
Vlassova 164, 195
Voesenek 68, 288
Vorren 208
Vuilluemier 366

Wagner 139
Wal 41, 316
Walker 199, 205
Walter 15, 87, 350
Walther 129
Wang 417
Wardle 128, 130, 250, 258, 311
Weathers 95
Webb 212, 343
Weber 297
Weis 114, 324
Werner 89
White 34
Whitlock 187
Whittle 24, 386
Wickham-Jones 386
Wielgolaski 171, 319
Wieser 13
Wiessner 281
Wildi 356, 358, 359
Williams 93
Williamson 32
Willis 36, 266, 345
Wilmking 175, 177
Wilmshurst 248, 335, 337
Wilson 14, 36
Wisheu 51
Wolff 56
Woodward 71
Wookey 106, 107

Xiong 248

Yallop 43
Yamaguchi 155
Yurtsev 389

Species index

Page numbers in *italics* refer to figures

Abies alba 84
Abies concolor 58
Abies grandis 58
Abies lasiocarpa 170, 177, 191
Abies squamata 366
Acacia longifolia 124
Acacia macrocantha 56, 96
Acer negundo 80, *81*, *97*, 151
Acer pseudoplatanus 84
Acer saccharin 305
Achillea atrata 351
Achillea moschata 351
Aciphylla horrida *144*
Aciphylla spp. *143*
Aconitum septentrionale 332
Acorus calamus 88, *88*, 158, 277, 281, 287, 295–297, 298, 300, *300*
Aegiceras corniculatum 261, *261*
Aetoxilon punctatum 114
Agave americana 66
Alces alces 37
alders *see Alnus* spp.
Alkanna orientalis 56
Alnus fruticosa *193*, 194
Alnus glutinosa 103, 302
Alnus incana 103, *321*
Alnus japonica *229*
Alnus rubra 103
Alnus spp. 199, 301, 303, 312
Ammophila arenaria 130, *130*, 234, *235*, 238, 239, 250, 263, 264
Ammophila brevigulata 238, 264
Ammophila spp. 238, 239
Andromeda polifolia ssp. *glaucophylla* 325
Andromeda polifolia ssp. *polifolia* 325–326
Andromeda polifolia 84, 325–326, *325*, 329
Anemone spp. 46
Angelica sylvestris 255
Anser caerulens 241
Anser spp. 39
Antennaria alpina *138*
Anthus pratensis 43
Arabis alpina *138*
Araucaria araucana 20, 155

Araucaria bidwillii 20
Araucaria heterophylla 20
Araucaria spp. 20
Arbutus unedo 22
Arctostaphylos alpina *138*, 309, 312, *313*, 329
Arctostaphylos uva-ursi *228*
Arenaria bryophylla 342
Armeria maritima 255
Arnica spp. 102
Artemisia norvegica 111, *112*
Artemisia spp. *134*
ash *see Fraxinus* spp.
aspens *see Populus tremula, P. tremuloides*
Aster tripolium 254
Atriplex confertifolia 151
Atriplex littoralis 131
Atriplex portulacoides 41
Atriplex spp. 235, *236*, 256
Austrocedrus chilensis 55
Avena sativa 405
Avena strigosa 405, *405*
Avicennia germinans 261, *261*
Avicennia marina 258, *260*, 261, *261*
Avicennia resinifera 258
Azolla filiculoides 124
Azorella spp. 369

Baccharis spp. 369
Banksia prionotes 92
barley *see Hordeum* spp.
Bassia spp. 50
bearberry *see Arctostaphylos* spp.
Betula ermanii *170*
Betula glandulosa 114, 319, 323–324
Betula nana 105, *138*, 171, 319, *322*, 323–324
Betula intermedia (*B. pubescens* × *B. nana*) 429
Betula papyrifera 87, 319
Betula pendul 171, 319
Betula pubescens ssp. *Carpatica* 319
Betula pubescens ssp. *czerepanovii* (formerly ssp. *tortuosa*) 170, 171, 319, *320, 321, 322, 323*

Betula pubescens *160*, 166, 171, 188, 302, 319
Betula spp. 199, 312
Betula utilis 383
bilberry *see Vaccinium myrtillus*
Bison bonasus 37
birches *see Betula* spp.
bog cotton *see Eriophorum* spp.
bog myrtle *see Myrica gale*
bog rosemary *see Andromeda polifolia*
Bolboschoenus australis 250
Bolboschoenus maritimus 90, 277, 291
Bolboschoenus spp. 292
Bos primigenius 37
Branta bernicla 41
Branta leucopsis 81, 241
buckwheat *see Fagopyrum esculentum*
Burkholderia sp. 264
buttercups *see Ranunculus* spp.
butterwort *see Pinguicula* spp.

Cabomba caroliniana 124
Caesalpina tinctoria 56, 96
Cakile islandica 239
Cakile maritima 65, 235, *239*
Calamagrostis stricta 138
Calamagrostis spp. 365, 369
Calluna vulgaris *44*, 128, 193, 256, *306*, 309, 311, 312, 324, 329, *330*, 364, 396
Caltha leptosepela 357
Campanula rotundifolia *138*
Campanula uniflora *138*
campion *see Silene* spp.
Capreolus capreolus 37
Cardamine bellidifolia *138*
Carex aquatalis 41, 106
Carex aquatilis ssp. *stans* *138*
Carex arctisiberia 68
Carex atrata 136
Carex atrofusca *138*
Carex bigelowii 105, 136, *138*
Carex brunescens 136
Carex capillaries *138*
Carex curvula 139, 140, 153

Carex curvula ssp. *curvula* 351
Carex curvula ssp. *rosae* 351
Carex ensifolia 68
Carex firma 125, 139, 140
Carex fuliginosa *138*
Carex lachenalii *138*
Carex misandra 245
Carex nigra 136, *300*
Carex norvegica 136
Carex papyrus 277
Carex parallela *138*
Carex rariflora *138*
Carex rostrata 277
Carex rupestris *138*
Carex saxatilis *138*
Carex serotina 136
Carex subspathacea 41
Carex spp. 14, *139*, 292
Carica candens 56, 96
Carpobrotus edulis 250
Cassiope hypnoides *138*, 193, 329, 332
Cassiope lycopodioides 309
Cassiope tetragona 106, *107*, *138*, 213, 314, *318*, 329
Castanea sativa 129
Castilleja linariaefolia *134*
Casuarina spp. 96
Ceiba chodati 55
Celmisia haastii 372
Celmisia prorepens 372
Celmisia viscosa 372
Celmisia spp. 143
Cerastium uniflorum *138*, 361
Cervus elaphus 37
Chamaecyparis formosensis *229*
Chamaecyparis lawsoniana *229*
Chamaecyparis nootkatensis 21, *229*
Chamaecyparis obtusa *229*
Chamaecyparis pisifera *229*
Chamaecyparis taiwanensis *229*
Chamaecyparis thyoides *229*
Chamaecyparis spp. 21, 227
Cheiranthus himalaicus 342
Chenopodium quinoa 415
Chimantaea spp. 347
Chionochloa rubra *143*
Chionochloa spp. 143
Chrysosplenium oppositifolium 256, *256*
Cinnamomum glanduliferum 118, 129
Cistus spp. 49
club-rushes *see Bolboschoenus* spp., *Schoenoplectus* spp., *Scirpus* spp.
Cochlearia officinalis 256
Coleus spp. 345
Colobanthus quitensis 148, 246, *247*, 248, *249*, 360

Combretum 50
Coprosma acerosa 250
Corallorhiza trifida 428, *428*
Corophium volutator 250
Cortaderia spp. 369
Corylus avellana 38, 166, *396*, *397*, 426
couch grasses *see Elytrigia* spp.
Crambe maritima 24, *24*
cranberry *see Vaccinium oxycoccos*, *V. macrocarpon*, *V. microcarpum*
Crassula helmsii 124
Crassula spp. 98
Crithmum maritimum 74, *75*, 252, *253*
crowberry *see Empetrum* spp.
Cryptomeria japonica 94
Cytisus oromediterraneus 329

Darlingtonia spp. 346
Daucus spp. 369
deer
   red deer *see Cervus elaphus*
   reindeer *see Rangifer tarandus*
   roe deer *see Capreolus capreolus*
Dendrosenecio spp. 356
Deschampsia antarctica 148, 246–248, *247*, *249*, 360
Deschampsia berengensis 215, 277
Deschampsia caespitosa 101
Deschampsia flexuosa 319
Desmoschoenus spiralis 120, *130*, 250
Diapensia lapponica 213
Dictamnus albus 50
Dicymbe corymbosa 144, 145
Digitalis purpurea 256
Distichlis spicata 151
Donacia claviceps 301
Draba daurica *138*
Draba fladnizensis *138*
Draba lactea *138*
Draba nivalis *138*
Draba sibirica 228
Draba spp. 120, 219, 369
Dracophyllum longifolium 335–337, *336*
Dracophyllum muscoides 372
Dracophyllum spp. 334
Drosera spp. 346
Dryas integrifolia 135
Dryas octopetala 16, 17, *17*, *33*, 39, 59, *62*, 106, 107, *138*, 213, 214, *318*
Dupontia fischeri 106

Echinopsis spp. 369
edelweiss *see Leontopodium alpinum*
eel grasses *see Zostera* spp.
Eichhornia crassipes 124

Eleocharis palustris 277
Elodea canadensis 124
Elytrigia aetherica 118
Elytrigia juncea 118, 238, 264
Elytrigia repens 81, 118, *239*
Empetrum nigrum *138*, 256, 309, 314, *315*
Empetrum nigrum ssp. *hermaphroditum* 312, 324
Empetrum nigrum ssp. *nigrum* 312, *314*, 324
Empetrum rubrum 314, 329
Empetrum spp. 329
Ephedera viridis 151
Epirrata autumnalis 171, 319, 322
Equisetum arvense 81
Equus przewalski 37
Erica arborea 49
Erica cinerea 256, *311*
Erica melanthera 48
Erica tetralix 324
Erica spp. 312, 329
Erigeron spp. 102
Erigeron uniflorus 138
Eriophorum angustifolium 71, *72*, 106, *138*, 277, 291, *292*, *293*
Eriophorum scheuchzeri *138*, 291, *292*, *293*
Eriophorum vaginatum 71, *72*, 99, 105, 106, 277, 291, *292*, *293*
Eriophorum spp. 291, *292*
Eritrichium nanum 351, *377*
Erophila verna 263
Erotia lanata 373
Eritrichium nanum ssp. *jankkae* 351
Erythronium grandiflorum 357, *359*
Espeletia spp. 356
Eucalyptus spp. 92, 96
Eugenia spp. 56, 96
Euphrasia frigidas *138*
eyebright *see Euphrasia frigida*

Fagopyrum esculentum 405
Fagus sylvatica 84
Fallopia japonica 124
fescue grasses *see Festuca* spp.
Festuca organogenesis 329
Festuca orthophylla 369
Festuca vivipara 215, 277
Festuca spp. 365, 369
Filipendula ulmaria *138*, 277, 295, 298, *299*
firs *see Abies* spp.
Fraxinus pennsylvanica 305
Fraxinus spp. 129
Fritillaria meleagris 23, *23*
Fucus vesiculosus 236

Galium spp. 46
Gaultheria spp. 329
geese *see Branta* spp.
Gentiana acaulis ssp. *clusii* 351
Gentiana acaulis ssp. *kochiana* 351
Gentiana nivalis 138
Gentiana spp. 369
Gentianella tenella 138
Geranium spp. 369
Geum intermedium 120
Geum reptans 367
Geum rivale 120, *374*
Geum urbanum 120
Ginkgo biloba 13, 20, *21, 22*
Glaux maritima 267, *269*
Glyceria maxima 158, 277, 295, *296*,
    298–300, *299, 300*
Glyceria spp. 405
gorse, whin or furze *see Ulex*
    europaeus
grass of Parnassus *see Parnassia*
    palustris
Grevillea robusta 93

hazel *see Corylus avellana*
heather *see Calluna vulgaris*
heaths *see Erica* spp. *Cassiope* spp.,
    Phyllodoce caerulea
Hebe mora 372
Hedera helix 129
Heliamphora nutans 346, *346*
Heliamphora spp. 346
Helianthus annuus 423
Helichrysum newii 365
Herpotrichia sp. 328
Hierochloe alpina 138
Hippophae rhamnoides 103, *152*, 153, 270,
    270, 317, 328
holly *see Ilex aquifolium*
Holocarpus spp. 334
Homo sapiens 383
Homogyne alpina 359
Honckenya peploides 235, *236*
honeysuckle *see Lonicera periclymenum*
Hordeum murinum 131
Hordeum vulgare var. *tetrastichon* 405, *405*
Hutchinsia alpina 351
Hutchinsia brevicaulis 351
Hyacinthoides hispanica 127
Hyacinthoides non-scripta 114, 127, *127*

Ilex aquifolium 129
Iris fulva 134
Iris germanica 277
Iris giganticaerulea 120
Iris pseudacorus 88, *88*, 277, 287
Isoetes lacustris 282

Isoetes spp. 98
ivy *see Hedera helix*

Jamensonia spp. 369
Juncus biglumis 138
Juncus conglomeratus 277
Juncus effusus 277, 281, 286
Juncus spp. 292, 295, 305
Juncus triglumis 138
Juniperus communis 326–328
Juniperus communis ssp. *communis*
    326–328, *326*
Juniperus communis ssp. *nana* (syns.
    alpina, J. sibirica) 326–328, *327*, 329

Kalmia spp. 312
Kobresia myosuroides 361
Kobresia pygmaea 370
Koenigia islandica 66, *68*, 140

Laminaria digitata 266
Laminaria hyperborea 266
larch *see Larix* spp.
Larix dahurica 303
Larix decidua 190, *375*, *375*
Larix gmelinii 185
Larix laricina 185
Larix sibirica 185
Larrea divaricata 92
Lathraea clandestine 91, *91*
Laurelia phillipiana 114
Ledum spp. 329
Lemmus lemmus 39
Leontopodium alpinum *358*
Lepidium draba 230
Lepidothamnus laxifolius 335
Lepus europaeus 41
Leucaena leucophala 134
Leymus arenarius 75, *77*, 235, 239, 264
Leymus mollis 239–240, 264
Leymus racemosus 250
Ligusticum scoticum 73, *73*, *74*, 75, 235,
    237
lime trees (linden) *see Tilia* spp.
Limonium vulgare 74
Littorella spp. 98
Littorella uniflora 98, 276, 282, 290, 291,
    282
Lobelia deckenii 365
Lobelia dortmanna 281, *282*
Lobelia spp. 356
Loiseleuria procumbens 213, 329
Lonicera periclymenum 256
louseworts *see Pedicularis* spp.
Ludwigia arcuata 291
Lupinus albus 103, *104*
Lupinus arboreus 250

Lupinus spp. 369
Luzula arcuata 138
Luzula confusa 312, 342
Luzula parviflora 147
Luzula spicata 138
Luzula sylvatica 254, 256
Lychnis alpina 138
Lycopodiella spp. 369
Lycopodium annotinum 106

Magnolia grandiflora 129
mangroves *see Avicennia* spp.
maples *see Acer* spp.
Marram grasses *see Ammophila* spp.
Meadow grasses *see Poa* spp.
Medicago laciniata 120
Mentha aquatica 277
Mercurialis perennis 101
Mertensia maritima 24, 73, *73*, *74*
Metasequoia glyptostroboides 13, *13*, *23*
Microdracoides squamosus 347
Microtus nivalis 348
Microtus spp. 348
Minuartia biflora 138
Minuartia rubella 138
Minuartia sedoides 351
Minuartia stricta 138
Minuartia verna 351
mistletoe *see Viscum* spp.
Molinia caerulea 405
monkey-puzzle (Chile pine) *see Araucaria*
    araucana
mountain avens *see Dryas octopetala*
mountain laurels *see Kalmia* spp.
Myrica gale 103
Myrsine divaricata 336
Myrtus communis 49

Nelumbo nucifera 147
Nereis diversicolor 250, 251
Nothofagus antarctica 360
Nothofagus dombeyi 6, 52, 126
Nothofagus menziesii 128
Nothofagus pumilio 6, 114, *121*, 127, 155,
    156
Nothofagus rubra 128
Nothofagus solandri 423
Nothofagus solandri var. *cliffortiodes* 145,
    146
Nothofagus spp. 5, 114, 146, 428
Nototriche spp. 369
Nucifraga caryocatactes 114
Nucifraga columbiana 114, *116*
nutcrackers *see Nucifraga* spp.
Nymphaea alba 272, *283*
Nyssa aquatica 229, 305
Nyssa sylvatica 301

oak *see Quercus* spp.
oats *see Avena* spp.
Ochotona princeps 348
Operophtera brumata 171, 319, 320
Otanthus maritimus 131, *132*
Ovibus moschatus 39
Oxalis tuberosus 415
Oxyria digyna 16, 17, *138*, 202, 277,
     348
Oxytropis deflexa 228

Papaver radicatum 314
Parnassia palustris 256
Pedicularis dasyantha 62, 90
Pedicularis hirsuta 138
Pedicularis lapponica 138
Pedicularis spp. 90, 91
Phacidium infestans 194
Phalaris arundinacea 277, 295, *295*
Phippsia algida 136, *138*
Phragmites australis 137, 156, 250, 252,
     272, 277, 281, 282–284, *285*, 286,
     292, 293, 298, 299, 300–301, *300*,
     426
Phyllitis scolopendrium 257
Phyllodoce caerulea 16, *138*
Picea abies 15, *186*, 191, *230*
Picea abies ssp. *obovata* 166
Picea engelmannii 170
Picea glauca 175, 187
Picea glehnii 229
Picea mariana 170, *179*, 185, 188, *189*,
     194, 303, 323, 324, 325
Picea sithchensis 13, 75, 304
Pinguicula alpina 137, *137*
Pinguicula villosa 137
Pinguicula vulgaris 137, *138*
Pinguicula spp. 137
Pinus abicaulis 114, *116*
Pinus aristata 155, 170, *155*
Pinus canariensis 94, *95*
Pinus cembra 12, 13, 114, 191
Pinus engelmannii 86
Pinus glauca 170
Pinus longeava 155, 170, *155*
Pinus moticola 58
Pinus mugo 2, 170, *176*, 292, 423
Pinus mugo ssp. *pumilio* 292, *294*
Pinus muricata 229
Pinus palustris 86, *86*
Pinus pinaster 265
Pinus pinea 265
Pinus pumila 169, *170*, *173*, *193*, 194, *229*,
     303
Pinus radiata 229
Pinus roxburghii 351
Pinus serotina 229

Pinus sibirica 173
Pinus spp. subsection *Cembrae* 127
Pinus sylvestris 146, 171, 187, *192*, 193,
     228, *276*, *309*, *311*, *321*, 322, 402,
     430
Pinus uncinata 190, 375, *375*
Piriqueta caroliniana 120
Pitcairnia feliciana 347
Pittosporum spp. 334
Plantago coronopus 254
Plantago lanceolata 423
Plantago maritima 68, *69*, 254, 255, 267,
     423
Plettkea spp. 369
Poa alpina 138, 155, *155*
Poa alpina var. *alpina* 155
Poa alpina var. *vivipara* 88, 155
Poa arctica 138
Poa colensoi 372
Poa glauca 138
Poa pratensis 138
Podocarpus nivalis 335, *335*, 379
Podocarpus spp. 379
Polygonella spp. 58
Polygonum oxyspermum ssp. *rayii* 140,
     142, 149
Polygonum viviparum 17, 88, 137, *138*,
     154, 241, *244*
Polylepis spp. *13*, 366–369, *367*
Polylepis tarapacana 366, *368*, 369
poplar *see Populus* spp.
poppy *see Papaver* spp.
Populus balsimifera 150, 151
Populus fremontii 92
Populus spp. 397
Populus tremula 156, *157*, 256, *259*, 302
Populus tremuloides 68, 151, 177
Potamogeton nodosus 291
Potentilla crantzii *138*
Potentilla hyparctica 88
Potentilla nivea 138
Potentilla stipularis 228
Primula auricula 351
Primula hirsuta 351
Primula scotica 14, 23, *35*, 428
Primula vulgaris 255
Prosopis juliflora 54
Protea spp. 48
Prunus laurocerasus 120
Pseudotsuga menziesii 58, 87
Pteridium aquilinum 68, 81, 129, 131, *133*
Puccinellia phryganodes 41, *78*, 97, 100,
     100, 154, 217, *218*, 241
Puccinellia spp. 241
Puya spp. 356
Pyrola grandiflora 65, *67*
Pyrola norvegica 138

Quercus alba 305
Quercus petraea 38, 430, *430*
Quercus robur 38, 84
Quercus spp. 129
quillwort *see Isoetes* spp.

Rangifer tarandus 39
Rangifer tarandus platyrhynchus 342,
     395
Ranunculus acris 138, 373, *374*
Ranunculus alpestris 351
Ranunculus glacialis 66, *138*, 343, 348,
     351, 361, 362, 373, *374*, 377, *378*, *379*
Ranunculus nivalis 135, *138*
Ranunculus pygmaeus 16, 17, 88, *138*, 213
Ranunculus repens 277, 292–294
Ranunculus spp. 46
Rhizophora mangle 261, *261*
Rhizophora stylosa 261, *261*
Rhodiola rosea 138
Rhododendron catawbiens 124
Rhododendron caucasicum 120, *121*
Rhododendron ferrugineum 351, *351*
Rhododendron hirsutum 351, *352*
Rhododendron lapponicum 138, 332
Rhododendron luteum 121
Rhododendron ponticum 82, 118, 120, 124,
     129, 311
Rhododendron smirnowii 121
Rhododendron subsection *Pontica* 22
Rhododendron × sochadzeae 120, *121*,
     122
Rhododendron spp. 120, *121*, 312, 329
Rosmarinus officinalis 49
Rubus chamaemorus 202, *325*, 422
Rumex acetosa 256, 288
Rumex crispus 261
Rumex obtusifolius 261
Rumex palustris 288, 289, *290*
Rumex spp. *289*
rushes *see Juncus* spp.
rye *see Scirpus* spp. 292

Secale cereale 405
Sagittaria spp. 98
Salix alba 275
Salix arbuscula 318
Salix arctica 76, *108*, *149*, 314, 316
Salix caprea 129
Salix cinerea 118
Salix fragilis 275
Salix glauca 149, 314
Salix goodingii 92
Salix herbacea 314, 316, 317, 332, 351
Salix herbacea × *S. polaris* 102
Salix lanata 138, 277, 314, *318*
Salix lapponum 318

Salix myrsinifolia 149
Salix myrsinites 318
Salix phylicifolia 277
Salix polaris 17, 88, *89, 149,* 156, 191,
     193, 312, 314, 316, 317, *317,* 318, *318,*
     333, *333*
Salix polaris ssp. *pseudopolaris* 314
Salix repens 129, *149,* 153, 250, *428*
Salix reticulata 191, 309, 314, 316, *316,*
     318
Salix retusa 351
Salix spp. 199, 219, 301, 312, 396
saltcedar *see Tamarix* spp.
salt-marsh grasses *see Puccinellia* spp.
Salvinia molesta 125
sandworts *see Minuartia* spp.,
     Honckenya peploides
Saracenia spp. 346, 423
Saussurea gnaphalodes 342, *342*
Saxifraga aizoides 138
Saxifraga aizoon 351
Saxifraga biflora 343
Saxifraga caespitosa 277
Saxifraga cernua 138, 154, 277
Saxifraga cespitosa 138
Saxifraga cotyledon 351
Saxifraga flagellaris 75
Saxifraga flagellaris ssp. *platysepela 77*
Saxifraga hieracifolia 277
Saxifraga hyperborea 213
Saxifraga kochii 343
Saxifraga × kochii 343
Saxifraga nivalis 138
Saxifraga oppositifolia 7, 17, *17,* 20, 34,
     58, 76, *79,* 88, 97, 105, 112, 137, *138,*
     139, 200, *201,* 204, *211, 212,* 213, 214,
     216, 221, 241, *244, 277,* 312, 314,
     343, 351, 361–363, *362,* 378–379, 383,
     420
Saxifraga rivularis 138
Saxifraga rupestre 351
Saxifraga seguieri 361
Saxifraga spp. 120, 219
Saxifraga stellaris 138
Saxifraga tenuis 138, 213
Schinus molle 56
Schoenoplectus (Scirpus) californicus 406,
     407
Schoenoplectus lacustris 88, 277, 285, 286,
     289, 291, 297, 404
Schoenoplectus maritimus 291
Schoenoplectus tabernaemontani 291, 406
Schoenoplectus spp. 293
Scirpus spp. 292, *see also Bolboschoenus*
     spp., *Schoenoplectus* spp.
sea-buckthorn *see Hippophae rhamnoides*
sea-kale *see Crambe maritima*

Secale cereale 405
sedges *see Carex* spp.
Sedum atratum 351
Sedum montanum 351
Sedum rosea 254
Senecio aethensis 125
Senecio aquaticus 118, 119, *119,* 120,
     130
Senecio bellidiides 123
Senecio cambrensis 114, *126*
Senecio chrysanthemifolius 125
Senecio eboracensis 126, *126*
Senecio jacobea 118, *119,* 120
Senecio johnstonii ssp. *cottonii* 365
Senecio johnstonii ssp. *johnstonii* 365
Senecio leucophyllus 370, *371*
Senecio smithii 123
Senecio squalidus 123, 125, *126,* 139,
     140
Senecio vulgaris 125, 126, *126*
Senecio vulgaris var. *hibernicus* 126
Senecio spp. 123, 365
Sequoia sempervirens 20, 21, 227, *229*
Sequoiadendron giganteum 94
Seslaria coerulea 351
Seslaria disticha 351
shoreweed *see Littorella* spp.
Sibbaldia procumbens 88, *138*
Silene acaulis 135, *136, 138,* 241
Silene dioica 255
Silene rupestris 351
Silene uniflora ssp. *petraea* 351
Silene wahlbergella 138
Solanum acaule 415
Solanum tuberosum 277, 415
Soldanella alpina 351
Soldanella pusillia 351, *359*
Solidago virgaurea 256
Sonchus arvensis 81
Sorbus aucuparia 146, *176,* 397
southern beeches *see Nothofagus* spp.
Spartina alterniflora 122, 127, 298, 300
Spartina anglica 122, 123, *123,* 137, 251,
     252, *252,* 277
Spartina foliosa 127
Spartina maritima 122
Spartina townsendii 122
Spartina × neyrautii 122
Spartina spp. 258
Spergularia diandra 135
Spergularia marina 131
Sphagnum magellanicum 96
Sphagnum spp. 96, 103
Spinafex sericeus 250
spruce *see Picea* spp.
Stegolopsis spp. 347
Stellaria decumbens 342

Stipa spp. 369
strawberry tree *see Arbutus unedo*
Sueada vera 8
Sus scrofa 37
sweet flag *see Acorus calamus*
Syncarpha vestita 47

Tamarix spp. 92, 125, 311
Tanacetum gossypinum 357
Taraxacum 102
Taxodium distichum 229, 301
Taxodium spp. 21, 227
Taxus baccata 84
Tephroseris palustris ssp. *congesta* (syn.
     Senecio paludosus) 140, *141*
Teucrium scorodonia 256
Thalictrum alpinum 138
Thalictrum fendleri 151
thrift *see Armeria maritima*
Thuja orientalis 229
Thuja plicata 58
Thymus vulgaris 50
Tilia cordata 111, 113, *113*
Tilia spp. 129
Tillandsia brachycaulos 87
Tillandsia capillaries 56
Tillandsia landbeckii 56
Tillandsia latifolia 56, *57, 87*
Tillandsia marconae 56
Tillandsia purpurea 56
Tillandsia usneoides 87
Tillandsia werdermanii 56
Tillandsia spp. 56
Tofieldia pusilla 138
Tortula ruraliformis 263
Trapa natans 291
Triplospermum maritimum 256
Trisetum spicatum 138
Trisetum subalpestre 228
Tropaeolum tuberosum 415
Tussilago farfara 277
Typha latifolia 6, 7, 88, 98, 277, 281,
     285

Ulex europaeus 49
Ullucus tuberosus 415
Unioloa paniculata 264
Urtica dioica 101, 256

Vaccinium macrocarpon 288, 409, *410*
Vaccinium microcarpon 409
Vaccinium myrtillus 13, 71, *71,* 193, 319,
     329–330, *331,* 332, 372, 409
Vaccinium oxycoccos 409
Vaccinium uliginosum 73, *138,* 202, 319,
     324, 422
Vaccinium vitis-idaa 202, 324, 409

Vaccinium spp. 312, 329
Valeriana spp. 369
Vallisneria spp. 98
Verbascum blattaria 147
Veronica alpina 138
Veronica peregrina 292
Vicia faba 405
Vicia lathyroides 263
Victoria regina 284

Viscum spp. 90
Vochysia tucanorum 52

Werneria spp. 369
willows *see Salix* spp.
Wollemia nobilis 20
wood-rushes *see Luzula* spp.

Xanthoria parietina 254

yew *see Taxus baccata*
Yucca brevifolia 54, *55*

Zea mays 415
Zostera marina 250
Zostera noltii 251
Zostera spp. 122, 251, 258

Subject index

Page numbers in *italics* refer to figures

Åbisko – northern Sweden 319, 322
acetaldehyde 287
    post-anoxic generation 279, 297
    tolerance 280
achlorophyllous species 90
acorn production 145
acrotelm 105, 277
active layer *see also acrotelm* 205
adversity selection 83
aerenchyma 98
aerobic metabolism 215
age and area concept 36
agriculture 24
Alaska
    air temperatures 202
    reduced growth of black spruce 178
alcohol dehydrogenase induction by
        flooding 279
alder growth forms
    monocormic 301
    polycormic 301
Alexandra Fiord 204
alien species 82, 130
allelic frequencies 33
allelic heterozygosity/variation 33, 36
allopatric speciation 120
allopolyploid 35
alpine
    bogs 292
    lower alpine zone 311
    middle alpine zone 364
    upper alpine zone 364
    pastures 129
    sedges 139
    species 70, 332, 335
    treeline 190
    tundra 324
Altai Mountains 23, 212
alternative resources 90
altitude limits 42, 45
aluminium 51
Amazon forest 51
    flood pulses 59
    lakes 116

American marram grass 264
amino acids 99
    concentrations 99
ammonium 98
    rich soils 99
amphibious and aquatic species 217
    adaptations 290
    communities and species 98, 276, 279,
        286
    graminoid species 292
        Glyceria maxima 298–300
        Phragmites australis 300–301
    phenotypic plasticity 290
    seed germination 290
    winter flooding 292
anaerobic conditions *see also* anoxia 19
    anaerobic mud 19
    anaerobic respiration 74
    carbohydrate consumption 286
    metabolite leakage 286
anaerobiosis 123
    and pathogenic infection 286
    in tree stems 303
Andes 23, 54, 70
Andøya – Lofoten Islands 206
animal pollinated species 145
annual species
    cold climate/arctic species 135,
        140–141
    spring flowering 133
anoxia *see also* anaerobiosis 36, 88, 122,
    214
    acceleration of glycolysis 287
    metabolic down-regulation 280, 285,
        287, 289
    prolonged duration 295
    spring starvation 299
    survival 279
    tolerant species 277, 295
Antarctica 240
    Antarctic Peninsula 246
    flowering plants 148
    shores 246–248
antioxidants 46, 215, 263

aphids 50
apomictic species 34, 158
Appalachian forests 94
Arctic 39
    as a marginal area 204
    climatic oscillation 145
    defined 199
    European 239
    flora 224
    flora – ancient 39
    growing season 65
    habitat preferences 213–214
    low and high Arctic 199, 221
    margins – maps 204–205
    oases 219–220
    ocean 185, 203, 206, 212, 224
    pH boundary, Alaska 205
    radioactivity 203
    Russia 209
    salt marsh grasses 241
    scrub line 337
    shores 238–240, 241
    Siberia – increase in oceanicity 181
    signs of change 199–203
    species/ plant diversity 36, 57, 75, 76,
        213
    subarctic treelines 165
    temperatures 70
    tree fossils carbon dated 181
    tundra 37
    vegetation units 204
    vegetation zones/regions 34, 36, 206
    willows 148
Argentina 20
ascorbate reductase 287
asexual reproduction 153–155, 158
    bulbils 137
Asia Minor 25
assessing genetic variation 35
Atacama Desert 56, 95
Atlantic bogs *see also* bogs 277
Atlantic coasts 329
Auckland Island – New Zealand 330
aurochs 37

Australia 20, 103, 250
    Queensland 20
authochthonous (ancient) species 127,
        213, 216
    arctic flora 213, 216
autumn moths 171
Axel Heiberg Island 102

Baffin Island 211
Baltic coasts 328
Baltic heathlands 328
Bangladesh 262
Bathurst Inlet 219
Bay of Bengal 262
beach ridges 59
beaver 37
benzene 50
Beringia 209, 212
Beringian–Siberian glaciation 213
biodiversity 31–34, 42, 43, 57, 231, 234
    assessment 31
    conservation 269
    definitions 31, 32
        alpha diversity 32, 43
        beta diversity 32, 56
        delta diversity 32
        gamma diversity 32, 51
    floristic 51
    genetic see also genetic factors 35
    hotspot 44–47
    preservation 43
    reduction 252
    regional 48
biomass 31, 74, 81
    accumulation 77, 85, 187
biome 14
birch growth form
    monocormic see also alder 171, 318
    polycormic 171
bird species variability 58
bison 37
blanket bog 324–329
blue-green algae see also cyanobacteria
        107
bogs
    Atlantic bog 231
        bogs of Scotland and Ireland 324
        preservation 42
    barriers to forest advance 175
    climatic history 229
    growth 182
    homeostatic properties 183
    pool and hummock topography 229
    replacing tundra 231
    vs. forest 183–185
Bolshezemelskaya Tundra see West
        Siberian Lowlands

boreal forest 97, 163, 166, 185
    advance in northern America 177, 303
    advance in northern Russia 174, 303
    climatic limits 166
    climatic warming 193
    productivity 187–188
Borneo 260
bottomland forests 277, 301, 305
boundaries 31, 165
    biological 26
    physiological 37
Bowen ratio 179
Braunton Burrows – North Devon 266
breeding systems
    androdioecious 154
    gynodioecious and dioecious 140
    hermaphrodites 154
Britain/British Isles 114, 271, 303
British National Vegetation Classification
        324
Brøgger Peninsula 39
Bronze Age 25, 50
burning 36, 43
burning bush 50

C3 plants 79
C4 plants 79
calcicole species 34
calcifuge species 34
California – vernal pools 292
CAM and water use efficiency 98
Camancha (Chile) 56, 95
Camargue 41
Campbell Island 311, 335
camphor 50
Campo Cerrado 51
Canada 99, 163, 169, 241, 324
    Arctic 202, 204, 219, 325
    Arctic Archipelago 314
    forest fire 325
Canary Islands 94
capacity adaptation 69
Cape Peninsula – South Africa 45
carbohydrate
    availability 45
    depletion 74
    reserves 41, 84, 334
    under anoxia 305
    utilization 279
carbon balance 34, 39, 41, 42, 47, 73, 189
carbon deficits 73, 74, 190, 191, 314
carbon dioxide
    alternative sources 96
    atmospheric concentrations 25, 74, 84,
        323
    atmospheric effects at the treeline 190
    dissolved in water 98

enrichment 84, 96
    free-air carbon dioxide enrichment
        (FACE) 96
    stable isotope ratio 97
    substrate-derived 96
    utilization in aquatic species 98
carbon recycling 98
carbon sequestration 50
carbon sink 74, 97
carbon starvation 45
caribou 41
carrying capacity (K) 81
catalase 287
catotelm 105, 277
caves 256
Cenozoic temperature –decline 163, 165
Central Africa 219
Central Alps 329
Cerrado see also Campo Cerrado 42, 45
chalcolithic 49
Chernobyl Nuclear Reactor 103
Chile 20, 95
China 20
chloroplast DNA see DNA
chloroplasts in alder stems 302
Chukotsky Peninsula 209
circumpolar zones 205
cladistic methods 33
classification problems 34
cliffs 68, 227, 234, 235, 252–254
climate change 25, 31, 140, 216, 325, 329
    anomalies 204
    boreal –sub-boreal period 277
    cooling 330
    deterioration 25
    forest migration 174–178
    seasonal sensitivity 113
climatic oases 219
climatic optimum or hypsithermal 164
climatic oscillations/fluctuations 36
    boreal forest responses 174
climatic stability 48
climatic transition zones 31
climatic warming 24, 65, 74, 102,
        227–231, 323
    Antarctica 246
    Arctic 200, 337
    disturbance and invasion 130–131
    drought stress 178
    forest advance 175
    trends 129
    tundra–taiga interface 187, 193–195
clones
    age 68
    diversity 36
    DNA 153
    growth and reproduction 153

longevity 153, 316
proliferation 422
species 45
cluster roots 103
coasts
defences 250, 269
erosion 252
habitats/sites 41, 75, 118, 204, 271
pasture biodiversity 267
vegetation
northern hemisphere 235–240
physiological adaptations 263–268
southern hemisphere 246–250
cold climates 57
cold resistance – differing adaptations 248
cold temperatures 260
colonization 235
Colorado Rockies 324
competition 14, 36, 58, 65, 78, 81
absolute-intensity 77
consumptive 36, 85, 187
interference 14
relative-intensity 77
flooding – *Glyceria maxima* vs.
Filipendula ulmaria 295
interactions 57
intraspecific 85
Conrad's Index of Continentality 181, 183
conservation 130
biodiversity 269
coastal habitats 269–271
continental seasonality 114
core habitats 19, 85, 86
crassulacean acid metabolism (CAM) 54,
98
crop production 25
cryoprotective sugars 333
cryoturbation 165
cyanobacteria 107
cycad species 20
cytochromes 286
cytoplasmic acidosis 286, 305

Darwin (Charles Robert Darwin,
1809–1882) 34
Darwinian fitness 106, 131
day degrees 65
defining margins 31
deforestation 25
deleterious mutations 155
demography 26, 56
deprivation indifference 14, 39, 49, 68,
77, 86–87, 87–89
deserts
Africa 56
annual species 133
flowering events 133

hot 54, 57
polar 57, 163, 312
seed survival strategies 134–135
warm 131
desiccation injury 260, 263
Devon Island 219
dew 91, 93–94
diploid arctic species 33
disjunct distribution 111, 209
Disko – West Greenland 326
distribution – species 38
– limits *39*, 47
disturbance 51, 57, 83, 234, 256, 269, 335
divergent evolution 120
diversity *see* biodiversity
DNA chloroplast haplotypes (cpDNA)
51, 127, 129, 211, 212, 224
DNA fingerprinting 328
DNA variation 31, 33
dormancy 66
drought tolerance/resistance 52–57, 214,
263–264
drylands – plant diversity 52
dunes 118, 227
depth and accretion 264
dune grasses 264
dune heaths 309
dune slacks 267, 267
dune systems 238–240, 267
embryo dunes 238
migration 264
yellow/mobile 238
dwarf birch 341–332
biogeographical history 323–324
hybridization 323
North American limits 114
dwarf mammoth population 39
dwarf shrub zone 49

ecological release (competitive release) 14,
36, 58
ecological tolerance 48
ecosystem destabilization 36
ecotone *see also Limes convergens* 14, 120
ecotypes 34, 50, 76, 137
specialization 158
variation 19, 80, 215
ectomycorrhizal species 99, 144
edge effects 37
El Niño 252
electron transfer system 50
elk 37
Ellesmere Island 204, 219
embolism injury at the treeline 191
endangered species 23
endemic species/endemism 34, 47, 51, 58
energy capture 66

entropy 65
environmental phenomena
change 33, 34
extremes 31
fluctuations 85
monitoring 31–34
ephemeral migrations 323
epicuticular wax 214
essential oils 50
Estonia 271, 303
ethanol – production and accumulation
279, 286–318
eutrophic habitats 34
species 34
evapotranspiration rates 92
evolutionary strategies 34
units of evolution 19
extermination/extinction 20, 73

F_1 hybrids 120
Falkland Islands 312
fertility barriers 118
Finland 111, 202, 271, 319
Finnish Lapland 146
fire *see also forest fires* 42, 50, 79, 178, 324,
325
effect on juniper 328
post-fire-habitat degradation 178–179
subarctic Finland 179
fitness – immediate and long-term 14, 51,
215
flooding 75, 214, 267
abscisic acid 288
adaptations and plant hormones
288
ATP levels 288
flash floods 56
frequency 36, 118
habitats 34
high temperatures 260
induced ethylene and shoot extension
288
long-term flooding and aerenchyma
288
long-term oxygen deprivation 285–286
phenotypic plasticity 281
prolonged 118, 324
responses – long term 284–285, 305
responses – short term 287
seasonal responses 43
survival of overwintering organs 287
tolerance 71, 275–281
advantages and disadvantages 289
life form 277–281
tropical vs. temperate trees 301–305
unflooding – the post-anoxic
experience 286

floral herbivory 39
Florida 120, 258
flowering
    dual induction in cold climates 136
    early species 137
    initiation in cold climates 136
    late species 137
    phenology 137
    synchronous 143
fluctuating water levels 291
fog 95
    dependent species 96
foraging species 71
foreshore plant communities 235
forest
    altitude limits 322
    and agriculture north of the Arctic
        Circle 181
    fires 163, 178
    forest-tundra see also lesotundra 163
    gallery forests 51
    mesophytic 51
    regeneration and snow drifting 178
    retreat 14, 35, 174
    wetlands 277
France 75, 122
frost
    freezing injury – lethal temperatures
        248
    frost sensitive mangrove forests 260
    frost sensitivity 15, 107
    frost tolerance 333
fruits – delayed dehiscence 135
functional adjustment/adaptation 69,
    76
Fynbos 32, 36, 42–47, 50, 103, 329
    dry 48
    ericaceous 48
    graminoid 48
    proteoid 48
    restoid 48

Ganges 263
garrigue 49
Garua (Peru) 56, 95
gene flow 19, 119, 120, 246
    gene flow in the Arctic 153
genetic factors
    boundaries 51
    diversity in the High Arctic 243
    diversity/variation 19, 26, 31, 36, 57,
        326
    drift 31, 57, 216
    fitness 111
    genotypic alteration 106
    geography 33
    markers 51

memory 19
geographical factors
    barriers 56
    isolation 26
    parthenogenesis 154
germination
    establishment in marginal areas 114
    inhibitors 133
    phenology 279
    physiology 280
glacial relict species 35
global temperature rise 25
glycophytic species 34
Gondwana 46
grazing 37, 83, 314
    birds – Giant Moa 250
    deterrents
        chemical defence compounds 317
        metabolic 317
    geese 39, 41, 50, 241, 406
    intensity 105
    invertebrates 250
    ragworms 251
Great Lakes 238
Greenland 66, 75, 135, 211, 235, 241, 291,
        309, 311, 314, 324, 330
    earliest record of mountain birch 323
    hybrid birch 319
    Little Ice Age 341
    Norse settlement 323
    West Greenland 331
ground water – phreatic zone 92
growing season
    advance 316
    length 45, 73, 84, 202
growth rates
    increased 68
    low/reduced 68, 68
habitat
    fragility – physical vs. biological
        231–235
    productivity 77–78

halophytic species 34, 92, 131, 241, 254,
        260
hard coasts/shores 227, 234, 252–256
Hartig net 102
heat injury 75, 76
heat resistant species 50
heathlands 234, 329
    climate 329
    high arctic heaths 312
    historical ecology 334
    Mediterranean 46, 48
    muirburn north-west Europe 334
    regeneration 37
    soils 99

hemi-parasitism 91
herbivory/herbivores see also grazing 37,
        38, 231, 250, 316
hermaphrodite species 312
heterozygosity 35
High Arctic 34, 58, 66, 96, 107
hill land reclamation 24
Himalaya 66, 263, 329
Holocene 36, 37, 39, 165
    forest advance 174
    mid–late 337
    migrations 36
    treeline history 323
homeostasis and tree stability 185–187
horses 41
Hudson Bay 39, 168, 241
Humboldt Current 56, 95
hurricanes and storm surges 260
hybrid species 19
    swarm 118, 120
    swarm stability 120
    transient populations 120
    zones 118–126
hybridization 19, 158
hydraulic lift 93, 264
hydraulic conductivity 260
hyperoceanicity 25
    oceanic islands – Orkney 118
hypoxia 74, 98
hypsithermal 164, 165, 174, 181, 183

Iberian Peninsula 22
ice
    age see also Pleistocene and Weichselian
        255
    cover in polar regions – reassessment
        205–211
    encasement and prolonged anoxia 185,
        213, 214–215, 280
Iceland 25, 66, 206, 330
illumination compensation point 68
India 262
International Tundra Experiment
        (ITEX) 221
introduced species 129
invasion
    climatic warming 127–130
    genetic 126–131
    marshes 252
    theories 131
invasive species 78, 122, 124
Iona – Inner Hebrides 266
Ireland 114
Iron Age 334
iron precipitation 103
irrigation 25
island biogeography 36, 56

isomorphism 14
isoprene 50

Japan 20, 94
Jotunheimen Mts. 322
juniper *see also* fire, longevity 326

K-selected species 82
Kamchatka 328
Karoo biome 56
Knudsen diffusion 283
Kola Peninsula 103, 175, 330
Köppen's Rule 14, 44, 165, 167
krummholz 165, 309, 312
    carbon balance 190–191
    expansion – North America 177
    spruce *krummholz* 324
    spruce–lichen *krummholz* 324
    and treeline advance 169–172
    winter desiccation injury 191
Kwongan (Australia) 50

La Pérouse Bay 39
Labrador 168, 235
lactating mammals 219
land reclamation 250
landnám *see* pre-landnám
landscape degradation/deterioration 25
land-use changes 23
lapse rates (adiabatic) 25
Last Glacial Maximum 22, 163, 206, 209,
    216
Late Glacial/Holocene transition 25
latitude limits 42
Latvia 271, 303
Laurentide ice sheet 20
laurophylization 129
leaching 100
leaf
    area index (LAI) 165
    area ratios 84
    C:N ratio 316
    heterophylly and abscisic acid 298–300
    heterophylly and DNA 291
    nitrogen content 316
lemmings 39, 106
Leonie Islands – Antarctica 246
lesotundra *see also* forest tundra 163, 165,
    194
lesser snow goose 39
Levant 25
lichen woodland 165
life-forms 34, 37, 52
life-history traits/strategies 14, 34, 42, 59,
    65, 81–84
    2-class 81–83
    3-class 83

4-class 83, 83–84
light 90
light respiration 50
limits
    definitions
        Limes convergens see also ecotone 14,
            32
        Limes divergens see also ecocline 14, 32,
            205
    forest distribution 36, 111
        boreal forest – southern limit 32
        deciduous forest – northern limit 32
    physiological 26
    reproduction 19, 111–114
lipid peroxidation 287
litter decomposition 101
Little Ice Age 82, 153, 166, 178, 178–166,
    325
living fossils 20
lomas 56, 96
longevity
    juniper 328
    long-term environmental change 34
    persistence in marginal habitats 155
    trees 155
Louisiana swamps 301
Lüneburger Heide 334
Luskentyre Banks – Outer Hebrides 266

Macquarie Island 335
malic acid 292
mammoth fauna 39
mangroves 231, 256–263
    colonization 263
    frost sensitivity 258
    habitat 263
manipulation experiments – short term
    106
manuring with seaware 266
maquis 49
marginal habitats/areas 19, 31, 32, 33, 57,
    94
marginal plant diversity 43
maritime sedge heath 37
mast seeding 142–146
maternal effects 120, 135
meadow pipit 43
Mediterranean
    climate 45
    mountains 324
    plants/vegetation 50
Mesolithic sites 24
Mesopotamia (Iraq) 36
Mesozoic Period 20
metabolism
    metabolic adaptation 287
    metabolic oxygen demand 281

metabolic rate and temperature 76
metabolic rates and carbohydrate
    conservation 216
metabolic responses to anoxia
    280
metapopulations 14, 59, 215
microbial (soil) respiration 98
microhabitats 204
migration 32, 51, 128, 332–334, 337
migration by rafting 228
mineralization rates 100
Miocene 335
Mississippi River 120, 241, 301
Missouri River 241
mitochondria 70
Mohave Desert 107
Mongolian ass (kulan) 39
mono- and polycormic stems 166
Montgomery effect 14, 50
moorland 42
    Scottish 43
Moran effects 145
morphological variation 214
mountain birches 170, 171, 318–322,
    341
    biographical history 322
    in North Atlantic regions 323
mountain pine 170
Mt. Etna (Sicily) 123
muirburn (Moor burn) 42
muskoxen 39, 41, 202
mutualism 65, 114
    arctic species 215–216
mycorrhizal associations 145
    in the Arctic 102–103
    nutrient poor soils 102

Namaqualand (South Africa) 56
Narvik 208
Neolithic Age
    agriculture 25
    deforestation 337
    farming 334
    middle 49
    sites/settlements 24, 25, 37
Nepal 75
net primary production 312
Netherlands 119, 127
New Caledonia 250
New England 235
New Zealand 143, 248–250, 258
    hyperoceanicity 334–337
    heather 311
    southern oceanic islands 335
Newfoundland 185, 211, 235
niche 14
nitrate 98

nitrogen 98–100
  available to plants in the Arctic 202
  cycling 100
  fixation 107, 264
  mineralization 223
  sources 56
  turnover 100
nitrophile species 101
non-structural carbohydrate content –
    conifer needles 190
Normalized Difference Vegetation Index
    (NDVI) 167
Norse settlements in Greenland 103
North America 58, 206, 252
  black spruce 194
  polar front 167
North Atlantic 235
North Atlantic oscillation 181
North Cape (Norway) 239
northern limit birch trees in Russia 181
northern limits in Britain for *Tilia cordata*
    111
Norway 111, 166, 188, 303
Nova Scotia 235
Novaya Zemlya 209, 331
Nunataks 206, 213
nutrients (mineral)
  cycling 105, 107
  flowering 106
  frost sensitivity 106
  levels in soils 51
  requirements in the Arctic 314–318
  retention in marginal areas 103–106

oak – post-glacial history 51
oceanic climates 75, 185, 193, 303
  areas/regions/environments/habitats
      25, 37, 51, 168
  climate gradient – Scotland 324
  cooling of the growing season 181
  heathlands 21, 337
  niche 14, 80
  soil leaching 175
  species 129
  treelines 324
  woody plants 324–326
oligotrophic species 34
ombrotrophic peat 96
opposing and incompatible adaptations 75
opposing strategies 76, 85
optimum leaf temperature 312
Orkney 23, 25, 35, 37, 120
osmo–conformers 34
osmo–regulators 34
osmotic injury 260
overgrazing 25, 36, 39
overwinter survival 107

Oxford Botanic Garden 123
oxidative damage 50
oxygen 46, 74, 86
  debt 305
  deprivation 88, 276, 288
  deprivation and lipid metabolism 298
  deprivation extreme tolerance in *Acorus
      calamus* 295–298
  deprivation in winter 295
oxygen free radicals 263
ozone pollution 50

Pakistan 75
Palaeolithic hunting 209
palisade tissue 70
paludification 166, 185
  history 181–183
  Shetland 175
  Siberia 175
parapatric speciation 120
parasite 91
patchiness 56
Peary Land (north-east Greenland) 163,
    219, 314
peripheral habitats/areas 19, 36
permafrost 39, 183, 184, 187, 199,
    203
phenological time 316
phenology 34, 65, 116, 212
  clusters 19
  plasticity 70
phosphate 100–101, 145
  availability at high latitudes 101–102
  deficiency 101
  deficient soils 100
photosynthesis
  activity 88, 191
  C3 amphibious plants 98
  capacity 50, 85
  corticular 302
  efficiency 312
  low temperatures 248
  production 83
  under water 288
phreatophytes 51
  desert 92
  facultative 92
phytochrome 69
pioneer species 324
plaggen soils 334
plant biomass 58
plant functional types 78–81, 84
plant life strategies *see* life-history traits
plaque deposits on roots 25, 51, 244,
    282
Pleistocene 175, 271
  history – arctic flora 205–213

ice-sheets/glaciation 25, 36
  refugia 58
  speciation 335
  tabula rasa 206, 213
  tundra–steppe 39
Pliocene 224
Pliocene–Pleistocene 163
ploidy levels 137, 213
podzolization 25, 334
polar ice *see also* sea ice 206
polar regions 36
polar shores *see also* arctic shores 241, 241
pollen riskers *see also* risk reduction 137
pollination 111, 137
  wind 19
pollution at high latitudes 314
polycormic growth – causes *see also* alder,
    birch 303, 319
polymorphic populations/species 50, 216,
    326
polymorphism 19, 50
  balance 50
  somatic 135
polyploidy 19
  high latitudes 216–219
Pomors (Russian Arctic hunters) 103
populations
  decline 57
  growth 19
  margins 56
  peripheral 31, 36
post-anoxia
  injury 191, 215, 286
  lipid destruction 279
potential water deficit 229
prairies 129
pre- and post-zygotic limitations to seed
    production 111–114
pre-adaptation 15
precipitation 91
predation 86, 256
predator satiation 143
pre-landnám (pre-agricultural settlement
    – Iceland) 37, 38
pre-Neolithic 38
pulse periods 85

Québec 171, 178, 235, 324
Quechua Indians 70

r- and K-selection 81, 137
radial oxygen loss 281–282
radiative evolution 43
rain belts – Russia 32
raised bogs *see also* bogs 281
reactive oxygen species (ROS) 286
recalcitrant seeds 118

recruitment vs. mortality 33
red data books 23
red deer 37, 39
red Grouse 42
Red Sea 258
redox balance – root system 305
refugium *pl.* refugia 15, 48, 163, 212
regeneration 231
    episodic 114
    strategies 83
reindeer 39, 106, *139*
    population dynamics 145
    Spitsbergen 316
    starvation through tundra
        ice encasement 185
relict species 19–24
    climatic and evolutionary 15, 20
    populations 111
reproduction
    asexual 19
    experiments 78
    hot deserts 131–135
    in flood-prone habitats 116
    sexual *see also* asexual reproduction 68
    strategies 15, 59
    vegetative 82
resources *see also* alternative resources
    access 48
    acquisition 35, 85, 86
    allocation 84–85
    availability 31, 38
    availability and climatic warming in the
        Arctic 106–107
    definitions 65
    deprivation 86, 89
    fluctuations 128
    foraging 71
    limitations 68
    requirements 47, 65
    tracking 143
respiration *46*, 76
    dark respiration 73, 84
    rates 73, 314
    temperature responses *see also*
        physiological entries
    fluctuations 134
    phenological 221–224
    phenotypic 221
Rhine river 301
rhizosphere redox state 281
ribosomal DNA 129
ribulose bisphosphate carboxylase
    (RuB is CO) 70
riparian forests 92
risk reduction 25
Rockies 23
rodents 51

roe deer 37
Roman farming 34
root meristems and carbohydrate
        depletion 191
root to shoot ratio 83
ruderal plants 58
ruderal species (definition) 15
Russia 163

Sahara 56
salinity 241, 252
    saline habitats 260
    salt flooding 122
    salt marshes – arctic 217
    salt marshes/mudflats 39, 118, 122,
        227, 250–251
sap flow – birch trees 305
satellite observations of the Arctic 202
savannas – African 51
scale 34
scaling-up 38, 80
Scandinavia 24, 330
Schiermonnikoog 41
Scotland 23, 24, 35, 42, 73, 111, 252
    Atlantic coastline 256
sea ice 241, 314, 318
sea level rise 227, 250, 251–252, 269
seasonality 25, 329, 332, 337
seed
    dormancy – post maturation 133
    longevity 147
    polymorphism 118
    production 19
    reserves 279
    scarification 141
    seed bank – aerial 135
    seed bank – polar 147–148
    seed bank – semi-desert 148
    seed bank – soil 57, 66, 134, 146–148
    seed bank – warm desert 148
    seed dispersal 115
        by fish (ichthyochory) 116
        by floods (hydrochory) 116
    seed 'rain' 148
    seed 'riskers' *see also* pollen riskers 137
    viability records
        longest running trial 147
        weight and viability 146
        world-record 147
selection – habitat mediated 120
semelparous species 140
serotiny (seed retention) 135
sex ratios
    biased 85, 148, 317, 318
    female bias 151, 317
sexual reproduction in marginal habitats
        111, 241

shade
    detection 69
    tolerance 68, 84
Shetland Islands 25, 101, 312
shingle beaches 24
shore as long distance migration pathway
        228
Siberia 39, 163, 166, 174, 209–213
Siberian Arctic 68
    lowlands 36
    rain shadow 209
Signy Island – Antarctica 246
Sinai Peninsula 56
slugs 50
snow
    beds/banks 59, 140
    blight damage 194
    cover 334
    melt 97, 139, 316
soft coasts/shores 227, 235
soil
    atmosphere 96
    carbon 97
    impoverishment 25
    pH 51, 103
    temperature 43
South African Cape Flora 219
South Aucklands – New Zealand 335
South Carolina – swamps 301
Spain 127
Spartina anglica – dieback 123
speciation
    history 34
    rapid 51
species
    accumulation curve (species area curve)
        31–32
    richness 37, 43, 56, 86
    sub-specific variation 58
    variability 36
Spiekeroog – East Friesian Islands
        312
Spitsbergen *see also*
    Svalbard 36, 39, 75, 99, 103, 106, 148,
        156, 200, 204, 208, 243, 291, 312,
        314, 316
spring starvation 75
St Catherine Desert 56
St Columba 266
St Lawrence River 238
stoloniferous plants 71
stomata 70
stomatal density 71
stress
    avoidance or tolerance 68, 85
    environmental 70
    tolerant species 85

succulents 54
   Succulent Karoo *46*, 47
sulphide
   detoxification 281
   injury 281
Sumerians 25
summer grazing *see also* transhumance 334
Sunderbans 263
sun-tracking 76
super-cooling 248
superoxide dismutase 287
survival
   burial 264–267
   strategies – avoiders and tolerators
      89–90, 300
   strategies – incompatibilities 214, 216
suspended speciation 216
Svalbard, 216 *see also*
   Spitsbergen
Sweden 102, 332
Swedish Scandes 323
synchronous mass flowering 145

tarpan (European wild horse) 37
Taymyr Peninsula 14, 164, 209, 212
terra firme (flood-free) 118
Tertiary forests 271
Tertiary Period 21
tetraploid cytotypes 212
Texas 258
Theophrastus 78
thermal effects
   energy 65
   indicators 43
   oasis 102
thermo-osmosis 98, 282–285
therophytes 146
Tierra del Fuego 66, 140, 312
timberline *see also* treeline species –
   North America and Eurasia 114
time 88
   as a resource 65
tissue lignification 309
tolerance – physiological and ecological
   15
tomillaraes 43
toolik Lake (Alaska) 102, 106, 177, 223
transhumance 411–413
transpiration rates 260, 263

treeline 45, 74, 94, 318, 323
   Alaska 100, 175–178, 187–189
   altitude decrease 175
   birch 188
   montane 31
   North-western Canada 175
   paludification 178–179
   polar/arctic 145, 167, 169
   retreat 182
   Scandinavia 175
   sea proximity 256–263
   soil temperature *167*
   subarctic 169
   summer insolation 181
trees
   colonization of peatlands 177, 185
   establishment 177
   growth trends – Alaska 188
   growth trends – Europe and North
      America 187
   growth trends – Kola Peninsula 187
   growth trends – Mongolia and Siberia
      187
   limits and forest migration 177
   limits – Glacier National Park,
      Montana 177
   seed dispersal 177
   soil temperatures *167*
   survival after localized injury – specific
      tissues 190
tropical marshlands 277
tropical rainforest 43, 133
tundra
   soils 98
   –steppe 163
tundra–taiga interface 23, 163–166, 169,
      199, 311
   fire and paludification 178
   interface and land use 188
   interface and physiological limits for
      tree survival 188–190
   interface and thunder storms 178
   migrational history 163
Turkey 120
turloughs (Irish – dry lakes) 292
Tyrell Sea 41

Upper Pleistocene glaciations 209
Ural Mountains 23

UV-B radiation 248

variation
   biological 33
   genotypic 68
   phenotypic 68
várzea 118
vegetative layering 158
venturi-convection 286
vesicular-arbuscular mycorrhizae (VAM)
      101, 102

water deficit 260
water meadows 23
water table levels 264, 275
waterlogging 131, 185
Weichselian ice age 36, 206
West Siberian Lowlands 168, 185
wetland species 88
wetlands 24
wild boar 37
willows – montane and arctic 314–318
wind-pruning 169
wind throw 75
winter climate effects
   cooling 330
   desiccation injury 191
   grazing 41, 334
   snow pack 328
   warming 75, 318, 329, 337
winter moth populations 145
woody plants
   as invasive species 311
   beyond the treeline 309–311
   in the tundra 311–314
   migration 323
   on salt marshes 258
Wrangel Island 39

xanthine oxidase 287
xerothermic (or hypsithermal) 164
xylem embolism 260
xylopodia (ligno-tubers) 51

Yakutia 171

zonation and flooding – species
   and population limits
      291–294